《21世纪理论物理及其交叉学科前沿丛书》编委会
（第二届）

主　　编：孙昌璞
执行主编：常　凯
编　　委：（按姓氏拼音排序）

蔡荣根	段文晖	方　忠	冯世平
李　定	李树深	梁作堂	刘玉鑫
卢建新	罗民兴	马余刚	任中洲
王　炜	王建国	王玉鹏	向　涛
谢心澄	邢志忠	许甫荣	尤　力
张伟平	郑　杭	朱少平	朱世琳
庄鹏飞	邹冰松		

国家自然科学基金
理论物理专款资助

"十三五"国家重点出版物出版规划项目
21世纪理论物理及其交叉学科前沿丛书

可积模型方法及其应用

杨文力　杨战营　杨　涛　等　编著

科学出版社
北　京

内 容 简 介

本书从可积模型的基本概念出发,系统介绍了求解可积模型的典型方法及其在超冷原子和低维凝聚态理论等非线性物理系统中的应用. 全书共 6 章, 分别讲述了四种求解量子可积模型的方法; 介绍如何基于可积模型的精确解研究量子多体模型的物理性质; 二维共形场论的基本理论; 类非线性薛定谔可积系统中的怪波物理等问题; 具有长程相互作用的量子多体系统 (如 Calogero-Sutherland-Moser 系统) 和具有差分性质的相关联系统 (如 Ruijsenaars-Schneider-van Diejen-Macdonald-Koornwinder 系统) 的精确解及其相关的各类正交多项式; 准可积多体模型的基本理论.

本书可供高等学校理论物理和凝聚态物理及相关专业研究生、高年级本科生了解和学习精确可解模型时使用, 也可作为相关领域科研人员的参考书.

图书在版编目(CIP)数据

可积模型方法及其应用/杨文力等编著. —北京: 科学出版社, 2019.4
(21 世纪理论物理及其交叉学科前沿丛书)
ISBN 978-7-03-061069-0

Ⅰ. ①可··· Ⅱ. ①杨··· Ⅲ. ①物理学-研究 Ⅳ. O4

中国版本图书馆 CIP 数据核字 (2019) 第 074393 号

责任编辑: 钱 俊 陈艳峰 / 责任校对: 彭珍珍
责任印制: 吴兆东 / 封面设计: 无极书装

科学出版社 出版
北京东黄城根北街 16 号
邮政编码: 100717
http://www.sciencep.com

北京虎彩文化传播有限公司 印刷
科学出版社发行 各地新华书店经销
*
2019 年 4 月第 一 版 开本: 720×1000 1/16
2019 年 6 月第二次印刷 印张: 25 1/4
字数: 500 000
定价: 168.00 元
(如有印装质量问题, 我社负责调换)

《21 世纪理论物理及其交叉学科前沿丛书》
出版前言

物理学是研究物质及其运动规律的基础科学. 其研究内容可以概括为两个方面: 第一, 在更高的能量标度和更小的时空尺度上, 探索物质世界的深层次结构及其相互作用规律; 第二, 面对由大量个体组元构成的复杂体系, 探索超越个体特性"演生"出来的有序和合作现象. 这两个方面代表了两种基本的科学观 —— 还原论 (reductionism) 和演生论 (emergence). 前者把物质性质归结为其微观组元间的相互作用, 旨在建立从微观出发的终极统一理论, 是一代又一代物理学家的科学梦想; 后者强调多体系统的整体有序和合作效应, 把不同层次"演生"出来的规律当成自然界的基本规律加以探索. 它涉及从固体系统到生命软凝聚态等各种多体系统, 直接联系关乎日常生活的实际应用.

现代物理学通常从理论和实验两个角度探索以上的重大科学问题. 利用科学实验方法, 通过对自然界的主动观测, 辅以理论模型或哲学上思考, 先提出初步的科学理论假设, 然后借助进一步的实验对此进行判定性检验. 最后, 据此用严格的数学语言精确、定量表达一般的科学规律, 并由此预言更多新的、可以被实验再检验的物理效应. 当现有的理论无法解释一批新的实验发现时, 物理学就要面临前所未有的挑战, 有可能产生重大突破, 诞生新理论. 新的理论在解释已有实验结果的同时, 还将给出更一般的理论预言, 引发新的实验研究. 物理学研究这些内禀特征, 决定了理论物理学作为一门独立学科存在的必要性以及在当代自然科学中的核心地位.

理论物理学立足于科学实验和观察, 借助数学工具、逻辑推理和观念思辨, 研究物质的时空存在形式及其相互作用规律, 从中概括和归纳出具有普遍意义的基本理论. 由此不仅可以描述和解释自然界已知的各种物理现象, 而且还能够预言此前未知的物理效应. 需要指出, 理论物理学通过当代数学语言和思想框架, 使得物理定律得到更为准确的描述. 沿循这个规律, 作为理论物理学最基础的部分, 20 世纪初诞生的相对论和量子力学今天业已成为当代自然科学的两大支柱, 奠定了理论物理学在现代科学中的核心地位. 统计物理学基于概率统计和随机性的思想处理多粒子体系的运动, 是二者的必要补充. 量子规范场论从对称性的角度描述微观粒子的基本相互作用, 为自然界四种基本相互作用的统一提供坚实的基础.

关于理论物理的重要作用和学科发展趋势, 我们分六点简述.

(1) 理论物理研究纵深且广泛, 其理论立足于全部实验的总和之上. 由于物质结构是分层次的, 每个层次上都有自己的基本规律, 不同层次上的规律又是互相联系的. 物质层次结构及其运动规律的基础性、多样性和复杂性不仅为理论物理学提供了丰富的研究对象, 而且对理论物理学家提出了巨大的智力挑战, 激发出人类探索自然的强大动力. 因此, 理论物理这种高度概括的综合性研究具有显著的多学科交叉与知识原创的特点. 在理论物理中, 有的学科 (诸如粒子物理、凝聚态物理等) 与实验研究关系十分密切, 但还有一些更加基础的领域 (如统计物理、引力理论和量子基础理论), 它们一时并不直接涉及实验. 虽然物理学本身是一门实验科学, 但物理理论是立足于长时间全部实验总和之上的, 而不是只针对个别实验. 虽然理论正确与否必须落实到实验检验上, 但在物理学发展过程中, 有的阶段性理论研究和纯理论探索性研究, 开始不必过分强调具体的实验检验. 其实, 产生重大科学突破甚至科学革命的广义相对论、规范场论和玻色–爱因斯坦凝聚就是这方面的典型例证, 它们从纯理论出发, 实验验证却等待了几十年, 甚至近百年. 近百年前爱因斯坦广义相对论预言了一种以光速传播的时空波动 —— 引力波. 直到 2016 年 2 月, 美国科学家才宣布人类首次直接探测到引力波. 引力波的预言是理论物理发展的里程碑, 它的观察发现将开创一个崭新的引力波天文学研究领域, 更深刻地揭示宇宙奥秘.

(2) 面对当代实验科学日趋复杂的技术挑战和巨大经费需求, 理论物理对物理学的引领作用必不可少. 第二次世界大战后, 基于大型加速器的粒子物理学开创了大科学工程的新时代, 也使得物理学发展面临经费需求的巨大挑战. 因此, 伴随着实验和理论对物理学发展发挥的作用有了明显的变化, 理论物理高屋建瓴的指导作用日趋重要. 在高能物理领域, 轻子和夸克只能有三代是纯理论的结果, 顶夸克和最近在大型强子对撞机 (LHC) 发现的 Higgs 粒子首先来自理论预言. 当今高能物理实验基本上都是在理论指导下设计进行的, 没有理论上的动机和指导, 高能物理实验如同大海捞针, 无从下手. 可以说, 每一个大型粒子对撞机和其他大型实验装置, 都与一个具体理论密切相关. 天体宇宙学的观测更是如此. 天文观测只会给出一些初步的宇宙信息, 但其物理解释必须依赖于具体的理论模型. 宇宙的演化只有一次, 其初态和末态迄今都是未知的. 宇宙学的研究不能像通常的物理实验那样, 不可能为获得其演化的信息任意调整其初末态. 因此, 仅仅基于观测, 不可能构造完全合理的宇宙模型. 要对宇宙的演化有真正的了解, 建立自洽的宇宙学模型和理论, 就必须立足于粒子物理和广义相对论等物理理论.

(3) 理论物理学本质上是一门交叉综合科学. 大家知道, 量子力学作为 20 世纪的奠基性科学理论之一, 是人们理解微观世界运动规律的现代物理基础. 它的建立, 带来了以激光、半导体和核能为代表的新技术革命, 深刻地影响了人类的物质和精神生活, 已成为社会经济发展的原动力之一. 然而, 量子力学基础却存在诸多的争议, 哥本哈根学派对量子力学的 "标准" 诠释遭遇诸多挑战. 不过这些学术争论不仅促进了量子理论自身发展, 而且促使量子力学走向交叉科学领域, 使得量子物理从观测解释阶段进入自主调控的新时代, 从此量子世界从自在之物变成为我之物. 近二十年来, 理论物理学在综合交叉方面的重要进展是量子物理与信息计算科学的交叉, 由此形成了以量子计算、量子通信和量子精密测量为主体的量子信息科学. 它充分利用量子力学基本原理, 基于独特的量子相干进行计算、编码、信息传输和精密测量, 探索突破芯片极限、保证信息安全的新概念和新思路. 统计物理学为理论物理研究开拓了跨度更大的交叉综合领域, 如生物物理和软凝聚态物理. 统计物理的思想和方法不断地被应用到各种新的领域, 对其基本理论和自身发展提出了更高的要求. 软物质是在自然界中存在的最广泛的复杂凝聚态物质, 它处于固体和理想流体之间, 与人们的日常生活及工业技术密切相关. 例如, 水是一种软凝聚态物质, 其研究涉及的基础科学问题关乎人类社会今天面对的水资源危机.

(4) 理论物理学在具体系统应用中实现创新发展, 并在基本层次上回馈自身. 从量子力学和统计物理对固体系统的具体应用开始, 近半个世纪以来凝聚态物理学业已发展成当代物理学最大的一个分支. 它不仅是材料、信息和能源科学的基础, 也与化学和生物等学科交叉与融合, 而其中发现的新现象、新效应, 都有可能导致凝聚态物理一个新的学科方向或领域的诞生, 为理论物理研究展现了更加广阔的前景. 一方面, 凝聚态物理自身理论发展异常迅猛和广泛, 描述半导体和金属的能带论和费米液体理论为电子学、计算机和信息等学科的发展奠定了理论基础; 另一方面, 从凝聚态理论研究提炼出来的普适的概念和方法, 对包括高能物理在内的其他物理学科的发展也起到了重要的推动作用. BCS 超导理论中的自发对称破缺概念, 被应用到描述电弱相互作用统一的 Yang-Mills 规范场论, 导致了中间玻色子质量演生的 Higgs 机制, 这是理论物理学发展的又一个重要里程碑. 近二十年来, 在凝聚态物理领域, 有大量新型低维材料的合成和发现, 有特殊功能的量子器件的设计和实现, 有高温超导和拓扑绝缘体等大量新奇量子现象的展示. 这些现象不能在以单体近似为前提的费米液体理论框架下得到解释, 新的理论框架建立已迫在眉睫, 如果能使凝聚态物理的基础及应用研究成功跨上一个新的历史台阶, 也将理论物理的引领作用发挥到极致.

(5) 理论物理的一个重要发展趋势是理论模型与强大的现代计算手段相结合. 面对纷繁复杂的物质世界 (如强关联物质和复杂系统), 简单可解析求解的理论物理模型不足以涵盖复杂物质结构的全部特征, 如非微扰和高度非线性. 现代计算机的发明和快速发展提供了解决这些复杂问题的强大工具. 辅以面向对象的科学计算方法 (如第一原理计算、蒙特卡罗方法和精确对角化技术), 复杂理论模型的近似求解将达到极高的精度, 可以逐渐逼近真实的物质运动规律. 因此, 在解析手段无法胜任解决复杂问题任务时, 理论物理必须通过数值分析和模拟的办法, 使得理论预言进一步定量化和精密化. 这方面的研究导致了计算物理这一重要学科分支的形成, 成为连接物理实验和理论模型必不可少的纽带.

(6) 理论物理学将在国防安全等国家重大需求上发挥更多作用. 大家知道, 无论决胜第二次世界大战、冷战时代的战略平衡, 还是中国国家战略地位提升, 理论物理学在满足国家重大战略需求方面发挥了不可替代的作用. 爱因斯坦、奥本海默、费米、彭桓武、于敏、周光召等理论物理学家也因此彪炳史册. 与战略武器发展息息相关, 第二次世界大战后开启了物理学大科学工程的新时代, 基于大型加速器的重大科学发现反过来为理论物理学提供广阔的用武之地, 如标准模型的建立. 国防安全方面等国家重大需求往往会提出自由探索不易提出的基础科学问题, 在对理论物理提出新挑战的同时, 也为理论物理研究提供了源头创新的平台. 因此, 理论物理也要针对国民经济发展和国防安全方面等国家重大需求, 凝练和发掘自己能够发挥关键作用的科学问题, 在实践应用和理论原始创新方面取得重大突破.

为了全方位支持我国理论物理事业长足发展, 1993 年国家自然科学基金委员会设立 "理论物理专款", 并成立学术领导小组 (首届组长是我国著名理论物理学家彭桓武先生). 多年来, 这个学术领导小组凝聚了我国理论物理学家集体智慧, 不断探索符合理论物理特点和发展规律的资助模式, 培养理论物理优秀创新人才并做出杰出的研究成果, 对国民经济和科技战略决策提供指导和咨询. 为了更全面地支持我国的理论物理事业, "理论物理专款" 持续资助我们编辑出版这套《21 世纪理论物理及其交叉学科前沿丛书》, 目的是系统全面地介绍现代理论物理及其交叉学科前沿领域的基本内容、最新进展和发展前景, 以及中国理论物理学家在这些领域中的科学贡献和所取得的主要进展. 希望这套丛书能帮助大学生、研究生、博士后、青年教师和研究人员全面了解理论物理学研究进展, 培养其对物理学研究的兴趣, 迅速进入理论物理前沿研究领域, 同时吸引更多的年轻人献身理论物理学事业, 为我国的科学研究在国际上占有一席之地做出自己的贡献.

孙昌璞

中国科学院院士, 发展中国家科学院院士

国家自然科学基金委员会 "理论物理专款" 学术领导小组组长

前　　言

随着现代材料科学和微加工技术的迅速发展, 低维量子多体系统已经成为当前量子物理研究的前沿领域之一. 由于维度降低和尺寸缩小, 量子效应和关联效应更加凸显, 此类系统呈现出许多内涵极为丰富的新的物理特性. 由于缺乏普适的理论方法, 精确可解的非线性强关联量子多体模型的严格解已成为研究低维量子多体系统一个非常好的出发点. 精确可解模型是数学物理领域的一个重要分支, 这些模型不但具有优美的数学结构, 同时具有丰富的物理内涵, 在物理学和数学的多个领域 (如场论和超弦理论、统计物理和凝聚态物理以及纯数学领域 (如代数几何和量子群)) 中都扮演着非常重要的角色, 尤其是在精确揭示高度非线性和强关联物理系统在远离微扰区的特征和临界行为等方面扮演着不可替代的角色, 为某些重要的物理概念提供基准. 几个著名的例子有: ① 二维伊辛模型的精确解为热力学相变理论提供了无可置疑的佐证, 从而结束了人们关于有无热力学相变的纷争; ② 一维 Hubbard 模型的精确解证明了 Mott 绝缘体及 Hubbard 能隙的存在; ③ 海森伯自旋链的精确解直接给出了严格的自旋子元激发, 从而阐明了多体系统中分数元激发的物理机制.

目前, 求解量子可积模型的典型方法大致有四种: ① H.A. Bethe 于 1931 年提出的坐标 Bethe Ansatz 方法; ② 20 世纪 70 年代初 R.J. Baxter 发展的 T-Q 关系方法; ③ 20 世纪 70 年代末 L.D. Faddeev 学派基于 Yang-Baxter 方程提出的量子逆散射方法 (或代数 Bethe Ansatz 方法); ④ 近期发展起来的非对角 Bethe Ansatz 方法. 前三种方法在处理粒子数守恒 (或 $U(1)$-对称) 可积模型中取得了很大成功, 但在面对粒子数不守恒 (或 $U(1)$-对称破缺) 模型时却遇到了巨大困难, 原因是这类方法是通过构造本征态来获取转移矩阵的本征值, 因而强烈依赖于赝真空态–参照态 (reference state) 的存在. 第四种方法将可积模型的谱问题纳入统一的理论框架 —— 非均匀 (inhomogeneous) T-Q 关系, 不仅可以成功求解长期困扰可积模型领域的 "可积但不可解" 的几个典型粒子数不守恒模型, 而且还大大简化了粒子数守恒模型的求解过程.

本书的目的是向读者介绍目前基于 Yang-Baxter 方程的解 R 矩阵求解量子可积模型的四种主要的方法及其在统计物理、凝聚态物理和量子规范理论中的应用. 它是基于 2016 年暑假在西安西北大学举办的国家自然科学基金委员会 "第九期理论物理前沿暑期讲习班 —— 可积模型方法及其应用" 中几位授课教授的讲稿进行整理和适当地扩充而成的. 本书由杨文力、杨战营和杨涛组织编写, 共 6 章, 第 1

章由中国科学院物理研究所的王玉鹏研究员、曹俊鹏研究员和西北大学杨文力教授撰写，较为系统地介绍了坐标 Bethe Ansatz、(嵌套) 代数 Bethe Ansatz、T-Q 关系和非对角 Bethe Ansatz 方法，并将它们应用于海森伯自旋链、自旋梯子模型和高自旋玻色气体等模型；第 2 章由中国科学院武汉物理与数学研究所的管习文研究员和彭黎博士撰写，基于 Leib-Liniger 模型的 Bethe Ansatz 解，系统地介绍了该模型的基态和低能激发态的物理性质以及在弱相互作用和强相互作用下的热力学性质；第 3 章由中国科学院数学与系统科学研究院的丁祥茂研究员撰写，主要介绍二维共形场的基本理论和相关算子代数；第 4 章由西北大学的杨战营教授、赵立臣教授和刘冲副教授撰写，主要介绍不同非线性物理系统中平面波背景上非线性局域波精确解的构造、动力学特征以及物理机制的分析；第 5 章由日本京都大学汤川理论物理研究所的 R. Sasaki 教授撰写，主要介绍一维相互作用量子多粒子可积体系的薛定谔方程及相联系的差分薛定谔方程的精确解，以及相关的正交多项式；第 6 章由澳大利亚昆士兰大学的张耀中教授撰写，主要介绍一维准精确可积量子多粒子体系的基本理论和几类典型准精确可解模型及相关的差分模型的严格解.

本书在国家自然科学基金委员会"理论物理专款"(项目编号 11547607)、国家自然科学基金委员会杰出青年基金和重点项目 (项目编号 11434013 和 11425522) 以及西北大学理论物理学科建设经费的资助下得以完成，在此谨向国家自然科学基金委员会，特别是数理科学部"理论物理专款"和西北大学的长期支持表示感谢. 感谢"第九期理论物理前沿暑期讲习班"授课教师以及组委会老师、同学的辛勤付出；感谢中国科学院半导体研究所夏建白院士和常凯研究员在本书编写过程中给予的支持和鼓励；感谢《21 世纪理论物理及其交叉学科前沿丛书》编委会的大力支持与帮助；本书在编写和出版过程中还得到了科学出版社的钱俊编辑的大力支持和帮助，对此表示诚挚的感谢.

编　者
2019 年 3 月于西北大学

目 录

前言
第 1 章 Bethe Ansatz 方法简介 ································· 1
  1.1 可积性概述 ································· 1
  1.2 自旋链模型 ································· 4
    1.2.1 坐标 Bethe Ansatz ································· 4
    1.2.2 基态及低能元激发 ································· 9
    1.2.3 热力学性质 ································· 15
    1.2.4 代数 Bethe Ansatz ································· 20
    1.2.5 开边界问题 ································· 26
    1.2.6 反射代数 Bethe Ansatz ································· 31
    1.2.7 非对角 Bethe Ansatz ································· 39
    1.2.8 近藤问题 ································· 52
    1.2.9 各向异性 ································· 56
  1.3 嵌套的代数 Bethe Ansatz ································· 60
  1.4 $SU(4)$ 对称自旋梯子模型 ································· 66
  1.5 自旋为 1 的玻色气体 ································· 70
  参考文献 ································· 77
第 2 章 Lieb-Liniger 模型：多体物理之美 ································· 79
  2.1 引言 ································· 79
  2.2 Bethe 假设 ································· 80
  2.3 基态行为 ································· 86
  2.4 弱相互作用：半圆律 ································· 88
  2.5 强相互作用：费米化 ································· 90
  2.6 元激发：集体运动 ································· 92
  2.7 Yang-Yang 热力学方法 ································· 96
  2.8 Lieb-Liniger 模型中的量子统计 ································· 99
  2.9 普适的热力学行为 ································· 104
  2.10 Luttinger 液体理论 ································· 108
  2.11 量子临界性 ································· 111
  2.12 关联函数 ································· 114

2.13　关于 Lieb-Liniger 玻色气体的实验发展 · 120
参考文献 · 124

第 3 章　共形场论入门 · 126
3.1　共形变换 · 126
3.1.1　d 维共形变换 · 127
3.1.2　二维共形变换 · 132
3.1.3　Witt 代数 · 133
3.1.4　共形子代数 · 135
3.1.5　元场的多点函数 · 136
3.1.6　二维共形代数的中心扩张 · 137
3.1.7　守恒量与能-动张量 · 138
3.2　几个常用概念 · 140
3.2.1　径向积 · 141
3.2.2　算子积展开 · 143
3.2.3　正规积 · 145
3.2.4　能-动张量展开 · 146
3.3　共形场：举例 · 147
3.3.1　自由玻色子 · 148
3.3.2　顶点算子 · 150
3.3.3　$\widehat{su(2)}_1$ 代数 · 152
3.3.4　自由费米子 · 152
3.3.5　鬼系统 · 155
3.3.6　中心荷 · 157
3.4　态 · 158
3.4.1　最高权态 · 159
3.4.2　嗣场 · 161
3.4.3　Kac 行列式和酉表示 · 163
3.4.4　极小模型 · 168
3.4.5　Virasoro 特征标 · 170
3.5　有理共形场 · 171
3.5.1　伊辛模型 · 171
3.5.2　融合代数 · 173
3.5.3　共形块的交换关系 · 176
3.5.4　流代数 · 177
3.5.5　$\mathcal{W}$-代数 · 178

    3.5.6 结语 ··· 180
 参考文献 ··· 181
第 4 章 类非线性薛定谔可积系统中光怪波物理 ······························ 184
 4.1 光怪波物理简介 ··· 184
   4.1.1 怪波现象 ··· 184
   4.1.2 理论解释 ··· 186
   4.1.3 研究进展 ··· 188
 4.2 类非线性薛定谔可积模型方法 ··· 192
   4.2.1 达布变换 ··· 193
   4.2.2 相似变换 ··· 200
   4.2.3 调制不稳定性 ··· 203
 4.3 高斯背景上光怪波的激发 ·· 207
   4.3.1 高斯背景上光怪波精确解 ······································· 207
   4.3.2 怪波激发性质 ··· 210
 4.4 高阶效应诱发光学局域波态转换 ······································ 214
   4.4.1 调制不稳定性分析 ·· 216
   4.4.2 局域波精确解构造 ·· 219
   4.4.3 一阶怪波与孤子的态转换 ······································· 221
   4.4.4 二阶怪波与孤子的态转换 ······································· 225
   4.4.5 呼吸子与其他非线性波的态转换 ··························· 229
   4.4.6 Kuznetsov-Ma 呼吸子与单峰孤子的态转换 ············· 232
   4.4.7 一般呼吸子与多峰孤子的态转换 ··························· 233
 参考文献 ··· 235
第 5 章 Introduction to Exactly Solvable Quantum Many-body Systems
   (精确可解量子多体系统导论) ··· 245
 5.1 Introduction (引言) ··· 246
 5.2 1-Degree of Freedom System (单自由度系统) ···················· 251
   5.2.1 Factorised Hamiltonian (因式化哈密顿量) ··············· 251
   5.2.2 Intertwining Relations: Crum's Theorem (交互关系: Crum's 定理) ···· 253
   5.2.3 Five Typical Solvable Potentials (五种典型的可解势) ···· 255
   5.2.4 Shape Invariance: Sufficient Condition of Exact Solvability
      (形状不变性: 完全可解性的充分条件) ················· 261
   5.2.5 Solvability in the Heisenberg Picture (海森伯绘景下的可解性) ······· 264
   5.2.6 Difference Schrödinger Equations (差分薛定谔方程) ······· 269
   5.2.7 From One Particle to Many Particles (从单粒子到多粒子) ·········· 278

  5.2.8 Reflection Groups and Root Systems (反射群和根系) ········· 279
 5.3 Calogero-Sutherland Systems (Calogero-Sutherland 系统) ········· 283
  5.3.1 Simplest Cases (Based on A_{r-1} Root System) (几个简单的例子
   (基于 A_{r-1} 根系)) ········· 283
  5.3.2 Universal Formalism (普适形式) ········· 287
  5.3.3 Jack Polynomials (Jack 多项式) ········· 299
 5.4 Multi-Particle QM with Difference Schrödinger Equations
  (差分薛定谔方程中的多粒子量子力学) ········· 305
  5.4.1 Ruijsenaars-Schneider Systems (Ruijsenaars-Schneider 系统) ········· 307
  5.4.2 Macdonald Polynomials (Macdonald 多项式) ········· 308
 5.5 Comments and Discussion (总结和讨论) ········· 313
 5.6 Appendix: Symbols, Definitions & Formulas
  (附录: 符号、定义和公式) ········· 314
 参考文献 ········· 315

第 6 章 Quasi-Exactly Solvable Systems (准精确可解系统) ········· 322
 6.1 Exact Solvability Versus Quasi-Exact Solvability
  (精确可解性与准精确可解性) ········· 322
 6.2 Generalities: Characterization of QES Operators (概论: 准精确可解算符
  的特性) ········· 323
 6.3 Lie Algebra Approach (李代数方法) ········· 324
 6.4 Relationship Between 2nd-order Differential Operator and Schrödinger
  Operator (二阶差分算符和薛定谔算符的关系) ········· 328
 6.5 Examples of QES Systems with Lie Algebraization
  (准精确可解系统李代数化的例子) ········· 329
  6.5.1 Sextic Potential (六次势) ········· 329
  6.5.2 Harmonic Oscillator (谐振子) ········· 331
  6.5.3 Lamé Equation (Lamé 方程) ········· 331
  6.5.4 QES Quartic Potential (准精确可解的四次势) ········· 333
  6.5.5 Quantum (Driven) Rabi Model (量子驱动的 Rabi 模型) ········· 335
 6.6 Stäckel Transform and Coupling Constant Metamorphosis
  (Stäckel 变换和耦合常数变形) ········· 338
  6.6.1 Two Electrons in External Oscillator Potential (谐振外势中的两
   电子体系) ········· 338
  6.6.2 2D Hydrogen in Uniform Magnetic Field (均匀磁场中的二维
   氢原子) ········· 340

6.6.3 Hooke Atom: Two Planar Charged Particles in Uniform Magnetic Field
(Hooke 原子: 均匀磁场中的两个平面带电粒子) ······················ 341

6.6.4 Two Coulombically Repelling Electrons on a Sphere
(球面上的两个具有库仑排斥势的电子) ······························ 343

6.6.5 Inverse Sextic Power Potential (逆六次势) ························ 344

6.7 Solutions to QES: Bender-Dunne Polynomials
(精确解: Bender-Dunne 多项式) ·· 346

6.8 3-term Recurrence Relation and Continued Fractions
(3 项递归关系和连分式) ·· 348

6.8.1 Bargmann-Hilbert Spaces (Bargmann-Hilbert 空间) ············· 351

6.8.2 2-photon Quantum Rabi Model (两光子量子 Rabi 模型) ········ 352

6.8.3 Two-mode Quantum Rabi Model (双模量子 Rabi 模型) ········ 358

6.9 Solutions to QES Differential Equations: Heine-Stieltjes Polynomials
(准精确可解差分方程的解: Heine-Stieltjes 多项式) ····················· 362

6.10 Solutions to QES Systems: Functional Bethe Ansatz Method
(准精确可解系统的解: Bethe Ansatz 方法) ································ 365

6.11 Examples of Solutions of QES Models (准精确可解模型的解的一些
例子) ·· 370

6.11.1 Bose-Hubbard Dimer with Local M-body Interaction (Bose-Hubbard
二聚体中的局域多体相互作用) ····································· 370

6.11.2 BA Solutions of the Driven Rabi Model (驱动 Rabi 模型的
Bethe Ansatz 解) ·· 372

6.11.3 BA Solutions of Two Electrons on a Sphere (球面上两电子的
Bethe Ansatz 解) ·· 374

6.11.4 BA Solutions of the Inverse Sextic Power Potential (逆六次势的
Bethe Ansatz 解) ·· 375

6.11.5 Kink Stability Analysis of the ϕ^6-type Field Theory (ϕ^6 型场论的扭结
稳定性) ·· 377

6.12 Realization of $gl(M)$ in Fock Spaces and Bargmann-Hilbert Spaces
($gl(M)$ 在 Fock 空间和 Bargmann-Hilbert 空间中的表示) ··········· 380

参考文献 ·· 382

第 1 章 Bethe Ansatz 方法简介

王玉鹏　杨文力　曹俊鹏

1.1　可积性概述

在多体物理中, 有一类模型是可以严格求解的, 这类模型通常被称为可积模型. 可积概念来源于经典力学, 一个经典力学系统一般可以用一组微分方程来描述, 这组方程的解就是它们的积分, 因此, 可积就是可解. 另外, 一组微分方程的积分又伴随着一系列积分常数, 这组常数由系统的初始条件和边界条件所决定, 它们不随时间的改变而改变, 因而又被称为运动积分或守恒量. 如果一个具有 N 个自由度的力学系统同时又具有 N 个独立的守恒量, 则我们说这个系统是完全可积的.

为了初步理解可积系统, 我们首先考虑一个简单模型. 假定有 N 个质量相等的刚球在一条直线上或一个环上运动, 它们之间的碰撞是弹性的; 再假定在初始时刻, 它们分别携带动量 $k_1, k_2, \cdots, k_N$. 在刚球的两两碰撞过程中, 刚球的动量只会互相交换而不会变值. 考虑第 i 个刚球和第 j 个刚球的碰撞过程, 假定它们在碰撞前分别携带动量 k_i 和 k_j, 碰撞后携带动量 k'_i 和 k'_j, 由于它们具有相同质量并且碰撞是弹性的, 则碰撞过程中的动量守恒定律和能量守恒定律分别给出

$$k_i + k_j = k'_i + k'_j, \tag{1.1.1}$$

$$k_i^2 + k_j^2 = k'^2_i + k'^2_j, \tag{1.1.2}$$

上述方程只有两种可能的解, 即 $k'_i = k_j$, $k'_j = k_i$ 或 $k'_i = k_i$, $k'_j = k_j$. 由于经典刚球是不可互相穿越的, 只有第一组解在物理上是允许的, 即刚球在碰撞后互换动量. 由此可以看出, 初始动量组 $\{k_1, k_2, \cdots, k_N\}$ 在碰撞过程中只会重组, 不会变值. 显然, 这样一个系统具有 N 个独立的守恒量:

$$I_n = \sum_{j=1}^{N} k_j^n, \quad n = 1, 2, \cdots, N. \tag{1.1.3}$$

因此, 它是一个完全可积系统. 在以上的考虑中, 我们实际上已假定了周期性边界条件. 如果我们将刚球的运动限制在一个有限的区间内, 则刚球在边界上会被反弹回来, 同时其动量变号. 这时原始动量不再是守恒的, 但 $\{|k_1|, |k_2|, \cdots, |k_n|\}$ 仍然保

持不变, 因此我们仍然有下述守恒量:

$$I_n = \sum_{j=1}^{N} |k_j|^n, \quad n = 1, 2, \cdots, N. \tag{1.1.4}$$

量子可积性要比经典可积性复杂得多. 考虑 N 个全同粒子 (允许携带内部自由度, 如自旋等) 在一维空间运动. 假定在初始时刻其渐近波函数为

$$\Psi_{\text{in}} \sim e^{i \sum_{j=1}^{N} k_j x_j}, \quad x_1 \ll x_2 \ll \cdots \ll x_N. \tag{1.1.5}$$

第一个粒子通过与其他每个粒子散射后到达终态的波函数可以写为

$$\Psi_{\text{out}} \sim S_{12\cdots N} e^{i \sum_{j=1}^{N} k_j x_j}, \quad x_2 \ll x_3 \ll \cdots \ll x_N \ll x_1, \tag{1.1.6}$$

其中, $S_{12\cdots N}$ 是散射矩阵或 S 矩阵. 在一定的边界条件下, 如果散射矩阵可以写为二体散射矩阵的乘积:

$$S_{12\cdots N} = S_{1N} S_{1N-1} \cdots S_{12}, \tag{1.1.7}$$

则我们称这个系统是可因式化的. 下面我们将看到, 可因式化保证了系统的可积性. 考虑一个三粒子系统, 假定它的初态为 (123), 终态为 (321), 由图 1.1 可以看出, 系统从初态到终态有两种途径, 由于波函数的唯一性, 如果系统是可因式化的, 则我们有下列方程:

$$S_{12} S_{13} S_{23} = S_{23} S_{13} S_{12}. \tag{1.1.8}$$

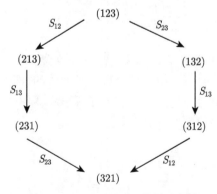

图 1.1 Yang-Baxter 方程示意图

上述方程就是著名的 Yang-Baxter 方程. 它是由杨振宁和 Baxter 在研究一维 δ 势费米气体模型[1] 及二维经典格子统计模型[2] 时分别独立提出的. 杨振宁曾经证

明, 方程 (1.1.8) 是可因式化条件 (1.1.7), 也就是周期边条件下可积性的充分必要条件. 可因式化实际上保证了量子系统与经典系统的可比性, 因为这时我们只需考虑二体散射. 在二体散射过程中, 由于粒子的全同性, 它们在碰撞前后只会互换动量或保持原有动量不变, 这就保证了系统的本征波函数可以由一组参数 $\{k_1,\cdots,k_N\}$ 来描述, 同时我们也总能够构造出 N 个独立的守恒量, 使得它们具有本征值 (1.1.3).

二体散射矩阵 S_{ij} 在可积模型中是一个非常重要的物理量, 它除了满足 Yang-Baxter 方程 (1.1.8) 外, 还具有下述性质:

$$S_{ij}S_{ji}=1, \tag{1.1.9}$$

$$[S_{ij},S_{kl}]=0, \quad i\neq k,l, \ j\neq k,l. \tag{1.1.10}$$

一般来说, 当粒子具有内部对称性时, S_{ij} 和 S_{kl} 是不可对易的.

当系统具有反射边界时, 可积性除了要求 S_{ij} 满足 (1.1.8) 之外, 还要求在边界上满足一定的条件. 考虑一个二粒子系统, 假定它的初态为 $(12)B$, 终态为 $(\bar{1}\,\bar{2})B$, 则系统从初态到终态同样有两种途径 (图 1.2). 以上记号中 $\bar{i}$ 表示第 i 个粒子的反射波, 即携带有动量 $-k_i$; B 表示开边界; $\overline{S}_{ij}$ 表示第 i 个粒子与第 j 个粒子的反散波之间的散射矩阵; K_j 表示第 j 个粒子在边界上的反射矩阵, 即入射波振幅与反射波振幅之比. 从图 1.2 我们可以得出下列方程:

$$S_{12}K_1\overline{S}_{12}K_2 = K_2\overline{S}_{12}K_1S_{12}. \tag{1.1.11}$$

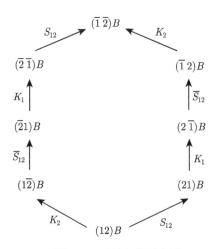

图 1.2 反射方程示意图

方程 (1.1.8) 与 (1.1.11) 是开边界可积系统的充分必要条件. 式 (1.1.11) 通常称为反射方程, 它是由 Cherednik[3] 和 Sklyanin[4] 首先提出的. 值得指出的是, 可积条件是非常严格的. 目前所发现的可积边界只有以上两种, 即周期性边条件 (允许有位相损失, 即扭曲边条件, 但无反射) 和全反射边条件. 这类边条件类似于光学中的全透射或全反射, 因此又被称为无衍射条件. 另外, 如果一个粒子具有与其他粒子不同的质量, 则初始动量组 $k_1, \cdots, k_N$ 不再是碰撞不变的, 这也会破坏可积性.

常用的求解可积模型的方法大致有两类, 即坐标 Bethe Ansatz 和代数 Bethe Ansatz(又称为量子反散射方法) 以及其他衍生的代数方法, 分别由 Bethe[5] 在研究一维 Heisenberg 反铁磁链时和 Faddeev 学派[6, 7] 基于 Baxter 的工作提出. 量子反散射方法因经典反散射方法[8] 而得名. 坐标 Bethe Ansatz 的思想非常简单. 假定第 j 个粒子从系统的最左边 ($x_j = 0$) 运动到系统的最右边 ($x_j = L$), 其间它将和每个其他粒子散射一次. 利用波函数的周期性边条件 $\Psi(\cdots, x_j = 0, \cdots) = \Psi(\cdots, x_j = L, \cdots)$, 我们可以得到下列方程:

$$S_{jN}S_{jN-1}\cdots S_{jj+1}S_{jj-1}\cdots S_{j1}\mathrm{e}^{\mathrm{i}k_j L}\xi_0 = \xi_0. \tag{1.1.12}$$

上式即所谓的 Bethe Ansatz 方程, 其中 ξ_0 表示初态波函数的振幅. 方程 (1.1.12) 确定了动量组 $\{k_1, \cdots, k_N\}$ 的可能取值. 如果粒子具有内部自由度, 则 (1.1.12) 是一个算子方程, 需要进一步求解. 这些方法我们将在以下章节中仔细介绍.

尽管可积模型非常特殊, 但对它们的研究是非常有意义的. 用重整化群的语言来说, 每一个物理模型都对应于参数空间中的一个点. 在重整化群变换下, 如果某些点流向同一个固定点, 则我们说它们属于同一个普适类, 一个普适类中的模型或物理系统具有相同或类似的物理性质. 因此, 只要我们在一个普适类中找到一个可积模型, 就可以理解整个普适类的物理行为. 另外, 某些可积模型本身就具有明确的物理意义, 它们在理解低维强关联系统的努力中起到了不可替代的作用[9–13].

1.2 自旋链模型

1.2.1 坐标 Bethe Ansatz

Heisenberg 模型是 Heisenberg 于 1928 年在研究绝缘体的磁性时提出的. 在这里, 我们考虑自旋为 $\frac{1}{2}$ 的一维 Heisenberg 模型. 该模型由 Bethe 于 1931 年精确求解[5]. 他所提出的方法被称为 Bethe Ansatz(德语词, 初始假定的意思). 此后, 杨振宁和杨振平[14]、Takhtajan 和 Faddeev[7] 对此模型的严格解做了更具体的讨论. 考虑 N 个自旋为 $\frac{1}{2}$ 的原子等距排列在一条直线上, 每个原子的自旋用 $\boldsymbol{S}_j = \frac{1}{2}\boldsymbol{\sigma}_j$ 来

1.2 自旋链模型

表示. 假定近邻原子之间存在交换相互作用, 我们考虑如下模型:

$$H = \frac{1}{2}J\sum_{j=1}^{N}\boldsymbol{\sigma}_j \cdot \boldsymbol{\sigma}_{j+1} - \frac{1}{2}h\sum_{j=1}^{N}\sigma_j^z, \tag{1.2.1}$$

其中, J 为耦合常数, $J>0$ 表示反铁磁相互作用, $J<0$ 表示铁磁相互作用; h 为加在 z 方向上的外磁场. 注意, 以上我们用到了周期性边界条件 $\boldsymbol{\sigma}_{N+1} \equiv \boldsymbol{\sigma}_1$. 为了方便起见, 我们引入自旋升降算子

$$S_j^+ = \frac{1}{2}(\sigma_j^x + \mathrm{i}\sigma_j^y), \quad S_j^- = \frac{1}{2}(\sigma_j^x - \mathrm{i}\sigma_j^y). \tag{1.2.2}$$

利用自旋算子的对易性质可以得到

$$[S_j^z, S_i^+] = \delta_{ij}S_j^+, \quad [S_j^z, S_i^-] = -\delta_{ij}S_j^-, \quad [S_j^+, S_i^-] = 2\delta_{ij}S_j^z. \tag{1.2.3}$$

哈密顿量 (1.2.1) 可以重新写为

$$H = J\sum_{j=1}^{N}(S_j^+ S_{j+1}^- + S_{j+1}^+ S_j^- + 2S_j^z S_{j+1}^z) - h\sum_{j=1}^{N}S_j^z. \tag{1.2.4}$$

定义第 j 个格点上的两个量子态 $|\uparrow\rangle_j$ 和 $|\downarrow\rangle_j$, 分别对应于自旋沿 z 方向向上和向下的态, 则 $\boldsymbol{S}_j$ 作用到它们上面有如下性质:

$$\begin{cases} S_j^+|\uparrow\rangle_j = 0, \quad S_j^-|\uparrow\rangle_j = |\downarrow\rangle_j, \quad S_j^z|\uparrow\rangle_j = \frac{1}{2}|\uparrow\rangle_j, \\ S_j^+|\downarrow\rangle_j = |\uparrow\rangle_j, \quad S_j^-|\downarrow\rangle_j = 0, \quad S_j^z|\downarrow\rangle_j = -\frac{1}{2}|\downarrow\rangle_j. \end{cases} \tag{1.2.5}$$

再定义初始态

$$|0\rangle = |\uparrow\rangle_1 \otimes |\uparrow\rangle_2 \otimes \cdots \otimes |\uparrow\rangle_N, \tag{1.2.6}$$

利用性质 (1.2.5) 很容易验证态 $|0\rangle$ 是哈密顿量 (1.2.1) 或 (1.2.4) 的本征态, 对应的本征方程为

$$H|0\rangle = \frac{1}{2}N(J-h)|0\rangle \equiv E_0|0\rangle. \tag{1.2.7}$$

当有一个自旋反转朝下时, 哈密顿量的本征态可以写为

$$|k\rangle = \sum_{x=1}^{N}\psi(x)S_x^-|0\rangle. \tag{1.2.8}$$

将 H 作用到态 $|k\rangle$ 上可以得到

$$H|k\rangle = J\sum_{x=1}^{N}\psi(x)[(S_{x+1}^{-}+S_{x-1}^{-})2S_x^z - 2]S_x^{-}|0\rangle + (E_0+h)\sum_{x=1}^{N}\psi(x)S_x^{-}|0\rangle. \quad (1.2.9)$$

利用本征方程

$$H|k\rangle = E|k\rangle = E\sum_{x=1}^{N}\psi(x)S_x^{-}|0\rangle, \quad (1.2.10)$$

其中, E 表示本征能量. 再比较式 (1.2.9) 与式 (1.2.10) 中的 $S_x^{-}|0\rangle$ 分量, 我们得到

$$J[\psi(x+1)+\psi(x-1)-2\psi(x)] + h\psi(x) = (E-E_0)\psi(x). \quad (1.2.11)$$

方程 (1.2.11) 可以用 Fourier 变换求解. 设 $\psi(x) = Ae^{ikx}$, 其中 A 是一个任意常数, 则 $\psi(x)$ 是一个本征函数, 本征能量可以用动量 k 表达为

$$E = E_0 + 2J[\cos k - 1] + h. \quad (1.2.12)$$

上述解是一个简单的单体问题, 它即通常所谓的自旋波. 利用波函数的周期性边条件 $\psi(x) = \psi(x+N)$, 可以得到动量或自旋波波数 k 的可能取值

$$k = \frac{2\pi I}{N}, \quad I = 0, \pm 1, \pm 2, \cdots. \quad (1.2.13)$$

注意 I 的取值应限制在第一布里渊区, 即 $|I| \leqslant N/2$. 当有两个自旋反转时, H 的本征态可以表示为

$$|k_1, k_2\rangle = \sum_{x_1, x_2=1}^{N} \psi(x_1, x_2) S_{x_1}^{-} S_{x_2}^{-} |0\rangle, \quad (1.2.14)$$

将 H 作用到态 $|k_1, k_2\rangle$ 上, 我们同样可以得到下述本征方程:

$$J[\psi(x_1+1, x_2)+\psi(x_1-1, x_2)+\psi(x_1, x_2+1)+\psi(x_1, x_2-1)-4\psi(x_1, x_2)]$$
$$+2h\psi(x_1, x_2) + 2J[\delta_{x_1, x_2+1}+\delta_{x_1, x_2-1}]\psi(x_1, x_2) = (E-E_0)\psi(x_1, x_2). \quad (1.2.15)$$

与单个自旋反转时不同, 当两个反转自旋处在相邻位置时, 它们之间存在相互作用. Bethe 提出, 对于二体波函数可以做 Ansatz:

$$\psi(x_1, x_2) = [A_{12}e^{ik_1x_1+ik_2x_2} + A_{21}e^{ik_2x_1+ik_1x_2}]\theta(x_2-x_1)$$
$$+ [A_{21}e^{ik_1x_1+ik_2x_2} + A_{12}e^{ik_2x_1+ik_1x_2}]\theta(x_1-x_2), \quad (1.2.16)$$

1.2 自旋链模型

其中, A_{12} 和 A_{21} 是与 k_1, k_2 有关的两个常数; $\theta(x)$ 是阶梯函数:

$$\theta(x) = \begin{cases} 1, & x > 0, \\ 0, & x < 0. \end{cases} \tag{1.2.17}$$

显然, 波函数是对称的, 即 $\psi(x_1, x_2) = \psi(x_2, x_1)$, 不失一般性, 我们可以只考虑 $x_1 < x_2$ 情形. 当 x_1 和 x_2 不相邻时, 将式 (1.2.16) 代入式 (1.2.15) 可以得到

$$E - E_0 = 2J(\cos k_1 + \cos k_2 - 2) + 2h. \tag{1.2.18}$$

当 $x_2 = x_1 + 1$, 即两个反转自旋相邻时, 本征方程 (1.2.15) 化为

$$J[\psi(x_1 - 1, x_1 + 1) + \psi(x_1, x_1 + 2) - 2\psi(x_1, x_1 + 1)]$$
$$+ 2h\psi(x_1, x_1 + 1) = (E - E_0)\psi(x_1, x_2), \tag{1.2.19}$$

注意两个反转自旋不能处于同一格点, 即 $\psi(x, x) \equiv 0$.

再将式 (1.2.16) 代入式 (1.2.19) 可以得到

$$S_{12} \equiv \frac{A_{21}}{A_{12}} = -\frac{\mathrm{e}^{\mathrm{i}(k_1+k_2)} + 1 - 2\mathrm{e}^{\mathrm{i}k_2}}{\mathrm{e}^{\mathrm{i}(k_1+k_2)} + 1 - 2\mathrm{e}^{\mathrm{i}k_1}}. \tag{1.2.20}$$

上式即所谓的二体散射矩阵. 它唯一确定了二体波函数的形式, 对波函数加上周期性边条件 $\psi(x_1 + N, x_2) = \psi(x_1, x_2) = \psi(x_1, x_2 + N)$, 则我们可以确定 k_1, k_2 的可能取值

$$\mathrm{e}^{\mathrm{i}k_1 N} = S_{12}^{-1}, \quad \mathrm{e}^{\mathrm{i}k_2 N} = S_{21}^{-1}, \tag{1.2.21}$$

这就是二体情形的Bethe Ansatz方程.

对于任意自旋反转数 M 情形, 系统的本征矢可以写为

$$|k_1, \cdots, k_M\rangle = \sum_{\{x_1, \cdots, x_M\}=1}^{N} \psi(x_1, \cdots, x_M) S_{x_1}^- \cdots S_{x_M}^- |0\rangle. \tag{1.2.22}$$

对于 $\psi(x_1, \cdots, x_M)$, 我们仍然有限制条件

$$\psi(x_1, \cdots, x_M)|_{x_i = x_j} \equiv 0. \tag{1.2.23}$$

将 H 作用到 $|k_1, \cdots, k_M\rangle$ 上, 我们得到本征方程

$$J \sum_{j=1}^{M} \sum_{\delta = \pm 1} \psi(x_1, \cdots, x_j + \delta, \cdots, x_M) + (h - 2J)M\psi(x_1, \cdots, x_M)$$
$$+ 2J \sum_{i<j} \psi(x_1, \cdots, x_M)[\delta_{x_i, x_j+1} + \delta_{x_i, x_j-1}] = (E - E_0)\psi(x_1, \cdots, x_M). \tag{1.2.24}$$

上述差分方程在每个反转自旋都不相邻时是平庸的,这时它的解应为平面波形式. 因此,我们可以对系统的波函数做如下 Ansatz:

$$\psi(x_1,\cdots,x_M) = \sum_{P,Q} A_P \mathrm{e}^{\mathrm{i}\sum_{j=1}^M k_{P_j} x_{Q_j}} \theta(x_{Q_1} < \cdots < x_{Q_M}), \quad (1.2.25)$$

其中, $P \equiv (P_1,\cdots,P_M)$, $Q \equiv (Q_1,\cdots,Q_M)$ 表示 $(1,2,\cdots,M)$ 的置换; $\theta(x_{Q_1} < \cdots < x_{Q_M})$ 是广义阶梯函数 $\theta(x_{QM}-x_{QM-1})\cdots\theta(x_{Q_2}-x_{Q_1})$. 不难看出, 式 (1.2.25) 是一个全对称波函数, 因此我们可以只考虑一个特殊区域 $x_{Q_1} < x_{Q_2} < \cdots < x_{Q_M}$. 对本征方程 (1.2.24) 的求解包括: ①求得本征值 E; ②确定常数 A_P; ③利用边条件确定动量 k_j 的可能取值. 当所有的反转自旋都不相邻时, 将式 (1.2.25) 代入式 (1.2.24) 中可以得到

$$E - E_0 = 2J \sum_{j=1}^M (\cos k_j - 1) + Mh. \quad (1.2.26)$$

假定两个反转自旋相邻而其余的都不相邻, 写出与这两个自旋的有关项

$$\psi(\cdots,x_{Q_i}-1,x_{Q_i}+1,\cdots) + \psi(\cdots,x_{Q_i},x_{Q_i}+2,\cdots) - 2\psi(\cdots,x_{Q_i},x_{Q_i}+1,\cdots)$$
$$= 2(\cos k_{P_i} + \cos k_{P_{i+1}})\psi(\cdots,x_{Q_i},x_{Q_i}+1,\cdots). \quad (1.2.27)$$

与二粒子情形相同, 我们可以得到

$$\frac{A_{P_1\cdots P_{i+1}P_i\cdots P_M}}{A_{P_1\cdots P_i P_{i+1}\cdots P_M}} = -\frac{\mathrm{e}^{\mathrm{i}(k_{P_i}+k_{P_{i+1}})} + 1 - 2\mathrm{e}^{\mathrm{i}k_{P_{i+1}}}}{\mathrm{e}^{\mathrm{i}(k_{P_i}+k_{P_{i+1}})} + 1 - 2\mathrm{e}^{\mathrm{i}k_{P_i}}}. \quad (1.2.28)$$

可以验证, 当若干反转自旋相邻时, 式 (1.2.28) 仍然保证式 (1.2.25) 是 H 的本征波函数, 因此波函数被唯一确定. 这时 A_P 中只有一个是独立的. 利用波函数的周期性边条件 $\psi(\cdots,x_j+N,\cdots) = \psi(\cdots,x_j,\cdots)$, 我们很容易得到

$$A_{P_1\cdots P_M} = A_{P_2\cdots P_M P_1} \mathrm{e}^{\mathrm{i}k_{P_1}N}. \quad (1.2.29)$$

再利用关系式 (1.2.28):

$$\frac{A_{P_1\cdots P_M}}{A_{P_2\cdots P_M P_1}} = \frac{A_{P_1\cdots P_M}}{A_{P_2 P_1\cdots P_M}} \frac{A_{P_2 P_1\cdots P_M}}{A_{P_2 P_3 P_1\cdots P_M}} \cdots \frac{A_{P_2\cdots P_1 P_M}}{A_{P_2\cdots P_M P_1}}, \quad (1.2.30)$$

上式右边正是二体散射矩阵倒数的连乘. 令 $P_1 = j$, 则我们得到

$$\mathrm{e}^{\mathrm{i}k_j N} = \prod_{l \neq j}^M S_{jl}^{-1}, \quad j = 1,\cdots,M. \quad (1.2.31)$$

1.2 自旋链模型

这就是著名的 Bethe Ansatz 方程, 它确定了 k_j 的可能取值. 为了方便起见, 我们将 k_j 参数化为

$$\mathrm{e}^{\mathrm{i}k_j} = \frac{\lambda_j - \frac{\mathrm{i}}{2}}{\lambda_j + \frac{\mathrm{i}}{2}}, \tag{1.2.32}$$

则二体散射矩阵可以化为

$$S_{ij} = \frac{\lambda_i - \lambda_j + \mathrm{i}}{\lambda_i - \lambda_j - \mathrm{i}}. \tag{1.2.33}$$

最后, 我们得到 Bethe Ansatz 方程及本征能量:

$$\left(\frac{\lambda_j - \frac{\mathrm{i}}{2}}{\lambda_j + \frac{\mathrm{i}}{2}}\right)^N = -\prod_{l=1}^{M} \frac{\lambda_j - \lambda_l - \mathrm{i}}{\lambda_j - \lambda_l + \mathrm{i}}, \tag{1.2.34}$$

$$E(\lambda_1, \cdots, \lambda_M) = -\sum_{j=1}^{M} \left(\frac{J}{\lambda_j^2 + \frac{1}{4}} - h\right) + E_0. \tag{1.2.35}$$

1.2.2 基态及低能元激发

在研究 Heisenberg 反铁磁链的基态性质之前, 我们首先证明下述定理: 如果 $\{\lambda_1, \cdots, \lambda_M\}$ 是 Bethe Ansatz 方程 (1.2.34) 的一组解, 则 λ_j 两两不等. 给定一个坐标区间 $x_{Q_1} < \cdots < x_{Q_M}$, 考虑波函数 $\psi(x_1, \cdots, x_M)$ 的两项:

$$A_{\cdots P_i \cdots P_j \cdots} \mathrm{e}^{\cdots + \mathrm{i}k_{P_i} x_{Q_i} + \mathrm{i}k_{P_j} x_{Q_j} + \cdots} + A_{\cdots P_j \cdots P_i \cdots} \mathrm{e}^{\cdots + \mathrm{i}k_{P_j} x_{Q_i} + \mathrm{i}k_{P_i} x_{Q_j} + \cdots}. \tag{1.2.36}$$

当 $k_{P_j} = k_{P_i}$ 时, 利用式 (1.2.28) 可以得到

$$A_{\cdots P_j \cdots P_i \cdots} = -A_{\cdots P_i \cdots P_j \cdots}. \tag{1.2.37}$$

因此式 (1.2.36) 为零. 由于 P_i, P_j 的任意性, 当 $k_j = k_i$ 时, $\psi(x_1, \cdots, x_M) = 0$, 即 $k_i = k_j$ 或 $\lambda_i = \lambda_j$ 是不允许的.

对式 (1.2.34) 取对数可得

$$\theta_1(\lambda_j) = \frac{2\pi I_j}{N} + \frac{1}{N} \sum_{l=1}^{M} \theta_2(\lambda_j - \lambda_l), \tag{1.2.38}$$

其中, $\theta_n(x) = 2\arctan(2x/n)$; I_j 取整数 (当 $N - M$ 为奇数) 或半整数 (当 $N - M$ 为偶数). 由于 λ_j 两两不等, 所以 I_j 两两不等. 注意, $k_j = \theta_1(\lambda_j)$, $2\pi I_j/N$ 正是无

相互作用时自旋波波矢 k_j 的量子化取值,因此相互作用系统的波矢 k_j 与自由系统的波矢有着一一对应关系. 定义

$$\begin{cases} Z(\lambda) = \dfrac{1}{2\pi}\left[\theta_1(\lambda) - \dfrac{1}{N}\sum_{l=1}^{M}\theta_2(\lambda-\lambda_l)\right], \\ \sigma(\lambda) = \dfrac{\mathrm{d}Z(\lambda)}{\mathrm{d}\lambda}. \end{cases} \quad (1.2.39)$$

显然, $Z(\lambda_i) = I_j/N$. 当 $N \to \infty, M \to \infty, M/N$ 有限时, λ_j 趋于连续分布. 对于反铁磁 $(J > 0)$ 基态, I_j 也应该取连续值, 这时 $\sigma(\lambda)$ 可以被看成翻转自旋在 λ 点上的数密度. 对式 (1.2.39) 求导并假定 λ 的取值限于两个费米点 $\pm\Lambda$ 之间, 可以得到

$$\sigma(\lambda) = a_1(\lambda) - \int_{-\Lambda}^{\Lambda} \sigma(\mu) a_2(\lambda-\mu)\mathrm{d}\mu, \quad (1.2.40)$$

其中

$$a_n(\lambda) = \frac{1}{2\pi}\frac{n}{\lambda^2 + n^2/4}. \quad (1.2.41)$$

当没有外磁场时, 由式 (1.2.35) 我们可以看出每个实 λ_j 都贡献一个负值, 因此 Λ 在基态尽可能取大, 即 $\Lambda = \infty$, 这时式 (1.2.40) 可以用 Fourier 变换求解. 令 $f(\lambda)$ 的 Fourier 变换为 $\widetilde{f}(\omega)$,

$$\widetilde{f}(\omega) = \int \mathrm{e}^{\mathrm{i}\omega\lambda}f(\lambda)\mathrm{d}\lambda, \quad f(\lambda) = \frac{1}{2\pi}\int \mathrm{e}^{-\mathrm{i}\omega\lambda}\widetilde{f}(\omega)\mathrm{d}w, \quad (1.2.42)$$

则

$$\widetilde{a}_n(\omega) = \mathrm{e}^{-\frac{n}{2}|\omega|}. \quad (1.2.43)$$

从式 (1.2.40) 可以解得

$$\widetilde{\sigma}(\omega) = \frac{\widetilde{a}_1(\omega)}{1+\widetilde{a}_2(\omega)} = \frac{1}{2\cosh\dfrac{\omega}{2}}, \quad (1.2.44)$$

因此, λ 的基态分布为

$$\sigma(\lambda) = \frac{1}{2\cosh(\pi\lambda)}. \quad (1.2.45)$$

基态的自旋反转数密度为

$$\frac{M}{N} = \int \sigma(\lambda)\mathrm{d}\lambda = \widetilde{\sigma}(0) = \frac{1}{2}. \quad (1.2.46)$$

这正是我们期望的结果. 这时 $\frac{N}{2} - M = 0$, 即在 $h = 0$ 时没有自发磁化. 注意, 当 N 为奇数时, 基态存在一个自由的剩余自旋, 它对应于 λ 海中无穷远处的一个空穴. 同样, 式 (1.2.35) 的热力学极限给出基态能量密度:

$$\begin{aligned}E/N &= -2\pi J \int a_1(\lambda)\sigma(\lambda)\mathrm{d}\lambda + E_0/N \\ &= -J \int \tilde{a}_1(\omega)\tilde{\sigma}(\omega)\mathrm{d}\omega + \frac{1}{2}J = \left(\frac{1}{2} - 2\ln 2\right)J.\end{aligned} \quad (1.2.47)$$

显然, 式 (1.2.47) 小于经典 Neel 态 $|\uparrow\rangle_1 \otimes |\downarrow\rangle_2 \otimes \cdots \otimes |\uparrow\rangle_{N-1} \otimes |\downarrow\rangle_N$ 的能量期望值 $-J/2$. 这说明量子涨落会进一步降低能量.

在讨论元激发之前, 我们首先定义准粒子能量 $\varepsilon(\lambda)$:

$$\varepsilon(\lambda) = \varepsilon_0(\lambda) - \int a_2(\lambda - \mu)\varepsilon(\mu)\mathrm{d}\mu, \quad (1.2.48)$$

其中 $\varepsilon_0(\lambda) = 2\pi a_1(\lambda)$. 利用 Fourier 变换可以求得

$$\varepsilon(\lambda) = \frac{1}{2\cosh(\pi\lambda)}. \quad (1.2.49)$$

系统最简单的元激发是将一个向下的自旋反转向上, 这时表示 I_j 序列中的其中一个不被占据或在 λ 海中产生空穴. 考虑 N 为偶数, 其基态构型如图 1.3 所示. 翻转一个自旋的激发态如图 1.4 所示. 由图 1.4 可以看出, 翻转一个自旋相当于在 λ 海中加入两个空穴. 设这两个空穴在 λ 海中的位置分别为 λ_1^h 和 λ_2^h, 则 $Z(\lambda_j^h) = I_j^h/N$ 是式 (1.2.39) 的解, 因此我们有

$$\sigma(\lambda) + \sigma_h(\lambda) = \frac{\mathrm{d}Z(\lambda)}{\mathrm{d}\lambda} = a_1(\lambda) - \int a_2(\lambda - \mu)\sigma(\mu)\mathrm{d}\mu, \quad (1.2.50)$$

其中, $\sigma_h(\lambda)$ 表示空穴密度,

$$\sigma_h(\lambda) = \frac{1}{N}[\delta(\lambda - \lambda_1^h) + \delta(\lambda - \lambda_2^h)]. \quad (1.2.51)$$

图 1.3 基态构型

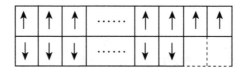

图 1.4 翻转一个自旋的激发态

对式 (1.2.50) 做 Fourier 变换可以求得 $\tilde{\sigma}(\omega)$ 相对于基态态密度的变化 $\delta\tilde{\sigma}(\omega)$:

$$\delta\tilde{\sigma}(\omega) = -\frac{1}{N}\frac{\mathrm{e}^{\mathrm{i}\lambda_1^h\omega} + \mathrm{e}^{\mathrm{i}\lambda_2^h\omega}}{1 + \mathrm{e}^{-|\omega|}}. \qquad (1.2.52)$$

元激发能量则为

$$\begin{aligned}\delta E &= -2\pi N\int \delta\sigma(\lambda)a_1(\lambda)\mathrm{d}\lambda\\ &= -N\int \delta\tilde{\sigma}(\omega)\tilde{a}_1(\omega)\mathrm{d}\omega = \varepsilon(\lambda_1^h) + \varepsilon(\lambda_2^h).\end{aligned} \qquad (1.2.53)$$

因此, 元激发能量可以表达为两个空穴的准粒子能量之和. 进一步我们可以求得元激发所携带的自旋为

$$S = -N\int \delta\sigma(\lambda)\mathrm{d}\lambda = -N\delta\tilde{\sigma}(0) = 1. \qquad (1.2.54)$$

显然, 这是一个三重态元激发. 与自旋波概念不同, 在一维中, 三重态元激发劈裂为两个独立的自旋为 $\frac{1}{2}$ 的准粒子, 通常被称为自旋子. 这种现象首先由 Takhtajan 和 Faddeev 观察到[15]. 实际上, 这两个准粒子对应于一个扭折和反扭折对, 分别以不同的速度独立运动, 如图 1.5 所示. 扭折和反扭折之间所含有的自旋数为单数, 保证了激发态总自旋为 1.

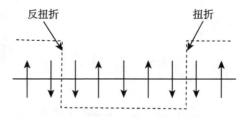

图 1.5 翻转一个自旋的激发态

在以上的讨论中, 我们只用到了 Bethe Ansatz 方程的实解. 实际上, 式 (1.2.34) 允许有复数解. 一般的复解可以表示为

$$\lambda_{j,\alpha}^{(n)} = \lambda_\alpha^{(n)} - \frac{\mathrm{i}}{2}(n+1-2j) + o(\mathrm{e}^{-\delta N}), \quad j = 1, 2, \cdots, n. \qquad (1.2.55)$$

式 (1.2.55) 即著名的弦假设[16], 其中 $\lambda_\alpha^{(n)}$ 表示第 α 个 n 弦在实轴上的位置, δ 是某个正数. 在热力学极限下, 弦对称于实轴沿虚轴方向等距排列. $n=1$ 时则对应于实解. 假定 Bethe Ansatz 方程的解由一系列这样的弦构成, 将式 (1.2.55) 代入式 (1.2.34) 中并对 j 连乘, 我们得到

$$\prod_{j=1}^{n}\left(\frac{\lambda_{j,\alpha}^{n}-\frac{\mathrm{i}}{2}}{\lambda_{j,\alpha}+\frac{\mathrm{i}}{2}}\right)^{N}=\prod_{j=1}^{n}\prod_{m=1}^{\infty}\prod_{l,\beta\neq j,\alpha}\frac{\lambda_{j,\alpha}^{(n)}-\lambda_{l,\beta}^{(m)}-\mathrm{i}}{\lambda_{j,\alpha}^{(n)}-\lambda_{l,\beta}^{(m)}+\mathrm{i}}, \tag{1.2.56}$$

略去无穷小量 $o(\mathrm{e}^{-\delta N})$ 可得

$$\left(\frac{\lambda_\alpha^{(n)}-\frac{\mathrm{i}}{2}n}{\lambda_\alpha^{(n)}+\frac{\mathrm{i}}{2}n}\right)^N$$

$$=-\prod_{m=1}^{\infty}\prod_{\beta}\frac{\lambda_\alpha^{(n)}-\lambda_\beta^{(m)}-\frac{\mathrm{i}}{2}(m+n)}{\lambda_\alpha-\lambda_\beta^{(m)}+\frac{\mathrm{i}}{2}(m+n)}\left[\frac{\lambda_\alpha^{(n)}-\lambda_\beta^{(m)}-\frac{\mathrm{i}}{2}(m+n-2)}{\lambda_\alpha^{(n)}-\lambda_\beta^{(m)}+\frac{\mathrm{i}}{2}(m+n-2)}\right]^2$$

$$\times\cdots\times\left[\frac{\lambda_\alpha^{(n)}-\lambda_\beta^{(m)}-\frac{\mathrm{i}}{2}(|m-n|+2)}{\lambda_\alpha^{(n)}-\lambda_\beta^{(m)}+\frac{\mathrm{i}}{2}(|m-n|+2)}\right]^2\frac{\lambda_\alpha^{(n)}-\lambda_\beta^{(m)}-\frac{\mathrm{i}}{2}|m-n|}{\lambda_\alpha^{(n)}-\lambda_\beta^{(m)}+\frac{\mathrm{i}}{2}|m-n|}. \tag{1.2.57}$$

对上式取对数可得

$$\theta_n(\lambda_\alpha^{(n)})=\frac{2\pi I_\alpha^n}{N}+\frac{1}{N}\sum_{m,\beta}\theta'_{m,n}(\lambda_\alpha^{(n)}-\lambda_\beta^{(m)}), \tag{1.2.58}$$

其中, I_α^n 取整数或半整数,

$$\theta'_{m,n}(\lambda)=\theta_{m+n}(\lambda)+2\theta_{m+n-2}(\lambda)+\cdots+2\theta_{|m-n|+2}(\lambda)+(1+\delta_{m,n})\theta_{|m-n|}(\lambda). \tag{1.2.59}$$

定义

$$Z_n(\lambda)=\frac{1}{2\pi}\left[\theta_n(\lambda)-\frac{1}{N}\sum_{m,\beta}\theta'_{m,n}(\lambda-\lambda_\beta^{(m)})\right]. \tag{1.2.60}$$

显然 $Z_n(\lambda_\alpha^{(n)})=I_\alpha^n/N$ 对应于 Bethe Ansatz 方程的解. 在热力学极限下,

$$\frac{\mathrm{d}Z_n(\lambda)}{\mathrm{d}\lambda}=\sigma_n(\lambda)+\sigma_n^h(\lambda), \tag{1.2.61}$$

其中, $\sigma_n(\lambda)$ 表示 n 弦在 λ 空间的密度; $\sigma_n^h(\lambda)$ 则表示 n 弦空穴在 λ 空间的密度. 翻转自旋密度数则由下式给出:

$$\frac{M}{N} = \sum_{n=1}^{\infty} n \int \sigma_n(\lambda) \mathrm{d}\lambda. \tag{1.2.62}$$

对式 (1.2.60) 取导数, 我们得到 $\sigma_n^h(\lambda)$ 与 $\sigma_m(\lambda)$ 的关系:

$$\sigma_n^h(\lambda) = a_n(\lambda) - \sum_{m=1}^{\infty} \int A_{m,n}(\lambda-\mu)\sigma_m(\mu)\mathrm{d}\mu, \tag{1.2.63}$$

其中

$$\begin{cases} A_{m,n}(\lambda) = a_{m+n}(\lambda) + 2a_{m+n-2}(\lambda) + \cdots + 2a_{|m-n|+2}(\lambda) + a_{|m-n|}(\lambda), \\ a_0(\lambda) \equiv \delta(\lambda). \end{cases} \tag{1.2.64}$$

式 (1.2.63) 是一个非常重要的公式, 它是讨论所有元激发和热力学行为的基础. 为了进一步给出元激发的图像, 我们考虑一个最简单的弦激发, 即 2 弦激发, 这种元激发相当于在实轴上产生两个空穴 λ_1^h 和 λ_2^h 并加入一个 2 弦 $\lambda_s \pm \mathrm{i}/2$, 则

$$\sigma_1^h(\lambda) = \frac{1}{N}[\delta(\lambda-\lambda_1^h) + \delta(\lambda-\lambda_2^h)], \tag{1.2.65}$$

$$\sigma_2(\lambda) = \frac{1}{N}\delta(\lambda-\lambda_s). \tag{1.2.66}$$

取 (1.2.63) 中 $n=1$ 情形:

$$\sigma_1(\lambda) + \sigma_1^h(\lambda) = a_1(\lambda) - \int a_2(\lambda-\mu)\sigma_1(\mu)\mathrm{d}\mu$$
$$- \int [a_1(\lambda-\mu) + a_3(\lambda-\mu)]\sigma_2(\mu)\mathrm{d}\mu. \tag{1.2.67}$$

将式 (1.2.65) 和式 (1.2.66) 代入式 (1.2.67), 可以求得 $\sigma_1(\lambda)$ 相对于基态密度变化:

$$\delta\widetilde{\sigma}_1(\omega) = -\frac{\mathrm{e}^{\mathrm{i}\lambda_1^h\omega} + \mathrm{e}^{\mathrm{i}\lambda_2^h\omega}}{N(1+\mathrm{e}^{-|\omega|})} - \frac{\mathrm{e}^{-\frac{1}{2}|\omega|} + \mathrm{e}^{-\frac{3}{2}|\omega|}}{N(1+\mathrm{e}^{-|\omega|})}\mathrm{e}^{\mathrm{i}\lambda_s\omega}, \tag{1.2.68}$$

由此可得元激发能量为

$$\delta E = -N\int \delta\widetilde{\sigma}_1(\omega)\widetilde{a}_1(\omega)\mathrm{d}\omega - 2\pi\left[a_1\left(\lambda_s+\frac{\mathrm{i}}{2}\right) + a_1\left(\lambda_s-\frac{\mathrm{i}}{2}\right)\right]$$
$$= \varepsilon(\lambda_1^h) + \varepsilon(\lambda_2^h). \tag{1.2.69}$$

由上式我们清楚地看到, 激发能的表达式与式 (1.2.53) 完全相同, 即只依赖于两个空穴而似乎与 2 弦无关. 2 弦对能量的直接贡献与其所引起的 λ 海重组对能量的

影响相互抵消. 实际上, 当考虑两个空穴之间的散射时, 它们的 S-矩阵是与弦有关的. 再考察 2 弦激发的磁化:

$$S = \frac{1}{2}\left(N - 2N\int \sigma_1(\lambda)\mathrm{d}\lambda - 4\int \sigma_2(\lambda)\mathrm{d}\lambda\right) = 0. \tag{1.2.70}$$

因此, 这种激发是一种自旋单态元激发, 具体图像可用图 1.6 表征. 注意, 这时扭折和反扭折畴中的自旋个数为偶数.

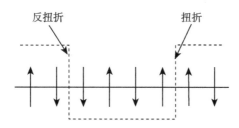

图 1.6 翻转零个自旋的激发态

1.2.3 热力学性质

可积模型的热力学研究方法是 1969 年由杨振宁和杨振平[17] 提出的. 随后, Takahashi 基于弦假设, 研究了一系列模型的热力学[16]. 在有限温度下, 系统的自由能可以表达为

$$F = E - TS, \tag{1.2.71}$$

其中, E 是系统的能量; S 是系统的熵. 注意一个 n 弦所携带的能量为

$$\begin{aligned}\varepsilon_n^0(\lambda) &= J\sum_{j=1}^n\left[\frac{\lambda + \frac{\mathrm{i}}{2}(n+1-2j) - \frac{\mathrm{i}}{2}}{\lambda + \frac{\mathrm{i}}{2}(n+1-2j) + \frac{\mathrm{i}}{2}} + \frac{\lambda + \frac{\mathrm{i}}{2}(n+1-2j) + \frac{\mathrm{i}}{2}}{\lambda + \frac{\mathrm{i}}{2}(n+1-2j) - \frac{\mathrm{i}}{2}} - 2\right] + nh \\ &= -2\pi J a_n(\lambda) + nh. \end{aligned} \tag{1.2.72}$$

其中, nh 是外磁场的贡献, 我们得到

$$E/N - \frac{1}{2}(J-h) = -2\pi J\sum_{n=1}^{\infty}\int \varepsilon_n^0(\lambda)\sigma_n(\lambda)\mathrm{d}\lambda. \tag{1.2.73}$$

系统的熵可由状态数求得. 考虑一个小区间 $[\lambda, \lambda+\mathrm{d}\lambda]$, 其中 n-弦的可占据数为 $N[\sigma_n(\lambda)+\sigma_n^h(\lambda)]\mathrm{d}\lambda$, 占据数和空位数则分别为 $N\sigma_n(\lambda)\mathrm{d}\lambda$, $N\sigma_n^h(\lambda)\mathrm{d}\lambda$, 因此这个小区间内可能的状态数为

$$\mathrm{d}\Omega(\lambda) = \prod_{n=1}^{\infty}\frac{[N(\sigma_n(\lambda)+\sigma_n^h(\lambda))\mathrm{d}\lambda]!}{[N\sigma_n(\lambda)\mathrm{d}\lambda]![N\sigma_n^h(\lambda)\mathrm{d}\lambda]!}. \tag{1.2.74}$$

利用 Sterring 公式 $\ln N! \approx N \ln N$，我们可以得到小区间内的熵为

$$dS(\lambda) = \ln d\Omega(\lambda) \approx N \sum_n \{[\sigma_n(\lambda) + \sigma_n^h(\lambda)] \ln[\sigma_n(\lambda) + \sigma_n^h(\lambda)]$$
$$- \sigma_n(\lambda) \ln \sigma_n(\lambda) - \sigma_n^h(\lambda) \ln \sigma_n^h(\lambda)\} d\lambda. \tag{1.2.75}$$

定义相对自由能密度

$$f = F/N - \frac{1}{2}(J - h). \tag{1.2.76}$$

将式 (1.2.73) 和式 (1.2.75) 代入式 (1.2.71) 可得

$$f = \sum_n \int \varepsilon_n^0(\lambda) \sigma_n(\lambda) d\lambda - T \sum_n \int \{[\sigma_n(\lambda) + \sigma_n^h(\lambda)] \ln[\sigma_n(\lambda) + \sigma_n^h(\lambda)]$$
$$- \sigma_n(\lambda) \ln \sigma_n(\lambda) - \sigma_n^h(\lambda) \ln \sigma_n^h(\lambda)\} d\lambda. \tag{1.2.77}$$

热力学平衡态由自由能极小即 $\delta f/\delta \sigma_n(\lambda) = 0$ 得到. 对式 (1.2.77) 取变分可得

$$\sum_n \int \{\varepsilon_n^0(\lambda) \delta\sigma_n(\lambda) - T \ln[1 + \eta_n(\lambda)] \delta\sigma_n(\lambda) - T \ln[1 + \eta_n^{-1}(\lambda)] \delta\sigma_n^h(\lambda)\} d\lambda = 0, \tag{1.2.78}$$

其中, $\eta_n(\lambda) \equiv \sigma_n^h(\lambda)/\sigma_n(\lambda)$. 注意 $\delta\sigma_n(\lambda)$ 与 $\delta\sigma_m^h(\lambda)$ 互相不独立. 由式 (1.2.63) 可知

$$\delta\sigma_n^h(\lambda) = -\sum_{m=1}^{\infty} \int A_{m,n}(\lambda - \mu) \delta\sigma_m(\mu) d\mu. \tag{1.2.79}$$

将式 (1.2.79) 代入式 (1.2.78) 中并令 $\delta\sigma_n(\lambda)$ 的系数为零可得

$$\ln[1 + \eta_n(\lambda)] = \frac{\varepsilon_n^0(\lambda)}{T} + \sum_{m=1}^{\infty} \int A_{m,n}(\lambda - \mu) \ln[1 + \eta_m^{-1}(\mu)] d\mu. \tag{1.2.80}$$

为了方便起见，我们引入积分算子 $[n]$, $\hat{A}_{m,n}$:

$$[n]F(\lambda) \equiv \int a_n(\lambda - \mu) F(\mu) d\mu, \tag{1.2.81}$$

$$\hat{A}_{m,n} F(\lambda) \equiv \int A_{m,n}(\lambda - \mu) F(\mu) d\mu, \tag{1.2.82}$$

其中, $F(\lambda)$ 是任意函数, 在 Fourier 变换下, 卷积变为乘积. 因此, $[n]$ 的作用相当于一个乘子 $\exp(-|\omega|n/2)$. 可以证明下述关系成立：

$$[m][n] = [m+n]. \tag{1.2.83}$$

1.2 自旋链模型

再定义算子 $\hat{G} = [1]/([0]+[2])$, 则它的积分核为

$$g(\lambda) = \frac{1}{2\pi}\int e^{-i\lambda\omega}\frac{e^{-\frac{1}{2}|\omega|}}{1+e^{-|\omega|}}d\omega = \frac{1}{2\cosh(\pi\lambda)}. \quad (1.2.84)$$

同样可以证明

$$\begin{cases} \hat{G}[\hat{A}_{m,n+1} + \hat{A}_{m,n-1}] = -\delta_{m,n} + \hat{A}_{m,n}, & n > 1, \\ \hat{G}\hat{A}_{m,2} = -\delta_{1,m} + \hat{A}_{1,m}. \end{cases} \quad (1.2.85)$$

式 (1.2.80) 可以重新写为

$$\ln(1+\eta_n) = \frac{\varepsilon_n^0(\lambda)}{T} + \sum_m \hat{A}_{n,m}\ln(1+\eta_m^{-1}). \quad (1.2.86)$$

取 $n+1$ 和 $n-1$ 时式 (1.2.86) 之和并作用以 $\hat{G}$, 当 $n > 1$ 时得到

$$\begin{aligned}&\hat{G}[\ln(1+\eta_{n+1}) + \ln(1+\eta_{n-1})]\\&= \frac{1}{T}\hat{G}(\varepsilon_{n+1}^0 + \varepsilon_{n-1}^0) + \sum_m \hat{G}(\hat{A}_{n+1,m} + \hat{A}_{n-1,m})\ln(1+\eta_m^{-1})\\&= \frac{\varepsilon_n^0}{T} - \ln(1+\eta_n^{-1}) + \sum_m \hat{A}_{n,m}\ln(1+\eta_m^{-1}).\end{aligned} \quad (1.2.87)$$

利用式 (1.2.86) 和式 (1.2.87) 可得

$$\ln\eta_n = \hat{G}[\ln(1+\eta_{n+1}) + \ln(1+\eta_{n-1})]. \quad (1.2.88)$$

再将 $\hat{G}$ 作用到式 (1.2.86) 并且取 $n = 2$, 可得

$$\ln\eta_1 = -\frac{2\pi g(\lambda)}{T} + \hat{G}\ln(1+\eta_2). \quad (1.2.89)$$

当 $n \to \infty$ 时, 由式 (1.2.86) 可以看出

$$\lim_{n\to\infty}\frac{\ln\eta_n}{n} = \frac{h}{T}. \quad (1.2.90)$$

式 (1.2.88)~ 式 (1.2.90) 构成了热力学量 η_n 的封闭方程.

再来考察自由能密度 f 并将之用 η_n 表示出来:

$$\begin{aligned}f = \sum_n \int \varepsilon_n^0(\lambda)\sigma_n(\lambda)d\lambda - T\sum_n \bigg\{&\int \sigma_n(\lambda)\ln[1+\eta_n(\lambda)]d\lambda\\&+ \int \sigma_n^h(\lambda)\ln[1+\eta_n^{-1}(\lambda)]d\lambda\bigg\}.\end{aligned} \quad (1.2.91)$$

将式 (1.2.63) 代入式 (1.2.91) 并利用式 (1.2.78), 可得

$$f = -T \sum_n \int a_n^0(\lambda) \ln(1 + \eta_n^{-1}) d\lambda. \tag{1.2.92}$$

注意, $\hat{A}_{1,m} = [m+1] + [m-1]$, 由式 (1.2.86) 可得

$$\hat{G} \ln(1 + \eta_1) = \frac{1}{T} \hat{G} \varepsilon_1^0 + \sum_{m=1}^{\infty} [m] \ln(1 + \eta_m^{-1}). \tag{1.2.93}$$

上式中令宗量 $\lambda = 0$ 可得

$$\sum_{m=1}^{\infty} \int a_m(\lambda) \ln[1 + \eta_m^{-1}(\lambda)] d\lambda = \frac{2\ln 2 J - \frac{1}{2} h}{T} + \int g(\lambda) \ln[1 + \eta_1(\lambda)] d\lambda. \tag{1.2.94}$$

将式 (1.2.94) 代入式 (1.2.92), 我们得到自由能的最终表达形式:

$$F/N = e_0 - T \int g(\lambda) \ln[1 + \eta_1(\lambda)] d\lambda, \tag{1.2.95}$$

其中, $e_0 = \left(-2\ln 2 + \frac{1}{2}\right) J$ 是无外磁场时的基态能量密度. 尽管我们开始时利用了弦假设, 但自由能只与实解的分布有关. 一般说来, 热力学 Bethe Ansatz 方程 (1.2.88)~(1.2.90) 不能严格求解, 但是利用数值方法, 原则上我们可以在任意精度下求解. 下面我们讨论两个极限情况, 即 $T \to 0$ 和 $T \to \infty$ 时的渐近解.

当 $T \to 0$, $h \to 0$ 时, 由于 $g(\lambda) > 0$, 方程 (1.2.89) 中的驱动项趋于负无穷, 因此 $\eta_1(\lambda) \to 0$, 这说明在零温下 $\sigma_1^h(\lambda) = 0$, 即所有实解都被填满, 它对应于我们在 1.2 节所讨论的基态. 这时所有的 $\eta_n(\lambda)$ 变为与 λ 无关的常数, 热力学方程退化为下列代数方程:

$$\eta_n^2 = (1 + \eta_{n+1})(1 + \eta_{n-1}), \quad n > 1. \tag{1.2.96}$$

这组方程的通解为

$$\eta_n = \left(\frac{az^n - a^{-1}z^{-n}}{z - z^{-1}}\right)^2 - 1, \tag{1.2.97}$$

其中, 参数 a 和 z 由边界条件 $\eta_1 = 0$ 和式 (1.2.90) 决定, $a = 1$, $z = e^{\frac{h}{2T}}$, 因此当 $T \to 0$ 时,

$$\eta_n = \frac{\sinh^2 \frac{nh}{2T}}{\sinh^2 \frac{h}{2T}} - 1. \tag{1.2.98}$$

在无外磁场时, $\eta_n(h=0) = n^2 - 1$. 令 $\eta_n = \exp(-\varepsilon_n/T)$, 则 $\varepsilon_1 \sim T^0$, 取式 (1.2.86) 中 $n=1$ 情形并比较主导项 T^{-1} 的系数可得

$$\varepsilon_1(\lambda) = -\varepsilon_0(\lambda) - \int \varepsilon_1(\mu) a_2(\lambda - \mu) \mathrm{d}\mu. \tag{1.2.99}$$

比较式 (1.2.99) 与式 (1.2.48), 我们得到 $\varepsilon_1(\lambda) = \varepsilon(\lambda)$, 即准粒子的能量. 注意 $\varepsilon(\lambda) = g(\lambda) = \sigma_0(\lambda)$. 以上的讨论表明, 利用基态的性质我们可以讨论自由能的低温展开. 基于 Landau 的费米液体理论, 准粒子动量对应于无相互作用时的裸粒子动量, 因此在连续情形, 我们可以定义准粒子动量 $p(\lambda) = 2\pi Z(\lambda)$. 准粒子的态密度为

$$N(\lambda) = \frac{1}{2\pi}\frac{\mathrm{d}p(\lambda)}{\mathrm{d}\varepsilon(\lambda)} = \frac{g(\lambda)}{\varepsilon'(\lambda)}. \tag{1.2.100}$$

基态时 $N(\lambda) = \pi^{-1}|\coth(\pi\lambda)|$, 在费米面上的取值为 $N(\infty) = \pi^{-1}$. 因此, 自由能可写为

$$(F - E_0)/N = -T \int N(\lambda) \ln\left[1 + \mathrm{e}^{-\frac{\varepsilon(\lambda)}{T}}\right] \mathrm{d}\varepsilon(\lambda). \tag{1.2.101}$$

当 $T \to 0$ 时, 积分核呈指数衰减, 这时只有费米面附近 ($\lambda \sim \pm\infty$) 的贡献才是重要的. 因此, $N(\lambda)$ 可近似取为 $N(\infty)$. 令 $x = \varepsilon/T$, 则

$$(F - E_0)/N \approx -\frac{T^2}{\pi}\int_{-\infty}^{\infty}\ln(1 + \mathrm{e}^{-x})\mathrm{d}x = -\frac{\pi}{6}T^2. \tag{1.2.102}$$

由此可求得反铁磁自旋链的低温比热为

$$C(T) = \frac{\pi}{3}T + o(T^2). \tag{1.2.103}$$

当 $T \to \infty$ 时, 式 (1.2.89) 中的驱动项趋近于零, η_n 也可近似为常数并满足方程 (1.2.96), 包括 $n=1$ 情形. 利用 $\eta_1 = 0$ 可以求得

$$\eta_n = \frac{\sinh^2\frac{(n+1)h}{2T}}{\sinh^2\frac{h}{2T}} - 1, \tag{1.2.104}$$

因此

$$\frac{F}{N} \to e_0 - T\ln\left(2\cosh\frac{h}{2T}\right). \tag{1.2.105}$$

利用上式可以求得高温磁化率为

$$\chi = -\frac{1}{N}\frac{\partial^2 F}{\partial h^2} \to \frac{1}{4T\cosh^2\frac{h}{2T}}. \tag{1.2.106}$$

当 $h \to 0$ 时, 上式正是 Curie 定律, 即自旋链的经典极限. 由式 (1.2.95) 可知, 系统的熵密度为

$$\frac{S}{N} = \int g(\lambda) \ln[1 + \eta_1(\lambda)] d\lambda. \tag{1.2.107}$$

将 η_1 的高温极限解代入上式可得 $S/N = \ln 2$, 这表示高温下所有的自旋退耦而自旋自由度冻结消失, 同样可以求得 $T = 0$ 时的熵密度为零, 表示所有自旋自由度冻结.

1.2.4 代数 Bethe Ansatz

反散射方法是在研究经典可积非线性方程时提出的. 众所周知, KdV 方程的解 $u(x,t)$ 可由散射方程

$$\left[-\frac{\partial^2}{\partial x^2} + u(x,t)\right]\psi(x,t,\lambda) = \lambda^2\psi(x,t,\lambda) \tag{1.2.108}$$

得到. 假定我们知道了散射数据 $\psi(\pm\infty, 0, \lambda)$, 则 $u(x,t)$ 可由这组数据表达出来, 反散射因此而得名. 另外, Lax[18] 曾经提出, 如果一组微分方程存在共轭对 $L(x,t)$, $M(x,t)$, 并且满足本征方程

$$\frac{\partial}{\partial x}\psi(x,t) = L(x,t)\psi(x,t), \tag{1.2.109}$$

$$\frac{\partial}{\partial t}\psi(x,t) = M(x,t)\psi(x,t), \tag{1.2.110}$$

则相容条件

$$\frac{\partial}{\partial t}L(x,t) - \frac{\partial}{\partial x}M(x,t) + [L(x,t), M(x,t)] = 0 \tag{1.2.111}$$

保证了这组方程的可积性. 对于格点模型, 相应的本征方程为

$$T_N(\lambda) = L_N(\lambda) T_{N-1}(\lambda), \tag{1.2.112}$$

$$\frac{\partial}{\partial t} T_N(\lambda) = M_N(\lambda) T_N(\lambda). \tag{1.2.113}$$

显然可令 $T_N = L_N L_{N-1} \cdots L_1$, 这时相容条件为

$$\frac{\partial}{\partial t} L_N + L_N M_{N-1} - M_N L_N = 0. \tag{1.2.114}$$

对于一个 N 格点系统, T_N 通常被称为单值矩阵, 它的迹 $\tau(\lambda) = \mathrm{tr} T_N(\lambda)$ 则被称为转移矩阵, 是可积统计模型和量子可积模型中一个非常重要的物理量. 利用相容性条件 (1.2.114) 可得

$$\begin{aligned}\frac{\partial \tau(\lambda)}{\partial t} &= \mathrm{tr} \sum_{j=1}^{N} L_N \cdots L_{j+1}[M_j \cdot L_j - L_j \cdot M_{j-1}]L_{j-1} \cdots L_1 \\ &= \mathrm{tr}(M_N T_N - T_N M_0),\end{aligned} \tag{1.2.115}$$

加上周期性边条件 $M_{j+N} = M_j$, 可得

$$\frac{\partial \tau(\lambda)}{\partial t} = 0. \qquad (1.2.116)$$

因此 $\tau(\lambda)$ 可以作为守恒量的产生泛函. 转移矩阵概念是在研究伊辛模型时提出的并在 Baxter 的工作[19] 中得到了发展. 考虑一个二维格子系统由图 1.7 中的顶角组成. 假定它的 Boltzmann 权重为 $L_{l,l+1}^{\alpha,\beta}(\lambda)$, 转移矩阵定义为

$$\tau^{\{\alpha\}\{\beta\}}(\lambda) = \sum_{\{l\}} L_{l_1,l_2}^{\alpha_1,\beta_1} L_{l_2,l_3}^{\alpha_2,\beta_2} \cdots L_{l_{N-1},l_N}^{\alpha_{N-1},\beta_{N-1}} L_{l_N,l_1}^{\alpha_N,\beta_N}, \qquad (1.2.117)$$

则系统的配分函数可以写为

$$Z = \mathrm{tr}\tau^M(\lambda), \qquad (1.2.118)$$

其中, N 和 M 分别为系统纵横方向的格点数. 因此, 得到了转移矩阵的本征值, 就可得到系统的热力学性质. 利用 Yang-Baxter 方程即 $L_{12}(\lambda - \mu)L_{13}(\lambda)L_{23}(\mu) = L_{23}(\mu)L_{13}(\lambda)L_{12}(\lambda - \mu)$, 可以严格对角化转移矩阵 $\tau(\lambda)$. Faddeev 等在经典反散射方法和 Baxter 工作的基础上, 发展了量子反散射方法[6, 7], 其主要宗旨是将 Lax 算子 $L_{l,m}^{\alpha,\beta}(\lambda)$ 中的 l,m 看为辅助矩阵元标号, 而将 α,β 看成是量子算子的矩阵元标号, 这样在转移矩阵 $\tau(\lambda)$ 的展开中可以构造出量子模型的哈密顿量. 量子反散射主要是沿用了经典反散射的名称. 由于它多用来求解可积模型的本征值问题, 因此又被称为代数 Bethe Ansatz.

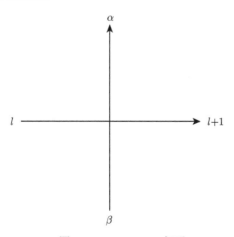

图 1.7 Boltzmann 权重

对于 Heisenberg 反铁磁链, 我们定义 Lax 算子

$$L_{n\tau}(\lambda) = \lambda + P_{n\tau} \equiv \lambda + \frac{1}{2}(1 + \boldsymbol{\sigma}_n \cdot \boldsymbol{\tau}), \qquad (1.2.119)$$

其中, τ 是辅助 Pauli 矩阵; σ_n 则是 n 格点上的 Pauli 自旋算子; $P_{n\tau}$ 是置换算子, 它满足下列关系:

$$P_{n\tau}^2 = 1, \quad P_{n\tau}\sigma_n = \tau P_{n\tau}, \quad P_{n\tau}\tau = \sigma_n P_{n\tau}. \tag{1.2.120}$$

可以验证, $L_{n\tau}(\lambda)$ 满足 Yang-Baxter 关系:

$$L_{\tau\tau'}(\lambda-\mu)L_{n\tau}(\lambda)L_{n\tau'}(\mu) = L_{n\tau'}(\mu)L_{n\tau}(\lambda)L_{\tau\tau'}(\lambda-\mu). \tag{1.2.121}$$

由于 τ, τ', σ_n 分别作用于不同的空间, 满足

$$[L_{n\tau}(\lambda), L_{m\tau'}(\mu)] = 0, \quad m \neq n. \tag{1.2.122}$$

定义单值矩阵 $T_N(\lambda)$ 为

$$T_{N\tau}(\lambda) = L_{N\tau}(\lambda)L_{N-1\tau}(\lambda)\cdots L_{1\tau}(\lambda). \tag{1.2.123}$$

利用式 (1.2.121) 和式 (1.2.122), 可得

$$\begin{aligned}
&L_{\tau\tau'}(\lambda-\mu)T_{N\tau}(\lambda)T_{N\tau'}(\mu) \\
&= L_{\tau\tau'}(\lambda-\mu)L_{N\tau}(\lambda)L_{N\tau'}(\mu)L_{N-1\tau}(\lambda)L_{N-1\tau'}(\mu)\cdots L_{1\tau}(\lambda)L_{1\tau'}(\mu) \\
&= L_{N\tau'}(\mu)L_{N\tau}(\lambda)L_{\tau\tau'}(\lambda-\mu)L_{N-1\tau}(\lambda)L_{N-1\tau'}(\mu)\cdots L_{1\tau}(\lambda)L_{1\tau'}(\mu) \\
&= L_{N\tau'}(\mu)L_{N\tau}(\lambda)L_{N-1\tau'}(\mu)L_{N-1\tau}(\lambda)\cdots L_{1\tau'}(\mu)L_{1\tau}(\lambda)L_{\tau\tau'}(\lambda-\mu) \\
&= T_{N\tau'}(\mu)T_{N\tau}(\lambda)L_{\tau\tau'}(\lambda-\mu).
\end{aligned} \tag{1.2.124}$$

式 (1.2.124) 是推广的 Yang-Baxter 关系, 将 $L_{\tau\tau'}$ 看成数矩阵, 我们可以求得单值矩阵矩阵元之间的对易关系. 对式 (1.2.123) 中的辅助空间 τ 求迹可得

$$[\tau(\lambda), \tau(\mu)] = 0, \tag{1.2.125}$$

即转移矩阵对不同的参量是相互对易的, 做 $\tau(\lambda)$ 的泰勒展开可知转移矩阵的系数是互相对易的. 因此, 只要其中一个系数或几个系数的组合是哈密顿量, 我们就得到了一系列相互对易的守恒量. 系统的哈密顿量可由 $\ln\tau(\lambda)$ 的一阶导数得到

$$\begin{aligned}
\left.\frac{\partial \ln\tau(\lambda)}{\partial\lambda}\right|_{\lambda=0} &= \text{tr}_\tau \sum_{j=1}^{N}[P_{12}P_{13}\cdots P_{1N}P_{N\tau}\cdots P_{j+1\tau}P_{j-1\tau}\cdots P_{1\tau}] \\
&= \sum_{j=1}^{N} P_{jj-1} = \frac{1}{2}\sum_{j=1}^{N}(1 + \sigma_j \cdot \sigma_{j-1}),
\end{aligned} \tag{1.2.126}$$

1.2 自旋链模型

因此
$$H = J \frac{\partial \ln \tau(\lambda)}{\partial \lambda}\bigg|_{\lambda=0} - \frac{1}{2}JN. \tag{1.2.127}$$

为了求得转移矩阵元之间的对易关系, 将 $T_{N\tau}(\lambda)$, $T_{N\tau'}(\mu)$ 和 $L_{\tau\tau'}(\lambda-\mu)$ 在 τ 和 τ' 的直积空间展开

$$T_{N\tau}(\lambda) = \begin{pmatrix} A(\lambda) & B(\lambda) \\ C(\lambda) & D(\lambda) \end{pmatrix} \otimes I_{\tau'} = \begin{pmatrix} A(\lambda) & 0 & B(\lambda) & 0 \\ 0 & A(\lambda) & 0 & B(\lambda) \\ C(\lambda) & 0 & D(\lambda) & 0 \\ 0 & C(\lambda) & 0 & D(\lambda) \end{pmatrix}, \tag{1.2.128}$$

$$T_{N\tau'}(\mu) = I_\tau \otimes \begin{pmatrix} A(\mu) & B(\mu) \\ C(\mu) & D(\mu) \end{pmatrix} = \begin{pmatrix} A(\mu) & B(\mu) & 0 & 0 \\ C(\mu) & D(\mu) & 0 & 0 \\ 0 & 0 & A(\mu) & B(\mu) \\ 0 & 0 & C(\mu) & D(\mu) \end{pmatrix}, \tag{1.2.129}$$

$$L_{\tau\tau'}(\lambda-\mu) = \begin{pmatrix} \lambda-\mu+1 & 0 & 0 & 0 \\ 0 & \lambda-\mu & 1 & 0 \\ 0 & 1 & \lambda-\mu & 0 \\ 0 & 0 & 0 & \lambda-\mu+1 \end{pmatrix}, \tag{1.2.130}$$

其中, I_τ 和 $I_{\tau'}$ 分别代表 τ 和 τ' 空间的单位矩阵. 将式 (1.2.128)~式 (1.2.130) 代入式 (1.2.124) 中, 可以求得

$$\begin{cases} [A(\lambda), A(\mu)] = [D(\lambda), D(\mu)] = 0, \\ A(\lambda)B(\mu) = \dfrac{\lambda-\mu-1}{\lambda-\mu}B(\mu)A(\lambda) + \dfrac{1}{\lambda-\mu}B(\lambda)A(\mu), \\ D(\lambda)B(\mu) = \dfrac{\lambda-\mu+1}{\lambda-\mu}B(\mu)D(\lambda) - \dfrac{1}{\lambda-\mu}B(\lambda)D(\mu), \\ [B(\lambda), B(\mu)] = [C(\lambda), C(\mu)] = 0, \\ [B(\lambda), C(\mu)] = \dfrac{1}{\lambda-\mu}(D(\mu)A(\lambda) - D(\lambda)A(\mu)). \end{cases} \tag{1.2.131}$$

与坐标 Bethe Ansatz 相同, 选择真空态为全部自旋向上的态, 则

$$L_{n\tau}(\lambda)|0\rangle = \begin{pmatrix} \lambda+\frac{1}{2}(1+\sigma_n^z) & S_n^- \\ S_n^+ & \lambda+\frac{1}{2}(1-\sigma_n^z) \end{pmatrix}|0\rangle$$

$$= \begin{pmatrix} \lambda+1 & S_n^- \\ 0 & \lambda \end{pmatrix}|0\rangle. \tag{1.2.132}$$

利用上式可得

$$\begin{cases} A(\lambda)|0\rangle = a(\lambda)|0\rangle = (\lambda+1)^N|0\rangle, \\ D(\lambda)|0\rangle = d(\lambda)|0\rangle = \lambda^N|0\rangle, \\ C(\lambda)|0\rangle = 0. \end{cases} \quad (1.2.133)$$

$B(\lambda)$ 作用于 $|0\rangle$ 相当于一个翻转自旋的叠加态, 因此可以作为本征态的产生算子, 定义态

$$|\mu_1,\cdots,\mu_M\rangle = B(\mu_1)\cdots B(\mu_M)|0\rangle, \quad (1.2.134)$$

我们将证明它在某些限制条件下是转移矩阵 $\tau(\lambda) \equiv A(\lambda) + D(\lambda)$ 的本征态.

首先我们证明下列对易关系:

$$A(\lambda)B(\mu_1)\cdots B(\mu_M) = \prod_{j=1}^{M} \frac{\lambda-\mu_j-1}{\lambda-\mu_j} B(\mu_1)\cdots B(\mu_M)A(\lambda)$$
$$+ \sum_{j=1}^{M} \frac{1}{\lambda-\mu_j} \prod_{l\neq j} \frac{\mu_j-\mu_l-1}{\mu_j-\mu_l} B(\mu_1)\cdots B(\mu_{j-1})B(\lambda)B(\mu_{j+1})\cdots B(\mu_M)A(\mu_j),$$
$$(1.2.135)$$

$$D(\lambda)B(\mu_1)\cdots B(\mu_M) = \prod_{j=1}^{M} \frac{\lambda-\mu_j+1}{\lambda-\mu_j} B(\mu_1)\cdots B(\mu_M)D(\lambda)$$
$$- \sum_{j=1}^{M} \frac{1}{\lambda-\mu_j} \prod_{l\neq i} \frac{\mu_j-\mu_l+1}{\mu_j-\mu_l} B(\mu_1)\cdots B(\mu_{j-1})B(\lambda)B(\mu_{j+1})\cdots B(\mu_M)D(\mu_j).$$
$$(1.2.136)$$

式 (1.2.135) 和式 (1.2.136) 可由数学归纳法得到证明. 由式 (1.2.131) 可知, 当 $M=1$ 时, 式 (1.2.135) 成立. 设在给定 M 时式 (1.2.135) 成立, 则

$$A(\lambda)B(\mu_{M+1})B(\mu_1)\cdots B(\mu_M)$$
$$= \frac{\lambda-\mu_{M+1}-1}{\lambda-\mu_{M+1}} B(\mu_{M+1})A(\lambda)B(\mu_1)\cdots B(\mu_M)$$
$$+ \frac{1}{\lambda-\mu_{M+1}} B(\lambda)A(\mu_{M+1})B(\mu_1)\cdots B(\mu_M)$$
$$= \prod_{j=1}^{M+1} \frac{\lambda-\mu_j-1}{\lambda-\mu_j} B(\mu_1)\cdots B(\mu_M)B(\mu_{M+1})A(\lambda)$$

$$+ \frac{\lambda - \mu_{M+1} - 1}{\lambda - \mu_{M+1}} \sum_{j=1}^{M} \frac{1}{\lambda - \mu_j} \prod_{l \neq j, M+1} \frac{\mu_j - \mu_l - 1}{\mu_j - \mu_l}$$
$$\times B(\mu_1) \cdots B(\mu_{j-1}) B(\lambda) B(\mu_{j+1}) \cdots B(\mu_{M+1}) A(\mu_j)$$
$$+ \frac{1}{\lambda - \mu_{M+1}} \prod_{j=1}^{M} \frac{\mu_{M+1} - \mu_j - 1}{\mu_{M+1} - \mu_j} B(\mu_1) \cdots B(\mu_M) B(\lambda) A(\mu_{M+1})$$
$$+ \frac{1}{\lambda - \mu_{M+1}} \sum_{j=1}^{M} \frac{1}{\mu_{M+1} - \mu_j} \prod_{l \neq j, M+1} \frac{\mu_j - \mu_l - 1}{\mu_j - \mu_l}$$
$$\times B(\mu_1) \cdots B(\mu_{j-1}) B(\lambda) B(\mu_{j+1}) \cdots B(\mu_{M+1}) A(\mu_j). \tag{1.2.137}$$

合并上式中的第二项和第四项可得

$$A(\lambda) B(\mu_1) \cdots B(\mu_{M+1}) = \prod_{j=1}^{M+1} \frac{\lambda - \mu_j - 1}{\lambda - \mu_j} B(\mu_1) \cdots B(\mu_{M+1}) A(\lambda)$$
$$+ \sum_{j=1}^{M+1} \frac{1}{\lambda - \mu_j} \prod_{l \neq j}^{M+1} \frac{\mu_j - \mu_l - 1}{\mu_j - \mu_l} B(\mu_1) \cdots B(\mu_{j-1})$$
$$\times B(\lambda) B(\mu_{j+1}) \cdots B(\mu_{M+1}) A(\mu_j). \tag{1.2.138}$$

因此式 (1.2.135) 对 $M+1$ 情形也成立, 即对任意 M 成立. 同样, 可以证明式 (1.2.136). 利用式 (1.2.135) 和式 (1.2.136) 可以求得

$$\tau(\lambda) |\mu_1, \cdots, \mu_M\rangle = \Lambda(\lambda; \mu_1, \cdots, \mu_M) |\mu_1, \cdots, \mu_M\rangle$$
$$+ \sum_{j=1}^{M} \Lambda_j(\lambda; \mu_1, \cdots, \mu_M) B(\mu_1) \cdots$$
$$B(\mu_{j-1}) B(\lambda) B(\mu_{j+1}) \cdots B(\mu_M) |0\rangle, \tag{1.2.139}$$

其中

$$\Lambda(\lambda; \mu_1, \cdots, \mu_M) = (\lambda+1)^N \prod_{j=1}^{M} \frac{\lambda - \mu_j - 1}{\lambda - \mu_j} + \lambda^N \prod_{j=1}^{M} \frac{\lambda - \mu_j + 1}{\lambda - \mu_j}, \tag{1.2.140}$$

$$\Lambda_j(\lambda; \mu_1, \cdots, \mu_M) = \frac{(\mu_j + 1)^N}{\lambda - \mu_j} \prod_{l \neq j}^{M} \frac{\mu_j - \mu_l - 1}{\mu_j - \mu_l} - \frac{\mu_j^N}{\lambda - \mu_j} \prod_{l \neq j}^{M} \frac{\mu_j - \mu_l + 1}{\mu_j - \mu_l}. \tag{1.2.141}$$

如果 $|\mu_1, \cdots, \mu_M\rangle$ 是 $\tau(\lambda)$ 的本征态, 则 $\Lambda_j(\lambda; \mu_1, \cdots, \mu_M) = 0$, 由此得出 Bethe Ansatz 方程:

$$\left(1 + \frac{1}{\mu_j}\right)^N = \prod_{l \neq j}^{M} \frac{\mu_j - \mu_l + 1}{\mu_j - \mu_l - 1}. \tag{1.2.142}$$

做参数变换 $\mu_j = i\lambda_j - \frac{1}{2}$, 则式 (1.2.142) 变为与式 (1.2.34) 相同的形式. $\tau(\lambda)$ 的本征值即 $\Lambda(\lambda; \mu_1, \cdots, \mu_M)$. 利用式 (1.2.127), 可得哈密顿量的本征值为

$$E(\mu_1, \cdots, \mu_M) = J \left. \frac{\partial \ln \Lambda(\lambda; \mu_1, \cdots, \mu_M)}{\partial \lambda} \right|_{\lambda=0} - \frac{1}{2} JN$$

$$= J \sum_{j=1}^{M} \frac{1}{\mu_j(\mu_j+1)} + \frac{1}{2} JN. \qquad (1.2.143)$$

同样做参量变换 $\mu_j = i\lambda_j - \frac{1}{2}$, 上式转化为式 (1.2.35). 从以上的讨论中可以看出, 量子反散射方法为求解可积模型提供了一种非常优美的方法. 在这种方法中, 本征矢的明确表达式并不需要, 它只用到散射数据 $A(\lambda)$, $B(\lambda)$, $D(\lambda)$ 的简单代数关系, 导致这些代数关系的基础就是 Yang-Baxter 方程. 因此, 从原则上讲, 只要找到 Yang-Baxter 方程的解, 就可以构建一个可积模型. 近年来, 求解 Yang-Baxter 方程以及在此基础上发展起来的量子群理论已成为数学研究的一个重要领域.

1.2.5 开边界问题

开边界问题可以用坐标方法[20, 21] 或代数方法[4] 求解. 我们考虑以下模型:

$$H = \frac{1}{2} J \sum_{j=1}^{N-1} \boldsymbol{\sigma}_j \cdot \boldsymbol{\sigma}_{j+1} - \frac{1}{2} h \sum_{j=1}^{N} \sigma_j^z + h_1 \sigma_1^z + h_N \sigma_N^z. \qquad (1.2.144)$$

注意在该模型中, 第一个格点和第 N 个格点没有耦合, 因此是一条开链. 同时, 在边界的两个格点上加上了边界磁场 h_1 和 h_N. 显然, 这一系统不具有平移不变性. 哈密顿量 (1.2.144) 可以重新写为

$$H = J \sum_{j=1}^{N-1} (S_j^+ S_{j+1}^- + S_{j+1}^+ S_j^- + 2 S_j^z S_{j+1}^z)$$

$$- h \sum_{j=1}^{N} S_j^z + 2 h_1 S_1^z + 2 h_N S_N^z. \qquad (1.2.145)$$

利用自旋算子的性质, 我们仍然可以验证初始态 $|0\rangle$ 是式 (1.2.144) 的本征态

$$\begin{cases} H|0\rangle = E_0 |0\rangle, \\ E_0 = \frac{1}{2}(N-1)J - \frac{1}{2}Nh + h_1 + h_N. \end{cases} \qquad (1.2.146)$$

当一个自旋反转时, 哈密顿量的本征态可以写为

$$|k\rangle = \sum_{x=1}^{N} \psi(x) S_x^- |0\rangle. \qquad (1.2.147)$$

1.2 自旋链模型

将 H 作用到 $|k\rangle$ 上, 当 $x \neq 1, N$ 时, 可得

$$J[\psi(x+1) + \psi(x-1) - 2\psi(x)] + (E_0 + h)\psi(x) = E\psi(x), \tag{1.2.148}$$

其中, E 仍然表示本征能量; k 是准动量. 由于系统具有开边界, 我们设

$$\psi(x) = A_+ e^{ikx} + A_- e^{-ikx}, \tag{1.2.149}$$

$A_\pm$ 是两个待定系数. 将式 (1.2.149) 代入式 (1.2.148) 可得

$$E(k) = 2J(\cos k - 1) + E_0 + h. \tag{1.2.150}$$

当 $x = 1$ 时, 由于第 1 个格点与第 N 个格点不再相连, 本征方程变为

$$J[\psi(2) - \psi(1)] + (E_0 + h - 2h_1)\psi(1) = E\psi(1). \tag{1.2.151}$$

再将式 (1.2.149) 代入上式, 可得到 $A_\pm$ 之间的关系为

$$\frac{A_+}{A_-} = -\frac{1 - (1 - 2h_1/J)e^{-ik}}{1 - (1 - 2h_1/J)e^{ik}}. \tag{1.2.152}$$

当 $x = N$ 时, 本征方程变为

$$J[\psi(N-1) - \psi(N)] + (E_0 + h - 2h_N)\psi(N) = E\psi(N). \tag{1.2.153}$$

同样可以得到

$$\frac{A_+}{A_-} = -e^{-2iNk}\frac{e^{-ik} - (1 - 2h_N/J)}{e^{ik} - (1 - 2h_N/J)}, \tag{1.2.154}$$

联立式 (1.2.152) 和式 (1.2.154) 可得到 Bethe Ansatz 方程:

$$e^{2iNk} = \frac{1 - (1 - 2h_1/J)e^{ik}}{1 - (1 - 2h_1/J)e^{-ik}} \frac{e^{-ik} - (1 - 2h_N/J)}{e^{ik} - (1 - 2h_N/J)}. \tag{1.2.155}$$

对于 M 个自旋反转情形, 系统的本征矢可以写为

$$|k_1, \cdots, k_M\rangle = \sum_{x_1, \cdots, x_M = 1}^{N} \psi(x_1, \cdots, x_M) S_{x_1}^- \cdots S_{x_M}^- |0\rangle. \tag{1.2.156}$$

将式 (1.2.145) 作用到该本征矢上, 可以得到本征方程

$$\begin{aligned} &J \sum_{j=1}^{M} \sum_{\delta = \pm 1} \psi(x_1, \cdots, x_j + \delta, \cdots, x_M) + (h - 2J)M\psi(x_1, \cdots, x_M) \\ &+ 2J \sum_{i<j} \psi(x_1, \cdots, x_M)[\delta_{x_i, x_j+1} + \delta_{x_i, x_j-1}] \\ &+ \sum_{j=1}^{M}[(J - 2h_1)\delta_{x_j, 1} + (J - 2h_N)\delta_{x_j, N}]\psi(x_1, \cdots, x_M) \\ &= (E - E_0)\psi(x_1, \cdots, x_M). \end{aligned} \tag{1.2.157}$$

上式中对波函数有以下限制条件:

$$\begin{cases} \psi(x_1,\cdots,x_M)\,|_{x_i=x_j}\equiv 0, \\ \psi(x_1,\cdots,x_M)\,|_{x_j<1}=\psi(x_1,\cdots,x_M)\,|_{x_j>N}\equiv 0. \end{cases} \quad (1.2.158)$$

与周期性边界情形类似, 对波函数做下述假定:

$$\psi(x_1,\cdots,x_M)=\sum_{P,Q}\sum_{r_j=\pm}A_{r,P}\mathrm{e}^{\mathrm{i}\sum_{j=1}^M r_{P_j}k_{P_j}x_{Q_j}}\theta(x_{Q_1}<\cdots<x_{Q_M}). \quad (1.2.159)$$

与周期系统不同, 上式中对每个 k_j 都包含了其相应的反射项. 当所有的 x_j 不相邻且不在边界点 1, N 时, 将式 (1.2.159) 代入式 (1.2.157) 中可以得到

$$E-E_0=2J\sum_{j=1}^M(\cos k_j-1)+Mh. \quad (1.2.160)$$

考虑两个相邻的反转自旋 $2<x_{Q_{j+1}}=x_{Q_j}+1<N$, 写出上式中与这两个自旋的相关项

$$\psi(\cdots,x_{Q_j}-1,x_{Q_j}+1,\cdots)+\psi(\cdots,x_{Q_j},x_{Q_j}+2,\cdots)-2\psi(\cdots,x_{Q_j},x_{Q_j}+1,\cdots)$$
$$=2(\cos k_{P_j}+\cos k_{P_{j+1}})\psi(\cdots,x_{Q_j},x_{Q_j}+1,\cdots). \quad (1.2.161)$$

利用式 (1.2.159) 比较 $\exp[\mathrm{i}r_{P_j}k_{P_j}x_{Q_j}+\mathrm{i}r_{P_{j+1}}k_{P_{j+1}}x_{Q_j}]$ 项可得

$$\frac{A_{P_1\cdots P_{j+1}P_j\cdots P_M}(\cdots r_{P_{j+1}}r_{P_j}\cdots)}{A_{P_1\cdots P_j P_{j+1}\cdots P_M}(\cdots r_{P_j}r_{P_{j+1}}\cdots)}=-\frac{\mathrm{e}^{\mathrm{i}(r_{P_j}k_{P_j}+r_{P_{j+1}}k_{P_{j+1}})}+1-2\mathrm{e}^{\mathrm{i}r_{P_{j+1}}k_{P_{j+1}}}}{\mathrm{e}^{\mathrm{i}(r_{P_j}k_{P_j}+r_{P_{j+1}}k_{P_{j+1}})}+1-2\mathrm{e}^{\mathrm{i}r_{P_j}k_{P_j}}}.$$
$$(1.2.162)$$

由于 P_j 具有任意性, 上式给出了 $A_{r,P}$ 的相互关系. 注意, 上式中交换 P_j 和 P_{j+1} 时, r_{P_j} 和 $r_{P_{j+1}}$ 也同时交换了. 当 $x_{Q_1}=1$ 时, 从本征方程 (1.2.161) 可得

$$J[\psi(2,x_{Q_2}\cdots x_{Q_M})+\psi(1,x_{Q_2}\cdots x_{Q_M})]-2h_1\psi(1,x_{Q_2}\cdots x_{Q_M})$$
$$=2J\cos k_{P_1}\psi(1,x_{Q_2}\cdots x_{Q_M}), \quad (1.2.163)$$

比较同类项可得

$$\frac{A_P(+,\cdots)}{A_P(-,\cdots)}=-\frac{1-(1-2h_1/J)\mathrm{e}^{-\mathrm{i}k_{P_1}}}{1-(1-2h_1/J)\mathrm{e}^{\mathrm{i}k_{P_1}}}. \quad (1.2.164)$$

同理, 当 $x_{Q_M}=N$ 时, 式 (1.2.157) 给出

$$J[\psi(\cdots,N-1)+\psi(\cdots,N)]=2(\cos k_{P_N}+h_N)\psi(\cdots,N), \quad (1.2.165)$$

1.2 自旋链模型

再比较同类项可得

$$\frac{A_P(\cdots,+)}{A_P(\cdots,-)} = -\mathrm{e}^{-2\mathrm{i}Nk_{P_M}}\frac{\mathrm{e}^{-\mathrm{i}k_{P_M}} - (1-2h_N/J)}{\mathrm{e}^{\mathrm{i}k_{P_M}} - (1-2h_N/J)}. \tag{1.2.166}$$

由式 (1.2.164) 可知

$$\frac{A_{P_2\cdots P_M P_1}(r_{P_2}\cdots r_{P_M}r_{P_1})}{A_{P_1 P_2\cdots P_M}(r_{P_1}r_{P_2}\cdots r_{P_M})} = (-1)^{M-1}\prod_{j=2}^{M}\frac{\mathrm{e}^{\mathrm{i}(r_{P_1}k_{P_1}+r_{P_j}k_{P_j})}+1-2\mathrm{e}^{\mathrm{i}r_{P_j}k_{P_j}}}{\mathrm{e}^{\mathrm{i}(r_{P_1}k_{P_1}+r_{P_j}k_{P_j})}+1-2\mathrm{e}^{\mathrm{i}r_{P_1}k_{P_1}}}. \tag{1.2.167}$$

再利用式 (1.2.165) 和式 (1.2.167) 并取 $P_1 = j$，可得 Bethe Ansatz 方程:

$$\mathrm{e}^{2\mathrm{i}Nk_j} = \frac{1-(1-2h_1/J)\mathrm{e}^{\mathrm{i}k_j}}{1-(1-2h_1/J)\mathrm{e}^{-\mathrm{i}k_j}}\frac{\mathrm{e}^{-\mathrm{i}k_j} - (1-2h_N/J)}{\mathrm{e}^{\mathrm{i}k_j} - (1-2h_N/J)}$$
$$\times \prod_{l\neq j}^{M}\prod_{r=\pm}\frac{\mathrm{e}^{\mathrm{i}(k_j+rk_l)}+1-2\mathrm{e}^{\mathrm{i}k_j}}{\mathrm{e}^{\mathrm{i}(k_j+rk_l)}+1-2\mathrm{e}^{\mathrm{i}rk_l}}. \tag{1.2.168}$$

利用参数化 (1.2.32)，上式可简化为

$$\left(\frac{\lambda_j-\frac{\mathrm{i}}{2}}{\lambda_j+\frac{\mathrm{i}}{2}}\right)^{2(N+1)\mathrm{or}2N} = \frac{\lambda_j-\frac{\mathrm{i}}{2}\mu_1}{\lambda_j+\frac{\mathrm{i}}{2}\mu_1}\frac{\lambda_j-\frac{\mathrm{i}}{2}\mu_N}{\lambda_j+\frac{\mathrm{i}}{2}\mu_N}\prod_{l\neq j}^{M}\frac{\lambda_j-\lambda_l-\mathrm{i}}{\lambda_j-\lambda_l+\mathrm{i}}\frac{\lambda_j+\lambda_l-\mathrm{i}}{\lambda_j+\lambda_l+\mathrm{i}}, \tag{1.2.169}$$

其中

$$\mu_1 = 1-\frac{J}{h_1}, \quad \mu_N = 1-\frac{J}{h_N}. \tag{1.2.170}$$

本征能量可用 λ_j 表示

$$E = -\sum_{j=1}^{M}\left(\frac{J}{\lambda_j^2+\frac{1}{4}} - h\right) + E_0. \tag{1.2.171}$$

很容易看出，式 (1.2.169) 具有反射不变性，即 $\lambda_j \to -\lambda_j$ 时方程不变. 利用与周期系统类似的方法，我们可以验证，当 $\lambda_j = \pm\lambda_l$ 时，波函数为零，即 $\lambda_j = \pm\lambda_l$ 不是允许解. 同样，当 $\lambda_j = 0$ 时，波函数也为零. 先考虑 $h_1 = h_N = 0$ 的情形，这时的 Bethe Ansatz 方程为

$$\left(\frac{\lambda_j-\frac{\mathrm{i}}{2}}{\lambda_j+\frac{\mathrm{i}}{2}}\right)^{2N} = \prod_{l\neq j}^{M}\frac{\lambda_j-\lambda_l-\mathrm{i}}{\lambda_j-\lambda_l+\mathrm{i}}\frac{\lambda_j+\lambda_l-\mathrm{i}}{\lambda_j+\lambda_l+\mathrm{i}}. \tag{1.2.172}$$

定义

$$Z_0(\lambda) = \frac{1}{2\pi}\left\{\theta_1(\lambda) - \frac{1}{2N}\sum_{l=1}^{M}[\theta_2(\lambda-\lambda_l) + \theta_2(\lambda+\lambda_l)]\right\} + \frac{1}{4N\pi}[\theta_1(\lambda) + \pi]. \tag{1.2.173}$$

可以验证

$$Z_0(\lambda_j) = \frac{I_j}{N} \tag{1.2.174}$$

给出式 (1.2.172) 的对数形式,其中 I_j 取整数. 注意, 以上用到 $\theta_n(0) = \pi$. 再定义

$$\sigma_0(\lambda) = \frac{dZ_0(\lambda)}{d\lambda} - \frac{1}{2N}\delta(\lambda), \tag{1.2.175}$$

则在热力学极限 $N\to\infty, M/N$ 取有限值时, 无外磁场基态 $\sigma_0(\lambda)$ 满足

$$\sigma_0(\lambda) + \frac{1}{2N}\delta(\lambda) = a_1(\lambda) - [2]\sigma_0(\lambda) + \frac{1}{2N}a_1(\lambda). \tag{1.2.176}$$

对上式做 Fourier 变换可得

$$\tilde\sigma_0(\omega) = \frac{1}{2\cosh\frac{\omega}{2}} + \frac{1}{N}\tilde\sigma_b(\omega), \tag{1.2.177}$$

其中, $\tilde\sigma_b(\omega)$ 是边界项的贡献:

$$\tilde\sigma_b(\omega) = \frac{1}{4\cosh\frac{\omega}{2}} - \frac{1}{2(1+e^{-|\omega|})}. \tag{1.2.178}$$

由 $\tilde\sigma_0(0) = \int\sigma(\lambda)d\lambda = \frac{1}{2}$ 可知, 系统的基态仍然是自旋单态. 当 h_1 或 h_N 不等于零时, Bethe Ansatz 方程 (1.2.172) 存在边界弦解. 对于反铁磁耦合情形, 在热力学极限下边界弦所产生的能量与其导致的 λ 海的移动所引起的能量相抵消, 这与体内弦激发的结果类似, 因此基态存在某种隐藏的简并性. 为了更清楚地认识边界弦, 我们讨论铁磁情形的基态及边界弦激发. 对于 $h_1 \neq 0$ 且 $h_N = 0$, 我们总可以选择 $h_1 < 0$, 这时的基态即 $|0\rangle$ 态. 显然, 当 $N\to\infty$ 且 $|h_1| > |J|$ 时,

$$\lambda_1^b = \frac{i}{2}\mu_1 \tag{1.2.179}$$

是式 (1.2.172) 的一个解. 实际上,

$$\lambda_n^b = \frac{i}{2}\mu_1 + in, \quad n = 1, 2, \cdots, M-1 \tag{1.2.180}$$

是热力学极限下式 (1.2.172) 的一个解. 它所携带的能量为

$$E_{b,1}^M = |J| \sum_{h=0}^{M-1} \frac{1}{-\left(\frac{1}{2}\mu_1 + n\right)^2 + \frac{1}{4}} = \frac{-2|J|}{\mu_1 - 1} + \frac{2|J|}{\mu_1 + 2M - 1}. \qquad (1.2.181)$$

由于 $\mu_1 < 1$, 所以 $E_b^M > 0$, 并且随 M 增大而降低. 当 $|h_1| \leqslant |J|$ 时, $\mu_1 \leqslant 0$, 式 (1.2.172) 不存在边界弦解. 当存在两个边界磁场时, 假定 $|h_1| > |h_N|$, 且 $h_1 < 0$, $h_N > 0$, 因此 $\mu_N > 1$. 这时式 (1.2.180) 仍是 Bethe Ansatz 方程 (1.2.172) 的解并给出正能量, 但在左端存在另一个边界弦

$$\lambda_m^b = \frac{\mathrm{i}}{2}\mu_N + \mathrm{i}m, \quad m = 0, 1, \cdots, M - 1. \qquad (1.2.182)$$

它所携带的能量为

$$E_{b,N}^M = -2|J| \left(\frac{1}{\mu_N - 1} - \frac{1}{\mu_N + 2M - 1} \right). \qquad (1.2.183)$$

注意 $E_{b,N}^M < 0$, 因此它将存在于基态. 这说明在两个边界场反平行时, 基态的自旋构型会发生变化, 在热力学极限下, $M \to \infty$, 代表反转自旋的个数. 在有限链情形, 由于式 (1.2.182) 存在 $o(\mathrm{e}^{-\delta N})$ 修正, M 取有限值. 这时基态自旋构型可粗略地用两个反向铁磁畴来描述.

1.2.6 反射代数 Bethe Ansatz

前文我们提到, 对于自由边界系统, 可积的充分条件是 Yang-Baxter 方程 (1.1.8) 和反射方程 (1.1.11). 对于海森伯自旋链, 我们定义双行单值矩阵

$$\mathcal{T}_0(\lambda) = (1 - \lambda^2)^N T_0^{-1}(-\lambda) K_0^-(\lambda) T_0(\lambda). \qquad (1.2.184)$$

其中, $K_0^-(\lambda)$ 是反射方程

$$R_{12}(\lambda-\mu)K_1^-(\lambda)R_{12}(\lambda+\mu)K_2^-(\mu) = K_2^-(\mu)R_{12}(\lambda+\mu)K_1^-(\lambda)R_{12}(\lambda-\mu) \qquad (1.2.185)$$

的解, 并且

$$R_{12}(\lambda) = \frac{1}{1 - \lambda^2} L_{12}(\lambda). \qquad (1.2.186)$$

下面我们证明 $\mathcal{T}_0(\lambda)$ 也满足反射方程 (1.2.185):

$$R_{12}(\lambda-\mu)\mathcal{T}_1(\lambda)R_{12}(\lambda+\mu)\mathcal{T}_2(\mu)$$
$$=(1-\lambda^2)^N(1-\mu^2)^N R_{12}(\lambda-\mu)T_1^{-1}(-\lambda)K_1^-(\lambda)T_1(\lambda)R_{12}(\lambda+\mu)T_2^{-1}(-\mu)K_2^-(\mu)T_2(\mu)$$
$$=(1-\lambda^2)^N(1-\mu^2)^N R_{12}(\lambda-\mu)T_1^{-1}(-\lambda)K_1^-(\lambda)T_2^{-1}(-\mu)R_{12}(\lambda+\mu)T_1(\lambda)K_2^-(\mu)T_2(\mu)$$
$$=(1-\lambda^2)^N(1-\mu^2)^N R_{12}(\lambda-\mu)T_1^{-1}(-\lambda)T_2^{-1}(-\mu)K_1^-(\lambda)R_{12}(\lambda+\mu)T_1(\lambda)K_2^-(\mu)T_2(\mu)$$
$$=(1-\lambda^2)^N(1-\mu^2)^N T_2^{-1}(-\mu)T_1^{-1}(-\lambda)R_{12}(\lambda-\mu)K_1^-(\lambda)R_{12}(\lambda+\mu)K_2^-(\mu)T_1(\lambda)T_2(\mu)$$
$$=(1-\lambda^2)^N(1-\mu^2)^N T_2^{-1}(-\mu)T_1^{-1}(-\lambda)K_2^-(\mu)R_{12}(\lambda+\mu)K_1^-(\lambda)R_{12}(\lambda-\mu)T_1(\lambda)T_2(\mu)$$
$$=(1-\lambda^2)^N(1-\mu^2)^N T_2^{-1}(-\mu)K_2^-(\mu)T_1^{-1}(-\lambda)R_{12}(\lambda+\mu)K_1^-(\lambda)T_2(\mu)T_1(\lambda)R_{12}(\lambda-\mu)$$
$$=(1-\lambda^2)^N(1-\mu^2)^N T_2^{-1}(-\mu)K_2^-(\mu)T_1^{-1}(-\lambda)R_{12}(\lambda+\mu)T_2(\mu)K_1^-(\lambda)T_1(\lambda)R_{12}(\lambda-\mu)$$
$$=(1-\lambda^2)^N(1-\mu^2)^N T_2^{-1}(-\mu)K_2^-(\mu)T_2(\mu)R_{12}(\lambda+\mu)T_1^{-1}(-\lambda)K_1^-(\lambda)T_1(\lambda)R_{12}(\lambda-\mu)$$
$$=\mathcal{T}_2(\mu)R_{12}(\lambda+\mu)\mathcal{T}_1(\lambda)R_{12}(\lambda-\mu), \qquad (1.2.187)$$

其中用到

$$T_1(\lambda)R_{12}(\lambda+\mu)T_2^{-1}(-\mu)=T_2^{-1}(-\mu)R_{12}(\lambda+\mu)T_1(\lambda), \qquad (1.2.188)$$
$$R_{12}^{-1}(-\lambda+\mu)=R_{12}(\lambda-\mu), \qquad (1.2.189)$$
$$R_{12}(\lambda-\mu)T_1^{-1}(-\lambda)T_2^{-1}(-\mu)=T_2^{-1}(-\mu)T_1^{-1}(-\lambda)R_{12}(\lambda-\mu), \qquad (1.2.190)$$

和反射方程 (1.2.185).

再定义双行转移矩阵

$$\tau(\lambda)=\text{tr}_0 K_0^+(\lambda)\mathcal{T}_0(\lambda), \qquad (1.2.191)$$

其中, $K_0^+(\lambda)$ 满足对偶反射方程:

$$R_{12}(-\lambda+\mu)K_1^{+t_1}(\lambda)R_{12}(-\lambda-\mu-2)K_2^{+t_2}(\mu)$$
$$=K_2^{+t_2}(\mu)R_{12}(-\lambda-\mu-2)K_1^{+t_1}(\lambda)R_{12}(-\lambda+\mu), \qquad (1.2.192)$$

其中, t_1, t_2 代表在 1, 2 空间的转置. 不难验证 $K_2^{+t_2}(-\lambda+1)$ 同调于 $K_1^-(\lambda)$. 利用式 (1.2.185) 和式 (1.2.192) 可以证明双行转移矩阵对于不同的谱参数是彼此对易的, 即

$$[\tau(\lambda),\tau(\mu)]=0. \qquad (1.2.193)$$

1.2 自旋链模型

在证明式 (1.2.193) 之前,我们先介绍 $R_{12}(u)$ 矩阵具有以下性质:

$$P_{12}R_{12}(u)P_{12} = R_{12}(u), \tag{1.2.194}$$

$$R_{12}^{t_1}(u) = R_{12}^{t_2}(u), \tag{1.2.195}$$

$$R_{12}(u)R_{12}(-u) = 1, \tag{1.2.196}$$

$$R_{12}^{t_1}(u)R_{12}^{t_1}(-u-2) = R_{12}^{t_1}(-u-2)R_{12}^{t_1}(u) = \frac{u(u+2)}{u^2-1} \equiv \rho(u), \tag{1.2.197}$$

式 (1.2.196) 称为幺正性而式 (1.2.197) 称为交叉幺正性.

$$\begin{aligned}
\tau(\lambda)\tau(\mu) &= \text{tr}_1 K_1^+(\lambda)\mathcal{T}_1(\lambda)\text{tr}_2 K_2^+(\mu)\mathcal{T}_2(\mu) \\
&= \text{tr}_1 K_1^{+t_1}(\lambda)\mathcal{T}_1^{t_1}(\lambda)\text{tr}_2 K_2^+(\mu)\mathcal{T}_2(\mu) \\
&= \text{tr}_{12} K_1^{+t_1}(\lambda)\mathcal{T}_1^{t_1}(\lambda)K_2^+(\mu)\mathcal{T}_2(\mu) \\
&= \text{tr}_{12} K_1^{+t_1}(\lambda)K_2^+(\mu)\mathcal{T}_1^{t_1}(\lambda)\mathcal{T}_2(\mu) \\
&= \frac{1}{\rho(\lambda+\mu)}\text{tr}_{12} K_1^{+t_1}(\lambda)K_2^+(\mu)R_{12}^{t_1}(-\lambda-\mu-2)R_{12}^{t_1}(\lambda+\mu)\mathcal{T}_1^{t_1}(\lambda)\mathcal{T}_2(\mu) \\
&= \frac{1}{\rho(\lambda+\mu)}\text{tr}_{12} K_1^{+t_1}(\lambda)K_2^+(\mu)R_{12}^{t_2}(-\lambda-\mu-2)R_{12}^{t_1}(\lambda+\mu)\mathcal{T}_1^{t_1}(\lambda)\mathcal{T}_2(\mu) \\
&= \frac{1}{\rho(\lambda+\mu)}\text{tr}_{12}\left[K_1^{+t_1}(\lambda)R_{12}(-\lambda-\mu-2)K_2^{+t_2}(\mu)\right]^{t_2}\left[\mathcal{T}_1(\lambda)R_{12}(\lambda+\mu)\mathcal{T}_2(\mu)\right]^{t_1} \\
&= \frac{1}{\rho(\lambda+\mu)}\text{tr}_{12}\left[K_1^{+t_1}(\lambda)R_{12}(-\lambda-\mu-2)K_2^{+t_2}(\mu)\right]^{t_{12}}\mathcal{T}_1(\lambda)R_{12}(\lambda+\mu)\mathcal{T}_2(\mu) \\
&= \frac{1}{\rho(\lambda+\mu)}\text{tr}_{12}\left[K_1^{+t_1}(\lambda)R_{12}(-\lambda-\mu-2)K_2^{+t_2}(\mu)\right]^{t_{12}} \\
&\quad \times R_{12}(-\lambda+\mu)R_{12}(\lambda-\mu)\mathcal{T}_1(\lambda)R_{12}(\lambda+\mu)\mathcal{T}_2(\mu) \\
&= \frac{1}{\rho(\lambda+\mu)}\text{tr}_{12}\left[R_{12}(-\lambda+\mu)K_1^{+t_1}(\lambda)R_{12}(-\lambda-\mu-2)K_2^{+t_2}(\mu)\right]^{t_{12}} \\
&\quad \times R_{12}(\lambda-\mu)\mathcal{T}_1(\lambda)R_{12}(\lambda+\mu)\mathcal{T}_2(\mu) \\
&= \frac{1}{\rho(\lambda+\mu)}\text{tr}_{12}\left[K_2^{+t_2}(\mu)R_{12}(-\lambda-\mu-2)K_1^{+t_1}(\lambda)R_{12}(-\lambda+\mu)\right]^{t_{12}} \\
&\quad \times \mathcal{T}_2(\mu)R_{12}(\lambda+\mu)\mathcal{T}_1(\lambda)R_{12}(\lambda-\mu) \\
&= \frac{1}{\rho(\lambda+\mu)}\text{tr}_{12} R_{12}(-\lambda+\mu)\left[K_2^{+t_2}(\mu)R_{12}(-\lambda-\mu-2)K_1^{+t_1}(\lambda)\right]^{t_{12}} \\
&\quad \times \mathcal{T}_2(\mu)R_{12}(\lambda+\mu)\mathcal{T}_1(\lambda)R_{12}(\lambda-\mu) \\
&= \frac{1}{\rho(\lambda+\mu)}\text{tr}_{12}\left[K_2^{+t_2}(\mu)R_{12}(-\lambda-\mu-2)K_1^{+t_1}(\lambda)\right]^{t_{12}} \\
&\quad \times \mathcal{T}_2(\mu)R_{12}(\lambda+\mu)\mathcal{T}_1(\lambda) \\
&= \frac{1}{\rho(\lambda+\mu)}\text{tr}_{12}\left[R_{12}^{t_2}(-\lambda-\mu-2)K_2^+(\mu)K_1^{+t_1}(\lambda)\right]^{t_1}
\end{aligned}$$

$$\begin{aligned}
&\times \mathcal{T}_2(\mu)\mathcal{T}_1^{t_1}(\lambda)R_{12}^{t_1}(\lambda+\mu)\\
&= \frac{1}{\rho(\lambda+\mu)}\text{tr}_{12}\left[R_{12}^{t_1}(-\lambda-\mu-2)K_2^+(\mu)K_1^{+t_1}(\lambda)\right]^{t_1}\mathcal{T}_2(\mu)\mathcal{T}_1^{t_1}(\lambda)R_{12}^{t_1}(\lambda+\mu)\\
&= \frac{1}{\rho(\lambda+\mu)}\text{tr}_{12}R_{12}^{t_1}(-\lambda-\mu-2)K_2^+(\mu)K_1^{+t_1}(\lambda)\mathcal{T}_2(\mu)\mathcal{T}_1^{t_1}(\lambda)R_{12}^{t_1}(\lambda+\mu)\\
&= \text{tr}_{12}K_2^+(\mu)\mathcal{T}_2(\mu)K_1^{+t_1}(\lambda)\mathcal{T}_1^{t_1}(\lambda)\\
&= \text{tr}_2 K_2^+(\mu)\mathcal{T}_2(\mu)\text{tr}_1 K_1^+(\lambda)\mathcal{T}_1(\lambda)\\
&= \tau(\mu)\tau(\lambda).
\end{aligned} \tag{1.2.198}$$

因此, 双行转移矩阵 $\tau(\lambda)$ 可以作为守恒量的产生元. 选取

$$K_0^-(\lambda) = p + \lambda\sigma_0^z, \quad K_0^+(\lambda) = q + (\lambda+1)\sigma_0^z, \tag{1.2.199}$$

则

$$\left.\frac{\mathrm{d}\tau(\lambda)}{\mathrm{d}\lambda}\right|_{\lambda=0} = 2pK_N^+(0) + 4pq\sum_{j=1}^{N-1}P_{jj+1} + 2qK_1^{-1}(0). \tag{1.2.200}$$

哈密顿量 (1.2.144) 可由下式给出:

$$H = \frac{1}{4pq}\left.\frac{\mathrm{d}\tau(\lambda)}{\mathrm{d}\lambda}\right|_{\lambda=0} - \frac{1}{2}N. \tag{1.2.201}$$

参数 p 和 q 分别由边界场 h_1 和 h_N 决定:

$$p = \frac{1}{2h_1}, \quad q = \frac{1}{2h_N}. \tag{1.2.202}$$

利用反射方程

$$R_{12}(\lambda-\mu)\mathcal{T}_1(\lambda)R_{12}(\lambda+\mu)\mathcal{T}_2(\mu) = \mathcal{T}_2(\mu)R_{12}(\lambda+\mu)\mathcal{T}_1(\lambda)R_{12}(\lambda-\mu), \tag{1.2.203}$$

我们可以求得 $\mathcal{T}_1(\lambda)$ 矩阵元之间的对易关系. 将 $\mathcal{T}_1(\lambda)$ 在 V_1 空间展开成矩阵形式:

$$\mathcal{T}_1(\lambda) \equiv \begin{pmatrix} \alpha(\lambda) & \beta(\lambda) \\ \gamma(\lambda) & \delta(\lambda) \end{pmatrix}, \tag{1.2.204}$$

再将 $\mathcal{T}_1(\lambda)$ 和 $\mathcal{T}_2(\lambda)$ 在量子空间 $V_1 \otimes V_2$ 展开为

$$\mathcal{T}_1(\lambda) \equiv \begin{pmatrix} \alpha(\lambda) & 0 & \beta(\lambda) & 0 \\ 0 & \alpha(\lambda) & 0 & \beta(\lambda) \\ \gamma(\lambda) & 0 & \delta(\lambda) & 0 \\ 0 & \gamma(\lambda) & 0 & \delta(\lambda) \end{pmatrix}, \quad \mathcal{T}_2(\lambda) \equiv \begin{pmatrix} \alpha(\lambda) & \beta(\lambda) & 0 & 0 \\ \gamma(\lambda) & \delta(\lambda) & 0 & 0 \\ 0 & 0 & \alpha(\lambda) & \beta(\lambda) \\ 0 & 0 & \gamma(\lambda) & \delta(\lambda) \end{pmatrix}. \tag{1.2.205}$$

R 矩阵的定义为

$$R_{12}(\lambda) = \begin{pmatrix} \lambda+1 & 0 & 0 & 0 \\ 0 & \lambda & 1 & 0 \\ 0 & 1 & \lambda & 0 \\ 0 & 0 & 0 & \lambda+1 \end{pmatrix}. \tag{1.2.206}$$

将式 (1.2.205) 和式 (1.2.206) 代入反射方程 (1.2.203) 中，令等式两边第一行第一列的矩阵元 [为方便起见，第 n 行第 m 列的矩阵元标记为 (nm)] 相等，可以得到

$$[\alpha(\lambda), \alpha(\mu)] = 0. \tag{1.2.207}$$

由等式两边 (44) 矩阵元相等可得

$$[\delta(\lambda), \delta(\mu)] = 0. \tag{1.2.208}$$

由等式两边 (14) 和 (41) 矩阵元分别相等给出

$$[\beta(\lambda), \beta(\mu)] = 0, \quad [\gamma(\lambda), \gamma(\mu)] = 0. \tag{1.2.209}$$

由等式两边 (13) 矩阵元和 (34) 矩阵元分别相等给出

$$\begin{aligned} \alpha(\lambda)\beta(\mu) &= \frac{(\lambda+\mu)(\lambda-\mu-1)}{(\lambda-\mu)(\lambda+\mu+1)}\beta(\mu)\alpha(\lambda) + \frac{\lambda+\mu}{(\lambda-\mu)(\lambda+\mu+1)}\beta(\lambda)\alpha(\mu) \\ &\quad -\frac{1}{\lambda+\mu+1}\beta(\lambda)\delta(\mu), \end{aligned} \tag{1.2.210}$$

$$\begin{aligned} \delta(\lambda)\beta(\mu) &= \frac{(\lambda-\mu+1)(\lambda+\mu+2)}{(\lambda-\mu)(\lambda+\mu+1)}\beta(\mu)\delta(\lambda) - \frac{(\lambda+\mu+2)}{(\lambda-\mu)(\lambda+\mu+1)}\beta(\lambda)\delta(\mu) \\ &\quad -\frac{2}{(\lambda-\mu)(\lambda+\mu+1)}\beta(\mu)\alpha(\lambda) + \frac{(\lambda-\mu+2)}{(\lambda-\mu)(\lambda+\mu+1)}\beta(\lambda)\alpha(\mu). \end{aligned} \tag{1.2.211}$$

为了方便，我们引入

$$\bar{\delta}(\lambda) = (2\lambda+1)\delta(\lambda) - \alpha(\lambda). \tag{1.2.212}$$

这是为了使式 (1.2.211) 中不显示出 $\beta(\mu)\alpha(\lambda)$ 项，则新的对易关系为

$$\begin{aligned} \bar{\delta}(\lambda)\beta(\mu) &= \frac{(\lambda-\mu+1)(\lambda+\mu+2)}{(\lambda-\mu)(\lambda+\mu+1)}\beta(\mu)\bar{\delta}(\lambda) - \frac{2(\lambda+1)}{(\lambda-\mu)(2\mu+1)}\beta(\lambda)\bar{\delta}(\mu) \\ &\quad +\frac{4(\lambda+1)u}{(2\mu+1)(\lambda+\mu+1)}\beta(\lambda)\alpha(\mu), \end{aligned} \tag{1.2.213}$$

$$\begin{aligned} \alpha(\lambda)\beta(\mu) &= \frac{(\lambda+\mu)(\lambda-\mu-1)}{(\lambda-\mu)(\lambda+\mu+1)}\beta(\mu)\alpha(\lambda) - \frac{1}{(\lambda+\mu+1)(2\mu+1)}\beta(\lambda)\bar{\delta}(\mu) \\ &\quad +\frac{2\mu}{(\lambda-\mu)(2\mu+1)}\beta(\lambda)\alpha(\mu). \end{aligned} \tag{1.2.214}$$

式 (1.2.213) 和式 (1.2.214) 是代数 Bethe Ansatz 的关键公式. 利用关系式

$$\sigma_0^y L_{0j}^{t_0}(\lambda)\sigma_0^y = \begin{pmatrix} 0 & -i \\ i & 0 \end{pmatrix} \begin{pmatrix} \lambda + \frac{1}{2}(1+\sigma_j^z) & \sigma_j^+ \\ \sigma_j^- & \lambda + \frac{1}{2}(1-\sigma_j^z) \end{pmatrix} \begin{pmatrix} 0 & -i \\ i & 0 \end{pmatrix}$$

$$= \begin{pmatrix} \lambda + \frac{1}{2}(1-\sigma_j^z) & -\sigma_j^- \\ -\sigma_j^+ & \lambda + \frac{1}{2}(1+\sigma_j^z) \end{pmatrix}$$

$$= -L_{0j}(-\lambda - 1), \tag{1.2.215}$$

可以得到

$$\sigma_0[(1-\lambda^2)^N T_0^{-1}(-\lambda)]^{t_0}\sigma_0 = \sigma_0^y[L_{N0}(-\lambda)\cdots L_{10}(-\lambda)]^{t_0}\sigma_0^y$$

$$= \sigma_0^y L_{10}(-\lambda)^{t_0}\sigma_0^y \sigma_0^y L_{20}^{t_0}(-\lambda)\sigma_0^y \cdots \sigma_0^y L_{N0}^{t_0}(-\lambda)\sigma_0^y$$

$$= (-1)^N T_0(-\lambda - 1), \tag{1.2.216}$$

因此有

$$(\lambda^2 - 1)^N T_0^{-1}(-\lambda)$$
$$= \sigma_0^y T_0(-\lambda - 1)^{t_0} \sigma_0^y$$
$$= \begin{pmatrix} D(-\lambda-1) & -B(-\lambda-1) \\ -c(-\lambda-1) & A(-\lambda-1) \end{pmatrix} \begin{pmatrix} \alpha(\lambda) & \beta(\lambda) \\ \gamma(\lambda) & \delta(\lambda) \end{pmatrix}$$
$$= (-1)^N \begin{pmatrix} D(-\lambda-1) & -B(-\lambda-1) \\ -C(-\lambda-1) & A(-\lambda-1) \end{pmatrix} \times \begin{pmatrix} p+\lambda & 0 \\ 0 & p-\lambda \end{pmatrix} \begin{pmatrix} A(\lambda) & B(\lambda) \\ C(\lambda) & D(\lambda) \end{pmatrix}.$$
$$\tag{1.2.217}$$

再利用对易关系

$$[B(\lambda), C(\mu)] = \frac{1}{\lambda - \mu}[D(\mu)A(\lambda) - D(\lambda)A(\mu)], \tag{1.2.218}$$

得到

$$\begin{cases} \gamma(\lambda)|0\rangle = 0, \\ \alpha(\lambda)|0\rangle = (p+\lambda)(\lambda+1)^{2N}|0\rangle, \\ \bar{\delta}(\lambda)|0\rangle = 2(p-\lambda-1)\lambda^{2N+1}|0\rangle, \end{cases} \tag{1.2.219}$$

即初始态 $|0\rangle$ 是 $\alpha(\lambda)$ 和 $\bar{\delta}(\lambda)$ 的本征态, 而 $\beta(\lambda)$ 则可以作为 $\alpha(\lambda)$ 和 $\bar{\delta}(\lambda)$ 本征态的产生算子. 令

$$\begin{cases} B_M = \beta(\lambda_1)\cdots\beta(\lambda_M), \\ B_M^j = \beta_1(\lambda_1)\cdots\beta(\lambda_{j-1})\beta(\lambda)\beta(\lambda_{j+1})\cdots\beta(\lambda_M), \end{cases} \tag{1.2.220}$$

1.2 自旋链模型

则利用式 (1.2.213) 和式 (1.2.214) 可以证明

$$\alpha(\lambda)B_M = \prod_{j=1}^{M}\frac{(\lambda+\lambda_j)(\lambda-\lambda_j-1)}{(\lambda-\lambda_j)(\lambda+\lambda_j+1)}B_M\alpha(\lambda)$$

$$-\sum_{j=1}^{M}\frac{1}{(\lambda+\lambda_j+1)(2\lambda_j+1)}\prod_{l\neq j}\frac{(\lambda_j-\lambda_l+1)(\lambda_j+\lambda_l+2)}{(\lambda_j-\lambda_l)(\lambda_j+\lambda_l+1)}B_M^j\bar{\delta}(\lambda_j)$$

$$+\sum_{j=1}^{M}\frac{2\lambda_j}{(\lambda-\lambda_j)(2\lambda_j+1)}\prod_{l\neq j}\frac{(\lambda_j+\lambda_l)(\lambda_j-\lambda_l-1)}{(\lambda_j-\lambda_l)(\lambda_j+\lambda_l+1)}B_M^j\alpha(\lambda_j), \quad (1.2.221)$$

$$\bar{\delta}(\lambda)B_M = B_M\bar{\delta}(\lambda)\prod_{j=1}^{M}\frac{(\lambda+\lambda_j+2)(\lambda-\lambda_j+1)}{(\lambda+\lambda_j+1)(\lambda-\lambda_j)}$$

$$-\sum_{j=1}^{M}\frac{2(\lambda+1)}{(\lambda-\lambda_j)(2\lambda_j+1)}\prod_{l\neq j}\frac{(\lambda_j-\lambda_l+1)(\lambda_j+\lambda_l+2)}{(\lambda_j-\lambda_l)(\lambda_j+\lambda_l+1)}B_M^j\bar{\delta}(\lambda_j)$$

$$+\sum_{j=1}^{M}\frac{4\lambda_j(\lambda+1)}{(\lambda+\lambda_j+1)(2\lambda_j+1)}\prod_{l\neq j}\frac{(\lambda_j+\lambda_l)(\lambda_j-\lambda_l-1)}{(\lambda_j-\lambda_l)(\lambda_j+\lambda_l+1)}B_M^j\alpha(\lambda_j).$$

$$(1.2.222)$$

式 (1.2.221) 和式 (1.2.222) 的证明可以用数学归纳法.

双行转移矩阵 $\tau(\lambda)$ 可以表达为

$$\tau(\lambda) = (q+\lambda+1)\alpha(\lambda) + (q-\lambda-1)\delta(\lambda)$$
$$= \frac{q-\lambda-1}{2\lambda+1}\bar{\delta}(\lambda) + \left(\frac{q-\lambda-1}{2\lambda+1}+q+\lambda+1\right)\alpha(\lambda). \quad (1.2.223)$$

假定 $\tau(\lambda)$ 的本征态为

$$|\lambda_1,\cdots,\lambda_M\rangle = B_M|0\rangle. \quad (1.2.224)$$

由式 (1.2.221) 和式 (1.2.222) 可知, $\tau(\lambda)$ 作用于该本征态上可以得到两种项, 一种为含 B_M 的本征项, 另一种为含 B_M^j 的非本征项. 本征项的系数给出 $\tau(\lambda)$ 的本征值, 而令非本征项的系数为零则给出对 λ_j 取值的限制, 即 Bethe Ansatz 方程. 也就是说,

$$\tau(\lambda)|\lambda_1,\cdots,\lambda_M\rangle = \Lambda(\lambda;\lambda_1,\cdots,\lambda_M)B_M|0\rangle + \sum_{j=1}^{M}\Lambda_j(\lambda;\lambda_1\cdots,\lambda_M)B_j^M|0\rangle,$$

$$(1.2.225)$$

其中

$$\Lambda(\lambda;\lambda_1,\cdots,\lambda_M) = \left(\frac{q-\lambda-1}{2\lambda+1} + q + \lambda + 1\right)$$

$$\times (p+\lambda)(\lambda+1)^{2N} \prod_{j=1}^{M} \frac{(\lambda+\lambda_j)(-\lambda-\lambda_j-1)}{(\lambda-\lambda_j)(\lambda-\lambda_j+1)}$$

$$+ 2\frac{q-\lambda-1}{2\lambda+1}(p-\lambda-1)\lambda^{2N+1} \prod_{j=1}^{M} \frac{(\lambda-\lambda_j+1)(\lambda+\lambda_j+2)}{(\lambda-\lambda_j)(\lambda+\lambda_j+1)},$$
(1.2.226)

$$\Lambda_j(\lambda;\lambda_1,\cdots,\lambda_M)$$
$$= \left(\frac{q-\lambda-1}{2\lambda+1} + q + \lambda + 1\right)(p+\lambda_j)(\lambda_j+1)^{2N}\frac{2\lambda_j}{(\lambda-\lambda_j)(2\lambda_j+1)}$$
$$\times \prod_{l\neq j} \frac{(\lambda_j+\lambda_l)(\lambda_j-\lambda_l-1)}{(\lambda_j-\lambda_l)(\lambda_j+\lambda_l+1)}$$
$$- 2\left(\frac{q-\lambda-1}{2\lambda+1} + q + \lambda + 1\right)(p-\lambda_j-1)\lambda_j^{2N+1}$$
$$\times \frac{1}{(\lambda-\lambda_j+1)(2\lambda_j+1)}\prod_{l\neq j}\frac{(\lambda_j-\lambda_l+1)(\lambda_j+\lambda_l+2)}{(\lambda_j-\lambda_l)(\lambda_j+\lambda_l+1)}$$
$$- \frac{q-\lambda-1}{2\lambda+1}2(p-\lambda_j-1)\lambda_j^{2N+1}\frac{2(\lambda+1)}{(\lambda-\lambda_j)(2\lambda_j+1)}\prod_{l\neq j}\frac{(\lambda_j-\lambda_l+1)(\lambda_j+\lambda_l+2)}{(\lambda_j-\lambda_l)(\lambda_j+\lambda_l+1)}$$
$$+ \frac{q-\lambda-1}{2\lambda+1}\frac{4(\lambda+1)\lambda_j}{(2\lambda_{j+1})(\lambda+\lambda_{j+1})}\prod_{l\neq j}\frac{(\lambda_j+\lambda_l)(\lambda_j-\lambda_l-1)}{(\lambda_j-\lambda_l)(\lambda_j+\lambda_l+1)}(p+\lambda_j)(\lambda_j+1)^{2N}.$$
(1.2.227)

令 $\Lambda_j(\lambda;\lambda_1,\cdots,\lambda_M) = 0$ 得

$$\frac{(q+\lambda_j)(p+\lambda_j)}{(\lambda_j+1-q)(\lambda_j+1-p)}\left(1+\frac{1}{\lambda_j}\right)^{2N}\prod_{l\neq j}\frac{(\lambda_j+\lambda_l)(\lambda_j-\lambda_l-1)}{(\lambda_j-\lambda_l+1)(\lambda_j+\lambda_l+2)} = 1.$$
(1.2.228)

因此, 满足上述条件的 $|\lambda_1,\cdots,\lambda_M\rangle$ 是 $\tau(\lambda)$ 的本征态. 由式 (1.2.201) 和式 (1.2.226) 可以得到哈密顿量的本征值为

$$E(\lambda_1,\cdots,\lambda_M) = \frac{1}{4pq}\left.\frac{\mathrm{d}\Lambda(\lambda;\lambda_1,\cdots,\lambda_M)}{\mathrm{d}\lambda}\right|_{\lambda=0} - \frac{1}{2}N. \qquad (1.2.229)$$

为了方便起见, 令 $\lambda_j = \mathrm{i}\lambda_j' - \frac{1}{2}$ 并将 λ_j' 重新标记为 λ_j, 则 Bethe Ansatz 方程 (1.2.228) 变为 (1.2.172), 其中 $\mu_1 = 2-p$, $\mu_N = 2-q$, 本征能量 $E(\lambda_1,\cdots,\lambda_M)$ 仍

然取式 (1.2.171) 的形式.

1.2.7 非对角 Bethe Ansatz

最一般的具有不平行边界磁场的 $s=\frac{1}{2}$ 各向同性 XXX 自旋链的哈密顿量是

$$H = \sum_{j=1}^{N-1} \boldsymbol{\sigma}_j \cdot \boldsymbol{\sigma}_{j+1} + h_1^z \sigma_1^z + h_N^x \sigma_N^x + h_N^z \sigma_N^z, \tag{1.2.230}$$

其中, h_N^z, h_1^x 和 h_1^z 是边界磁场强度.

系统的 R 矩阵是

$$R(\lambda) = \begin{pmatrix} \lambda+1 & & & \\ & \lambda & 1 & \\ & 1 & \lambda & \\ & & & \lambda+1 \end{pmatrix}. \tag{1.2.231}$$

该矩阵具有以下性质:

初值: $\quad R_{12}(0) = P_{12},$ (1.2.232)

幺正性: $\quad R_{12}(\lambda)R_{21}(-\lambda) = -\xi(\lambda)\,\mathrm{id}, \quad \xi(\lambda) = (\lambda+1)(\lambda-1),$ (1.2.233)

交叉关系: $\quad R_{12}(\lambda) = V_1 R_{12}^{t_2}(-\lambda-1)V_1, \quad V = -\mathrm{i}\sigma^y,$ (1.2.234)

PT- 对称性: $\quad R_{12}(\lambda) = R_{21}(\lambda) = R_{12}^{t_1 t_2}(\lambda),$ (1.2.235)

反对称性: $\quad R_{12}(-1) = -(1-P) = -2P^{(-)}.$ (1.2.236)

这里, $R_{21}(\lambda) = P_{12}R_{12}(\lambda)P_{12}$. R 矩阵满足 Yang-Baxter 方程:

$$R_{12}(\lambda-u)R_{13}(\lambda-v)R_{23}(u-v) = R_{23}(u-v)R_{13}(\lambda-v)R_{12}(\lambda-u). \tag{1.2.237}$$

引入单行单值矩阵 $T(\lambda)$ 和 $\hat{T}(\lambda)$:

$$T_0(\lambda) = R_{0N}(\lambda-\theta_N)R_{0\,N-1}(\lambda-\theta_{N-1})\cdots R_{01}(\lambda-\theta_1), \tag{1.2.238}$$

$$\hat{T}_0(\lambda) = R_{01}(\lambda+\theta_1)R_{02}(\lambda+\theta_2)\cdots R_{0N}(\lambda+\theta_N), \tag{1.2.239}$$

其中, $\{\theta_j, j=1,\cdots,N\}$ 是非均匀参数. 与哈密顿量 (1.2.230) 对应的反射矩阵 $K^-(\lambda)$ 和对偶反射矩阵 $K^+(\lambda)$ 分别是

$$K^-(\lambda) = \begin{pmatrix} p+\lambda & 0 \\ 0 & p-\lambda \end{pmatrix}, \tag{1.2.240}$$

$$K^+(\lambda) = \begin{pmatrix} q+\lambda+1 & \xi(\lambda+1) \\ \xi(\lambda+1) & q-\lambda-1 \end{pmatrix}. \tag{1.2.241}$$

它们满足反射方程和对偶反射方程:

$$R_{12}(\lambda - u)K_1^-(\lambda)R_{21}(\lambda + u)K_2^-(u)$$
$$= K_2^-(u)R_{12}(\lambda + u)K_1^-(\lambda)R_{21}(\lambda - u), \qquad (1.2.242)$$

$$R_{12}(u - \lambda)K_1^+(\lambda)R_{21}(-\lambda - u - 2)K_2^+(u)$$
$$= K_2^+(u)R_{12}(-\lambda - u - 2)K_1^+(\lambda)R_{21}(u - \lambda). \qquad (1.2.243)$$

为了得到可积的开边界系统,引入双行单值矩阵 $\mathbb{T}(u)$:

$$\mathbb{T}(\lambda) = T(\lambda)K^-(\lambda)\hat{T}(\lambda), \qquad (1.2.244)$$

相应的转移矩阵是

$$t(\lambda) = \text{tr}(K^+(\lambda)\mathbb{T}(\lambda)). \qquad (1.2.245)$$

利用 Yang-Baxter 方程和反射方程,可以证明含有不同谱参数的转移矩阵彼此对易, $[t(\lambda), t(u)] = 0$, 因此系统 (1.2.230) 是可积的. 转移矩阵的对数的一阶导数给出哈密顿量 (1.2.230):

$$H = \frac{\partial \ln t(\lambda)}{\partial \lambda}\Big|_{\lambda=0, \theta_j=0} - N$$
$$= 2\sum_{j=1}^{N-1} P_{j,j+1} + \frac{K_N^{-\prime}(0)}{K_N^-(0)} + 2\frac{K_1^{+\prime}(0)}{K_1^+(0)} - N$$
$$= \sum_{j=1}^{N-1} \boldsymbol{\sigma}_j \cdot \boldsymbol{\sigma}_{j+1} + \frac{1}{p}\sigma_1^z + \frac{1}{q}(\sigma_N^z + \xi\sigma_N^x). \qquad (1.2.246)$$

因此 $h_N = 1/p$, $h_1^x = \xi/q$, $h_1^z = 1/q$.

利用 R 矩阵的交叉对称性 (1.2.234), 可以证明转移矩阵 $t(u)$ 具有以下性质:

交叉对称性: $\quad t(-\lambda - 1) = t(\lambda), \qquad (1.2.247)$

初值: $\quad t(0) = 2pq\prod_{j=1}^{N}(1-\theta_j)(1+\theta_j) \times \text{id}, \qquad (1.2.248)$

渐近行为: $\quad t(\lambda) \sim 2\lambda^{2N+2} \times \text{id} + \cdots, \quad \lambda \to \pm\infty. \qquad (1.2.249)$

相应地,转移矩阵的本征值 $\Lambda(\lambda)$ 也应该满足:

交叉对称性: $\quad \Lambda(-\lambda - 1) = \Lambda(\lambda), \qquad (1.2.250)$

初值: $\quad \Lambda(0) = 2pq\prod_{j=1}^{N}(1-\theta_j)(1+\theta_j) = \Lambda(-1), \qquad (1.2.251)$

渐近行为: $\quad \Lambda(\lambda) \sim 2\lambda^{2N+2} + \cdots, \quad \lambda \to \pm\infty. \qquad (1.2.252)$

1.2 自旋链模型

本征值 $\Lambda(\lambda)$ 是一个关于变量 λ 的 $2N+2$ 阶多项式, 因此需要它在 $2N+3$ 个点上的值来确定它的具体形式. 事实上, 我们已经知道 $\Lambda(u)$ 在 $\lambda = 0, -1, \infty$ 上的取值, 因此还需要另外的 $2N$ 个点, 譬如由 $u = \theta_j$ 和 $-\theta_j - 1$ 时的关系来确定 $\Lambda(\lambda)$.

现在, 我们计算转移矩阵 $t(\lambda)$ 在 θ_j 点和 $\theta_j - 1$ 点的值. 直接的计算给出:

$$t(\theta_j) = R_{j\,j-1}(\theta_j - \theta_{j-1}) \cdots R_{j1}(\theta_j - \theta_1) K_j^-(\theta_j) R_{1j}(\theta_1 + \theta_j) \cdots R_{j-1\,j}(\theta_{j-1} + \theta_j)$$
$$\times \mathrm{tr}_0 \left\{ K_0^+(\theta_j) R_{0N}(\theta_j - \theta_N) \cdots R_{0\,j+2}(\theta_j - \theta_{j+2}) R_{0\,j+1}(\theta_j - \theta_{j+1}) \right.$$
$$\left. \times P_{0j} R_{j0}(2\theta_j) R_{j+1\,0}(\theta_{j+1} + \theta_j) R_{j+2\,0}(\theta_{j+2} + \theta_j) \cdots R_{N0}(\theta_N + \theta_j) \right\}.$$

利用 Yang-Baxter 方程 (1.2.237), 可以得到

$$R_{0\,j+1}(\theta_j - \theta_{j+1}) P_{0j} R_{j0}(2\theta_j) R_{j+1\,0}(\theta_{j+1} + \theta_j)$$
$$= R_{0\,j+1}(\theta_j - \theta_{j+1}) R_{0j}(2\theta_j) R_{j+1\,j}(\theta_{j+1} + \theta_j) P_{0j}$$
$$= R_{j+1\,j}(\theta_{j+1} + \theta_j) R_{0j}(2\theta_j) R_{0\,j+1}(\theta_j - \theta_{j+1}) P_{0j}$$
$$= R_{j+1\,j}(\theta_{j+1} + \theta_j) P_{0j} R_{j0}(2\theta_j) R_{j\,j+1}(\theta_j - \theta_{j+1}).$$

因此

$$t(\theta_j) = R_{j\,j-1}(\theta_j - \theta_{j-1}) \cdots R_{j1}(\theta_j - \theta_1) K_j^-(\theta_j) R_{1j}(\theta_1 + \theta_j) \cdots R_{j-1\,j}(\theta_{j-1} + \theta_j)$$
$$\times R_{j+1\,j}(\theta_{j+1} + \theta_j) \cdots R_{Nj}(\theta_N + \theta_j) \mathrm{tr}_0 \{ K_0^+(\theta_j) P_{0j} R_{j0}(2\theta_j) \}$$
$$\times R_{jN}(\theta_j - \theta_N) \cdots R_{j\,j+1}(\theta_j - \theta_{j+1}). \qquad (1.2.253)$$

另外, 利用 R 矩阵的交叉关系式 (1.2.234), 可以得到

$$t(\theta_j - 1) = \mathrm{tr}_0 \left\{ V_0 K_0^+(\theta_j - 1) V_0 R_{0N}^{t_0}(-\theta_j + \theta_N) \cdots R_{01}^{t_0}(-\theta_j + \theta_1) \right.$$
$$\left. \times V_0 K^-(\theta_j - 1) V_0 R_{10}^{t_0}(-\theta_1 - \theta_j) \cdots R_{0N}^{t_0}(-\theta_N - \theta_j) \right\}$$
$$= \mathrm{tr}_0 \left\{ \left[V_0 K_0^+(\theta_j - 1) V_0 R_{0N}^{t_0}(-\theta_j + \theta_N) \cdots R_{01}^{t_0}(-\theta_j + \theta_1) \right]^{t_0} \right.$$
$$\left. \times \left[V_0 K^-(\theta_j - 1) V_0 R_{10}^{t_0}(-\theta_1 - \theta_j) \cdots R_{0N}^{t_0}(-\theta_N - \theta_j) \right]^{t_0} \right\}$$
$$= \mathrm{tr}_0 \left\{ \left[V_0 K_0^-(\theta_j - 1) V_0 \right]^{t_0} R_{01}(-\theta_j + \theta_1) \cdots R_{0N}(-\theta_j + \theta_N) \right.$$
$$\left. \times \left[V_0 K^+(\theta_j - 1) V_0 \right]^{t_0} R_{N0}(-\theta_N - \theta_j) \cdots R_{10}(-\theta_1 - \theta_j) \right\}$$
$$= R_{j\,j+1}(-\theta_j + \theta_{j+1}) \cdots R_{jN}(-\theta_j + \theta_N) \left\{ V_j K_j^+(\theta_j - 1) V_j \right\}^{t_j}$$
$$\times R_{Nj}(-\theta_N - \theta_j) \cdots R_{j+1\,j}(-\theta_{j+1} - \theta_j) R_{j-1\,j}(-\theta_{j-1} - \theta_j)$$
$$\times \cdots R_{1j}(-\theta_1 - \theta_j) \mathrm{tr}_0 \left\{ \left[V_0 K_0^-(\theta_j - 1) V_0 \right]^{t_0} P_{0j} R_{j0}(-2\theta_j) \right\}$$
$$\times R_{j1}(-\theta_j + \theta_1) \cdots R_{j\,j-1}(-\theta_j + \theta_{j-1}). \qquad (1.2.254)$$

再利用 R 矩阵的幺正关系式 (1.2.233), 可以得到转移矩阵满足以下关系:

$$t(\theta_j)t(\theta_j - 1) = -\frac{\Delta_q(\theta_j)}{(2\theta_j - 1)(2\theta_j + 1)}, \quad (1.2.255)$$

其中, 量子行列式 $\Delta_q(\lambda)$ 的具体表达式为

$$\Delta_q(\lambda) = \text{Det}\{T(\lambda)\} \text{Det}\{\hat{T}(\lambda)\} \text{Det}\{K^-(\lambda)\} \text{Det}\{K^+(\lambda)\}. \quad (1.2.256)$$

进一步的计算给出

$$\text{Det}\{T(\lambda)\}\,\text{id} = \text{tr}_{12}\left(P_{12}^{(-)}T_1(\lambda-1)T_2(\lambda)P_{12}^{(-)}\right) = \prod_{j=1}^N (\lambda - \theta_j + 1)(\lambda - \theta_j - 1)\,\text{id},$$

$$\text{Det}\{\hat{T}(\lambda)\}\,\text{id} = \text{tr}_{12}\left(P_{12}^{(-)}\hat{T}_1(\lambda-1)\hat{T}_2(\lambda)P_{12}^{(-)}\right) = \prod_{j=1}^N (\lambda + \theta_j + 1)(\lambda + \theta_j - 1)\,\text{id},$$

$$\text{Det}\{K^-(\lambda)\} = \text{tr}_{12}\left(P_{12}^{(-)}K_1^-(\lambda-1)R_{12}(2\lambda-1)K_2^-(\lambda)\right) = 2(\lambda-1)(p^2 - \lambda^2),$$

$$\text{Det}\{K^+(\lambda)\} = \text{tr}_{12}\left(P_{12}^{(-)}K_2^+(\lambda)R_{12}(-2\lambda-1)K_1^+(\lambda-1)\right) = 2(\lambda+1)[(1+\xi^2)\lambda^2 - q^2].$$

相应地, 转移矩阵的本征值 $\Lambda(\lambda)$ 满足算子恒等式:

$$\Lambda(\theta_j)\Lambda(\theta_j - 1) = \frac{2(\theta_j + 1)[q^2 - (1+\xi^2)\theta_j^2]}{(2\theta_j - 1)(2\theta_j + 1)} a(\theta_j) d(\theta_j - 1)$$

$$= \frac{\Delta_q(\theta_j)}{(1 - 2\theta_j)(1 + 2\theta_j)}, \quad j = 1, \cdots, N. \quad (1.2.257)$$

方程 (1.2.257) 建立了转移矩阵的本征值 $\Lambda(\lambda)$ 在特殊点 $\{\theta_j\}$ 的值和量子行列式 $\Delta_q(\lambda)$ 之间的关系. 由于矩阵的迹和行列式都不依赖于态, 因此非对角 Bethe Ansatz 解决了传统方法缺少真空态的困难. 而且, 方程 (1.2.250)~(1.2.252) 和 (1.2.257) 可以确定函数 $\Lambda(\lambda)$. 下面, 我们通过推广的 $T-Q$ 关系来定出函数 $\Lambda(\lambda)$.

首先, 我们定义 $\bar{a}(\lambda)$ 和 $\bar{d}(\lambda)$:

$$\bar{a}(\lambda) = \frac{2\lambda + 2}{2\lambda + 1}(\lambda + p)(\sqrt{1+\xi^2}\,\lambda + q)\prod_{j=1}^N (\lambda + \theta_j + 1)(\lambda - \theta_j + 1), \quad (1.2.258)$$

$$\bar{d}(\lambda) = \frac{2\lambda}{2\lambda + 1}(\lambda - p + 1)[\sqrt{1+\xi^2}\,(\lambda+1) - q]\prod_{j=1}^N (\lambda + \theta_j)(\lambda - \theta_j)$$

$$= \bar{a}(-\lambda - 1). \quad (1.2.259)$$

1.2 自旋链模型

方程 (1.2.250)~(1.2.252) 的试探解为

$$\Lambda(\lambda) = \bar{a}(\lambda)\frac{Q_1(\lambda-1)}{Q_2(\lambda)} + \bar{d}(\lambda)\frac{Q_2(\lambda+1)}{Q_1(\lambda)} + 2(1-\sqrt{1+\xi^2})\lambda^n(\lambda+1)^n$$

$$\times \frac{\prod_{j=1}^{N}(\lambda+\theta_j)(\lambda-\theta_j)(\lambda+\theta_j+1)(\lambda-\theta_j+1)}{Q_1(\lambda)Q_2(\lambda)}, \qquad (1.2.260)$$

其中, n 是整数; 函数 $Q_1(\lambda)$ 和 $Q_2(\lambda)$ 可以用 $2M$ 个 Bethe 根 $\{\mu_j|j=1,\cdots,2M\}$ 参数化为

$$Q_1(\lambda) = \prod_{j=1}^{2M}(\lambda - \mu_j), \qquad (1.2.261)$$

$$Q_2(\lambda) = \prod_{j=1}^{2M}(\lambda + \mu_j + 1). \qquad (1.2.262)$$

注意, $2M$ 个 Bethe 根彼此不等. 通过分析该 $T-Q$ 关系, 就可以得到 Bethe Ansatz 方程中根的分布, 从而研究系统的物理性质.

数值结果表明, 任意一个固定的 M 就可以给出转移矩阵或者哈密顿量的所有能谱, 所以不同的 M 仅仅给出本征值不同的参数化形式. 因此, 我们可以给出简单的 $T-Q$ 关系: 对于偶数格点 N, 令 $n=1$ 并且 $M=N/2$; 对于奇数格点, 令 $n=2$ 并且 $M=(N+1)/2$. 我们先考虑偶数格点. 为了叙述方便, 定义

$$c(\lambda) = 2(1-\sqrt{1+\xi^2})\lambda(\lambda+1)\frac{2\lambda+1}{(2\lambda+2)(\lambda+p)(\sqrt{1+\xi^2}\lambda+q)}$$

$$\times \frac{2\lambda+1}{2\lambda(\lambda-p+1)[\sqrt{1+\xi^2}(\lambda+1)-q]}. \qquad (1.2.263)$$

函数 $Q_1(\lambda)$ 的零点是 $\lambda = \mu_j$, 函数 $Q_2(\lambda)$ 的零点是 $\lambda = -\mu_j - 1$. 零点 μ_j 和 $-\mu_j-1$ 互为交叉对称. 首先考虑零点 $\lambda = \mu_j$. 因为 $\Lambda(\lambda)$ 是一个 $N+2$ 阶多项式, 它本身没有奇点, 所以 $T-Q$ 试探解 (1.2.260) 中右端在 $\lambda = \mu_j$ 处的留数应该为零. 据此可得 Bethe Ansatz 方程:

$$c(\mu_j)\bar{a}(\mu_j) = -Q_2(\mu_j)Q_2(\mu_j+1)$$

$$= -\prod_{l=1}^{2M}(\mu_j+\mu_l+1)(\mu_j+\mu_l+2), \quad j=1,\cdots,2M. \qquad (1.2.264)$$

对 $T-Q$ 试探解 (1.2.260) 求在 $\lambda = -\mu_j - 1$ 处的留数, 可得 Bethe Ansatz 方程:

$$c(-\mu_j - 1)\bar{d}(-\mu_j - 1) = -Q_1(-\mu_j - 1)Q_1(-\mu_j - 2)$$
$$= -\prod_{l=1}^{2M}(\mu_j + \mu_l + 1)(\mu_j + \mu_l + 2), \quad j = 1, \cdots, 2M. \quad (1.2.265)$$

很明显, Bethe Ansatz 方程 (1.2.264) 和 (1.2.265) 等价. 在均匀极限下, $\{\theta_j = 0, j = 1, \cdots, N\}$, 系统的 Bethe Ansatz 方程是

$$(1 - \sqrt{1 + \xi^2})(2\mu_j + 1)(\mu_j + 1)^{2N+1}$$
$$= -(\mu_j - p + 1)[\sqrt{1 + \xi^2}(\mu_j + 1) - q]$$
$$\times \prod_{l=1}^{2M}(\mu_j + \mu_l + 1)(\mu_j + \mu_l + 2), \quad j = 1, \cdots, 2M. \quad (1.2.266)$$

利用关系式 (1.2.246), 哈密顿量 (1.2.230) 的本征值为

$$E = N - 1 + \frac{1}{p} + \frac{\sqrt{1 + \xi^2}}{q} - 2\sum_{j=1}^{2M}\frac{1}{\mu_j + 1}. \quad (1.2.267)$$

对于奇数 N, 令 $M = (N+1)/2$, $\Lambda(u)$ 可以参数化为

$$\Lambda(\lambda) = \bar{a}(\lambda)\frac{Q_1(\lambda - 1)}{Q_2(\lambda)} + \bar{d}(\lambda)\frac{Q_2(\lambda + 1)}{Q_1(\lambda)} + 2(1 - \sqrt{1 + \xi^2})\lambda^2(\lambda + 1)^2$$
$$\times \frac{\prod_{j=1}^{N}(\lambda + \theta_j)(\lambda - \theta_j)(\lambda + \theta_j + 1)(\lambda - \theta_j + 1)}{Q_1(\lambda)Q_2(\lambda)}. \quad (1.2.268)$$

由留数定理可得在均匀极限下的 Bethe Ansatz 方程:

$$(1 - \sqrt{1 + \xi^2})\mu_j(2\mu_j + 1)(\mu_j + 1)^{2N+2}$$
$$= -(\mu_j - p + 1)[\sqrt{1 + \xi^2}(\mu_j + 1) - q]$$
$$\times \prod_{l=1}^{2M}(\mu_j + \mu_l + 1)(\mu_j + \mu_l + 2), \quad j = 1, \cdots, 2M. \quad (1.2.269)$$

注意, Bethe 根两两不等, $\mu_j \neq \mu_l$. 哈密顿量的本征值仍然由方程 (1.2.267) 给出, 其中 $M = (N+1)/2$.

接下来, 我们构造系统的本征态. 为方便起见, 定义

$$a(\lambda) = \prod_{l=1}^{N}(\lambda - \theta_l + 1), \quad d(\lambda) = a(\lambda - 1) = \prod_{l=1}^{N}(\lambda - \theta_l). \quad (1.2.270)$$

1.2 自旋链模型

系统的转移矩阵的本征值参数化为以下非齐次 $T-Q$ 关系：

$$\Lambda(\lambda) = (-1)^N \frac{2\lambda+2}{2\lambda+1}(\lambda+p)(\sqrt{1+\xi^2}\lambda+q)a(\lambda)d(-\lambda-1)\frac{Q(\lambda-1)}{Q(\lambda)}$$

$$+(-1)^N \frac{2\lambda}{2\lambda+1}(\lambda-p+1)[\sqrt{1+\xi^2}(\lambda+1)-q]a(-\lambda-1)d(\lambda)\frac{Q(\lambda+1)}{Q(\lambda)}$$

$$+2(1-\sqrt{1+\xi^2})\lambda(\lambda+1)\frac{a(\lambda)a(-\lambda-1)d(\lambda)d(-\lambda-1)}{Q(\lambda)}, \quad (1.2.271)$$

其中，Q 函数为

$$Q(\lambda) = \prod_{j=1}^{N}(\lambda-\lambda_j)(\lambda+\lambda_j+1). \quad (1.2.272)$$

参数 $\{\lambda_j\}$ 满足以下 Bethe Ansatz 方程：

$$1+\frac{\lambda_j(\lambda_j-p+1)[\sqrt{1+\xi^2}(\lambda_j+1)-q]a(-\lambda_j-1)d(\lambda_j)Q(\lambda_j+1)}{(\lambda_j+1)(\lambda_j+p)(\sqrt{1+\xi^2}\lambda_j+q)a(\lambda_j)d(-\lambda_j-1)Q(\lambda_j-1)}$$

$$=(-1)^N\frac{(\sqrt{1+\xi^2}-1)\lambda_j(2\lambda_j+1)a(-\lambda_j-1)d(\lambda_j)}{(\lambda_j+p)(\sqrt{1+\xi^2}\lambda_j+q)Q(\lambda_j-1)}, \quad j=1,\cdots,N. \quad (1.2.273)$$

首先对角化对偶反射矩阵 K^+：

$$\bar{K}^+(\lambda) = UK^+(\lambda)U^{-1} = \begin{pmatrix} q+\sqrt{1+\xi^2}(\lambda+1) & 0 \\ 0 & q-\sqrt{1+\xi^2}(\lambda+1) \end{pmatrix}$$

$$= \begin{pmatrix} \bar{K}^+_{11}(\lambda) & 0 \\ 0 & \bar{K}^+_{22}(\lambda) \end{pmatrix}, \quad (1.2.274)$$

其中规范变换矩阵 U 定义为

$$U = \begin{pmatrix} \xi & \sqrt{1+\xi^2}-1 \\ \xi & -\sqrt{1+\xi^2}-1 \end{pmatrix}. \quad (1.2.275)$$

在此变换下，新的反射矩阵 $\bar{K}^-(\lambda)$ 为

$$\bar{K}^-(\lambda) = UK^-(\lambda)U^{-1} = \begin{pmatrix} p+\dfrac{1}{\sqrt{1+\xi^2}}\lambda & \dfrac{\sqrt{1+\xi^2}-1}{\sqrt{1+\xi^2}}\lambda \\ \dfrac{\sqrt{1+\xi^2}+1}{\sqrt{1+\xi^2}}\lambda & p-\dfrac{1}{\sqrt{1+\xi^2}}\lambda \end{pmatrix}$$

$$= \begin{pmatrix} \bar{K}^-_{11}(\lambda) & \bar{K}^-_{12}(\lambda) \\ \bar{K}^-_{21}(\lambda) & \bar{K}^-_{22}(\lambda) \end{pmatrix}. \quad (1.2.276)$$

定义两个局域参考态：

$$|1\rangle_n = \frac{\sqrt{1+\xi^2}+1}{2\xi\sqrt{1+\xi^2}}|\uparrow\rangle_n + \frac{1}{2\sqrt{1+\xi^2}}|\downarrow\rangle_n, \quad n=1,\cdots,N, \qquad (1.2.277)$$

$$|2\rangle_n = \frac{\sqrt{1+\xi^2}-1}{2\xi\sqrt{1+\xi^2}}|\uparrow\rangle_n - \frac{1}{2\sqrt{1+\xi^2}}|\downarrow\rangle_n, \quad n=1,\cdots,N, \qquad (1.2.278)$$

和它们的对偶态：

$$\begin{cases} \langle 1|_n = \xi\langle\uparrow|_n + (\sqrt{1+\xi^2}-1)\langle\downarrow|_n, \\ \langle 2|_n = \xi\langle\uparrow|_n - (\sqrt{1+\xi^2}+1)\langle\downarrow|_n, \quad n=1,\cdots,N. \end{cases} \qquad (1.2.279)$$

这些态满足以下正交关系：

$$\langle a|_j b\rangle_k = \delta_{a,b}\,\delta_{j,k}, \quad a,b=1,2,\ j,k=1,\cdots,N.$$

系统的参考态定义为

$$|\Omega\rangle_\xi = \otimes_{j=1}^N |1\rangle_j, \quad {}_\xi\langle\bar{\Omega}| = \otimes_{j=1}^N \langle 2|_j. \qquad (1.2.280)$$

双行单值矩阵

$$\mathbb{T}(\lambda) = T(\lambda)\,K^-(\lambda)\,\hat{T}(\lambda) = \begin{pmatrix} \mathcal{A}(\lambda) & \mathcal{B}(\lambda) \\ \mathcal{C}(\lambda) & \mathcal{D}(\lambda) \end{pmatrix}, \qquad (1.2.281)$$

在规范变换之后变为

$$\bar{\mathbb{T}}(\lambda) = U\,T(\lambda)\,K^-(\lambda)\,\hat{T}(\lambda)U^{-1} = UT(\lambda)U^{-1}\,UK^-(\lambda)U^{-1}\,U\hat{T}(\lambda)U^{-1}$$
$$= \bar{T}(\lambda)\,\bar{K}^-(\lambda)\hat{\bar{T}}(\lambda) = \begin{pmatrix} \bar{\mathcal{A}}(\lambda) & \bar{\mathcal{B}}(\lambda) \\ \bar{\mathcal{C}}(\lambda) & \bar{\mathcal{D}}(\lambda) \end{pmatrix}. \qquad (1.2.282)$$

规范的双行单值矩阵满足反射方程：

$$R_{12}(\lambda-u)\bar{\mathbb{T}}_1(\lambda)R_{21}(\lambda+u)\bar{\mathbb{T}}_2(u) = \bar{\mathbb{T}}_2(u)R_{12}(\lambda+u)\bar{\mathbb{T}}_1(\lambda)R_{21}(\lambda-u). \qquad (1.2.283)$$

1.2 自旋链模型

据此, 可以得到矩阵元 $\bar{\mathcal{A}}(\lambda), \bar{\mathcal{B}}(\lambda), \bar{\mathcal{C}}(\lambda)$ 和 $\bar{\mathcal{D}}(\lambda)$ 满足的对易关系:

$$\bar{\mathcal{C}}(\lambda)\bar{\mathcal{A}}(u) = \frac{(\lambda+u)(\lambda-u+1)}{(\lambda-u)(\lambda+u+1)}\bar{\mathcal{A}}(u)\bar{\mathcal{C}}(\lambda) - \frac{1}{\lambda+u+1}\bar{\mathcal{D}}(\lambda)\bar{\mathcal{C}}(u)$$
$$-\frac{(\lambda+u)}{(\lambda-u)(\lambda+u+1)}\bar{\mathcal{A}}(\lambda)\bar{\mathcal{C}}(u), \qquad (1.2.284)$$

$$\bar{\mathcal{D}}(u)\bar{\mathcal{C}}(\lambda) = \frac{(\lambda+u)(\lambda-u+1)}{(\lambda-u)(\lambda+u+1)}\bar{\mathcal{C}}(\lambda)\bar{\mathcal{D}}(u) - \frac{1}{\lambda+u+1}\bar{\mathcal{C}}(u)\bar{\mathcal{A}}(\lambda)$$
$$-\frac{(\lambda+u)}{(\lambda-u)(\lambda+u+1)}\bar{\mathcal{C}}(u)\bar{\mathcal{D}}(\lambda), \qquad (1.2.285)$$

$$\bar{\mathcal{A}}(\lambda)\bar{\mathcal{A}}(u) = \bar{\mathcal{A}}(u)\bar{\mathcal{A}}(\lambda) + \frac{1}{\lambda+u+1}\bar{\mathcal{B}}(u)\bar{\mathcal{C}}(\lambda) - \frac{1}{\lambda+u+1}\bar{\mathcal{B}}(\lambda)\bar{\mathcal{C}}(u), \qquad (1.2.286)$$

$$\bar{\mathcal{D}}(\lambda)\bar{\mathcal{D}}(u) = \bar{\mathcal{D}}(u)\bar{\mathcal{D}}(\lambda) + \frac{1}{\lambda+u+1}\bar{\mathcal{C}}(u)\bar{\mathcal{B}}(\lambda) - \frac{1}{\lambda+u+1}\bar{\mathcal{C}}(\lambda)\bar{\mathcal{B}}(u), \qquad (1.2.287)$$

$$\bar{\mathcal{D}}(\lambda)\bar{\mathcal{A}}(u) = \bar{\mathcal{A}}(u)\bar{\mathcal{D}}(\lambda) - \frac{(\lambda+u+2)}{(\lambda-u)(\lambda+u+1)}\bar{\mathcal{B}}(\lambda)\bar{\mathcal{C}}(u)$$
$$+\frac{(\lambda+u+2)}{(\lambda-u)(\lambda+u+1)}\bar{\mathcal{B}}(u)\bar{\mathcal{C}}(\lambda). \qquad (1.2.288)$$

转移矩阵可以表示为

$$t(\lambda) = K_{11}^+(\lambda)\mathcal{A}(\lambda) + K_{12}^+(\lambda)\mathcal{C}(\lambda) + K_{21}^+(\lambda)\mathcal{B}(\lambda) + K_{22}^+(\lambda)\mathcal{D}(\lambda)$$
$$= \bar{K}_{11}^+(\lambda)\bar{\mathcal{A}}(\lambda) + \bar{K}_{22}^+(\lambda)\bar{\mathcal{D}}(\lambda). \qquad (1.2.289)$$

注意到算子 $\bar{\mathcal{C}}(u)$ 彼此对易, $[\bar{\mathcal{C}}(u), \bar{\mathcal{C}}(v)] = 0$. 我们将利用算子 $\bar{\mathcal{C}}(u)$ 的共同本征态来构造系统希尔伯特空间的完备基. 为此, 我们引入以下由 N 个非均匀参数 $\{\theta_j\}$ 确定的右态和左态

$$|\theta_{p_1}, \cdots, \theta_{p_n}\rangle = \bar{\mathcal{A}}(\theta_{p_1}) \cdots \bar{\mathcal{A}}(\theta_{p_n})|\Omega\rangle_\xi, \quad 1 \leqslant p_1 < p_2 < \cdots < p_n \leqslant N, \qquad (1.2.290)$$

$$\langle -\theta_{q_1}, \cdots, -\theta_{q_n}| = {}_\xi\langle\bar{\Omega}|\bar{\mathcal{D}}(-\theta_{q_1}) \cdots \bar{\mathcal{D}}(-\theta_{q_n}), \quad 1 \leqslant q_1 < q_2 < \cdots < q_n \leqslant N. \qquad (1.2.291)$$

它们是算子 $\bar{\mathcal{C}}(\lambda)$ 的本征态:

$$\bar{\mathcal{C}}(\lambda)|\theta_{p_1}, \cdots, \theta_{p_n}\rangle = h(\lambda, \{\theta_{p_1}, \cdots, \theta_{p_n}\})|\theta_{p_1}, \cdots, \theta_{p_n}\rangle, \qquad (1.2.292)$$

$$\langle -\theta_{p_1}, \cdots, -\theta_{p_n}|\bar{\mathcal{C}}(\lambda) = h'(\lambda, \{-\theta_{p_1}, \cdots, -\theta_{p_n}\})\langle -\theta_{p_1}, \cdots, -\theta_{p_n}|, \qquad (1.2.293)$$

对应的本征值是

$$h(\lambda,\{\theta_{p_1},\cdots,\theta_{p_n}\}) = (-1)^N \bar{K}_{21}^-(\lambda) d(\lambda) d(-\lambda-1) \prod_{j=1}^{n} \frac{(\lambda+\theta_{p_j})(\lambda-\theta_{p_j}+1)}{(\lambda-\theta_{p_j})(\lambda+\theta_{p_j}+1)},$$
(1.2.294)

$$h'(\lambda,\{-\theta_{p_1},\cdots,-\theta_{p_n}\}) = (-1)^N \bar{K}_{21}^-(\lambda) a(\lambda) a(-\lambda-1) \prod_{j=1}^{n} \frac{(\lambda-\theta_{p_j})(\lambda+\theta_{p_j}+1)}{(\lambda+\theta_{p_j})(\lambda-\theta_{p_j}+1)}.$$
(1.2.295)

根据右态 (1.2.290) 中参数 p_n 和左态 (1.2.291) 中参数 q_n 的取值, 可得

$$\sum_{n=0}^{N} \frac{N!}{(N-n)!n!} = 2^N,$$
(1.2.296)

这正是系统希尔伯特空间的维数, 因此右态 (1.2.290) 和左态 (1.2.291) 都是完备的.

对于任意的非均匀参数 $\{\theta_j\}$, 左态和右态满足正交关系:

$$\langle-\theta_{q_1},\cdots,-\theta_{q_m}|\theta_{p_1},\cdots,\theta_{p_n}\rangle = f_n(\theta_{p_1},\cdots,\theta_{p_n})\delta_{m+n,N}\delta_{\{q_1,\cdots,q_m\};\{p_1,\cdots,p_n\}},$$
(1.2.297)

其中, $\delta_{\{q_1,\cdots,q_m\};\{p_1,\cdots,p_n\}}$ 的定义是

$$\delta_{\{q_1,\cdots,q_m\};\{p_1,\cdots,p_n\}} = \begin{cases} 1, & \{q_1,\cdots,q_m,p_1,\cdots,p_n\} = \{1,\cdots,N\}, \\ 0, & \text{其他}, \end{cases}$$
(1.2.298)

归一化系数 $f_n(\theta_{p_1},\cdots,\theta_{p_n})$ 是

$$\begin{aligned}
f_n(\theta_{p_1},\cdots,\theta_{p_n}) &= \langle-\theta_{p_{n+1}},\cdots,-\theta_{p_N}|\theta_{p_1},\cdots,\theta_{p_n}\rangle \\
&= \prod_{j=1}^{n} (-1)^N \bar{K}_{21}^-(\theta_{p_j}) d(-\theta_{p_j}-\eta) a(\theta_{p_j}) \prod_{k=n+1}^{N} (-1)^N \bar{K}_{21}^-(-\theta_{p_k}) \\
&\quad \times a(-\theta_{p_k}) d(\theta_{p_k}-\eta) \prod_{j=1}^{n}\prod_{l>j}^{n} \frac{\theta_{p_j}+\theta_{p_l}}{\theta_{p_j}+\theta_{p_l}+\eta} \\
&\quad \times \prod_{j=n+1}^{N}\prod_{l>j}^{N} \frac{\theta_{p_j}+\theta_{p_l}}{\theta_{p_j}+\theta_{p_l}-\eta} \prod_{j=1}^{n}\prod_{l=n+1}^{N} \frac{\theta_{p_l}-\theta_{p_j}}{\theta_{p_l}-\theta_{p_j}-\eta}.
\end{aligned}$$
(1.2.299)

由于右态 $\{|\theta_{p_1},\cdots,\theta_{p_n}\rangle\}$ 和左态 $\{\langle-\theta_{p_1},\cdots,-\theta_{p_n}|\}$ 是系统希尔伯特空间的完备基, 因此系统的本征态都可以按照这组完备基来展开. 设系统转移矩阵本征值 $\Lambda(\lambda)$ 对应的本征态为 $\langle\Psi|$, 即

$$\langle\Psi|t(\lambda) = \langle\Psi|\Lambda(\lambda),$$

1.2 自旋链模型

为简单起见，引入标记

$$\bar{F}_n(\theta_{p_1},\cdots,\theta_{p_n}) = \langle\Psi|\theta_{p_1},\cdots,\theta_{p_n}\rangle, \quad n=0,\cdots,N,\ 1\leqslant p_1<p_2<\cdots<p_n\leqslant N. \tag{1.2.300}$$

它唯一地确定了系统的本征态. 接下来考虑物理量 $\langle\Psi|t(\theta_{p_{n+1}})|\theta_{p_1},\cdots,\theta_{p_n}\rangle$. 经过复杂的计算，我们得到 F_n 满足的递推方程:

$$\begin{aligned}&\Lambda(\theta_{p_{n+1}})\bar{F}_n(\theta_{p_1},\cdots,\theta_{p_n})\\&=\frac{(2\theta_{p_{n+1}}+\eta)\bar{K}_{11}^+(\theta_{p_{n+1}})+\eta\bar{K}_{22}^+(\theta_{p_{n+1}})}{2\theta_{p_{n+1}}+\eta}\bar{F}_{n+1}(\theta_{p_1},\cdots,\theta_{p_{n+1}}).\end{aligned} \tag{1.2.301}$$

它的解 $\{\bar{F}_n(\theta_{p_1},\cdots,\theta_{p_n})\}$ 为

$$\bar{F}_n(\theta_{p_1},\cdots,\theta_{p_n}) = \left\{\prod_{j=1}^n \frac{(2\theta_{p_j}+\eta)\Lambda(\theta_{p_j})}{(2\theta_{p_j}+\eta)\bar{K}_{11}^+(\theta_{p_j})+\eta\bar{K}_{22}^+(\theta_{p_j})}\right\}\bar{F}_0,$$

其中，$\bar{F}_0 = \langle\Psi|\Omega\rangle_\xi$ 是标量因子. 代入本征值的具体形式，可得

$$\begin{aligned}\bar{F}_n(\theta_{p_1},\cdots,\theta_{p_n}) &= \langle\Psi|\theta_{p_1},\cdots,\theta_{p_n}\rangle\\&=\left\{\prod_{j=1}^n (-1)^N(\theta_{p_j}+p)\,a(\theta_{p_j})d(-\theta_{p_j}-\eta)\frac{Q(\theta_{p_j}-\eta)}{Q(\theta_{p_j})}\right\}\bar{F}_0,\\&n=0,\cdots,N,\quad 1\leqslant p_1<p_2<\cdots<p_n\leqslant N.\end{aligned} \tag{1.2.302}$$

另外，根据自旋向上态 $|0\rangle$ 和它的对偶态 $\langle 0|$ 的定义:

$$|0\rangle = \otimes_{j=1}^N|\uparrow\rangle_j, \quad \langle 0| = \langle\uparrow|_j\otimes_{j=1}^N, \tag{1.2.303}$$

我们得到

$$\langle 0|A(\lambda) = a(\lambda)\langle 0|,\quad \langle 0|D(\lambda)=d(\lambda)\langle 0|,\quad \langle 0|B(\lambda)=0,\quad \langle 0|C(\lambda)\neq 0, \tag{1.2.304}$$

其中，函数 $a(\lambda)$ 和 $d(\lambda)$ 的定义参见式 (1.2.270). 双行单值矩阵的矩阵元作用到态 $\langle 0|$ 上可得

$$\langle 0|\mathcal{A}(\lambda) = (-1)^N K_{11}^-(\lambda)\,a(\lambda)\,d(-\lambda-1)\,\langle 0|, \tag{1.2.305}$$

$$\langle 0|\mathcal{D}(\lambda) = (-1)^N \frac{1}{2\lambda+1} K_{11}^-(\lambda) a(\lambda) d(-\lambda-1) \langle 0|$$
$$+ (-1)^N \frac{(2\lambda+1)K_{22}^-(\lambda) - K_{11}^-(\lambda)}{2\lambda+1} d(\lambda) a(-\lambda-1) \langle 0|, \quad (1.2.306)$$

$$\langle 0|\mathcal{B}(\lambda) = 0, \quad (1.2.307)$$

$$\langle 0|\mathcal{C}(\lambda) = (-1)^N \frac{2\lambda}{2\lambda+1} K_{11}^-(\lambda) d(-\lambda-1) \langle 0| C(\lambda)$$
$$+ (-1)^N \frac{K_{11}^-(\lambda) - (2\lambda+1)K_{22}^-(\lambda)}{2\lambda+1} d(\lambda) \langle 0| C(-\lambda-1). \quad (1.2.308)$$

规范变换前后双行单值矩阵的矩阵元满足关系:

$$\bar{\mathcal{A}}(\lambda) = \frac{1}{2\xi\sqrt{1+\xi^2}} \left\{ \xi(1+\sqrt{1+\xi^2})\mathcal{A}(\lambda) + \xi^2 \mathcal{C}(\lambda) \right.$$
$$\left. + \xi^2 \mathcal{B}(\lambda) - \xi(1-\sqrt{1+\xi^2})\mathcal{D}(\lambda) \right\}, \quad (1.2.309)$$

$$\bar{\mathcal{C}}(\lambda) = \frac{1}{2\xi\sqrt{1+\xi^2}} \left\{ \xi(1+\sqrt{1+\xi^2})\mathcal{A}(\lambda) \right.$$
$$- (1+\sqrt{1+\xi^2})^2 \mathcal{C}(\lambda)$$
$$\left. + \xi^2 \mathcal{B}(\lambda) - \xi(1+\sqrt{1+\xi^2})\mathcal{D}(\lambda) \right\}, \quad (1.2.310)$$

$$\bar{\mathcal{D}}(\lambda) = \frac{1}{2\xi\sqrt{1+\xi^2}} \left\{ \xi(\sqrt{1+\xi^2}-1)\mathcal{A}(\lambda) - \xi^2 \mathcal{C}(\lambda) \right.$$
$$\left. - \xi^2 \mathcal{B}(\lambda) + \xi(1+\sqrt{1+\xi^2})\mathcal{D}(\lambda) \right\}. \quad (1.2.311)$$

考虑物理量 $\langle 0|\bar{\mathcal{C}}(\theta_{p_{n+1}})|\theta_{p_1},\cdots,\theta_{p_n}\rangle$. 把算子 $\bar{\mathcal{C}}(\theta_{p_{n+1}})$ 向左作用, 根据关系 (1.2.292) 和 (1.2.294), 可得 $\langle 0|\bar{\mathcal{C}}(\theta_{p_{n+1}})|\theta_{p_1},\cdots,\theta_{p_n}\rangle$ 等于零. 把算子 $\bar{\mathcal{C}}(\theta_{p_{n+1}})$ 向右作用, 利用关系式 (1.2.305)∼(1.2.308), 以及式 (1.2.310), 我们得到

$$\langle 0|C(\theta_{p_{n+1}})|\theta_{p_1},\cdots,\theta_{p_n}\rangle = \frac{\xi}{1+\sqrt{1+\xi^2}} a(\theta_{p_{n+1}}) \langle 0|\theta_{p_1},\cdots,\theta_{p_n}\rangle, \quad j=1,\cdots,N.$$
$$(1.2.312)$$

根据定义式 (1.2.290), 可得

$$\langle 0|\theta_{p_1},\cdots,\theta_{p_{n+1}}\rangle = \langle 0|\bar{\mathcal{A}}(\theta_{p_{n+1}})|\theta_{p_1},\cdots,\theta_{p_n}\rangle.$$

把算子 $\bar{\mathcal{A}}(\theta_{p_{n+1}})$ 向左作用, 利用关系式 (1.2.305)∼(1.2.308), 以及式 (1.2.309), 得到

1.2 自旋链模型

$$\begin{aligned}
&\langle 0|\theta_{p_1},\cdots,\theta_{p_{n+1}}\rangle \\
&= \frac{(-1)^N}{2\sqrt{1+\xi^2}}\Big\{(1+\sqrt{1+\xi^2})K_{11}^-(\theta_{p_{n+1}})a(\theta_{p_{n+1}})d(-\theta_{p_{n+1}}-\eta)\langle 0|\theta_{p_1},\cdots,\theta_{p_n}\rangle \\
&\quad -(1-\sqrt{1+\xi^2})\frac{\eta}{2\theta_{p_{n+1}}+\eta}K_{11}^-(\theta_{p_{n+1}})a(\theta_{p_{n+1}})d(-\theta_{p_{n+1}}-\eta)\langle 0|\theta_{p_1},\cdots,\theta_{p_n}\rangle \\
&\quad +\xi\frac{2\theta_{p_{n+1}}}{2\theta_{p_{n+1}}+\eta}K_{11}^-(\theta_{p_{n+1}})d(-\theta_{p_{n+1}}-\eta)\langle 0|C(\theta_{p_{n+1}})|\theta_{p_1},\cdots,\theta_{p_n}\rangle\Big\} \\
&= (-1)^N K_{11}^-(\theta_{p_{n+1}})a(\theta_{p_{n+1}})d(-\theta_{p_{n+1}}-\eta)\langle 0|\theta_{p_1},\cdots,\theta_{p_n}\rangle, \quad n=0,\cdots,N-1,
\end{aligned}$$
(1.2.313)

整理可得

$$\langle 0|\theta_{p_1},\cdots,\theta_{p_n}\rangle = \left\{\prod_{j=1}^{n}(-1)^N(\theta_{p_j}+p)\,a(\theta_{p_j})d(-\theta_{p_j}-\eta)\right\}\langle 0|\Omega\rangle_\xi,$$

$$n=0,\cdots,N,\quad 1\leqslant p_1<p_2<\cdots<p_n\leqslant N. \tag{1.2.314}$$

至此, 系统的本征态已经完全确定了.

接下来反演系统的 Bethe 态. 根据式 (1.2.277) 和式 (1.2.280), 对任意的非零参数 ξ, 整体标量因子 $\langle 0|\Omega\rangle_\xi$ 都不为零. 对于 Bethe Ansatz 方程 (1.2.273) 的每一组解, 我们引入如下的左 Bethe 态:

$$_B\langle\lambda_1,\cdots,\lambda_N| = \langle 0|\left\{\prod_{j=1}^{N}\frac{\bar{C}(\lambda_j)}{(-1)^N\bar{K}_{21}^-(\lambda_j)d(\lambda_j)d(-\lambda_j-\eta)}\right\}. \tag{1.2.315}$$

由关系式 (1.2.292)、(1.2.294) 和 (1.2.314), 可得

$$\begin{aligned}
&_B\langle\lambda_1,\cdots,\lambda_N|\theta_{p_1},\cdots,\theta_{p_n}\rangle \\
&= \left\{\prod_{j=1}^{n}(-1)^N(\theta_{p_j}+p)\,a(\theta_{p_j})d(-\theta_{p_j}-\eta)\frac{Q(\theta_{p_j}-\eta)}{Q(\theta_{p_j})}\right\}\langle 0|\Omega\rangle_\xi, \\
&n=0,\cdots,N,\quad 1\leqslant p_1<p_2<\cdots<p_n\leqslant N.
\end{aligned}$$
(1.2.316)

上式和式 (1.2.302) 仅相差一个标量因子. 因此, 由式 (1.2.315) 给出的 Bethe 态 $_B\langle\lambda_1,\cdots,\lambda_N|$ 是转移矩阵和哈密顿量的本征态, 其中的 Bethe 根 $\{\lambda_j\}$ 要满足 Bethe Ansatz 方程 (1.2.273). 经过相似的处理过程, 右 Bethe 态可以构造为

$$|\lambda_1,\cdots,\lambda_N\rangle_B = \prod_{j=1}^{N}\bar{\mathcal{B}}(\lambda_j)|0\rangle. \tag{1.2.317}$$

需要说明的是, Bethe 态 $_B\langle\lambda_1,\cdots,\lambda_N|$ (或者 $|\bar{\lambda}_1,\cdots,\bar{\lambda}_N\rangle_B$) 具有好的均匀极限. 有关非对角 Bethe Ansatz 方法的进一步讨论和应用, 可参看文献 [22].

1.2.8 近藤问题

金属中的磁性杂质问题 (近藤问题) 是强关联电子领域最基本的问题之一, 在许多物理系统 (如重费米子系统、耦合量子点系统) 都有重要应用. 许多处理强关联电子体系的重要方法都源于对此问题的研究. 例如, Wilson 研究此问题时所提出的数值重整化群理论曾获 1982 年诺贝尔物理学奖. 在传统的近藤问题中, 基于费米液体理论, 电子关联基本被忽略, 主要是由于在三维情形下, 弱电子–电子相互作用仅将无相互作用的电子转化为准粒子, 而费米液体的准粒子态同自由费米体系的电子态一一对应, 在低温时, 通常可近似认为准粒子之间无相互作用. 20 世纪 80 年代初, 美国和苏联科学家严格求解了传统的近藤问题[23, 24], 曾引起很大的轰动. 正是这一严格解的发现才从理论上最终解决了费米液体中单杂质近藤问题, 其中一个重要物理效应即费米液体中杂质磁矩会被电子云屏蔽而不呈现磁性. 但是, 对于低维关联体系, 近藤杂质会呈现出复杂性和新的普适类.

近藤问题的自旋动力学行为等价于具有边界杂质的自旋链系统. 周期 Heisenberg 模型中的可积杂质问题最早是由 Andrei 和 Johannesson 提出[25] 并被 Lee 和 Schlottmann 推广到任意杂质自旋情形的[26]. 考虑一维开边界自旋 1/2 Heisenberg 模型, 其中两个端点各放有一个磁性杂质, 杂质携带的自旋是任意的. 模型哈密顿量为[27]

$$H = \frac{1}{2}\sum_{j=1}^{N-1}\boldsymbol{\sigma}_j \cdot \boldsymbol{\sigma}_{j+1} + J_R\boldsymbol{\sigma}_N \cdot \boldsymbol{S}_R + J_L\boldsymbol{\sigma}_1 \cdot \boldsymbol{S}_L, \qquad (1.2.318)$$

其中, $\boldsymbol{\sigma}_j$ 是第 j 个格点的 Pauli 矩阵; $\boldsymbol{S}_{R,L}$ 分别是右左两个端点处磁性杂质的自旋, 它们是任意的; N 是体内的总格点数; $J_{R,L}$ 是杂质和体内自旋的耦合常数.

系统 (1.2.318) 可以严格求解. 系统的转移矩阵定义为单值矩阵 $U(\lambda)$ 的部分迹,

$$\tau(\lambda) = \mathrm{tr}_0 K_0^+ T_0(\lambda) K_0^- \tilde{T}_0(\lambda), \qquad (1.2.319)$$

其中单值矩阵是

$$T_0(\lambda) = L_{R0}^+(\lambda)L_{N0}(\lambda)L_{N-10}(\lambda)\cdots L_{10}(\lambda), \qquad (1.2.320)$$

$$\tilde{T}_0(\lambda) = L_{10}(\lambda)L_{20}(\lambda)\cdots L_{N0}(\lambda)L_{R0}^-(\lambda), \qquad (1.2.321)$$

L 算子为

1.2 自旋链模型

$$L_{j0}(\lambda) = \mathrm{i}\lambda + 1/2(1 + \boldsymbol{\sigma}_j \cdot \boldsymbol{\tau}), \tag{1.2.322}$$

$$L_{R0}^{\pm}(\lambda) = \mathrm{i}\lambda \pm c_R + \frac{1}{2} + \boldsymbol{\tau} \cdot \boldsymbol{S}_R, \tag{1.2.323}$$

反射矩阵是

$$K_0^+(\lambda) = 1, \tag{1.2.324}$$

$$K_0^-(\lambda) = \left(\mathrm{i}\lambda + c_L + \frac{1}{2} + \boldsymbol{\tau} \cdot \boldsymbol{S}_L\right)\left(\mathrm{i}\lambda - c_L + \frac{1}{2} + \boldsymbol{\tau} \cdot \boldsymbol{S}_L\right), \tag{1.2.325}$$

其中, $c_{R,L}$ 是两个常数. 哈密顿量 (1.2.318) 可以由转移矩阵得到

$$H = \frac{-\mathrm{i}}{4\prod\limits_{r=R,L}\left[\left(S+\frac{1}{2}\right)^2 - c_r^2\right]} \frac{\mathrm{d}\tau(\lambda)}{\mathrm{d}\lambda}\bigg|_{\lambda=0} - \frac{N}{2} - \frac{1}{2}\sum_{r=R,L}\frac{1}{\left(S+\frac{1}{2}\right)^2 - c_r^2}, \tag{1.2.326}$$

其中, $J_{R,L}$ 参数化为 $J_{R,L} = 1/[(S+1/2)^2 - c_{R,L}^2]$.

对任意的 $J_{R,L}$, 模型 (1.2.318) 都是严格可解的. 为方便起见, 我们仅考虑 $c_R = c_L = c$ ($J_R = J_L = J_i$), 其他参数可以采用同样的处理过程得到. 采用以前介绍的量子反散射方法, 可以得到哈密顿量 (1.2.318) 的本征能量为

$$E(\lambda_1, \cdots, \lambda_M) = -\sum_{j=1}^{M} \frac{1}{\lambda_j^2 + \frac{1}{4}} + \frac{1}{2}(N-1)J + 2SJ_i, \tag{1.2.327}$$

其中, 谱参数 λ_j 应该满足 Bethe Ansatz 方程:

$$\left(\frac{\lambda_j + \frac{\mathrm{i}}{2}}{\lambda_j - \frac{\mathrm{i}}{2}}\right)^{2N}\left(\frac{\lambda_j - \mathrm{i}c + \mathrm{i}S}{\lambda_j - \mathrm{i}c - \mathrm{i}S}\right)^2\left(\frac{\lambda_j + \mathrm{i}c + \mathrm{i}S}{\lambda_j + \mathrm{i}c - \mathrm{i}S}\right)^2 = \prod_{r=\pm 1}\prod_{l\neq j}^{M}\frac{\lambda_j - r\lambda_l + \mathrm{i}}{\lambda_j - r\lambda_l - \mathrm{i}}. \tag{1.2.328}$$

下面考虑系统的基态性质. 根据 J_i 和 c^2 的依赖关系可知, c 可以是实数也可以是纯虚数.

(i) 当 $c > S + 1/2$ 时. 杂质和体内的耦合是铁磁性的. 基态时所有的 λ 都是实数. 对 Bethe Ansatz 方程取对数,

$$\frac{I_j}{N} = \frac{1}{\pi}\left\{\theta_1(\lambda_j) + \frac{1}{2N}\left[\phi(\lambda_j) - \sum_{l=-M}^{M}\theta_2(\lambda_j - \lambda_l)\right]\right\}, \tag{1.2.329}$$

其中, $\theta_n(\lambda) = 2\arctan(2\lambda/n)$; $\phi(\lambda) = 2\theta_{2(S+|c|)}(\lambda) - 2\theta_{2(|c|-S)}(\lambda) + \theta_2(\lambda) + \theta_1(\lambda)$; I_j 是整数. 注意在上式中, 通过定义 $\lambda_j = -\lambda_{-j}$ ($\lambda_0 = 0$), 允许有负数解. 定义

$$Z_N(\lambda) = \frac{1}{\pi}\left\{\theta_1(\lambda) + \frac{1}{2N}\left[\phi(\lambda) - \sum_{l=-M}^{M}\theta_2(\lambda - \lambda_l)\right]\right\}. \tag{1.2.330}$$

$Z_N(\lambda_j) = I_j/N$ 就是 Bethe Ansatz 方程. 在基态, 量子数 I_j 应该对称地连续地分布在原点两侧. 在热力学极限下定义基态的密度函数为

$$\rho_N(\lambda) = \frac{\mathrm{d}Z_N(\lambda)}{\mathrm{d}\lambda}, \tag{1.2.331}$$

$\rho_N(\lambda)$ 满足

$$\int_{-\Lambda}^{\Lambda}\rho_N(\lambda)\mathrm{d}\lambda = \frac{2M+1}{N}, \tag{1.2.332}$$

其中, Λ 是 λ 的切断. 当 $\Lambda = \infty$ 时系统的能量最小. 此时, $M = N/2$, 这给出系统的自发磁化强度为

$$M_g = \frac{1}{2}N - M + 2S = 2S. \tag{1.2.333}$$

该结果说明, 由于杂质和体内是铁磁耦合, 杂质的磁矩不能被屏蔽.

(ii) 当 $S < |c| < S + 1/2$ 时. $J_i > 0$, 杂质和体内的交换作用进入反铁磁耦合区域. 基态时在 $\lambda = \mathrm{i}(|c| - S)$ 点可以存在两个虚模, 它们携带的能量是 $\epsilon_i = -1/[1/4 - (|c| - S)^2]$, 小于实 λ 携带的能量. 实 λ 满足如下 Bethe Ansatz 方程:

$$\left(\frac{\lambda_j + \frac{\mathrm{i}}{2}}{\lambda_j - \frac{\mathrm{i}}{2}}\right)^{2N}\left(\frac{\lambda_j - \mathrm{i}c + \mathrm{i}S}{\lambda_j - \mathrm{i}c - \mathrm{i}S}\right)^2\left(\frac{\lambda_j + \mathrm{i}c + \mathrm{i}S}{\lambda_j + \mathrm{i}c - \mathrm{i}S}\right)^2$$
$$= \left(\frac{\lambda_j - \mathrm{i}|c| + \mathrm{i}S + \mathrm{i}}{\lambda_j - \mathrm{i}|c| + \mathrm{i}S - \mathrm{i}}\right)^2 \times \left(\frac{\lambda_j + \mathrm{i}|c| - \mathrm{i}S + \mathrm{i}}{\lambda_j + \mathrm{i}|c| - \mathrm{i}S - \mathrm{i}}\right)^2 \prod_{r=\pm 1}\prod_{l\neq j}^{M-2}\frac{\lambda_j - r\lambda_l + \mathrm{i}}{\lambda_j - r\lambda_l - \mathrm{i}}. \tag{1.2.334}$$

采用和情形 (i) 相同的步骤得到 $M = (N+2)/2$, 基态自发的磁化强度为 $M_g = 2S - 1$. 这说明杂质磁矩是部分屏蔽的.

(iii) 当 $0 < |c| < S$ 时. 杂质和体内依然是反铁磁耦合. 基态中不存在束缚态. 函数 $Z_N(\lambda)$ 的定义和情形 (i) 相同, 但是 $\phi(\lambda)$ 变为 $\phi(\lambda) = 2\theta_{2(S+c)}(\lambda) + 2\theta_{2(S-c)}(\lambda) + \theta_2(\lambda) + \theta_1(\lambda)$. 在限制 $Z_N(\pm\infty) = \pm(M + 1/2)/N$ 下可得 $M = (N+2)/2$, 自发磁化强度是 $M_g = 2S - 1$.

1.2 自旋链模型

(iv) 当 c 取纯虚数时, J_i 恒正. 设 $c = \mathrm{i}b$. Bethe Ansatz 方程变为

$$\left(\frac{\lambda_j + \frac{\mathrm{i}}{2}}{\lambda_j - \frac{\mathrm{i}}{2}}\right)^{2N} \left(\frac{\lambda_j - b + \mathrm{i}S}{\lambda_j - b - \mathrm{i}S}\right)^2 \left(\frac{\lambda_j + b + \mathrm{i}S}{\lambda_j + b - \mathrm{i}S}\right)^2 = \prod_{r=\pm 1}\prod_{l\neq j}^{M} \frac{\lambda_j - r\lambda_l + \mathrm{i}}{\lambda_j - r\lambda_l - \mathrm{i}}. \tag{1.2.335}$$

基态所有的 λ 都是实数, 在热力学极限下它们的切断 Λ 仍然趋于无穷大. 在函数 $Z_N(\lambda)$ 中的 $\phi(\lambda)$ 需要重新改写为 $\phi(\lambda) = 2\theta_{2S}(\lambda - b) + 2\theta_{2S}(\lambda + b) + \theta_2(\lambda) + \theta_1(\lambda)$. 因此, $M = (N+2)/2$, 自发磁化强度是 $M_g = 2S - 1$.

综上分析, 杂质仅仅当 $J_i > 0$ 时可以被屏蔽 (当 $S > 1/2$ 时是部分屏蔽), 而当 $J_i < 0$ 时, 杂质磁矩不能被屏蔽. 这与 Furusaki 和 Nagaosa 预测的在 Luttinger 液体中的 Kondo 问题很不一样.

低温下, 系统的热力学性质可以利用描述 Kondo 问题的局域费米液体理论来研究. 系统的能谱由量子数 I_j 来刻画. 根据可积模型的 Luttinger-Fermi 液体图像, $p_N(\lambda_j) = \pi Z_N(\lambda_j)$ 可以看成是准粒子的动量. 热力学极限下的基态能量是

$$E_g = \frac{1}{2} N \int \epsilon_0(\lambda) \rho_N(\lambda) \mathrm{d}\lambda + \mathrm{const.}, \tag{1.2.336}$$

其中, $\epsilon_0(\lambda) = -1/(\lambda^2 + 1/4)$. 上式又可以改写成

$$E_g = \frac{N}{4\pi} \int \epsilon(\lambda) \rho_N^{(0)}(\lambda) \mathrm{d}\lambda + \mathrm{const.}, \tag{1.2.337}$$

其中, $\epsilon(\lambda) = -\pi/\cosh(\pi\lambda)$, 它表示准粒子的能量. 注意, 当 c 是纯虚数时, 精确到 $o(N^{-1})$ 项, $\rho_N(\lambda)$ 的解为

$$\rho_N(\lambda) = \rho_0(\lambda) + \frac{1}{N}\rho_i(\lambda) + \frac{1}{N}\rho_b(\lambda), \tag{1.2.338}$$

$$\rho_0(\lambda) = \frac{1}{\cosh(\pi\lambda)}, \tag{1.2.339}$$

$$\rho_b(\lambda) = \frac{1}{2\cosh(\pi\lambda)} + \int \frac{\mathrm{e}^{-\frac{1}{2}|\omega|}\mathrm{e}^{-\mathrm{i}\lambda\omega}}{4\pi\cosh\left(\frac{1}{2}\omega\right)}\mathrm{d}\omega, \tag{1.2.340}$$

$$\rho_i(\lambda) = \frac{1}{\cosh(\lambda - b)} + \frac{1}{\cosh(\lambda + b)}, \tag{1.2.341}$$

其中, $\rho_0(\lambda)$, $\rho_i(\lambda)/N$ 和 $\rho_b(\lambda)/N$ 分别为体内、杂质和开边界对态密度的贡献. 根据费米液体理论图像, 费米面上的态密度是

$$N(0) = \frac{1}{2\pi} \frac{\mathrm{d}p_N(\lambda)}{\mathrm{d}\epsilon(\lambda)}\Big|_{\lambda=\infty} = \frac{\rho_N(\lambda)}{2\epsilon'(\lambda)}\Big|_{\lambda=\infty}, \tag{1.2.342}$$

杂质对低温比热的贡献是

$$C_i = \frac{2\pi}{3N}\cosh(\pi b)T. \tag{1.2.343}$$

当 $|c| < 1/2$ 时，$\rho_i(\lambda)$ 的解为

$$\rho_i(\lambda) = \frac{1}{\cosh(\lambda - \mathrm{i}c)} + \frac{1}{\cosh(\lambda + \mathrm{i}c)}. \tag{1.2.344}$$

杂质对低温比热的贡献是

$$C_i = \frac{\pi}{3N}\cos(\pi c)T. \tag{1.2.345}$$

Kondo 温度是

$$T_k = \frac{\pi}{\cos(\pi c)}. \tag{1.2.346}$$

该结果显示了当 c 从虚数变为实数时的指数衰减和幂次衰减的过渡.

1.2.9 各向异性

自旋与轨道或其他自由度发生耦合时，会导致自旋耦合的各向异性，如会出现易磁化轴、易磁化面等. 本节我们考虑如下模型:

$$H = \frac{1}{2}\sum_{j=1}^{N}(\sigma_j^x\sigma_{j+1}^x + \sigma_j^y\sigma_{j+1}^y + \cos\eta\,\sigma_j^z\sigma_{j+1}^z). \tag{1.2.347}$$

该模型因其耦合形式又被称为 XXZ 模型，其中 η 为表征各向异性的参数. 当 $\eta = 0$ 时，式 (1.2.347) 退化为各向同性反铁磁链模型. 与 $\eta = 0$ 情形类似，式 (1.2.347) 的本征态仍可以写为式 (1.2.14)，但其二体散射矩阵则为

$$S_{12} = -\frac{\mathrm{e}^{\mathrm{i}(k_1+k_2)} + 1 - 2\cos\eta\,\mathrm{e}^{\mathrm{i}k_2}}{\mathrm{e}^{\mathrm{i}(k_1+k_2)} + 1 - 2\cos\eta\,\mathrm{e}^{\mathrm{i}k_1}}. \tag{1.2.348}$$

本征能量则表达为

$$E = 2\sum_{j=1}^{M}(\cos k_j - \cos\eta) + \frac{1}{2}N\cos\eta. \tag{1.2.349}$$

为了方便起见，对 $|\cos\eta| < 1$ 我们引入以下表示:

$$\mathrm{e}^{\mathrm{i}k_j} = \frac{\sinh\left(\lambda_j + \frac{\mathrm{i}}{2}\eta\right)}{\sinh\left(\lambda_j - \frac{\mathrm{i}}{2}\eta\right)}. \tag{1.2.350}$$

1.2 自旋链模型

把上式代入式 (1.2.348) 得到散射矩阵的表示形式:

$$S_{12} = \frac{\sinh(\lambda_1 - \lambda_2 - i\eta)}{\sinh(\lambda_1 - \lambda_2 + i\eta)}. \tag{1.2.351}$$

此时, 系统的 Bethe Ansatz 方程是

$$\frac{\sinh^N\left(\lambda_j - \frac{i}{2}\eta\right)}{\sinh^N\left(\lambda_j + \frac{i}{2}\eta\right)} = -\prod_{l=1}^{M} \frac{\sinh(\lambda_j - \lambda_l - i\eta)}{\sinh(\lambda_j - \lambda_l + i\eta)}. \tag{1.2.352}$$

相应的本征能量 (1.2.349) 可重新写为

$$E(\lambda_1, \cdots, \lambda_M) = -\sum_{j=1}^{M} \frac{2\sin^2 \frac{\eta}{2}}{\cosh^2 \lambda_j - \cos^2 \frac{\eta}{2}} + \frac{1}{2} N \cos \eta. \tag{1.2.353}$$

与各向同性情形类似, 定义

$$a_n(\lambda) = \frac{1}{\pi} \frac{\sin \frac{n\eta}{2}}{\cosh^2 \lambda_j - \cos^2 \frac{n\eta}{2}}. \tag{1.2.354}$$

在基态, 所有 λ_j 充满整个实轴, 其密度分布满足

$$a_1(\lambda) = \sigma_0(\lambda) + \int a_2(\mu) \sigma_0(\lambda - \mu) d\lambda. \tag{1.2.355}$$

在 Fourier 变换下

$$\tilde{a}_n(\omega) = e^{-\frac{1}{2}n\eta|\omega|}, \quad \tilde{\sigma}_0(\omega) = \frac{1}{2\cosh\left(\frac{1}{2}\eta|\omega|\right)}. \tag{1.2.356}$$

因此系统的总磁化为

$$S = \frac{1}{2} - \int \sigma_0(\lambda) d\lambda = \frac{1}{2} - \tilde{\sigma}_0(0) = 0, \tag{1.2.357}$$

即系统的基态是自旋单态. 基态的能量密度为

$$e_0 = -2\pi \sin \frac{\eta}{2} \int a_1(\lambda) \sigma_0(\lambda) d\lambda + \frac{1}{2} \cos \eta$$
$$= \frac{1}{2} \cos \eta - \frac{4}{\eta} \sin \frac{\eta}{2} \ln 2. \tag{1.2.358}$$

定义准粒子的能量函数 $\varepsilon(\lambda)$ 为

$$\varepsilon(\lambda) = 2\pi a_1(\lambda) - \int a_2(\mu) \varepsilon(\lambda - \mu) d\mu. \tag{1.2.359}$$

利用 Fourier 变换可得

$$\varepsilon(\lambda) = \frac{1}{2\cosh\dfrac{\pi\lambda}{\eta}}. \tag{1.2.360}$$

显然，最低能量在 $\lambda = \pm\infty$ 处. 利用同样的方法可以构造系统的自旋子元激发，其能量为 $\varepsilon(\lambda_j^h)$ 之和，其中 λ_j^h 是第 j 个空穴的位置. 不难看出，对于 $|\cos\eta| < 1$，系统的元激发是无能隙的.

对于 $\cos\eta > 1$ 的情形，令 $\cos\eta = \cosh\theta$，其中 θ 为实数. 做参数化

$$e^{ik_j} = \frac{\sin\left(\lambda_j + \dfrac{i}{2}\theta\right)}{\sin\left(\lambda_j - \dfrac{i}{2}\theta\right)}, \tag{1.2.361}$$

则系统的 Bethe Ansatz 方程和本征能量可以表述为

$$\frac{\sin^N\left(\lambda_j - \dfrac{i}{2}\theta\right)}{\sin^N\left(\lambda_j + \dfrac{i}{2}\theta\right)} = -\prod_{l=1}^{M} \frac{\sin(\lambda_j - \lambda_l - i\theta)}{\sin(\lambda_j - \lambda_l + i\theta)}, \tag{1.2.362}$$

$$E(\lambda_1, \cdots, \lambda_M) = -\sum_{j=1}^{M} \frac{2\sinh^2\dfrac{\theta}{2}}{\cosh^2\dfrac{\theta}{2} - \cosh^2\lambda_j} + \frac{1}{2}N\cosh\theta. \tag{1.2.363}$$

注意在式 (1.2.361) 中，k_j 是 λ_j 的周期函数，因此 λ_j 的取值范围被限定在 $(-\pi, \pi]$ 区间中. 定义

$$a_n(\lambda) = \frac{1}{\pi} \frac{\sinh\dfrac{n\theta}{2}}{\cosh^2\dfrac{n\theta}{2} - \cos^2\lambda}, \tag{1.2.364}$$

则

$$\tilde{a}_n(\omega) = \int_{-\pi}^{\pi} e^{i\omega\lambda} \frac{\sin\dfrac{n\theta}{2}}{\cosh^2\dfrac{n\theta}{2} - \cos^2\lambda} d\lambda = e^{-\dfrac{n\theta}{2}|\omega|}. \tag{1.2.365}$$

这时 ω 取分立值 $\omega = m$，其中 m 是整数. λ 的基态密度分布满足：

$$\sigma_0(\lambda) = a_1(\lambda) - \int_{-\pi}^{\pi} a_2(\mu)\sigma_0(\lambda - \mu)d\mu. \tag{1.2.366}$$

1.2 自旋链模型

利用 Fourier 变换可以求得

$$\tilde{\sigma}_0(\omega) = \frac{1}{2\cosh\left(\frac{1}{2}\theta|\omega|\right)}. \tag{1.2.367}$$

因此, 系统的基态总磁化为

$$S = \frac{1}{2} - \int_{-\pi}^{\pi} \sigma_0(\lambda) \mathrm{d}\lambda = 0. \tag{1.2.368}$$

同样, 该情形的准粒子能量函数满足以下方程:

$$\varepsilon(\lambda) = 2\pi a_1(\lambda) - \int_{-\pi}^{\pi} a_2(\lambda - \mu)\varepsilon(\lambda - \mu)\mathrm{d}\mu. \tag{1.2.369}$$

因此

$$\varepsilon(\lambda) = \sum_{n=-\infty}^{\infty} \frac{\mathrm{e}^{-in\lambda}}{2\cosh\left(\frac{1}{2}\theta|n|\right)}. \tag{1.2.370}$$

由于 λ 限定在有限空间, $\varepsilon(\lambda) > 0$, 即系统的元激发存在一个能隙

$$\Delta = \sum_{n=-\infty}^{\infty} \frac{(-1)^n}{2\cosh\left(\frac{1}{2}n\theta\right)}. \tag{1.2.371}$$

当 $\cos\theta < -1$ 时, 式 (1.2.347) 的基态是一个有隙铁磁态. 它的元激发能隙是 $2(|\cos\theta| - 1)$. 哈密顿量 (1.2.347) 仍然可以用代数 Bethe Ansatz 求解. 相应的 Lax 算子为

$$L_{0n}(\lambda) = \frac{1}{2}\left[\frac{\sin(\lambda+\eta)}{\sin\eta}(1 + \sigma_n^z \sigma_0^z) + \frac{\sin\lambda}{\sin\eta}(1 - \sigma_n^z \sigma_0^z)\right] + \frac{1}{2}(\sigma_n^x \sigma_0^x + \sigma_n^y \sigma_0^y). \tag{1.2.372}$$

不难验证, $L_{0n}(\lambda)$ 满足 Yang-Baxter 方程. 定义单值矩阵

$$T_0(\lambda) = L_{01}(\lambda) \cdots L_{0N}(\lambda), \tag{1.2.373}$$

利用 Yang-Baxter 方程可以证明 $[\tau(\lambda), \tau(\mu)] = 0$. 哈密顿量 (1.2.347) 可以由 $\tau(\lambda) = \mathrm{tr}_0 T_0(\lambda)$ 生成

$$H = \sin\eta \frac{\mathrm{d}\ln\tau(\lambda)}{\mathrm{d}\lambda}\bigg|_{\lambda=0} - N\cos\eta. \tag{1.2.374}$$

XXZ 模型的代数 Bethe Ansatz 与各向同性情形基本相同, 这里不再重复.

1.3 嵌套的代数 Bethe Ansatz

对于高自旋系统, 当系统具有某些特殊的对称性时也是可以严格求解的. 最典型的方法是嵌套的代数 Bethe Ansatz. 下面我们介绍这种方法.

具有 $SU(N)$ 对称性的周期自旋链模型哈密顿量可以写成[28]

$$H = \sum_{j=1}^{L} P_{j,j+1}, \tag{1.3.1}$$

其中, L 是格点数目; $P_{j,j+1}$ 是具有 $SU(N)$ 对称性的置换算子. 对自旋 $1/2$ 系统, 模型具有 $SU(2)$ 对称性, 置换算子为

$$P_{j,j+1} = \frac{1}{2}(1 + \boldsymbol{\sigma}_j \cdot \boldsymbol{\sigma}_{j+1}), \tag{1.3.2}$$

对自旋 1 的系统, 模型具有 $SU(3)$ 对称性, 置换算子为

$$P_{j,j+1} = -1 + \boldsymbol{S}_j \cdot \boldsymbol{S}_{j+1} + (\boldsymbol{S}_j \cdot \boldsymbol{S}_{j+1})^2, \tag{1.3.3}$$

其中, $\boldsymbol{S}$ 是自旋 1 算子. 对于任意自旋, 置换算子可以用 Hubbard 算子表示为

$$P_{j,j+1} = \sum_{a,b=1}^{N} X_j^{a,b} X_{j+1}^{b,a}, \tag{1.3.4}$$

其中, $X_j^{a,b}$ 是作用在第 j 个空间上 $N \times N$ 的 Weyl 矩阵, 其定义为

$$X^{a,b} \equiv |a\rangle\langle b|, \tag{1.3.5}$$

它的矩阵元为 $(X^{a,b})_{c,d} = \langle c|X^{a,b}|d\rangle = \delta_{a,c}\delta_{b,d}$, a, b, c 和 d 的取值都是 $1 \sim N$.

具有 $SU(N)$ 对称性的自旋链系统的 R 矩阵是

$$\begin{aligned} R_{ij}(\lambda) &= \sum_{a,b=1}^{N} \frac{\lambda}{\lambda+1} X_i^{a,a} \otimes X_j^{b,b} + \sum_{a,b=1}^{N} \frac{1}{\lambda+1} X_i^{a,b} \otimes X_j^{b,a} \\ &= a(\lambda) + b(\lambda) P_{i,j}, \end{aligned} \tag{1.3.6}$$

其中

$$a(\lambda) = \frac{\lambda}{\lambda+1}, \quad b(\lambda) = \frac{1}{\lambda+1}. \tag{1.3.7}$$

R 矩阵满足 Yang-Baxter 方程

$$R_{12}(\lambda-\mu)R_{13}(\lambda)R_{23}(\mu) = R_{23}(\mu)R_{13}(\lambda)R_{12}(\lambda-\mu). \tag{1.3.8}$$

1.3 嵌套的代数 Bethe Ansatz

定义单值矩阵

$$T_0(\lambda) = R_{01}(\lambda)R_{02}(\lambda)\cdots R_{0L}(\lambda), \tag{1.3.9}$$

$T_0(\lambda)$ 满足 Yang-Baxter 关系：

$$R_{12}(\lambda - u)T_1(\lambda)T_2(u) = T_2(u)T_1(\lambda)R_{12}(\lambda - u). \tag{1.3.10}$$

对单值矩阵在辅助空间求迹可得转移矩阵：

$$t(\lambda) = \text{tr}_0 T_0(\lambda). \tag{1.3.11}$$

哈密顿量可以表示为

$$H = \frac{\partial \ln t(\lambda)}{\partial \lambda}|_{\lambda=0}. \tag{1.3.12}$$

同时，辫子型的 $\hat{R}$ 矩阵定义为

$$\hat{R}_{12}(\lambda) = P_{1,2}R_{12}(\lambda). \tag{1.3.13}$$

辫子型 $\hat{R}$ 矩阵满足辫子型 Yang-Baxter 方程：

$$\hat{R}_{12}(\lambda - u)\hat{R}_{23}(\lambda)\hat{R}_{12}(u) = \hat{R}_{23}(u)\hat{R}_{12}(\lambda)\hat{R}_{23}(\lambda - u). \tag{1.3.14}$$

特别指出，R 矩阵和辫子型 $\hat{R}$ 矩阵具有以下性质，R 矩阵的矩阵元只有在以下条件下才不为零：① $a_1 = a_2 = b_1 = b_2$；② $a_1 = b_1, a_2 = b_2$；③ $a_1 = b_2, a_2 = b_1$. 该性质在推导矩阵元间的对易关系时十分有用. 设单值矩阵的矩阵形式是

$$T_0(\lambda) = \begin{pmatrix} A_{11}(\lambda) & \cdots & A_{1\ N-1}(\lambda) & B_1(\lambda) \\ \vdots & & \vdots & \vdots \\ A_{N-1\ 1}(\lambda) & \cdots & A_{N-1\ N-1}(\lambda) & B_{N-1}(\lambda) \\ C_1(\lambda) & \cdots & C_{N-1}(\lambda) & D(\lambda) \end{pmatrix}, \tag{1.3.15}$$

利用方程 (1.3.14)，我们可以证明单值矩阵 (1.3.15) 满足辫子型 Yang-Baxter 关系：

$$\hat{R}_{12}(\lambda - u)[T(\lambda) \otimes T(u)] = [T(u) \otimes T(\lambda)]\hat{R}_{12}(\lambda - u), \tag{1.3.16}$$

其中，$\otimes$ 代表直乘，

$$[A \otimes B]^{bd}_{ac} = A_{ab}B_{cd}, \tag{1.3.17}$$

a 和 c 是行指标, b 和 d 是列指标. 辫子型 Yang-Baxter 关系 (1.3.16) 可以利用指标法写为

$$\hat{R}_{12}(\lambda-u)_{a_1 a_2}^{b_1 b_2} T(\lambda)_{b_1}^{c_1} T(u)_{b_2}^{c_2} = T(u)_{a_1}^{b_1} T(\lambda)_{a_2}^{b_2} \hat{R}_{12}(\lambda-u)_{b_1 b_2}^{c_1 c_2}, \quad (1.3.18)$$

这里和以后, 如果不做特殊说明, 要对重复指标进行求和. 转移矩阵 $t(\lambda)$ 是单值矩阵 (1.3.15) 在辅助空间的迹, 因此我们有

$$t(\lambda) = \mathrm{tr}_0 T_0(\lambda) = \sum_{a=1}^{N} T(\lambda)_a^a = A_{11}(\lambda) + A_{22}(\lambda) + \cdots + D(\lambda). \quad (1.3.19)$$

利用 Yang-Baxter 关系 (1.3.16), 我们可以证明具有不同谱参数的转移矩阵彼此对易,

$$[t(\lambda), t(u)] = 0. \quad (1.3.20)$$

因此系统具有无穷多守恒量, 说明系统是可积的.

选择第 j 个格点的局域真空态为 $|0\rangle_j = (0, 0, \cdots, 1)^t$, $|0\rangle_j$ 是一个 N 维列矢量, 其中前 $N-1$ 个元素为零, 最后一个为 1, t 代表转置. R 算子作用在真空态上给出

$$R_{0,j}(\lambda)|0\rangle_j = \begin{pmatrix} a(\lambda) & 0 & \cdots & 0 \\ 0 & a(\lambda) & \cdots & 0 \\ \vdots & \vdots & & 0 \\ b(\lambda)X_j^{1N} & b(\lambda)X_j^{2N} & \cdots & 1 \end{pmatrix} |0\rangle_j. \quad (1.3.21)$$

系统整体的真空态为各局域真空态的直积: $|0\rangle = \otimes_{j=1}^{L}|0\rangle_j$. 单值矩阵 (1.3.15) 作用在真空态上, 我们得到

$$T_0(\lambda)|0\rangle = \begin{pmatrix} a^L(\lambda) & 0 & \cdots & 0 \\ 0 & a^L(\lambda) & \cdots & 0 \\ \vdots & \vdots & & \vdots \\ C_1(\lambda) & C_2(\lambda) & \cdots & 1 \end{pmatrix} |0\rangle. \quad (1.3.22)$$

在方程 (1.3.22) 中, 我们看到: 单值矩阵的对角元 $A_{11}(\lambda), \cdots, D(\lambda)$ 作用在真空态上给出本征值; 非对角元 $B_a(\lambda)$ 作用在真空态上为零; 非对角元 $C_a(\lambda)$ 作用在真空态上产生一些新态, 因此矩阵元 $C_a(\lambda)$ 可以被看成是系统本征态的产生算符. 假设系统的本征态可以由产生算子 C_a 作用到真空态上生成

$$|\lambda_1^{(1)}, \cdots, \lambda_{L_1}^{(1)}|F_1\rangle = C_{a_1}(\lambda_1^{(1)}) \cdots C_{a_{L_1}}(\lambda_{L_1}^{(1)})|0\rangle F_1^{a_{L_1} \cdots a_1}, \quad (1.3.23)$$

1.3 嵌套的代数 Bethe Ansatz

其中, $F_1^{a_{L_1}\cdots a_1}$ 是关于谱参数 $\lambda_j^{(1)}$ 的函数; L_1 是产生算子的数目. 当转移矩阵作用到试探解 (1.3.23) 上时, 我们需要矩阵元 $A_{11}, \cdots, D$ 和 C_a 的对易关系. 利用 Yang-Baxter 关系 (1.3.18) 并且利用 R 矩阵的性质, 我们得到

$$D(\lambda)C_c(u) = \frac{1}{a(u-\lambda)}C_c(u)D(\lambda) - \frac{b(u-\lambda)}{a(u-\lambda)}C_c(\lambda)D(u), \quad (1.3.24)$$

$$A_{ab}(\lambda)C_c(u) = \frac{\hat{R}^{(1)}(\lambda-u)_{de}^{cb}}{a(\lambda-u)}C_e(u)A_{ad}(\lambda) - \frac{b(\lambda-u)}{a(\lambda-u)}C_b(\lambda)A_{ac}(u), \quad (1.3.25)$$

$$C_{a_1}(\lambda)C_{a_2}(u) = \hat{R}^{(1)}(\lambda-u)_{b_1 b_2}^{a_2 a_1}C_{b_2}(u)C_{b_1}(\lambda), \quad (1.3.26)$$

其中, 所有上下指标的取值都为 $1 \sim N-1$; $\hat{R}^{(1)}$ 是第一阶嵌套的 $\hat{R}$ 矩阵. 实际上, $\hat{R}^{(1)}$ 正是具有 $SU(N-1)$ 对称性的自旋链的辫子型 $\hat{R}$ 矩阵, 它的具体形式是

$$\hat{R}_{ij}^{(1)}(\lambda) = \sum_{a,b=1}^{N-1} b(\lambda) X_i^{a,a} \otimes X_j^{b,b} + \sum_{a,b=1}^{N-1} a(\lambda) X_i^{a,b} \otimes X_j^{b,a}$$

$$= b(\lambda) + a(\lambda) P_{i,j}^{(1)}, \quad (1.3.27)$$

其中, $P_{i,j}^{(1)}$ 是具有 $SU(N-1)$ 对称性的置换算子. $P_{i,j}^{(1)}{}_{ac}^{bd} = \delta_{ad}\delta_{bc}$, 指标 a, b, c 和 d 的取值为 $1 \sim N-1$. 注意在上式中对 a 和 b 的求和中, a 可以等于 b.

这里做一些说明, 在辫子型 Yang-Baxter 关系 (1.3.18) 中, 适当选取行和列指标, 就可以得到对易关系 (1.3.24)~(1.3.26). 例如, 令 $a_1 = N$, $a_2 = N$, $c_1 = 1$, $c_2 = 1$, 可得

$$C_1(\lambda)C_1(u) = C_1(u)C_1(\lambda), \quad (1.3.28)$$

令 $a_1 = N$, $a_2 = N$, $c_1 = 1$, $c_2 = l$, l 选为 $2 \sim N-1$, 我们得到

$$C_1(\lambda)C_l(u) = b(\lambda-u)C_1(u)C_l(\lambda) + a(\lambda-u)C_l(u)C_1(\lambda). \quad (1.3.29)$$

综合考虑, 可以把方程 (1.3.28) 和 (1.3.29) 写成统一的形式, 即式 (1.3.26). 如果令 $a_1 = N$, $a_2 = N$, $c_1 = l$, $c_2 = N$, l 选为 $2 \sim N-1$, 可得对易关系式 (1.3.24). 采用类似的方法可得式 (1.3.25).

转移矩阵 (1.3.19) 作用到试探解 (1.3.23) 上, 重复利用对易关系式 (1.3.24)~(1.3.26) 和结果 (1.3.22), 我们得到

$$t(\lambda)C_{a_1}(\lambda_1^{(1)})\cdots C_{a_{L_1}}(\lambda_{L_1}^{(1)})|0\rangle F_1^{a_{L_1}\cdots a_1}$$

$$= \left\{\prod_{j=1}^{L_1}\frac{a^L(\lambda)}{a(\lambda-\lambda_j^{(1)})}\Lambda^{(1)}(\lambda) + \prod_{j=1}^{L_1}\frac{1}{a(\lambda_j^{(1)}-\lambda)}\right\}|\lambda_1^{(1)},\cdots,\lambda_{L_1}^{(1)}|F_1\rangle + u.t., \quad (1.3.30)$$

其中, $u.t.$ 代表不需要项; $\Lambda^{(1)}(\lambda)$ 是第一阶嵌套即 $SU(N-1)$ 对称性的转移矩阵 $t^{(1)}(\lambda)$ 的本征值. 如果试探解 (1.3.23) 是转移矩阵的本征态, 则不需要项彼此相互抵消, 这给出第一组 Bethe Ansatz 方程:

$$\prod_{j=1,\neq\alpha}^{L_1} \frac{a(\lambda_\alpha^{(1)} - \lambda_j^{(1)})}{a(\lambda_j^{(1)} - \lambda_\alpha^{(1)})} \frac{1}{a^L(\lambda_\alpha^{(1)})} F_1^{b_{L_1}\cdots b_1} = t^{(1)}(\lambda_\alpha^{(1)})_{a_1\cdots a_{L_1}}^{b_1\cdots b_{N_1}} F_1^{a_{L_1}\cdots a_1}, \quad \alpha = 1, 2, \cdots, L_1. \tag{1.3.31}$$

相应的本征值为

$$\Lambda(\lambda) = \prod_{j=1}^{L_1} \frac{1}{a(\lambda - \lambda_j^{(1)})} a^L(\lambda) \Lambda^{(1)}(\lambda) + \prod_{j=1}^{L_1} \frac{1}{a(\lambda_j^{(1)} - \lambda)}. \tag{1.3.32}$$

现在, 寻求转移矩阵 $t(u)$ 的本征值问题转化为寻求第一阶嵌套转移矩阵 $t^{(1)}(\lambda)$ 的本征值问题. 第一阶嵌套单值矩阵可以由 Lax 算子构造为

$$T_0^{(1)}(\lambda) = R_{01}^{(1)}(\lambda - \lambda_1^{(1)}) R_{02}^{(1)}(\lambda - \lambda_2^{(1)}) \cdots R_{0L_1}^{(1)}(\lambda - \lambda_{L_1}^{(1)}), \tag{1.3.33}$$

它满足 Yang-Baxter 关系

$$\hat{R}_{12}^{(1)}(\lambda - u)[T^{(1)}(\lambda) \otimes T^{(1)}(u)] = [T^{(1)}(u) \otimes T^{(1)}(\lambda)] \hat{R}_{12}^{(1)}(\lambda - u). \tag{1.3.34}$$

相应的转移矩阵 $t^{(1)}(\lambda)$ 是第一阶嵌套单值矩阵的迹:

$$t^{(1)}(\lambda) = \text{tr}_0 T_0^{(1)}(\lambda). \tag{1.3.35}$$

注意, 这时单值矩阵中含有格点依赖于非均匀参数 $\lambda_j^{(1)}$. 依次类推, 第 r 阶嵌套转移矩阵的本征值由第 $r+1$ 阶转移矩阵的本征值问题决定. 定义单值矩阵

$$\begin{aligned} T_0^{(r)}(\lambda) &= R_{01}^{(r)}(\lambda - \lambda_1^{(r)}) R_{02}^{(r)}(\lambda - \lambda_2^{(r)}) \cdots R_{0L_r}^{(r)}(\lambda - \lambda_{L_r}^{(r)}), \\ &= \begin{pmatrix} A_{11}^{(r)}(\lambda) & \cdots & A_{1N-r-1}^{(r)}(\lambda) & B_1^{(r)}(\lambda) \\ \vdots & & \vdots & \vdots \\ A_{N-r-11}^{(r)}(\lambda) & \cdots & A_{N-r-1N-r-1}^{(1)}(\lambda) & B_{N-r-1}^{(r)}(\lambda) \\ C_1^{(r)}(\lambda) & \cdots & C_{N-r-1}^{(1)}(\lambda) & D^{(r)}(\lambda) \end{pmatrix}, \end{aligned} \tag{1.3.36}$$

选择 $|0\rangle_j^{(r)} = (0, 0, \cdots, 1)^t$ 为第 r 阶嵌套系统的局域真空态. $|0\rangle_j^{(r)}$ 是一个 $N-r$ 维列矢量, 其中前 $N-r-1$ 个元素为零, 最后一个为 1. 系统的参考态为各局域真空

1.3 嵌套的代数 Bethe Ansatz

态的直积 $|0\rangle^{(r)} = \otimes_{j=1}^{L_r}|0\rangle_j^{(r)}$. 单值矩阵 (1.3.36) 作用在真空态上给出:

$$T^{(r)}(\lambda)|0\rangle^{(r)} = \begin{pmatrix} \prod_{l=1}^{L_r}a(\lambda-\lambda_l^{(r)}) & 0 & \cdots & 0 \\ 0 & \prod_{l=1}^{L_r}a(\lambda-\lambda_l^{(r)}) & \cdots & 0 \\ \vdots & \vdots & & \vdots \\ C_1^{(r)}(\lambda) & C_2^{(r)}(\lambda) & \cdots & 1 \end{pmatrix}|0\rangle^{(r)}. \quad (1.3.37)$$

因此, 第 r 阶嵌套系统的本征态可以由产生算子 $C_l^{(r)}(l=1,2,\cdots,N-r-1)$ 作用到真空态上生成

$$|\lambda_1^{(r+1)},\cdots,\lambda_{L_r}^{(r+1)}|F_r\rangle = C_{b_1}^{(r)}(\lambda_1^{(r+1)})\cdots C_{b_{L_{r+1}}}^{(r)}(\lambda_{L_{r+1}}^{(r+1)})|0\rangle^{(r)}F_r^{b_{L_{r+1}}\cdots b_1}, \quad (1.3.38)$$

其中, L_{r+1} 是产生算子的个数; $F_r^{b_{L_{r+1}}\cdots b_1}$ 是关于辅助参数 $\lambda_l^{(r+1)}$ 的函数. 利用 Yang-Baxter 关系式 (1.3.34), 我们得到如下对易关系:

$$D^{(r)}(\lambda)C_c^{(r)}(u) = \frac{1}{a(u-\lambda)}C_c^{(r)}(u)D^{(r)}(\lambda) - \frac{b(u-\lambda)}{a(u-\lambda)}C_c^{(r)}(\lambda)D^{(r)}(u), \quad (1.3.39)$$

$$A_{ab}^{(r)}(\lambda)C_c^{(r)}(u) = \frac{\hat{R}^{(r+1)}(\lambda-u)_{de}^{cb}}{a(\lambda-u)}C_e^{(r)}(u)A_{ad}^{(r)}(\lambda) - \frac{b(\lambda-u)}{a(\lambda-u)}C_b^{(r)}(\lambda)A_{ac}^{(r)}(u), (1.3.40)$$

$$C_{b_1}^{(r)}(\lambda)C_{b_2}^{(r)}(u) = \hat{R}^{(r+1)}(\lambda-u)_{c_1c_2}^{b_2b_1}C_{c_2}^{(r)}(u)C_{c_1}^{(r)}(\lambda), \quad (1.3.41)$$

其中, 所有的行列指标的取值都是 $1\sim N-r-1$; $\hat{R}^{(r+1)}(\lambda)$ 是第 $r+1$ 阶嵌套的 R 矩阵.

第 r 次嵌套转移矩阵 (1.3.35) 作用到试探态 (1.3.38) 上, 得到

$$\begin{aligned} &t^{(r)}(\lambda)C_{b_1}^{(r)}\left(\lambda_1^{(r+1)}\right)\cdots C_{b_{L_{r+1}}}^{(r)}\left(\lambda_{L_{r+1}}^{(r+1)}\right)|0\rangle^{(r)}F_r^{b_{L_{r+1}}\cdots b_1} \\ &= \left\{\prod_{j=1}^{L_{r+1}}\frac{1}{a\left(\lambda-\lambda_j^{(r+1)}\right)}\prod_{l=1}^{L_r}a\left(\lambda-\lambda_l^{(r+1)}\right)\Lambda^{(r+1)}(\lambda)\right. \\ &\left.+\prod_{j=1}^{L_{r+1}}\frac{1}{a\left(\lambda_j^{(r+1)}-\lambda\right)}\right\}|\lambda_1^{(r+1)},\cdots,\lambda_{L_{r+1}}^{(r+1)}|F_r\rangle + u.t., \quad (1.3.42) \end{aligned}$$

由此得出如下 Bethe Ansatz 方程:

$$\prod_{j=1,\neq\alpha}^{L_{r+1}} \frac{a(\lambda_\alpha^{(r+1)} - \lambda_j^{(r+1)})}{a(\lambda_j^{(r+1)} - \lambda_\alpha^{(r+1)})} \prod_{l=1}^{L_r} \frac{1}{a(\lambda_\alpha^{(r+1)} - \lambda_l^{(r)})} F_r^{b_{L_{r+1}}\cdots b_1}$$
$$= \Lambda^{(r+1)}(\lambda_\alpha^{(r+1)})_{a_1\cdots a_{L_{r+1}}}^{b_1\cdots b_{L_{r+1}}} F_r^{a_{L_{r+1}}\cdots a_1},$$

$$\Lambda^{(r)}(\lambda)$$
$$= \prod_{j=1}^{L_{r+1}} \frac{1}{a(\lambda - \lambda_j^{(r+1)})} \prod_{l=1}^{L_r} a(\lambda - \lambda_l^{(r)}) \Lambda^{(r+1)}(\lambda) + \prod_{j=1}^{L_{r+1}} \frac{1}{a(\lambda_j^{(r+1)} - \lambda)}, \quad (1.3.43)$$

其中, $\alpha = 1, 2, \cdots, L_{r+1}$.

把函数 $a^{(r)}(u)$ 和 $b^{(r)}(u)$ 的具体表达式代入 Bethe Ansatz 方程 (1.3.31) 和 (1.3.43) 中, 得到

$$\prod_{j\neq\beta}^{L_r} \frac{\lambda_\beta^{(r)} - \lambda_j^{(r)} - 1}{\lambda_\beta^{(r)} - \lambda_j^{(r)} + 1} = \prod_{l=1}^{L_{r-1}} \frac{\lambda_\beta^{(r)} - \lambda_l^{(r-1)}}{\lambda_\beta^{(r)} - \lambda_l^{(r-1)} + 1} \prod_{k=1}^{L_{r+1}} \frac{\lambda_\beta^{(r)} - \lambda_k^{(r+1)} - 1}{\lambda_\beta^{(r)} - \lambda_k^{(r+1)}}, \quad r = 1, 2, \cdots, N-1.$$
(1.3.44)

其中, $L_0 \equiv L$, $\lambda_j^{(0)} \equiv 0$, $L_N \equiv 0$. 为对称起见, 令 $\lambda_j^{(r)} \to \lambda_j^{(r)} - r/2$, 则 Bethe Ansatz 方程化为

$$\prod_{j\neq\beta}^{L_r} \frac{\lambda_\beta^{(r)} - \lambda_j^{(r)} - 1}{\lambda_\beta^{(r)} - \lambda_j^{(r)} + 1} = \prod_{l=1}^{L_{r-1}} \frac{\lambda_\beta^{(r)} - \lambda_l^{(r-1)} - \frac{1}{2}}{\lambda_\beta^{(r)} - \lambda_l^{(r-1)} + \frac{1}{2}} \prod_{k=1}^{L_{r+1}} \frac{\lambda_\beta^{(r)} - \lambda_k^{(r+1)} - \frac{1}{2}}{\lambda_\beta^{(r)} - \lambda_k^{(r+1)} + \frac{1}{2}}, \quad r=1,2,\cdots,N-1.$$
(1.3.45)

转移矩阵的本征值为

$$\Lambda(\lambda) = a^L(\lambda) \sum_{m=1}^{N-1} \prod_{j=1}^{L_m} \frac{1}{a(\lambda - \lambda_j^{(m)})} \prod_{l=1}^{L_{m+1}} \frac{1}{a(\lambda_j^{(m+1)} - \lambda)} + \prod_{j=1}^{L} \frac{1}{a(\lambda_j^{(1)} - \lambda)}. \quad (1.3.46)$$

哈密顿量的本征值为

$$E = \frac{\partial \ln \Lambda(\lambda)}{\partial \lambda}|_{\lambda=0}. \quad (1.3.47)$$

1.4 $SU(4)$ 对称自旋梯子模型

梯子系统是从一维向二维过渡的系统. 它和很多准一维材料有着密切的联系. 在两条耦合的自旋链中, 有很多种可能性来连接两个近邻的自旋. 本节考虑一个可

1.4 $SU(4)$ 对称自旋梯子模型

积的两腿自旋梯子模型, 其哈密顿量是[29]

$$H = \frac{1}{2}\sum_{j=1}^{N}[\boldsymbol{\sigma}_j \cdot \boldsymbol{\sigma}_{j+1} + \boldsymbol{\tau}_j \cdot \boldsymbol{\tau}_{j+1}] + \frac{1}{2}J\sum_{j=1}^{N}\boldsymbol{\sigma}_j \cdot \boldsymbol{\tau}_j$$
$$+ \frac{1}{4}U\sum_{j=1}^{N}(\boldsymbol{\sigma}_j \cdot \boldsymbol{\sigma}_{j+1})(\boldsymbol{\tau}_j \cdot \boldsymbol{\tau}_{j+1}), \tag{1.4.1}$$

其中, $\boldsymbol{\sigma}_j$ 是上腿中第 j 个格点上的 Pauli 矩阵; $\boldsymbol{\tau}_j$ 是下腿中第 j 个格点上的 Pauli 矩阵; J 是横档上的自旋交换作用; U 是相互作用系数; N 是总粒子数或者系统的长度. 如果没有四自旋相互作用项的话, 模型 (1.4.1) 退化为通常的自旋梯子模型. 哈密顿量 (1.4.1) 中的 U 项可以通过自旋声子相互作用或者空穴间的库仑排斥等来调节. 对任意参数值 U, 系统 (1.4.1) 是不可积的. 但是对于 U 的特殊取值 $U=1$, 系统是可严格求解的. 此时, 哈密顿量可写为

$$H = \frac{1}{4}\sum_{j=1}^{N}(1+\boldsymbol{\sigma}_j \cdot \boldsymbol{\sigma}_{j+1})(1+\boldsymbol{\tau}_j \cdot \boldsymbol{\tau}_{j+1})$$
$$+ \frac{1}{2}J\sum_{j=1}^{N}(\boldsymbol{\sigma}_j \cdot \boldsymbol{\tau}_j - 1) + \frac{1}{2}(J-\frac{1}{2})N. \tag{1.4.2}$$

哈密顿量的可积性还是不太明显. 为此, 改写哈密顿量的第一项为 $\sum_{j=1}^{N}P_{j,j+1}$, 其中 $P_{j,j+1}$ 是两个最近邻横档间的置换算子. 因此, 哈密顿量 (1.4.1) 的第一项具有 $SU(4)$ 对称性, 就像自旋轨道耦合系统那样. 利用 Hubbard 算子, 置换算子 $P_{j,j+1}$ 可以表示为 $P_{j,j+1} = \sum_{\alpha,\beta}X_j^{\alpha,\beta}X_{j+1}^{\beta,\alpha}$, 其中 $X_j^{\alpha,\beta} \equiv |\alpha_j\rangle\langle\beta_j|$, Dirac 态 $|\alpha_j\rangle$ 张成第 j 个横档的 Hilbert 空间, 它们是正交的 ($\langle\alpha_j|\beta_j\rangle = \delta_{\alpha\beta}$). 这些量子态的一个基本表示是 $|\sigma_j^z,\tau_j^z\rangle$, 其中 $\sigma_j^z,\tau_j^z = \uparrow,\downarrow$. 然而, 这些态并不是算子 $\boldsymbol{\sigma}_j \cdot \boldsymbol{\tau}_j$ 的本征态. 选择另外一种基本表示:

$$|0\rangle = \frac{1}{\sqrt{2}}(|\uparrow,\downarrow\rangle - |\downarrow,\uparrow\rangle), \quad |1\rangle = |\uparrow,\uparrow\rangle,$$
$$|2\rangle = \frac{1}{\sqrt{2}}(|\uparrow,\downarrow\rangle + |\downarrow,\uparrow\rangle), \quad |3\rangle = |\downarrow,\downarrow\rangle. \tag{1.4.3}$$

第一个态是横档上的自旋单态, 后三个是横档上的自旋三重态. 利用这些分析, 哈密顿量 (1.4.2) 可以写为

$$H = \sum_{j=1}^{N}\sum_{\alpha,\beta=0}^{3}X_j^{\alpha\beta}X_{j+1}^{\beta\alpha} - 2J\sum_{j=1}^{N}X_j^{00}. \tag{1.4.4}$$

这里我们已经抹去了一些常数. 明显地, 粒子数算子 $N_\alpha \equiv \sum_{j=1}^{N} X_j^{\alpha,\alpha}$ 是守恒的. 方程 (1.4.4) 中最后一项的系数 $2J$ 表示以 N_0 为基的化学势, 它把系统的 $SU(4)$ 对称性约化为 $U(1) \times SU(3)$ 对称性. 因此, 我们把哈密顿量 (1.4.1) 约化为一个可积的具有 $SU(4)$ 对称性的自旋链, 或者一个 $SU(4)$ 的 $t-J$ 模型, 可以利用标准的 Bethe Ansatz 方法求解. 设系统的参考态为 $|0\rangle = |0_1\rangle \otimes |0_2\rangle \otimes \cdots \otimes |0_N\rangle$. 系统的 Bethe Ansatz 方程是

$$\left(\frac{\lambda_j - \frac{i}{2}}{\lambda_j + \frac{i}{2}}\right)^N = \prod_{l \neq j}^{M_1} \frac{\lambda_j - \lambda_l - i}{\lambda_j - \lambda_l + i} \prod_{\alpha=1}^{M_2} \frac{\lambda_j - \mu_\alpha + \frac{i}{2}}{\lambda_j - \mu_\alpha - \frac{i}{2}}, \tag{1.4.5}$$

$$\prod_{\beta \neq \alpha}^{M_2} \frac{\mu_\alpha - \mu_\beta - i}{\mu_\alpha - \mu_\beta + i} = \prod_{j=1}^{M_1} \frac{\mu_\alpha - \lambda_j - \frac{i}{2}}{\mu_\alpha - \lambda_j + \frac{i}{2}} \prod_{\delta=1}^{M_3} \frac{\mu_\alpha - \nu_\delta - \frac{i}{2}}{\mu_\alpha - \nu_\delta + \frac{i}{2}}, \tag{1.4.6}$$

$$\prod_{\gamma \neq \delta}^{M_3} \frac{\nu_\delta - \nu_\gamma - i}{\nu_\delta - \nu_\gamma + i} = \prod_{\alpha=1}^{M_2} \frac{\nu_\delta - \mu_\alpha - \frac{i}{2}}{\nu_\delta - \mu_\alpha + \frac{i}{2}}, \tag{1.4.7}$$

其中, $M_1 = N_1 + N_2 + N_3$, $M_2 = N_2 + N_3$, $M_3 = N_3$; λ_j, μ_α 和 ν_δ 是快度指标. 在我们的推导中用到了周期性边界条件 $X_{N+1}^{\alpha,\beta} \equiv X_1^{\alpha,\beta}$. 哈密顿量 (1.4.4) 的本征能量 (允许差一个常数) 是

$$E = -\sum_{j=1}^{M_1} \left(\frac{1}{\lambda_j^2 + \frac{1}{4}} - 2J\right). \tag{1.4.8}$$

很明显, 当 $0 < J < 2$ 时, 基态快度 λ_j, μ_α 和 ν_δ 的取值都为实数, 它们对称地分布在原点的两侧. 这意味着我们有三个费米海和三支无能隙激发. 当 $J > 2$ 时, 参考态就是系统的基态, 三支激发都是有能隙的. 基态是横档上的单态的直积态, 这说明在横档方向是二聚化态. 利用方程 (1.4.8) 很容易得到能隙为 $\Delta = 2(J-2)$. 因此, 系统中存在一个从二聚化态到无能隙态的量子相变, 相应的临界点为 $J_+^c = 2$. 在该临界点, 系统的低温热力学表现出非费米液体行为. 对 $J < 0$, 选择 $|1_1\rangle \otimes |1_2\rangle \otimes \cdots \otimes |1_N\rangle$ 作为参考态. Bethe Ansatz 方程的形式和方程 (1.4.5)~(1.4.7) 完全相同, 但是新的自旋反转数为 $M_1 = N_0 + N_2 + N_3$, $M_2 = N_3 + N_0$, $M_3 = N_0$. 系统的能量 (允许差一个常数) 为

$$E = -\sum_{j=1}^{M_1} \frac{1}{\lambda_j^2 + \frac{1}{4}} - 2JN_0. \tag{1.4.9}$$

1.4 $SU(4)$ 对称自旋梯子模型

因此, 系统中存在另一个量子临界点 J_-^c. 当 $J < J_-^c$ 时, 基态自旋构型中不存在横档的单重态, 横档单态元激发是有能隙的. 在这种情况下, 只有两支有能隙的激发, 系统的有效低能哈密顿量等价于具有 $SU(3)$ 对称性的自旋链. 基态由两个费米海 (λ 和 μ) 组成, 并且 $M_1 = 2N/3$, $M_2 = N/3$, $M_3 = 0$. 标记 λ 和 μ 在基态的分布为 $\rho_1(\lambda)$ 和 $\rho_2(\mu)$. 单态激发可以由一个 ν 模和一个空穴 μ_h 构造. 相应地, $\rho_1(\lambda)$ 和 $\rho_2(\mu)$ 的变化量可分别标记为 $\delta\rho_1(\lambda)$ 和 $\delta\rho_2(\mu)$. 利用 Bethe Ansatz 方程, 我们得

$$\delta\tilde{\rho}_1(\omega) = \frac{1}{4\cosh^2\frac{\omega}{2} - 1}[e^{-i\nu\omega} - e^{-\frac{1}{2}|\omega|}e^{i\mu_h\omega}], \tag{1.4.10}$$

其中, $\delta\tilde{\rho}_1(\omega)$ 是 $\delta\rho_1(\lambda)$ 的 Fourier 变换. 从方程 (1.4.9) 和 (1.4.10), 可得从基态激发一个 ν 模的最小能量 (对应于 $\nu \to 0$, $\mu_h \to \infty$)

$$\epsilon_{\min} = -\frac{1}{2}\int \delta\tilde{\rho}_1(\omega)e^{-\frac{1}{2}|\omega|}d\omega - 2J = 2|J| - \frac{\pi}{2\sqrt{3}} + \ln\sqrt{3}. \tag{1.4.11}$$

临界点的值 J_-^c 可以从条件 $\epsilon_{\min} = 0$ 得到 $J_-^c = -\pi/(4\sqrt{3}) + (\ln 3)/4$. 当 $J_-^c < J < 0$ 时, 系统的行为和 $0 < J < 2$ 时完全相同. 所以, 我们有三种量子态, 当 $J > J_+^c$ 时的横档二聚化态, 当 $J_+^c > J > J_-^c$ 时的无能隙态 (三支元激发都是无能隙的), 当 $J < J_-^c$ 时的无能隙态 (只有两支无能隙的元激发). 二聚化相具有长程序

$$\langle 0|X_i^{00}X_j^{00}|0\rangle = 1, \tag{1.4.12}$$

这诱导了横档单态的凝聚. 当掺入空穴时, 系统的行为和 $t-J$ 梯子模型一样, 此时, 横档单态可以被看成 Cooper 对. 空穴掺杂时的单态移动产生超导电性. 基于以上观察, 我们得到 $J_\pm^c$ 表示系统的两个不稳定固定点. 此外, 稳定的固定点也可以得到. 当 $J > J_+^c$ 时, 横向交换作用是主导项, 系统应该流向固定点 $J^* = +\infty$; 当 $J_-^c < J < J_+^c$ 时, 两个不稳定固定点 $J_\pm^c$ 诱导出一个中间稳定固定点 $J_-^c < J^* < J_+^c$; 当 $J < J_-^c$ 时, 在低能区域内单态激发被剔除, 系统应该流向稳定的固定点 $J^* = -\infty$, 这等价于一个 $SU(3)$ 不变的自旋链.

采用标准的热力学 Bethe Ansatz 方法可以研究系统的热力学行为. 在无能隙相区域, 系统可以用 Luttinger 液体来描述. 然而, 在量子临界点, 系统会呈现出非费米液体行为. 首先考虑当 $J = J_+^c$ 时的零温磁化率. 在没有外加磁场时, 基态是横档单态的凝聚体. 如果对系统施加一个非常弱的外磁场, 基态构型中会出现一些 $S_z = 1$ 的横档三态, 此时, N_2 和 N_3 仍然是零, 这是因为这两类横档的能级要么抬升 ($|3\rangle$) 要么保持不动 ($|2\rangle$). 在外加磁场 ($H > 0$) 下基态能量密度是

$$E/N = \int_{-\Lambda}^{\Lambda}\left(4 - \frac{1}{\lambda^2 + \frac{1}{4}} - H\right)\rho_1(\lambda)d\lambda, \tag{1.4.13}$$

其中 $\rho_1(\lambda)$ 满足

$$\rho_1(\lambda) + \int_{-\Lambda}^{\Lambda} a_2(\lambda - \lambda')\rho_1(\lambda')\mathrm{d}\lambda' = a_1(\lambda), \tag{1.4.14}$$

并且 $a_n(\lambda) = n/2\pi[\lambda^2 + (n/2)^2]$, $\Lambda^2 = 1/(4-H) - 1/4$. 当外磁场很小时 $(H << 1)$, 可得 $\Lambda \approx \sqrt{H}/4$. 方程 (1.4.14) 的解为

$$\rho_1(\lambda) = \frac{2}{\pi} - \frac{1}{\pi^2}\sqrt{H} + o(H^{3/2}, \lambda^2). \tag{1.4.15}$$

联立方程 (1.4.13) 和 (1.4.15) 可以得到系统的磁化率为

$$\chi = -\frac{\partial^2(E/N)}{\partial H^2} = \frac{1}{2\pi}H^{-\frac{1}{2}} + o(H^{\frac{1}{2}}). \tag{1.4.16}$$

通过热力学 Bethe Ansatz 方程的低温展开, 可得系统的低温比热和低温磁化率为

$$C \sim T^{\frac{1}{2}}, \quad \chi \sim T^{-\frac{1}{2}}, \tag{1.4.17}$$

这说明了一种典型的量子临界行为. 需要说明的是, 在有能隙区域, 磁场也可以诱导出量子相变. 在临界点 $H_c = 2(J-2)$, 可以得到相似的量子临界行为.

1.5 自旋为 1 的玻色气体

一维 δ 势相互作用玻色气体最早是由 Lieb 和 Liniger 求解的[30]. δ 势是描述量子气体非常好的模型, 因为在稀薄气体中, 粒子间只有在非常靠近的时候才有相互作用. 体系的哈密顿量是

$$H = \int_0^L \partial_x\phi^\dagger(x)\partial_x\phi(x)\mathrm{d}x + \int_0^L c\phi^\dagger(x)\phi^\dagger(x)\phi(x)\phi(x)\mathrm{d}x, \tag{1.5.1}$$

其中, $\phi(x)$ 是玻色型场算符, 它满足对易关系 $[\phi^\dagger(x), \phi(y)] = \delta_{xy}$; c 表征相互作用强度. 哈密顿量 (1.5.1) 又可以写成

$$H = -\sum_{i=1}^{N}\frac{\partial^2}{\partial x_i^2} + 2c\sum_{i<j}\delta(x_i - x_j). \tag{1.5.2}$$

本节我们考虑自旋为 1 的玻色子, 假设除了 δ 势之外, 粒子间还有自旋的交换作用[31]:

$$H = -\sum_{i=1}^{N}\frac{\partial^2}{\partial x_i^2} + \sum_{i<j}[c_0 + c_2\boldsymbol{S}_i \cdot \boldsymbol{S}_j]\delta(x_i - x_j), \tag{1.5.3}$$

1.5 自旋为 1 的玻色气体

其中, c 为相互作用系数; $\boldsymbol{S}_i$ 为第 i 个格点上的自旋算子, 它们的矩阵形式是

$$S_i^x = \frac{1}{\sqrt{2}} \begin{pmatrix} 0 & 1 & 0 \\ 1 & 0 & 1 \\ 0 & 1 & 0 \end{pmatrix}, \quad S_i^y = \frac{1}{\sqrt{2}} \begin{pmatrix} 0 & -\mathrm{i} & 0 \\ \mathrm{i} & 0 & -\mathrm{i} \\ 0 & \mathrm{i} & 0 \end{pmatrix}, \quad S_i^z = \begin{pmatrix} 1 & 0 & 0 \\ 0 & 0 & 0 \\ 0 & 0 & -1 \end{pmatrix}, \tag{1.5.4}$$

在二次量子化表象下, 我们定义粒子的产生和湮没算符分别为 $a_s^\dagger(x)$ 和 $a_s(x)$. 很明显, 模型 (1.5.3) 是玻色-Hubbard 模型的低密度极限, 注意要选择适当的耦合参数 c_0 和 c_2. 在两体散射过程中, 系统的总自旋是守恒的. 系统具有 $SU(2)$ 对称性. 非平庸的散射发生在总自旋为 0 和 2 的通道. 在自旋为 1 的通道, 当交换两个粒子时, 波函数的自旋部分是反对称的, 而空间部分由于 δ 相互作用是对称的, 所以总的波函数是反对称的, 这和系统是玻色型系统波函数要是对称的相矛盾, 因此自旋为 1 的通道是禁止的. 需要特别强调的是, 在自旋为 0 的通道, 两粒子发生散射时存在以下关系:

$$a_\uparrow^\dagger(x) + a_\downarrow^\dagger(x) \longrightarrow 2a_0^\dagger(x) \tag{1.5.5}$$

这使得总自旋的各个分量或者说携带特定自旋分量的粒子数不守恒, 破坏了系统的对称性.

很容易验证该模型的守恒量是

$$\begin{cases} N = \sum\limits_s \int a_s^\dagger(x) a_s(x) \mathrm{d}x, \\ S^z = \int [a_\uparrow^\dagger(x) a_\uparrow(x) - a_\downarrow^\dagger(x) a_\downarrow(x)] \mathrm{d}x, \end{cases} \tag{1.5.6}$$

其中, N 和 S^z 分别为总粒子数和总自旋的 z 分量. 因为系统还具有 $SU(2)$ 不变性, 下面两个量也是守恒的:

$$\begin{aligned} S^+ &= \sqrt{2} \int [a_\uparrow^\dagger(x) a_0(x) + a_0^\dagger(x) a_\downarrow(x)] \mathrm{d}x, \\ S^- &= \sqrt{2} \int [a_0^\dagger(x) a_\uparrow(x) + a_\downarrow^\dagger(x) a_0(x)] \mathrm{d}x. \end{aligned} \tag{1.5.7}$$

S^z 和 $S^\pm$ 构成 $SU(2)$ 李代数的生成元. 三个自旋算子和以下五个自旋四极矩算子:

$$\begin{cases} Q_0 = \int [a_\uparrow^\dagger(x) a_\uparrow(x) + a_\downarrow^\dagger(x) a_\downarrow(x) - 2a_0^\dagger(x) a_0(x)] \mathrm{d}x, \\ Q_2 = \int [a_\uparrow^\dagger(x) a_\downarrow(x) + a_\downarrow^\dagger(x) a_\uparrow(x)] \mathrm{d}x, \\ Q_{xy} = -\mathrm{i} \int [a_\uparrow^\dagger(x) a_\downarrow(x) - a_\downarrow^\dagger(x) a_\uparrow(x)] \mathrm{d}x, \\ Q_{xz} = \dfrac{1}{\sqrt{2}} \int [a_\uparrow^\dagger(x) a_0(x) - a_0^\dagger(x) a_\downarrow(x) + h.c.] \mathrm{d}x, \\ Q_{yz} = -\dfrac{\mathrm{i}}{\sqrt{2}} \int [a_\uparrow^\dagger(x) a_0(x) - a_0^\dagger(x) a_\downarrow(x) - h.c.] \mathrm{d}x, \end{cases} \tag{1.5.8}$$

构成了 $SU(3)$ 李代数的基础表示.

下面我们用 Bethe Ansatz 方法来求解系统 (1.5.3). 在模型 (1.5.3) 中, 在两种情况下是可积的, 一种是 $c_2 = 0$, 此时系统具有 $SU(3)$ 不变性; 另一种情况是 $c_0 = c_2 = c$, 此时系统具有 $SU(2)$ 不变性并且也是可积的. 以下我们讨论这种情形.

在坐标 Bethe Ansatz 的框架下, 系统的波函数可以用一套准动量 $\{k_j\}$ 来刻画. 假设系统的波函数可以表达成以下形式:

$$\Psi(x_1 s_1, \cdots, x_N s_N) = \sum_{Q,P} \theta(x_{Q_1} < \cdots < x_{Q_N}) A_{s_1 \cdots s_N}(Q, P) e^{i \sum_{l=1}^N k_{P_l} x_{Q_l}}, \quad (1.5.9)$$

其中, $Q = (Q_1, \cdots, Q_N)$ 和 $P = (P_1, \cdots, P_N)$ 是整数 $1 \sim N$ 的置换; k 是准动量, $\theta(x_{Q_1} < \cdots < x_{Q_N}) = \theta(x_{Q_N} - x_{Q_{N-1}}) \cdots \theta(x_{Q_2} - x_{Q_1})$, $\theta(x)$ 是阶梯函数. 当置换两个粒子的坐标和自旋后, 系统的波函数必须是对称的. 当两个粒子处在同一点时, 波函数要连续, 但是波函数的一阶导数不连续. 利用标准的坐标 Bethe Ansatz 方法, 可以得到两体散射矩阵:

$$S_{ij} = \frac{k_i - k_j - ic}{k_i - k_j + ic} P_{ij}^0 + P_{ij}^1 + \frac{k_i - k_j + 2ic}{k_i - k_j - 2ic} P_{ij}^2, \quad (1.5.10)$$

其中, $P_{ij}^S (S = 0, 1, 2)$ 是作用在总自旋 S 上的自旋投影算子; k_i 和 k_j 是准粒子携带的准动量. 两体散射矩阵满足 Yang-Baxter 方程, 这保证了系统的可积性.

利用波函数的周期性边界条件, 我们得到如下本征值方程:

$$S_{jN} S_{jN-1} \cdots S_{jj+1} S_{jj-1} \cdots S_{j1} e^{i k_j L} \xi_0 = \xi_0, \quad (1.5.11)$$

其中, ξ_0 是初始态波函数的振幅. $c_0 = c_2$ 的情形也可以利用标准的代数 Bethe Ansatz 方法求解. 实际上, 该模型的两体散射矩阵和 Takhtajan-Babujian 模型[32]的 R 矩阵具有相同的形式. 因此, 该模型的自旋动力学行为和 Takhtajan-Babujian 自旋链具有一定的相似性. 首先, 我们定义单值矩阵为

$$\begin{aligned} \mathcal{T}(\lambda) &= S_{0j} S_{0N} S_{0N-1} \cdots S_{0j+1} S_{0j-1} \cdots S_{01} \\ &= \begin{pmatrix} A_1(\lambda) & B_1(\lambda) & B_2(\lambda) \\ C_1(\lambda) & A_2(\lambda) & B_3(\lambda) \\ C_2(\lambda) & C_3(\lambda) & A_3(\lambda) \end{pmatrix}, \end{aligned} \quad (1.5.12)$$

其中, $S_{0l} \equiv S_{0l}(\lambda - k_l)$, $\lambda = -ik/c$ 是谱参数, 0 表征辅助空间而 n 表征量子空间. 单值矩阵满足 Yang-Baxter 关系:

$$S_{12}(\lambda - u) \mathcal{T}_1(\lambda) \mathcal{T}_2(u) = \mathcal{T}_2(u) \mathcal{T}_1(\lambda) S_{12}(\lambda - u). \quad (1.5.13)$$

1.5 自旋为 1 的玻色气体

转移矩阵定义为

$$t(\lambda) = \mathrm{tr}\mathcal{T}(\lambda), \tag{1.5.14}$$

含有不同参数的转移矩阵彼此之间对易, $[t(\lambda), t(u)] = 0$, 因此系统是可积的且有无穷多个守恒量. 如此一来, 本征值问题(1.5.11) 约化为

$$t(k_j)\mathrm{e}^{\mathrm{i}k_j L}\xi_0 = \xi_0. \tag{1.5.15}$$

进一步, 我们定义辅助的单值矩阵为

$$\mathcal{T}(\lambda) = S^{\sigma s}_{0j} S^{\sigma s}_{0N} \cdots S^{\sigma s}_{0j+1} S^{\sigma s}_{0j-1} S^{\sigma s}_{01} = \begin{pmatrix} A(\lambda) & B(\lambda) \\ C(\lambda) & D(\lambda) \end{pmatrix}, \tag{1.5.16}$$

其中

$$S^{\sigma s}(\lambda) = \frac{\lambda - k_l - \frac{1}{2}\mathrm{i}c - \mathrm{i}c\boldsymbol{\sigma}_0\cdot\boldsymbol{S}_l}{\lambda - k_l + \frac{3}{2}\mathrm{i}c}. \tag{1.5.17}$$

单值矩阵 (1.5.12) 和 (1.5.16) 满足 Yang-Baxter 关系:

$$\begin{cases} S^{\sigma s}_{12}(\lambda - u)\mathcal{T}_1(\lambda)\mathcal{T}_2(u) = \mathcal{T}_2(u)\mathcal{T}_1(\lambda)S^{\sigma s}_{12}(\lambda - u), \\ S^{\sigma\sigma}_{12}(\lambda - \mu)\mathcal{T}_1(\lambda)\mathcal{T}_2(\mu) = \mathcal{T}_2(\mu)\mathcal{T}_1(\lambda)S^{\sigma\sigma}_{12}(\lambda - \mu), \end{cases} \tag{1.5.18}$$

其中, $S^{\sigma\sigma}_{12}(\lambda) = (\lambda - \mathrm{i}c)^{-1}(\lambda - \mathrm{i}c/2 - \mathrm{i}c\boldsymbol{\sigma}_1\cdot\boldsymbol{\sigma}_2/2)$. 利用方程 (1.5.18), 我们得到如下的对易关系

$$A_1(\lambda)B(u) = \frac{\lambda - u + \mathrm{i}\frac{3}{2}c}{\lambda - u - \mathrm{i}\frac{1}{2}c}B(u)A_1(\lambda) - \frac{\mathrm{i}\sqrt{2}c}{\lambda - u - \mathrm{i}\frac{1}{2}c}B_1(\lambda)A(u), \tag{1.5.19}$$

$$\begin{aligned}A_2(\lambda)B(u) &= \frac{\left(\lambda - u + \mathrm{i}\frac{3}{2}c\right)\left(\lambda - u - \mathrm{i}\frac{3}{2}c\right)}{\left(\lambda - u + \mathrm{i}\frac{1}{2}c\right)\left(\lambda - u - \mathrm{i}\frac{1}{2}c\right)}B(u)A_2(\lambda) \\ &\quad + \frac{\mathrm{i}\sqrt{2}c}{\lambda - u - \mathrm{i}\frac{1}{2}c}B_1(\lambda)D(u) - \frac{\mathrm{i}\sqrt{2}c}{\lambda - u + \mathrm{i}\frac{1}{2}c}B_3(\lambda)A(u) \\ &\quad + \frac{2\mathrm{i}c}{\left(\lambda - u + \mathrm{i}\frac{1}{2}c\right)\left(\lambda - u - \mathrm{i}\frac{1}{2}c\right)}B_2(\lambda)C(u), \end{aligned} \tag{1.5.20}$$

$$A_3(\lambda)B(u) = \frac{\lambda - u - \mathrm{i}\frac{3}{2}c}{\lambda - u + \mathrm{i}\frac{1}{2}c}B(u)A_3(\lambda) + \frac{\mathrm{i}\sqrt{2}c}{\lambda - u + \mathrm{i}\frac{1}{2}c}B_3(\lambda)D(u). \tag{1.5.21}$$

同时，矩阵元 $A(\lambda)$, $D(\lambda)$ 和 $B(\lambda)$ 满足

$$A(\lambda)B(u) = \frac{\lambda - u + \mathrm{i}c}{\lambda - u}B(u)A(\lambda) - \frac{\mathrm{i}c}{\lambda - u}B(\lambda)A(u), \tag{1.5.22}$$

$$D(\lambda)B(u) = \frac{\lambda - u - \mathrm{i}c}{\lambda - u}B(u)D(\lambda) + \frac{\mathrm{i}c}{\lambda - u}B(\lambda)D(u). \tag{1.5.23}$$

系统的参考态或真空态定义为 $|\Omega\rangle = |\uparrow\rangle_1 \otimes \cdots \otimes |\uparrow\rangle_N$，它是 $A_1(\lambda)$, $A_2(\lambda)$, $A_3(\lambda)$, $A(\lambda)$ 和 $D(\lambda)$ 的共同本征态. 矩阵元 $C(\lambda)$ 作用到真空态上为零. 矩阵元 $B(\lambda)$ 作用到真空态上给出一个有限值, 它可以被认为是系统本征态的产生算子:

$$|\Psi\rangle = B(u_1)\cdots B(u_M)|\Omega\rangle. \tag{1.5.24}$$

转移矩阵 $t(k_j) \equiv \sum_{n=1}^{3} A_n(k_j)$ 作用到 Bethe 态 (1.5.24) 上给出两项, 一项是本征值, 另一项是所谓的不需要项. 如果 Bethe 态 (1.5.24) 是系统的本征态, 则不需要项必须相消. 利用不需要项为零的条件, 我们可以得到如下关于准动量 $\{k_j\}$ 的 Bethe Ansatz 方程:

$$\mathrm{e}^{\mathrm{i}k_j L} = \prod_{l=1, l\neq j}^{N} \frac{k_j - k_l + 2\mathrm{i}c}{k_j - k_l - 2\mathrm{i}c} \prod_{\alpha=1}^{M} \frac{k_j - \lambda_\alpha - \mathrm{i}c}{k_j - \lambda_\alpha + \mathrm{i}c}, \tag{1.5.25}$$

$$\prod_{l=1}^{N} \frac{\lambda_\alpha - k_l - \mathrm{i}c}{\lambda_\alpha - k_l + \mathrm{i}c} = -\prod_{\beta=1}^{M} \frac{\lambda_\alpha - \lambda_\beta - \mathrm{i}c}{\lambda_\alpha - \lambda_\beta + \mathrm{i}c}, \tag{1.5.26}$$

其中, $j, l = 1, \cdots, N$; $\alpha, \beta = 1, \cdots, M$, M 是自旋反转数目; λ_α 是自旋快度, 通过对 Bethe 根 u_α 做适当移动得到. 系统 (1.5.3) 的能谱是

$$E = \sum_{j=1}^{N} k_j^2. \tag{1.5.27}$$

原则上, 利用 Bethe Ansatz 方程可以解出准动量的值, 然后代入式 (1.5.27), 就可以求得系统的能量.

在以上的讨论中, 我们把粒子限制在长度为 L 的一维盒子中. 在 Bethe Ansatz 方程解的基础上, 我们可以研究系统在热力学极限下的基态和低温性质. 热力学极限要求体系的长度 L 和总粒子数 N 都趋于无穷大, 但是它们的比值有限 $N/L \to n$. Bethe Ansatz 方程的解有一点复杂. 除了实数外, 快度 $\{k_j\}$, $\{\lambda_\alpha\}$ 还可以取复数, 无论 $c > 0$ 还是 $c < 0$, 它们通常称为弦解. 当 $c > 0$ 时, 吸引相互作用只能在总自旋 $S = 0$ 通道发生. 这意味着粒子形成自旋单重态. 一般来讲, Bethe Ansatz 方程的复数解由 Bethe Ansatz 方程在热力学极限下的奇点或者零点来确定. 例如, 如

1.5 自旋为 1 的玻色气体

果一些 k_j 的取值在上半复平面上, 则方程 (1.5.25) 的左手端趋于零, 当系统尺寸 L 趋于无穷大时, 对应地, 则存在一个 λ_α 满足 $k_j - \lambda_\alpha - \mathrm{i}c \to 0$. 进一步, 从方程 (1.5.26), 我们知道存在另一个 λ_β, 它满足 $\lambda_\alpha - \lambda_\beta + \mathrm{i}c \to 0$. 利用 Bethe Ansatz 方程的共轭不变性, 我们得到配对的互为复共轭的 k_j 解:

$$\begin{aligned} k_j &= K_j + \mathrm{i}c/2 + o(\mathrm{e}^{-\delta L}), \\ k_j^* &= K_j - \mathrm{i}c/2 + o(\mathrm{e}^{-\delta L}), \end{aligned} \quad (1.5.28)$$

同时, 快度 Λ 也形成 2 弦, 它们的实部和 k_j 的 2 弦解的实部相同,

$$\begin{aligned} \lambda_j &= K_j + \mathrm{i}c/2 + o(\mathrm{e}^{-\delta' L}), \\ \lambda_j^* &= K_j - \mathrm{i}c/2 + o(\mathrm{e}^{-\delta' L}), \end{aligned} \quad (1.5.29)$$

其中, K_j 是一个实数; δ 和 δ' 是正的常数. 解 (1.5.28) 描述了在电荷扇区唯一可能的束缚态. 事实上, 由于波函数对称性的限制, 系统中不会出现两体以上的束缚态. 在热力学极限下, 每一个束缚的粒子对贡献的束缚能为 $\Delta = c^2/2$. 因此, 该体系的低能物理是由这些束缚的粒子对形成的量子液体来描述的. 在基态, 假设总粒子数为偶数, 则所有的粒子都形成这样的对. 当总粒子数为奇数时, 系统中会存在一个多余的未配对的粒子, 并且基态是三重简并的. 把这些 2 弦解代入方程 (1.5.25) 和 (1.5.26) 中, 并对方程 (1.5.25) 和 (1.5.26) 求导, 可得如下约化的 Bethe Ansatz 方程:

$$K_j L = \pi I_j - \sum_{l=1, l \neq j}^{N} \left[\arctan\left(\frac{2(K_j - K_l)}{3c}\right) \right. \\ \left. + \arctan\left(\frac{K_j - K_l}{c}\right) - \arctan\left(\frac{2(K_j - K_l)}{c}\right) \right], \quad (1.5.30)$$

其中, 当 $N/2$ 为奇数时, 量子数 I_j 取整数值, 而当 $N/2$ 为偶数时, 量子数 I_j 取半奇数. 在基态, 量子数 $\{I_j\}$ 关于原点对称排列. 在热力学极限下, 定义反转自旋的粒子数密度为 $\rho_0(K_j) = L^{-1}(I_{j+1} - I_j)/(K_{j+1} - K_j)$. 对方程 (1.5.30) 取导, 可得基态的 $\rho_0(K)$ 分布满足以下积分方程:

$$\rho_0(K) = \frac{1}{\pi} + \frac{1}{\pi} \int_{-Q}^{Q} \left[\frac{6|c|}{9c^2 + 4(K-K')^2} \right. \\ \left. + \frac{|c|}{c^2 + (K-K')^2} - \frac{2|c|}{4c^2 + (K-K')^2} \right] \rho_0(K') \mathrm{d}K', \quad (1.5.31)$$

其中, 赝费米点 Q 由以下式子确定:

$$n = 2 \int_{-Q}^{Q} \rho_0(K) \mathrm{d}K. \quad (1.5.32)$$

基态的能量密度为

$$\frac{E_0}{L} = \int_{-Q}^{Q} \left(2K^2 - \frac{c^2}{2}\right) \rho_0(K) \mathrm{d}K. \tag{1.5.33}$$

很明显, 基态总自旋的 z 分量为零, 基态是一个自旋单重态. 需要注意的是, 该基态不是 $SU(3)$ 单重态. 在该模型中, 每一个束缚对的波函数的自旋部分具有如下形式: $(|\uparrow,\downarrow\rangle + |\downarrow,\uparrow\rangle - |0,0\rangle)/\sqrt{3}$. 很容易推导单位长度上的四极矩算符的期望值为 $\langle Q_0 \rangle = -n/2$, $\langle Q_\alpha \rangle = 0$, 其中 $\alpha \neq 0$. 自旋激发有一个能隙 $\Delta = c^2/2$. 电荷激发是无能隙的. 这类激发可以在赝费米海中挖一个洞或是在赝费米点上放一个粒子. 一个空穴的激发能量 $\epsilon(K)$ 满足方程:

$$\begin{aligned}\epsilon(K) = {}& 2(K^2 - Q^2) + \frac{1}{\pi} \int_{-Q}^{Q} \left[\frac{6|c|}{9c^2 + 4(K-K')^2}\right. \\ & \left. + \frac{|c|}{c^2 + (K-K')^2} - \frac{2|c|}{4c^2 + (K-K')^2}\right] \epsilon(K') \mathrm{d}K'.\end{aligned} \tag{1.5.34}$$

其余那些无能隙激发诸如粒子-空穴激发和流激发等都可以用单粒子激发和单空穴激发的叠加来刻画. 费米速度为

$$v_{\mathrm{F}} = \frac{\epsilon'(Q)}{\pi \rho_0(Q)}. \tag{1.5.35}$$

如果系统的温度很低并且 $T \ll \Delta$, 则自旋自由度完全被冻结. 因此, 该系统的低温物理和 Lieb-Liniger 模型有一定的相似之处. 此时, 系统可以用 Luttinger 液体来描述, 其低温比热和磁化率满足

$$C(T) = \frac{\pi^2}{3v_{\mathrm{F}}} T + o(T^2), \quad \chi(T) \sim \mathrm{e}^{-\frac{c^2}{2T}}, \tag{1.5.36}$$

这里我们已经把 Boltzmann 常数 k_{B} 设为单位.

在有限温度时, 体系处在激发态. 该激发态由一组实数 $\{k_l\}$, 一组由方程 (1.5.28) 和 (1.5.29) 确定的 $\{k_j, \lambda_j\}$ 对和一组由快度 $\{\lambda_\alpha\}$ 形成的 n 弦来定量刻画, 其中 n 弦具有如下形式: $\lambda_{\alpha,m}^{(n)} = \lambda_\alpha^{(n)} + \mathrm{i}(n+1-2m)c/2 + o(\mathrm{e}^{-\delta L})$, $m = 1, \cdots, n$, $\alpha = 1, \cdots, M_n$, $n = 1, 2, \cdots$, $M = \sum_n n M_n$. 标记在热力学极限下的形成束缚态的粒子对密度为 σ', 未形成束缚对的粒子数密度为 ρ, n 弦密度为 σ_n. 相应地, σ'^h, ρ^h 和 σ_n^h 分别表示各自对应的空穴密度. 当系统达到热平衡时, 系统 Gibbs 自由能应该取最小值.

对自由能求变分, 可得如下耦合的积分方程:

$$\begin{cases} \ln\eta' = 2T^{-1}(k^2 - \frac{c^2}{4} - \mu) - (a_5 - a_1) * \ln(1+\xi^{-1}) \\ \qquad - (a_6 + a_4 - a_2) * \ln(1+\eta'^{-1}), \\ \ln\xi = T^{-1}(k^2 - \mu - 2h) - (a_5 - a_1) * \ln(1+\eta'^{-1}) \\ \qquad + \ln\eta_1 - (a_4 + a_2 + a_0) * \ln(1+\xi^{-1}), \\ \ln\eta_1 = G * [\ln(1+\eta_2) + \ln(1+\xi^{-1})], \\ \ln\eta_n = G * [\ln(1+\eta_{n+1}) + \ln(1+\eta_{n-1})], \quad n = 2, 3, \cdots \\ \lim_{n\to\infty} \frac{\ln\eta_n}{n} = \frac{h}{T}, \end{cases} \quad (1.5.37)$$

其中, $a_n(x) = 4n|c|/\{\pi[(nc)^2 + (4x)^2]\}$, $\eta' = \sigma'^h/\sigma'$, $\xi = \rho^h/\rho$, $\eta_n = \sigma_n^h/\sigma_n$, $G(x) = c^{-1}\text{sech}(2\pi x/c)$, $f*g = \int f(x-y)g(y)\mathrm{d}y$, h 是外加的磁场强度. 当 $T = 0$ 并且 $h = 0$ 时, 很容易得到 $\rho = \sigma_n = 0$, $\sigma' \equiv \rho_0$. 这和以前的结论自洽.

如果相互作用常数 c 小于零, 则在总自旋 $S = 2$ 通道是吸引相互作用, 而在总自旋 $S = 0$ 通道是排斥相互作用. 通过仔细分析 Bethe Ansatz 方程的解, 可知系统的基态是不可压缩的铁磁态, 所有的粒子形成一个大 N 弦

$$k_j = \mathrm{i}c(N+1-2j), \quad j = 1, 2\cdots, N. \quad (1.5.38)$$

以上结果也可以推广到具有任意自旋的相互作用量子气体[33].

参 考 文 献

[1] Yang C N. Phys. Rev. Lett., 1967, 19: 1312; Phys. Rev., 1968, 168: 1920.
[2] Baxter R J. Exactly Solved Models in Statistics Mechanics. Academic Press, 1982.
[3] Cherednik I. Funct. Anal. Appl., 1985, 19: 77.
[4] Sklyanin E K. J. Phys. A, 1988, 21: 2375.
[5] Bethe H A. Z. Physik, 1931, 71: 205.
[6] Sklyanin E K, Faddeev L D. Sov. Phys. Dokl., 1978, 23: 902.
[7] Takhtadzhan L A, Faddeev L D. Rush. Math. Surveys, 1979, 34: 11.
[8] Gardner C S, Greene J M, Kruskal M D, et al. Phys. Rev. Lett., 1967, 19: 1095.
[9] Korepin V E, Bogoliubov N M, Izergin A G. Quantum Inverse Scattering Method and Correlation Function. Cambridge University Press, 1993.
[10] Sutherland B. Beautiful Models: 70 Years of Exactly Solved Quantum Many-Body Problems. World Scientify Publishing, 2004.
[11] Chowdhury A R, Choudhury A G. Quantum Integrable Systems. Chapman and Hall/CRC. 2004.

[12] Samaj L, Bajnok Z. Introduction to the Statistical Physics of Integrable Many-body Systems. Cambridge University Press, 2013.

[13] Franchini F. An Introduction to Integrable Techniques for One-Dimensional Quantum Systems. Springer Press, 2017.

[14] Yang C N, Yang C P. Phys. Rev., 1966, 150: 321; Phys. Rev., 1966, 150: 327; Phys. Rev., 1966, 151: 258.

[15] Faddeev L D, Takhtajan L A. Phys. Lett. A, 1981, 85: 375.

[16] Takahashi M. Thermodynamics of One-Dimensional Solvable Models. Cambridge University Press, 1999.

[17] Yang C N, Yang C P. J Math. Phys., 1969, 10: 1115.

[18] Lax P D. Commun. Pure. Appl. Math., 1968, 21: 467.

[19] Baxter R J. Phys. Rev. Lett., 1971, 26: 832; Phys. Rev. Lett., 1971, 26: 834; Ann. Phys., 1972, 70: 193; Ann. Phys., 1972, 70: 323; Ann. Phys., 1973, 76: 1; Ann. Phys., 1973, 76: 25; Ann. Phys., 1973, 76: 48.

[20] Gaudin M. Phys. Rev. A, 1971, 4: 386.

[21] Gaudin M. The Bethe Wavefunction. Cambridge University Press, 2014.

[22] Wang Y, Yang W L, Cao J, et al. Off-Diagonal Bethe Ansatz for Exactly Solvable Models. Springer Press, 2015.

[23] Andrei N. Phys. Rev. Lett., 1980, 45: 379.

[24] Wiegmann P B. Sov. Phys. JETP Lett., 1980, 31: 392.

[25] Andrei N, Johannesson H. Phys. Lett. A, 1984, 100: 108.

[26] Lee K J B, Schlottmann P. Phys. Rev. B, 1987, 37: 379.

[27] Wang Y. Phys. Rev. B, 1997, 56: 14045.

[28] De Vega H J, Lopes E. Phys. Rev. Lett., 1991, 67: 489; Lopes E. Nucl. Phys. B, 1992, 370: 636.

[29] Wang Y. Phys. Rev. B, 1999, 60: 9236.

[30] Lieb E H, Liniger W. Phys. Rev., 1963, 130: 1605; Lieb E H. Phys. Rev., 1963, 130: 1616.

[31] Cao J, Jiang Y, Wang Y. EPL, 2007, 79: 30005.

[32] Takhtajan L A. Phys. Lett. A, 1982, 87: 479; Babujian H M. Phys. Lett. A, 1982, 90: 479; Babujian H M. Nucl. Phys. B, 1983, 215: 337.

[33] Jiang Y, Cao J, Wang Y. EPL, 2009, 87: 10006; Phys J A, 2011, 44: 345001.

第 2 章 Lieb-Liniger 模型: 多体物理之美

彭 黎 管习文

2.1 引 言

Bethe 在 1931 年为了精确求解一维 Heisenberg 自旋链模型, 提出了一种构造一维量子多体系统波函数的方法 ——Bethe Ansatz, 即将波函数假设成所有可能的平面波的叠加 [1]. Bethe 的方法提出来以后, 并未得到广泛的关注. 直到 1963 年, Lieb 和 Liniger 首次利用 Bethe 的假设精确求解了一维 δ 相互作用的均匀无自旋玻色气体 ——Lieb-Liniger 模型 [2]. 在此基础上, 杨振宁和杨振平于 1969 年成功解决了该模型的有限温热力学问题 [3], 由此建立了 Yang-Yang 热力学. Lieb-Liniger 模型在数学的可积性和量子多体物理方面具有丰富的内涵, 展现了多体物理之美. 尤其是, 该模型的 Yang-Yang 热力学描述能够使人们精确地理解多体物理中的量子统计、热力学、量子流体、量子关联及量子临界现象的本质. 本章是基于作者之一在第九期理论物理前沿暑期讲习班的部分讲义为基础, 并做了相应的整理和扩充. 我们希望通过系统地介绍一个简单的可积模型, 来深入理解多体物理的基本概念, 借此丰富研究生在量子多体物理方面的认知.

在多体物理中, 有一类模型是可以严格求解的, 这类模型通常被称为可积模型. 可积性的概念最早来源于经典力学, 一个经典力学系统一般可以用一组微分方程来描述, 通过求解这组微分方程, 如果我们能找到和系统自由度个数相同的独立守恒量, 则可以说这个系统是完全可积的.

对于量子可积性概念的理解, 要追溯到 1931 年, Bethe 为精确求解一维海森伯自旋链 (考虑近邻相互作用) 的能量本征值问题, 构造了一种特殊形式的波函数, 即将波函数写成一维自旋链上所有可能的平面波的叠加 [1]. Bethe 的假设提出来后, 并未引起大家的关注, 直到 30 多年后, Lieb 和 Liniger 利用 Bethe Ansatz(BA) 第一次精确求解了一维 δ 相互作用的玻色气体. 一维 δ 相互作用的玻色气体的精确解, 是由一系列满足 Bethe Ansatz 方程的波数 $k_i (i=1,2,\cdots,N)$ 给出的, 该方程称为 Lieb-Liniger 方程. 量子可积模型的重要性体现在, BA 方程的解可帮助人们精确地理解多体系统中诸多物理现象, 如量子统计、热力学、量子流体、量子关联及量子临界现象等. 通常能够精确求解的物理体系包括自旋链、相互作用量子气体、强关联电子系统、Kondo 杂质问题、Gaudin 磁性链等.

基于 Lieb 和 Liniger 1963 年的工作, 杨振宁和杨振平于 1969 年成功解决了该模型的有限温热力学问题, 首次给出了 Lieb-Liniger 模型在平衡态下的巨正则统计物理描述, 建立了 Yang-Yang 热力学. 这是第一次真正意义上严格求解了一个有相互作用的多体系统的热力学问题. 具体思想是, 在有限温情况下, 系统平衡态存在很多微观态, 通过引入熵, 并最小化吉布斯自由能给出系统满足的 Yang-Yang 方程, 即热力学 Bethe Ansatz(TBA) 方程. 由 Yang-Yang 热力学方程可解析地得到 Lieb-Liniger 模型的所有热力学量. Yang-Yang 方程描述了系统的普适热力学规律, 描述着不同相区的量子涨落和热涨落. Yang-Yang 热力学方法的提出, 标志着在处理有限温多体系统精确解方面迈进了重要的一步[3].

本章的结构如下: 第 2.2 节, 给出 Lieb-Liniger 玻色气体的 BA 方程的严格推导, 展示了如何从一个场论问题转化到量子力学多体系统的推导, 并系统地介绍了 BA 方法. 第 2.3 节, 对系统基态的性质进行了讨论, 包括基态能量、量子多体系统的协同和集体特性等. 第 2.4 节和第 2.5 节, 分别讨论了在弱相互作用和强相互作用极限下系统的准动量密度分布. 在弱相互作用极限下, 准动量分布函数满足半圆律; 在强相互作用极限下, 一维 Lieb-Liniger 玻色气体可以看成一维自由的费米气体, 也称为 Tonks-Girardeau 气体[4]. 第 2.6 节, 讨论了两种低能激发态的激发谱. 第 2.7 节, 详细介绍了 Yang-Yang 热力学方法在 Lieb-Liniger 模型中的应用, 并给出了玻色系统从 BA 方程得到 TBA 方程的过程. 第 2.8 节、第 2.9 节和第 2.10 节分别讨论了在零温、高温、强相互作用、弱相互作用情况下 TBA 方程的解, 以及低温时 Lieb-Liniger 玻色气体的量子临界性. 第 2.11 节, 给出了在强排斥相互作用时 Lieb-Liniger 玻色气体的多体关联函数. 第 2.12 节中, 简单地回顾最近关于 Lieb-Liniger 玻色气体的最新实验.

2.2 Bethe 假设

考虑粒子总数为 N、长度为 L 的一维无自旋均匀玻色气体, 哈密顿量的二次量子化形式[2](取 $\hbar = 1, 2m = 1$) 为

$$\hat{H} = \int_0^L \mathrm{d}x \partial_x \hat{\psi}^\dagger(x) \partial_x \hat{\psi}(x) + c \int_0^L \mathrm{d}x \hat{\psi}^\dagger(x) \hat{\psi}^\dagger(x) \hat{\psi}(x) \hat{\psi}(x), \tag{2.2.1}$$

其中, m 为玻色子的质量; c 为相互作用强度. 正则玻色场算符 $\hat{\psi}(x)$ 满足以下对易关系:

$$[\hat{\psi}(x), \hat{\psi}^\dagger(y)] = \delta(x-y), \quad [\hat{\psi}(x), \hat{\psi}(y)] = [\hat{\psi}^\dagger(x), \hat{\psi}^\dagger(y)] = 0. \tag{2.2.2}$$

2.2 Bethe 假设

为了得到 Lieb-Liniger 模型的解，我们首先在 Fock 空间定义真空态：$\hat{\psi}(x)|0\rangle = 0, x \in \mathbb{R}$，满足 $\langle 0|0 \rangle = 1$，场算符 $\hat{\psi}(x)$ 的运动方程由如下 Heisenberg 方程得到：

$$-\mathrm{i}\partial_t \hat{\psi}(x) = [\hat{H}, \hat{\psi}(x)],$$

其运动方程为

$$\mathrm{i}\partial_t \hat{\psi}(x) = -\partial_x^2 \hat{\psi}(x) + 2c\hat{\psi}^\dagger(x)\hat{\psi}(x)\hat{\psi}(x). \tag{2.2.3}$$

上式可直接由 Heisenberg 方程得到

$$-\mathrm{i}\partial_t \hat{\psi}(x) = [\hat{H}, \hat{\psi}(x)] = \int_0^L \mathrm{d}y [\partial_y \hat{\psi}^\dagger(y) \partial_y \hat{\psi}(y) + c\hat{\psi}^\dagger(y)\hat{\psi}^\dagger(y)\hat{\psi}(y)\hat{\psi}(y), \hat{\psi}(x)],$$

其中

$$\int_0^L \mathrm{d}y \partial_y \hat{\psi}^\dagger(y) \partial_y \hat{\psi}(y) = \int_0^L \partial_y(\hat{\psi}^\dagger(y) \partial_y \hat{\psi}(y))\mathrm{d}y - \int_0^L \hat{\psi}^\dagger(y) \partial_y^2 \hat{\psi}(y)\mathrm{d}y$$

$$= -\int_0^L \hat{\psi}^\dagger(y) \partial_y^2 \hat{\psi}(y)\mathrm{d}y,$$

于是

$$-\mathrm{i}\partial_t \hat{\psi}(x) = \int_0^L \mathrm{d}y[-\hat{\psi}^\dagger(y)\partial_y^2\hat{\psi}(y), \hat{\psi}(x)] + c\int_0^L \mathrm{d}y[\hat{\psi}^{\dagger 2}(y), \hat{\psi}(x)]\hat{\psi}^2(y)$$

$$= -\int_0^L \mathrm{d}y \delta(x-y)\partial_y^2 \hat{\psi}(y) + c\int_0^L \mathrm{d}y\{\hat{\psi}^\dagger(y)[\hat{\psi}^\dagger(y), \hat{\psi}(x)]$$

$$+ [\hat{\psi}^\dagger(y), \hat{\psi}(x)]\hat{\psi}^\dagger(y)\}\hat{\psi}^2(y)$$

$$= \partial_x^2 \hat{\psi}(x) - 2c\hat{\psi}^\dagger(x)\hat{\psi}^2(x).$$

所以

$$\mathrm{i}\partial_t \hat{\psi}(x) = -\partial_x^2 \hat{\psi}(x) + 2c\hat{\psi}^\dagger(x)\hat{\psi}(x)\hat{\psi}(x).$$

如果 $\hat{\psi}(x)$ 是一个经典场，则这个运动方程可以约化为一个经典场中非线性的薛定谔方程. 相应地，我们很容易得到粒子数算符 $\hat{N}$ 和动量算符 $\hat{P}$：

$$\hat{N} = \int_0^L \mathrm{d}x \hat{\psi}^\dagger(x)\hat{\psi}(x), \tag{2.2.4}$$

$$\hat{P} = -\frac{\mathrm{i}}{2}\int_0^L [\partial_x, \hat{\psi}^\dagger(x)]\hat{\psi}(x)\mathrm{d}x. \tag{2.2.5}$$

可以证明，这两个算符均和哈密顿量对易：$[\hat{H}, \hat{N}] = 0, [\hat{H}, \hat{P}] = 0$. 所以，粒子数和动量均是此体系的守恒量.

此模型的 N 粒子态的本征波函数可写为

$$|\psi\rangle = \frac{1}{\sqrt{N!}} \int_0^L d^N x \psi(x) |x\rangle, \tag{2.2.6}$$

其中, $x = \{x_1, x_2, \cdots, x_N\}$, $|x\rangle = \hat{\psi}^\dagger(x_1)\hat{\psi}^\dagger(x_2)\cdots\hat{\psi}^\dagger(x_N)|0\rangle$, 在这里 x_j 为第 j 个粒子的位置坐标. 对于玻色子而言, 波函数满足交换对称性, 即交换空间中任意两个粒子的坐标, 波函数不变, $\psi(\cdots, x_\xi, \cdots, x_\eta, \cdots) = \psi(\cdots, x_\eta, \cdots, x_\xi, \cdots)$. 我们将二次量子化的哈密顿量作用在本征波函数上, 并利用分部积分, 可将二次量子化的哈密顿量 (2.2.1) 写成一次量子化形式:

$$H = -\sum_{i=1}^N \frac{\partial^2}{\partial x_i^2} + 2c \sum_{i<j}^N \delta(x_i - x_j). \tag{2.2.7}$$

这个多体哈密顿量描述了一维 δ 相互作用的玻色子系统, 称为 Lieb-Liniger 模型.

Bethe 为精确求解一维海森伯自旋链 (考虑近邻相互作用) 的能量本征值问题, 构造了一种特殊形式的波函数, 即将波函数写成一维自旋链上所有可能的平面波的叠加:

$$\psi(x_1, \cdots, x_N) = \sum_{\mathcal{P}, Q} A_{\mathcal{P}} e^{i\sum_{j=1}^N k_{\mathcal{P}_j} x_{Q_j}} \theta(x_{Q_1} < \cdots < x_{Q_N}). \tag{2.2.8}$$

这里, N 为自旋反转的个数; $\mathcal{P}_1, \mathcal{P}_2, \cdots, \mathcal{P}_N$ 和 $Q_1, Q_2, \cdots, Q_n$ 代表对粒子标记 $(1, 2, \cdots, N)$ 的所有可能的置换.

下面我们利用这个假设的波函数, 分两种情况讨论 Lieb-Liniger 模型.

(1) 首先考虑一种简单的情况: 两粒子系统. 两粒子波函数可以写为

$$\begin{aligned}\psi(x_1, x_2) =\, & \theta(x_2 - x_1)[A_{12}e^{i(k_1 x_1 + k_2 x_2)} + A_{21}e^{i(k_2 x_1 + k_1 x_2)}] \\ & + \theta(x_1 - x_2)[A_{12}e^{i(k_1 x_2 + k_2 x_1)} + A_{21}e^{i(k_2 x_2 + k_1 x_1)}],\end{aligned} \tag{2.2.9}$$

其中, A_{12} 和 A_{21} 是与 k_1, k_2 有关的系数; $\theta(x)$ 是阶梯函数, 当 $x > 0$ 时, $\theta(x) = 1$, 当 $x < 0$ 时, $\theta(x) = 0$. 利用质心坐标, 令 $X = \dfrac{x_1 + x_2}{2}$, $Y = x_2 - x_1$, 可以得到 $x_1 = X - \dfrac{1}{2}Y$, $x_2 = X + \dfrac{1}{2}Y$, 于是有

$$\begin{aligned}\frac{\partial}{\partial x_1} &= \frac{\partial}{\partial X}\frac{\partial X}{\partial x_1} + \frac{\partial}{\partial Y}\frac{\partial Y}{\partial x_1} = \frac{1}{2}\frac{\partial}{\partial X} - \frac{\partial}{\partial Y}, \\ \frac{\partial^2}{\partial x_1^2} &= \frac{\partial}{\partial X}\left(\frac{1}{2}\frac{\partial}{\partial X} - \frac{\partial}{\partial Y}\right)\frac{\partial X}{\partial x_1} + \frac{\partial}{\partial Y}\left(\frac{1}{2}\frac{\partial}{\partial X} - \frac{\partial}{\partial Y}\right)\frac{\partial Y}{\partial x_1} \\ &= \frac{1}{4}\frac{\partial^2}{\partial X^2} - \frac{\partial^2}{\partial X \partial Y} + \frac{\partial^2}{\partial Y^2},\end{aligned}$$

2.2 Bethe 假设

$$\frac{\partial}{\partial x_2} = \frac{\partial}{\partial X}\frac{\partial X}{\partial x_2} + \frac{\partial}{\partial Y}\frac{\partial Y}{\partial x_2} = \frac{1}{2}\frac{\partial}{\partial X} + \frac{\partial}{\partial Y},$$

$$\frac{\partial^2}{\partial x_2^2} = \frac{\partial}{\partial X}\left(\frac{1}{2}\frac{\partial}{\partial X} + \frac{\partial}{\partial Y}\right)\frac{\partial X}{\partial x_2} + \frac{\partial}{\partial Y}\left(\frac{1}{2}\frac{\partial}{\partial X} + \frac{\partial}{\partial Y}\right)\frac{\partial Y}{\partial x_2}$$

$$= \frac{1}{4}\frac{\partial^2}{\partial X^2} + \frac{\partial^2}{\partial X \partial Y} + \frac{\partial^2}{\partial Y^2},$$

因此

$$\frac{\partial^2}{\partial x_1^2} + \frac{\partial^2}{\partial x_2^2} = \frac{1}{2}\frac{\partial^2}{\partial X^2} + 2\frac{\partial^2}{\partial Y^2}.$$

由薛定谔方程 $H\psi = E\psi$ 可得到

$$\left[-\frac{1}{2}\frac{\partial^2}{\partial X^2} - 2\frac{\partial^2}{\partial Y^2}\right]\psi(X,Y) + 2c\delta(Y)\psi(X,Y) = E\psi(X,Y).$$

对上式作积分

$$\int_{-\varepsilon}^{+\varepsilon}\frac{\partial^2\psi}{\partial Y^2}\mathrm{d}Y = \int_{-\varepsilon}^{+\varepsilon}c\delta(Y)\psi\mathrm{d}Y - \frac{1}{2}\int_{-\varepsilon}^{+\varepsilon}E\psi\mathrm{d}Y - \frac{1}{4}\int_{-\varepsilon}^{+\varepsilon}\frac{\partial^2\psi}{\partial X^2}\mathrm{d}Y(\varepsilon \to 0)$$

得

$$\frac{\partial\psi}{\partial Y}|_{Y=0^+} - \frac{\partial\psi}{\partial Y}|_{Y=0^-} = c\psi|_{Y=0}. \tag{2.2.10}$$

又因为 $\frac{\partial}{\partial Y} = \frac{\partial}{\partial x_1}\frac{\partial x_1}{\partial Y} + \frac{\partial}{\partial x_2}\frac{\partial x_2}{\partial Y} = -\frac{1}{2}\frac{\partial}{\partial x_1} + \frac{1}{2}\frac{\partial}{\partial x_2}$, 取 $\varepsilon \to 0$ 的极限, 式 (2.2.10) 可写为

$$\left(\frac{\partial}{\partial x_2} - \frac{\partial}{\partial x_1}\right)\psi|_{x_2=x_1^+} - \left(\frac{\partial}{\partial x_2} - \frac{\partial}{\partial x_1}\right)\psi|_{x_2=x_1^-} = 2c\psi|_{x_2=x_1}.$$

由于玻色子满足交换对称性 $\left(\frac{\partial}{\partial x_2} - \frac{\partial}{\partial x_1}\right)\psi|_{x_2=x_1^+} = \left(\frac{\partial}{\partial x_1} - \frac{\partial}{\partial x_2}\right)\psi|_{x_2=x_1^-}$, 所以

$$\left(\frac{\partial}{\partial x_2} - \frac{\partial}{\partial x_1}\right)\psi|_{x_2=x_1^+} = c\psi|_{x_2=x_1}. \tag{2.2.11}$$

可根据式 (2.2.9) 与式 (2.2.11) 得

$$\frac{A_{12}}{A_{21}} = \frac{\mathrm{i}k_2 - \mathrm{i}k_1 + c}{\mathrm{i}k_2 - \mathrm{i}k_1 - c} = -\frac{c + \mathrm{i}(k_2 - k_1)}{c - \mathrm{i}(k_2 - k_1)}$$

$$= -\mathrm{e}^{\mathrm{i}\theta(k_2-k_1)}. \tag{2.2.12}$$

这里, $\theta(x) = 2\arctan(x/c)$.

(2) 考虑 N 个粒子的系统, 我们有

$$\left[-\sum_{i=1}^{N}\frac{\partial^2}{\partial x_i^2}+2c\sum_{i<j}^{N}\delta(x_i-x_j)\right]\psi=E\psi, \qquad (2.2.13)$$

$$\psi=\sum_{\mathcal{P}}A_{\mathcal{P}_1\cdots\mathcal{P}_N}\exp[\mathrm{i}(k_{\mathcal{P}_1}x_1+\cdots+k_{\mathcal{P}_N}x_N)]. \qquad (2.2.14)$$

利用质心坐标

$$X=\frac{x_j+x_k}{2}, \quad Y=x_j-x_k, \qquad (2.2.15)$$

$$x_j=X-\frac{1}{2}Y, \quad x_k=X+\frac{1}{2}Y. \qquad (2.2.16)$$

同样我们可以得到

$$\frac{\partial^2}{\partial x_j^2}+\frac{\partial^2}{\partial x_k^2}=\frac{1}{2}\frac{\partial^2}{\partial X^2}+2\frac{\partial^2}{\partial Y^2}.$$

由薛定谔方程进一步可得

$$\left(-\frac{\partial^2}{\partial x_1^2}-\cdots-\frac{1}{2}\frac{\partial^2}{\partial X^2}-2\frac{\partial^2}{\partial Y^2}-\cdots-\frac{\partial^2}{\partial x_N^2}\right)\psi+\cdots+2c\delta(x_j-x_k)\psi+\cdots=E\psi. \qquad (2.2.17)$$

对 Y 作无穷小积分

$$\left(-\frac{\partial^2}{\partial x_1^2}-\cdots-\frac{1}{2}\frac{\partial^2}{\partial X^2}\cdots\frac{\partial^2}{\partial x_N^2}\right)\psi Y|_{-\varepsilon}^{+\varepsilon}-2\frac{\partial}{\partial Y}\psi|_{-\varepsilon}^{+\varepsilon}+2c\psi|_{Y=0}=E\psi Y|_{-\varepsilon}^{+\varepsilon}$$

如上述两个粒子系统处理方法一样, N 粒子系统满足的边界条件为

$$\left(\frac{\partial}{\partial x_j}-\frac{\partial}{\partial x_k}\right)\psi|_{x_j=x_k^+}-\left(\frac{\partial}{\partial x_j}-\frac{\partial}{\partial x_k}\right)\psi|_{x_j=x_k^-}=2c\psi|_{x_j=x_k}. \qquad (2.2.18)$$

由于玻色子的交换对称性

$$\left(\frac{\partial}{\partial x_j}-\frac{\partial}{\partial x_k}\right)\psi|_{x_j=x_k^+}=\left(\frac{\partial}{\partial x_k}-\frac{\partial}{\partial x_j}\right)\psi|_{x_j=x_k^-},$$

可以得到同样的边界条件:

$$\left(\frac{\partial}{\partial x_j}-\frac{\partial}{\partial x_k}\right)\psi|_{x_j=x_k^+}=c\psi|_{x_j=x_k}. \qquad (2.2.19)$$

现在, 我们需要找出粒子不同排列时对应的系数关系, 因此, 我们取定区域 $x_1<x_2<\cdots<x_k<x_j<\cdots<x_N$, 利用如下表示:

$$\phi(P)=A(P)\mathrm{e}^{\mathrm{i}(k_{\mathcal{P}_1}x_1+\cdots+k_{\mathcal{P}_k}x_k+k_{\mathcal{P}_j}x_j+\cdots+k_{\mathcal{P}_N}x_N)}, \qquad (2.2.20)$$

$$\phi(Q)=A(Q)\mathrm{e}^{\mathrm{i}(k_{\mathcal{P}_1}x_1+\cdots+k_{\mathcal{P}_j}x_k+k_{\mathcal{P}_k}x_j+\cdots+k_{\mathcal{P}_N}x_N)}. \qquad (2.2.21)$$

2.2 Bethe 假设

显然, 整个区域上的波函数可写为 $\phi(x_1, x_2, \cdots, x_N) = [\cdots + \phi(P) + \phi(Q) + \cdots]$, 代入边界条件 (2.2.19) 可得

$$\left(\frac{\partial}{\partial x_j} - \frac{\partial}{\partial x_k}\right)[\cdots + \phi(P) + \phi(Q) + \cdots]|_{x_j = x_k^+} = c[\cdots + \phi(P) + \phi(Q) + \cdots]|_{x_j = x_k}. \tag{2.2.22}$$

由此, 可以进一步得到

$$-\mathrm{i}(k_{\mathcal{P}_k} - k_{\mathcal{P}_j})[A(P) - A(Q)] = c[A(P) + A(Q)].$$

因此, 我们可以得到如下关系:

$$\begin{aligned} A(Q) &= -\frac{c - \mathrm{i}(k_{\mathcal{P}_j} - k_{\mathcal{P}_k})}{c + \mathrm{i}(k_{\mathcal{P}_j} - k_{\mathcal{P}_k})} A(P) = \frac{k_{\mathcal{P}_k} - k_{\mathcal{P}_j} - \mathrm{i}c}{k_{\mathcal{P}_k} - k_{\mathcal{P}_j} + \mathrm{i}c} A(P) \\ &= -\mathrm{e}^{-\mathrm{i}\theta(k_{\mathcal{P}_k} - k_{\mathcal{P}_j})} A(P). \end{aligned}$$

即

$$A(k_1, \cdots, k_{\mathcal{P}_j}, \cdots, k_{\mathcal{P}_k}, \cdots, k_N) = -\mathrm{e}^{-\mathrm{i}\theta(k_{\mathcal{P}_k} - k_{\mathcal{P}_j})} A(k_1, \cdots, k_{\mathcal{P}_k}, \cdots, k_{\mathcal{P}_j}, \cdots, k_N), \tag{2.2.23}$$

其中, $\theta(k_i - k_j) = 2\arctan\left(\dfrac{k_i - k_j}{c}\right)$. 为了方便, 我们取 $A(I) = 1$, 这里 $A(I) = A(k_1, \cdots, k_N)$. 考虑波函数满足的周期性边界条件:

$$\psi(0, x_2, \cdots, x_N) = \psi(x_2, \cdots, x_N, L). \tag{2.2.24}$$

于是, 可以得到

$$A_P \mathrm{e}^{\mathrm{i}(k_{\mathcal{P}_2} x_2 + k_{\mathcal{P}_3} x_3 + \cdots + k_{\mathcal{P}_N} x_N)} = A_{P'} \mathrm{e}^{\mathrm{i}(k_{\mathcal{P}_2} x_2 + k_{\mathcal{P}_3} x_3 + \cdots + k_{\mathcal{P}_N} x_N + k_{\mathcal{P}_1} L)}$$

其中 $P = (\mathcal{P}_1, \mathcal{P}_2, \cdots, \mathcal{P}_N)$ 和 $P' = (\mathcal{P}_2, \mathcal{P}_3, \cdots, \mathcal{P}_N, \mathcal{P}_1)$, 可得

$$A_P = A_{P'} \mathrm{e}^{\mathrm{i}k_{\mathcal{P}_1} L}. \tag{2.2.25}$$

继而

$$\begin{aligned} A_{\mathcal{P}_1 \mathcal{P}_2 \cdots \mathcal{P}_N} &= -\mathrm{e}^{\mathrm{i}\theta(k_{\mathcal{P}_1} - k_{\mathcal{P}_2})} A_{\mathcal{P}_2 \mathcal{P}_1 \mathcal{P}_3 \cdots} \\ &= \left[-\mathrm{e}^{\mathrm{i}\theta(k_{\mathcal{P}_1} - k_{\mathcal{P}_2})}\right]\left[-\mathrm{e}^{\mathrm{i}\theta(k_{\mathcal{P}_1} - k_{\mathcal{P}_3})}\right] A_{\mathcal{P}_2 \mathcal{P}_3 \mathcal{P}_1 \mathcal{P}_4 \cdots} \\ &\vdots \\ &= (-1)^{N-1} \mathrm{e}^{\sum_q \mathrm{i}\theta(k_{\mathcal{P}_1} - k_q)} A_{\mathcal{P}_2 \mathcal{P}_3 \cdots \mathcal{P}_N \mathcal{P}_1}. \end{aligned} \tag{2.2.26}$$

比较式 (2.2.25) 与式 (2.2.26), 可以得到

$$\mathrm{e}^{\mathrm{i}k_{\mathcal{P}_j} L} = (-1)^{N-1} \mathrm{e}^{\sum_q \mathrm{i}\theta(k_{\mathcal{P}_j} - k_q)}. \tag{2.2.27}$$

$$e^{ik_jL} = (-1)^{N-1}(-1)^{N-1}\prod_{q\neq 1}^{N}\frac{k_j-k_q+\mathrm{i}c}{k_j-k_q-\mathrm{i}c}$$

$$=\prod_{q\neq 1}^{N}\frac{k_j-k_q+\mathrm{i}c}{k_j-k_q-\mathrm{i}c}$$

$$=-\prod_{q=1}^{N}\frac{k_j-k_q+\mathrm{i}c}{k_j-k_q-\mathrm{i}c}.$$

即

$$e^{ik_jL} = -\prod_{l=1}^{N}\frac{k_j-k_l+\mathrm{i}c}{k_j-k_l-\mathrm{i}c}, \tag{2.2.28}$$

我们称上式为 Bethe Ansatz 方程 [2]. 通过求解上述 BA 方程, 我们便可以得到所有的准动量 $k_j(j=1,\cdots,N)$, 从而给出体系的动量和能量. 为此, 我们先对上述 BA 方程两边取对数得到

$$k_iL = 2\pi I_i - \sum_{l=1}^{N}2\arctan\frac{k_i-k_l}{c}, \tag{2.2.29}$$

这里, $\{I_i\}$ 是准动量 $\{k_i\}$ 对应的量子数, 其中

$$\begin{cases} I_i = \text{整数} & (N=\text{奇数}) \\ I_i + \dfrac{1}{2} = \text{整数} & (N=\text{偶数}) \end{cases} \tag{2.2.30}$$

对于基态, 量子数 $\{I_i\}$ 的取值为 $\{I_i\} = \left\{-\dfrac{N-1}{2}, -\dfrac{N-3}{2}, \cdots, \dfrac{N-1}{2}\right\}.$

2.3 基态行为

为了理解一维 δ 相互作用的玻色气体的基态特性, 我们考虑排斥相互作用情况 $(c>0)$, 取热力学极限: $N,L\to\infty$, 粒子数密度 $n=N/L$ 为常数, 则 BA 方程 (2.2.28) 可由密度分布函数来表示. 通过对 N 个 BA 方程 (2.2.28) 累乘的计算, 我们得到一个关系: $e^{\mathrm{i}\sum_i^N k_iL} = 1$. 因此, 总的动量为 $P = \sum_j^N k_j = 2\pi l/L(l=0,\pm 1,\cdots)$. 所以, 对于基态, $P=0$, 所有的 k_i 在均匀区间 $[-k_\mathrm{F}, k_\mathrm{F}]$ 内对称分布, k_F 为费米点.

我们引入区间 $[k,k+\Delta k]$ 内的准动量密度分布函数 $\rho(k) = \lim_{\Delta k\to 0}1/(L\Delta k)$, 则此区间内的粒子数可表示为 $L\rho(k)\Delta k$. 区间 $[I_k, I_k+\Delta I]$ 内的粒子数与区间 $[k,k+\Delta k]$ 内的粒子数相同 $\Delta I = L\rho(k)\Delta k$. 由式 (2.2.29), 我们得到

$$k = \frac{2\pi I}{L} - \frac{2}{L}\int_{-k_\mathrm{F}}^{k_\mathrm{F}}\theta(k-q)\rho(q)\mathrm{d}q \tag{2.3.1}$$

2.3 基态行为

通过变形可得

$$1 = 2\pi\rho(k) - 2\int_{-k_F}^{k_F} \frac{c}{c^2 + (k-q)^2}\rho(q)\mathrm{d}q, \tag{2.3.2}$$

所以密度分布函数表示为

$$\rho(k) = \frac{1}{2\pi} + \frac{1}{2\pi}\int_{-k_F}^{k_F} \frac{2c}{c^2 + (k-q)^2}\rho(q)\mathrm{d}q. \tag{2.3.3}$$

对上式做如下标度变换 [5]：

$$k = k_F x, \quad c = k_F \lambda, \quad \rho(k_F x) = g(x), \tag{2.3.4}$$

于是式 (2.3.3) 变为

$$1 + 2\lambda \int_{-1}^{1} \frac{g(x)\mathrm{d}x}{\lambda^2 + (x-y)^2} = 2\pi g(y), \tag{2.3.5}$$

整理得到非齐次的 Fredholm 方程:

$$g(y) = \frac{1}{2\pi} + \frac{\lambda}{\pi}\int_{-1}^{1} \frac{g(x)\mathrm{d}x}{\lambda^2 + (x-y)^2}, \tag{2.3.6}$$

这里, $g(y)$ 为密度分布函数, $\gamma = Lc/N = c/n$, 相应的粒子数密度 $n = \int_{-k_F}^{k_F}\rho(k)\mathrm{d}k$ 和基态能量 $E_0 = Nn^2 e(\gamma) = \frac{N}{n}\int_{-k_F}^{k_F}\rho(k)k^2\mathrm{d}k$ 均可由密度分布函数来表示, 即

$$\gamma \int_{-1}^{1} g(x)\mathrm{d}x = \lambda, \tag{2.3.7}$$

$$e(\gamma) = \frac{r^3}{\lambda^3}\int_{-1}^{1} g(x)x^2\mathrm{d}x. \tag{2.3.8}$$

所以

$$e(\gamma) = \frac{\gamma^3}{\lambda^3}\int_{-1}^{1} g(x)x^2\mathrm{d}x, \tag{2.3.9}$$

$$k_F(\gamma) = \frac{c}{\lambda} = n\gamma\lambda^{-1} = n\left[\int_{-1}^{1} g(x)\mathrm{d}x\right]^{-1}. \tag{2.3.10}$$

其中, $e(\gamma)$ 为能量密度函数, 可证明, $g(x)$、$e(\gamma)$ 和 $k_F(\gamma)$ 均为 γ 的解析函数 (除 $\gamma = 0$ 点外). 因此, 我们可以用数值的方法求解上述积分方程. 在强相互作用和弱相互作用极限下, 我们可以给出体系基态能量的解析表达式, 这部分我们将在下面章节讨论.

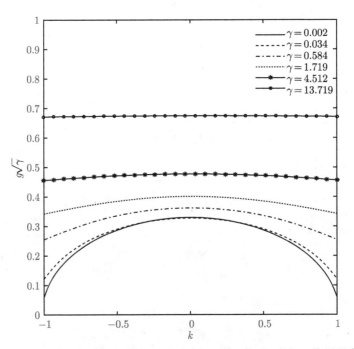

图 2.1 基态准动量密度: 准动量密度 $g(k)$ 由方程 (2.3.6) 得到. 当相互作用强度 γ 很小时, 分布函数 $g(k)$ 符合半圆律 (2.4.11); 随着相互作用强度 γ 的逐渐增大, 分布函数越来越平缓, 最终趋于 $g(k) \approx 1/2\pi$

2.4 弱相互作用: 半圆律

在弱相互作用极限下, 一维 δ 相互作用玻色气体的 Bethe Ansatz 方程的根满足半圆律 (semicircle law). 相应的非齐次积分方程 (2.3.6) 称为 Love 方程. 运用 Huston 的方法, Gaudin 发现密度分布函数和能量密度可以表示为[6]

$$g_0(t) \approx \frac{k_F}{2\pi c}(1-t^2)^{\frac{1}{2}} + \frac{1}{4\pi^2}(1-t^2)^{-\frac{1}{2}}\left(t\ln\frac{1-t}{1+t} + \ln\frac{16\pi e k_F}{c}\right), \quad (2.4.1)$$

$$e(\gamma) = n^3\left(\gamma - \frac{4}{3\pi}\gamma^{\frac{3}{2}}\right). \quad (2.4.2)$$

其中, k_F 为费米点. 下面我们应用一种简单的方法来证明弱相互作用极限下体系的准动量分布满足半圆律. 取极限条件: $c \to 0$ 或 $Lc \ll 1$, 线性化 BA 方程 (2.2.28), 得到

$$e^{ik_j L} \approx 1 - 2\sum_{l=1}^{N}\frac{c^2}{(k_j-k_l)^2} - 4\sum_{l=1}^{N-1}\sum_{l<l'=2}^{N}\frac{c^2}{(k_j-k_l)(k_j-k_{l'})} + i\sum_{l=1}^{N}\frac{2c}{k_j-k_l}, \quad (2.4.3)$$

2.4 弱相互作用：半圆律

上面的求和中 $l \neq j$ 且 $l' \neq j$，由此我们得到方程组：

$$\begin{cases} \cos k_j L \approx 1 - 2\sum_{l=1}^{N} \dfrac{c^2}{(k_j - k_l)^2} - 4\sum_{l=1}^{N-1}\sum_{l<l'=2}^{N} \dfrac{c^2}{(k_j-k_l)(k_j-k_{l'})} \\ \sin k_j L \approx \sum_{l=1}^{N} \dfrac{2c}{k_j - k_l} \end{cases} \quad (2.4.4)$$

显然，其中动量 k_j 满足如下关系：

$$k_j = \frac{2\pi d_j}{L} + \frac{2c}{L}\sum_{l\neq j}^{N}\frac{1}{k_j - k_l}, \quad j = 1, 2, \cdots, N, \quad (2.4.5)$$

这里 $d_j = \pm 1, \pm 2, \cdots$，代表激发态. $d_j = 0$ 对应体系的基态，则式 (2.4.5) 可写为

$$k_j = \frac{2c}{L}\sum_{l\neq j}^{N}\frac{1}{k_j - k_l}, \quad j = 1, 2, \cdots, N. \quad (2.4.6)$$

令 $q_j = k_j \sqrt{L/2c}$ [7]，则上式变为

$$q_j = \sum_{l\neq j}^{N}\frac{1}{q_j - q_l}, \quad j = 1, 2, \cdots, N. \quad (2.4.7)$$

如果我们定义 $H_N(q) = \prod_{i=1}^{N}(q - q_i)$，则有

$$\lim_{q\to q_j} \frac{H_N''(q_j)}{H_N'(q_j)} = \sum_{j\neq l}\frac{2}{q_j - q_l}, \quad (2.4.8)$$

我们定义 $F(q) = H_N''(q) - 2qH_N'(q)$，因此式 (2.4.8) 可以写为

$$F(q_j) \equiv H_N''(q_j) - 2q_j H_N'(q_j) = 0$$

则 $F(q)$ 和 $H_N(q)$ 都是 q 的 N 阶多项式，且满足 $F(q_j) = H_N(q_j) = 0$. $H_N(q)$ 和 $F(q)$ 的最高阶项分别为 $H_N(q) = q^N + \cdots$，$F(q) = -2Nq^N + \cdots$，因此 $F(q) = -2NH_N(q)$. 于是有

$$H_N''(q) - 2qH_N'(q) + 2NH_N(q) = 0. \quad (2.4.9)$$

上述微分方程的解为 q 的 N 次厄米多项式，$q_j(j = 1, 2, \cdots, N)$ 为该厄米多项式 $H_N(q) = 0$ 的根，如果我们令 $q_j < q_{j+1}$，则 q_j 满足

$$\frac{\sqrt{2N+1-q_{j+1}^2}}{\pi} < \frac{1}{q_{j+1} - q_j} < \frac{\sqrt{2N+1-q_j^2}}{\pi}, \quad q_j \geqslant 0, \quad (2.4.10)$$

根据定义，密度分布函数 $\rho(k) = \lim\limits_{L,N\to\infty} \dfrac{1}{L(k_{j+1} - k_j)}$，因此我们可以得到

$$\rho(k) = \dfrac{1}{2\pi c}\left(4cn + \dfrac{2c}{L} - k^2\right)^{\frac{1}{2}}$$

$$\approx \dfrac{1}{2\pi c}(4cn - k^2)^{\frac{1}{2}}$$

$$= \dfrac{1}{\pi}\sqrt{\dfrac{n}{c}}\left[1 - \left(\dfrac{k}{k_{\rm F}}\right)^2\right]^{\frac{1}{2}}$$

$$= \dfrac{1}{\pi\sqrt{\gamma}}\sqrt{1 - \dfrac{k^2}{4|\gamma|n^2}}. \tag{2.4.11}$$

上式为一维 δ 相互作用玻色气体的 Bethe Ansatz 方程的根满足的半圆律. 此式中 $k_{\rm F} = 2\sqrt{nc}$，由 $\displaystyle\int_{-k_{\rm F}}^{k_{\rm F}} \rho(k){\rm d}k = n$ 得到.

2.5 强相互作用: 费米化

在强排斥相互作用极限下，单粒子波函数之间重叠很少，因此，此极限下，Lieb-Liniger 玻色气体可以看成自由的费米气体，也称为 Tonks-Girardeau 气体 [4]. 在冷原子实验中，很多都观察到了强相互作用的量子简并玻色气体. 根据取对数后的 BA 方程 (2.2.29) 在强相互作用极限 $c \to \infty$ 时，量子数 I_j 满足

$$I_j = -\left(\dfrac{N-1}{2}\right) + j - 1, \quad j = 1, 2, \cdots, N. \tag{2.5.1}$$

利用泰勒展开 $\arctan(x) = x - \dfrac{1}{3}x^3 + o(x^5)$:

$$k_j L = 2\pi I_j - 2\sum_{l=1}^{N}\dfrac{k_j - k_l}{c} + 2\sum_{l=1}^{N}\dfrac{1}{3}\dfrac{(k_j - k_l)^3}{c^3} + o\left(\dfrac{1}{c^5}\right)$$

$$= 2\pi I_j - 2\dfrac{Nk_j}{c} + 2\sum_{l=1}^{N}\dfrac{1}{3c^3}(k_j - k_l)^3 + o\left(\dfrac{1}{c^5}\right)$$

$$= 2\pi I_j - 2\dfrac{Nk_j}{c} + \dfrac{16\pi^3}{3c^3 L^3}\sum_{l=1}^{N}(l - j)^3 + o\left(\dfrac{1}{c^5}\right).$$

在上述计算中有 $2\pi j/L = k_j$. 对于基态，$\sum\limits_{l=1}^{N} k_l/c = p/c = 0$. 通过如下近似处理:

$$\lim_{N\to +\infty}\sum_{l=1}^{N}(x-j)^3 = \int_{\frac{1}{2}}^{N+\frac{1}{2}}(x-j)^3 {\rm d}x = \dfrac{1}{4}\left[\left(N + \dfrac{1}{2} - j\right)^4 - \left(\dfrac{1}{2} - j\right)^4\right], \tag{2.5.2}$$

2.5 强相互作用：费米化

我们会得到准动量

$$k_jL \approx 2\pi I_j \left(1 - \frac{2}{\gamma} + \frac{4}{\gamma^2} - \frac{8}{\gamma^3}\right) + \frac{4\pi^3}{3c^3L^3}\left[\left(N+\frac{1}{2}-j\right)^4 - \left(\frac{1}{2}-j\right)^4\right] + o\left(\frac{1}{c^5}\right). \tag{2.5.3}$$

在得到式 (2.5.3) 的过程中，我们利用了迭代的方法：

$$\begin{aligned}k_jL &\approx 2\pi I_j - \frac{2N}{c}\left[2\pi I_j - \frac{2N}{c}\left(2\pi I_j - \frac{2N}{c}\right)\right] \\ &\quad + \frac{4\pi^3}{3c^3L^3}\left[\left(N+\frac{1}{2}-j\right)^4 - \left(\frac{1}{2}-j\right)^4\right] + o\left(\frac{1}{c^5}\right) \\ &= 2\pi I_j\left(1 - \frac{2}{\gamma} + \frac{4}{\gamma^2} - \frac{8}{\gamma^3}\right) + \frac{4\pi^3}{3c^3L^3}\left[\left(N+\frac{1}{2}-j\right)^4 - \left(\frac{1}{2}-j\right)^4\right] + o\left(\frac{1}{c^5}\right).\end{aligned}$$

于是在强排斥相互作用下玻色气体的基态能量为

$$\begin{aligned}E &= \sum_j k_j^2 \\ &\approx \frac{4\pi^2}{L^2}\sum_j I_j{}^2\left(1 - \frac{2}{\gamma} + \frac{4}{\gamma^2} - \frac{8}{\gamma^3}\right)^2 + \frac{16\pi^4}{3c^3L^5}\sum_j I_j\left[\left(N+\frac{1}{2}-j\right)^4 - \left(\frac{1}{2}-j\right)^4\right] \end{aligned} \tag{2.5.4}$$

$$\approx \frac{4\pi^2}{L^2}\sum_j I_j{}^2\left(1 - \frac{4}{\gamma} + \frac{12}{\gamma^2} - \frac{32}{\gamma^3}\right) + \frac{16\pi^4}{3c^3L^5}\sum_j I_j\left[\left(N+\frac{1}{2}-j\right)^4 - \left(\frac{1}{2}-j\right)^4\right], \tag{2.5.5}$$

这里，$\gamma = Lc/N$. 于是便可以得到

$$E = \sum_j k_j^2 = \frac{\pi^2 N^3}{3L^2}\left(1 - \frac{4}{\gamma} + \frac{12}{\gamma^2} - \frac{32}{\gamma^3} + \frac{32\pi^2}{15\gamma^3}\right). \tag{2.5.6}$$

即在强排斥相互作用下基态能为

$$\frac{E}{L} \approx \frac{\pi^2 n^3}{3}\left[1 - \frac{4}{\gamma} + \frac{12}{\gamma^2} + \frac{32}{\gamma^3}\left(\frac{\pi^2}{15} - 1\right)\right]. \tag{2.5.7}$$

另外，基态能量 (2.5.7) 也可以通过从 Fredholm 方程 (2.3.6) 做强相互作用展开获得. 从上面基态能量的表达式中我们可以看出，在强相互作用极限 $\gamma \to \infty$ 的条件下，基态能量一阶近似可表示成自由费米子的能量 $e_f = \frac{1}{3}\pi^2 n^3$. 根据基态能量，我们也可以得到一些其他的物理量，如压缩率、声速、Luttinger 参数等.

最近, Ristivojevic 给出了强相互作用极限条件下 [8] 一维玻色气体基态密度分布函数的解析表达式:

$$\rho(k) = \frac{1}{2\pi} + \frac{1}{\pi^2\lambda} + \frac{2}{\pi^3\lambda^2} + \frac{12-\pi^2}{3\pi^4\lambda^3} + \frac{8-2\pi^2}{\pi^5\lambda^4} - k^2\left(\frac{1}{\pi^2\lambda^3} + \frac{1}{\pi^3\lambda^4}\right) + o(\lambda^{-5}) \quad (2.5.8)$$

其中, $\lambda = c/Q$, Q 为费米点. 由此可得单位长度的能量和粒子数密度为

$$\begin{aligned}
\frac{E}{L} &= \int_{-Q}^{Q} \rho(k)k^2 \mathrm{d}k \\
&= \int_{-Q}^{Q} \left[\frac{1}{2\pi} + \frac{1}{\pi^2\lambda} + \frac{2}{\pi^3\lambda^2} + \frac{12-\pi^2}{3\pi^4\lambda^3} + \frac{8-2\pi^2}{\pi^5\lambda^4} - k^2\left(\frac{1}{\pi^2\lambda^3} + \frac{1}{\pi^3\lambda^4}\right)\right] k^2 \mathrm{d}k \\
&= \frac{1}{3\pi}\frac{c^3}{\lambda^3}\left[1 + \frac{2}{\pi\lambda} + \frac{4}{\pi^2\lambda^2} + \frac{120-28\pi^2}{15\pi^3\lambda^3} + \frac{80-26\pi^2}{5\pi^4\lambda^4}\right],
\end{aligned} \quad (2.5.9)$$

$$\begin{aligned}
n &= \int_{-Q}^{Q} \rho(k) \mathrm{d}k \\
&= \int_{-Q}^{Q} \left[\frac{1}{2\pi} + \frac{1}{\pi^2\lambda} + \frac{2}{\pi^3\lambda^2} + \frac{12-\pi^2}{3\pi^4\lambda^3} + \frac{8-2\pi^2}{\pi^5\lambda^4} - k^2\left(\frac{1}{\pi^2\lambda^3} + \frac{1}{\pi^3\lambda^4}\right)\right] \mathrm{d}k \\
&= \frac{1}{\pi}\frac{c}{\lambda}\left[1 + \frac{2}{\pi\lambda} + \frac{4}{\pi^2\lambda^2} + \frac{24-4\pi^2}{3\pi^3\lambda^3} + \frac{48-14\pi^2}{3\pi^4\lambda^4}\right].
\end{aligned} \quad (2.5.10)$$

图 2.2 在费米点附近的低能粒子空穴对激发: 一个粒子的准动量从动量 Q 激发到 $q > Q$, 其他粒子的准动量也将依次改变, 形成集体运动模式

2.6 元激发: 集体运动

Lieb-Liniger 玻色气体的元激发可以分为两种激发模式 [2]: 粒子激发和空穴激发.

Type 1 粒子激发

粒子激发是指在动量空间中将动量 Q(或 $-Q$) 的粒子激发到动量 $q > Q$(或 $q < -Q$) 的位置, 这种激发对应的能量和动量分别为

$$\varepsilon_1(q) = q^2 - Q^2, \quad (2.6.1)$$

$$p = \begin{cases} q - Q, & q > Q, \\ q + Q, & q < -Q, \end{cases} \quad (2.6.2)$$

2.6 元激发：集体运动

这里，p 为体系的动量. 进而体系的能量可以表示为 [5]

$$\varepsilon_1(p) = p^2 + 2\pi\rho|p|, \tag{2.6.3}$$

其中，$\rho = N/L$，对于粒子激发，体系动量的取值范围为 $-\infty < p < \infty$. 假设一个粒子从动量为 Q 的点激发到 q 处 $(q > Q)$，令

$$k_N = q, \quad k'_i = k_i + \Delta k_i, \quad i = 1, 2, \cdots, N-1, \tag{2.6.4}$$

这里，$\{k'_i\}$ 代表激发一个粒子后准动量的重新分布.

对于 $N-1$ 粒子，我们知道体系基态有

$$k_i L = 2\pi I_i - \sum_{l=1}^{N-1} 2\arctan\left(\frac{k_i - k_l}{c}\right). \tag{2.6.5}$$

对于新的激发态，我们有

$$k'_i L = 2\pi I'_i - \sum_{l=1}^{N-1} 2\arctan\left(\frac{k'_i - k'_l}{c}\right) - 2\arctan\left(\frac{k'_i - q}{c}\right),$$

$$I'_i = -\frac{N+1}{2} + i, \quad i = 1, 2, \cdots, N-1,$$

从而可以得到

$$(k'_i - k_i)L = \Delta k_i L$$
$$= -\pi - \sum_{l=1}^{N-1} \frac{2c}{c^2 + (k_i - k_l)^2}(\Delta k_i - \Delta k_l) - 2\arctan\left(\frac{k_i - q}{c}\right)$$

$$\Delta k_i L\left(1 + \frac{1}{L}\sum_{l=1}^{N-1}\frac{2c}{c^2+(k_i-k_l)^2}\right) = -\pi + \sum_{l=1}^{N-1}\frac{2c\Delta k_l}{c^2+(k_i-k_l)^2} - 2\arctan\left(\frac{k_i-q}{c}\right)$$

$$\Delta k L \rho(k) = -\frac{1}{2} + \frac{1}{2\pi}\int_{-Q}^{Q}\frac{2c\Delta k'\rho(k')\mathrm{d}k'}{c^2+(k-k')^2} - \frac{1}{\pi}\arctan\left(\frac{k-q}{c}\right).$$

令 $J(k) = L\rho(k)\Delta k$，于是

$$J(k) = \frac{1}{2\pi}\int_{-Q}^{Q}\frac{2c}{c^2+(k-k')^2}J(k')\mathrm{d}k' - \frac{1}{2} - \frac{1}{\pi}\arctan\frac{k-q}{c}. \tag{2.6.6}$$

此时，体系的动量为

$$p = \sum_{i=1}^{N-1} k'_i + q = \sum_{i=1}^{N-1}(k_i + \Delta k_i) + q$$

$$= q + \int_{-Q}^{Q} J(k)\mathrm{d}k. \tag{2.6.7}$$

能量为 [5]

$$\varepsilon_1(k) = \sum_{i=1}^{N-1} k_i'^2 + q^2 - E_0(N)$$
$$= -\mu + q^2 + 2\int_{-Q}^{Q} kJ(k)\mathrm{d}k, \qquad (2.6.8)$$

其中, $E_0(N)$ 为 N 粒子体系的基态能量. 实际上, 我们通过直接求解 Bethe Ansatz 方程可以很容易得到 Tonks - Girardeau 气体的激发能, 在强相互作用极限下, $c \gg 1$, 由离散的 Bethe Ansatz 方程可以得到如下根的分布:

$$\begin{cases} k_N \approx \left[\dfrac{(N-1)\pi}{L} + p + \dfrac{2p}{Lc}\right]\left(1 - \dfrac{2N}{Lc}\right) \\ k_j \approx \left(\dfrac{2n_j\pi}{L} + \dfrac{2p}{Lc}\right)\left(1 - \dfrac{2N}{Lc}\right) \end{cases} \qquad (2.6.9)$$

在这里我们取 $n_j = 0, \pm 1, \pm 2, \cdots, \pm\dfrac{N-1}{2}$ (假设 N 为奇数), 则激发能为

$$E(p) - E_0 \approx \dfrac{2(N-1)\pi}{L}\left(1 - \dfrac{2N}{Lc}\right)^2 p \approx v_s p, \qquad (2.6.10)$$

其中, $v_s = \dfrac{2N\pi}{L}\left(1 - \dfrac{4N}{Lc}\right)$ 为声速. 当 $c \to \infty$ 时, 声速即为无相互作用的费米速度. 在低能激发中, 声速是相互作用 γ 的函数 $\left(\gamma = \dfrac{c}{n}\right)$.

Type 2 空穴激发

空穴激发可以认为是将一个粒子从 $0 < q < Q$ 激发到 $Q + \dfrac{2\pi}{L}$ (或从 $-Q < q < 0$ 到 $-Q - \dfrac{2\pi}{L}$), 参见图 2.3. 这种激发的能量和动量分别为

$$\varepsilon_2(q) = Q^2 - q^2,$$
$$p = \begin{cases} +Q - q, & 0 < q < Q, \\ -Q - q, & -Q < q < 0, \end{cases}$$
$$\varepsilon_2(p) = 2\pi\rho|p| - p^2.$$

对于空穴激发, 其体系动量的取值范围为 $-\pi\rho \leqslant p \leqslant \pi\rho$. 定义:

$$\omega_i = (k_{i+1} - k_i)L$$
$$= 2\pi\Delta I_i - 2\sum_{l=1}^{N}\left[\arctan\left(\dfrac{k_{i+1} - k_l}{c}\right) - \arctan\left(\dfrac{k_i - k_l}{c}\right)\right]$$

其中, $\Delta I_i = I_{i+1} - I_i$ $(i = 1, 2, \cdots, N-1)$. 空穴激发相当于将第 k_{j+1} 个粒子移到费米面上, 在 $j+1$ 的位置上留下一个空穴.

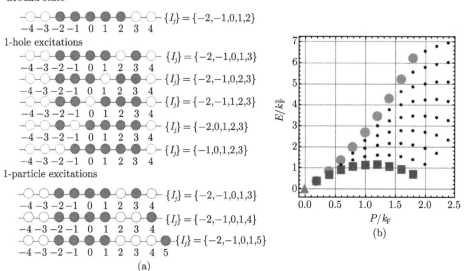

图 2.3 (a) 5 个粒子的基态, 1 个空穴激发和 1 个粒子激发的量子数的取值, 见方程 (2.2.29). 在基态的准动量分布中没有空穴; 空穴激发, 相当于 6 个粒子的准动量分布中, 在不同位置上挖掉一个粒子的动量; 粒子激发, 表示一个粒子从费米点附近激发到高准动量的态. (b) 粒子激发谱 (圆), 空穴激发谱 (正方块), 1 个粒子及 1 个空穴的激发 (圆点), 基态能量 (三角). 此图取自文献 [9]

在空穴激发情况下, 我们假设在 j 处存在一个空穴 $k_j = q$, 且 $0 < q < Q$. 分析可得

$$\begin{cases} k'_i = k_i + (1/L)w_i, & i \leqslant j, \\ k'_i = k_{i+1} + (1/L)w_i, & i > j, \end{cases} \quad (2.6.11)$$

这里, $\{k_i\}$ 为 $N+1$ 个粒子的基态; k'_i 为存在空穴激发时体系粒子的准动量值. 引入分布修正函数 $J(k) = \rho(k)\omega(k)$, 则

$$J(k) = \frac{1}{2\pi}\int_{-Q}^{Q} \frac{2c}{c^2 + (k-k')^2} J(k')\mathrm{d}k' + \frac{1}{2} + \frac{1}{\pi}\arctan\frac{k-q}{c}, \quad (2.6.12)$$

于是, 能量及动量分别为 [5]

$$\varepsilon_2(k) = \mu - q^2 + 2\int_{-Q}^{Q} kJ(k)\mathrm{d}k, \quad (2.6.13)$$

$$p = -q + \int_{-Q}^{Q} J(k)\mathrm{d}k. \quad (2.6.14)$$

2.7 Yang-Yang 热力学方法

在 1969 年，杨振宁与杨振平先生从 BA 方程的解出发推导出一维玻色气体的热力学，他们第一次利用巨正则系综描述了量子多体系统的热力学. 他们的工作开创了统计物理的新领域. 杨振宁先生在他的论文集 [10] 的后记中提到"何时和如何做下一个跳跃，是研究中一个经常出现的重要问题. 这个过程中所涉及的心理因素，如驱动力、气质、品味和自信心都十分重要，也许同样重要的是研究技巧". 这篇文章将玻尔兹曼统计概率的思想应用到极致 [3].

根据前面的分析我们知道，Lieb-Liniger 玻色气体的准动量 k_j 可由一组量子数 I_j 刻画，这组量子数 I_j 满足：

$$I_i = \begin{cases} 整数, & N = 奇数, \\ 半整数, & N = 偶数. \end{cases}$$

下面我们考虑热力学极限 $N, L \to \infty$，$n = N/L$ 为常数的情况. 当系统处于基态时，粒子会占据能量最低的能态. 相应的量子数 I_j/L 标志在 $[-n/2, n/2]$ 均匀分布的格点，准动量 k_j 在最小值与最大值之间非均匀分布.

系统激发态的能量比基态高，此时会有一些粒子占据其他的态，相应的量子数 I_j/L 排布在 $[-n/2, n/2]$ 之外. 因此，在 $[-n/2, n/2]$ 存在未被占据的态，即空穴.

我们定义一个总的粒子占有数函数：

$$h(p) = p + \frac{2}{L} \sum_{j=1}^{N} \arctan\left(\frac{p - k_j}{c}\right), \tag{2.7.1}$$

这是一个连续的单调递增函数. 在热力学极限下，原来不连续的准动量趋于连续分布，我们定义粒子数密度为 ρ，空穴密度为 ρ^h，则在区间 $[k, k + \mathrm{d}k]$ 内的粒子数和空穴数目分别为 $L\rho(k)\mathrm{d}k$ 和 $L\rho^h(k)\mathrm{d}k$，由此可得到关系：

$$L\mathrm{d}h(k) = 2\pi L[\rho(k) + \rho^h(k)]\mathrm{d}k. \tag{2.7.2}$$

热力学极限下，式 (2.7.1) 可写成积分的形式：

$$h(k) = k + \int_{-\infty}^{+\infty} 2\arctan\left(\frac{k - k'}{c}\right)\rho(k')\mathrm{d}k'. \tag{2.7.3}$$

由式 (2.7.2) 和式 (2.7.3) 可得

$$\rho(k) + \rho^h(k) = \frac{1}{2\pi} + \frac{c}{\pi}\int_{-\infty}^{+\infty} \frac{\rho(k')\mathrm{d}k}{c^2 + (k - k')^2}, \tag{2.7.4}$$

2.7 Yang-Yang 热力学方法

单位长度的能量和密度为

$$\frac{E}{L} = \int_{-\infty}^{+\infty} \rho(k)k^2 \mathrm{d}k, \qquad (2.7.5)$$

$$n = \int_{-\infty}^{+\infty} \rho(k)\mathrm{d}k. \qquad (2.7.6)$$

对于每一个区间 $\mathrm{d}k$, 由于能态的简并, 当能量相同时, 不同的粒子和空穴排布会给出不同的微观状态, 因此, 可以将 $\mathrm{d}k$ 区间内可能的微观态表示为

$$\mathrm{d}W = \frac{[L(\rho + \rho^h)\mathrm{d}k]!}{[L\rho \mathrm{d}k]![L\rho^h \mathrm{d}k]!}. \qquad (2.7.7)$$

对上式取对数, 并运用斯特林公式 ($\ln m! = m\ln m - m$) (取玻尔兹曼常数为 1), 由熵的可加性, 体系的熵 $S = \int \ln \mathrm{d}W$, 所以

$$s = \frac{S}{L} = \int_{-\infty}^{+\infty} [(\rho + \rho^h)\ln(\rho + \rho^h) - \rho\ln\rho - \rho^h\ln\rho^h]\mathrm{d}k \qquad (2.7.8)$$

在巨正则系综中, 单位长度的吉布斯自由能为

$$\frac{G}{L} = \frac{E}{L} - \mu n - Ts. \qquad (2.7.9)$$

当系统处于平衡态时, 满足 $\delta G = 0$, 于是

$$\frac{\delta G}{L} = \frac{\delta E}{L} - \mu \delta n - T\delta s = 0, \qquad (2.7.10)$$

$$\frac{\delta E}{L} = \int_{-\infty}^{+\infty} k^2 \delta\rho(k)\mathrm{d}k, \qquad (2.7.11)$$

$$\delta n = \int_{-\infty}^{+\infty} \delta\rho(k)\mathrm{d}k. \qquad (2.7.12)$$

我们对单位长度的熵做变分:

$$\delta s = \int_{-\infty}^{+\infty} \left[(\delta\rho + \delta\rho^h)\ln(\rho + \rho^h) + (\rho + \rho^h)\frac{\delta\rho + \delta\rho^h}{\rho + \rho^h} \right.$$
$$\left. -\delta\rho\ln\rho - \rho\frac{\delta\rho}{\rho} - \delta\rho^h\ln\rho^h - \rho^h\frac{\delta\rho^h}{\rho^h} \right]\mathrm{d}k$$
$$= \int_{-\infty}^{+\infty} \left[(\delta\rho + \delta\rho^h)\ln\left(1 + \frac{\rho}{\rho^h}\right) - \delta\rho\ln\frac{\rho}{\rho^h} \right]\mathrm{d}k,$$

由式 (2.7.4) 有

$$\delta\rho + \delta\rho^h = \frac{c}{\pi}\int_{-\infty}^{+\infty} \frac{\delta\rho(k')\mathrm{d}k'}{c^2 + (k-k')^2}, \qquad (2.7.13)$$

由 $\delta G = 0$, 得到

$$
\begin{aligned}
0 &= \frac{\delta G}{L} \\
&= \int_{-\infty}^{+\infty} k^2 \delta\rho(k)\mathrm{d}k - \int_{-\infty}^{+\infty} \mu\delta\rho(k)\mathrm{d}k - T\int_{-\infty}^{+\infty} \frac{c}{\pi}\int_{-\infty}^{+\infty} \frac{\mathrm{d}k'}{c^2+(k-k')^2} \\
&\quad \times \ln\left(1+\frac{\rho}{\rho^h}\right)\delta\rho \mathrm{d}k + T\int_{-\infty}^{+\infty} \ln\frac{\rho}{\rho^h}\delta\rho \mathrm{d}k \\
&= \int_{-\infty}^{+\infty} \left[k^2 - \mu + T\ln\left(\frac{\rho}{\rho^h}\right) - \frac{cT}{\pi}\int_{-\infty}^{+\infty} \frac{\mathrm{d}k'}{c^2+(k-k')^2}\ln\left(1+\frac{\rho}{\rho^h}\right)\right]\delta\rho \mathrm{d}k.
\end{aligned}
\tag{2.7.14}
$$

对于任意 $\delta\rho$, 上式成立需满足条件:

$$
k^2 - \mu + T\ln\frac{\rho}{\rho^h} - \frac{Tc}{\pi}\int_{-\infty}^{+\infty} \frac{\mathrm{d}k'}{c^2+(k-k')^2}\ln\left(1+\frac{\rho}{\rho^h}\right) = 0. \tag{2.7.15}
$$

定义系统的缀饰能 (dressed energy)$\varepsilon(k)$ 为

$$
\ln\frac{\rho^h}{\rho} = \frac{\varepsilon(k)}{T}. \tag{2.7.16}
$$

缀饰能 $\varepsilon(k)$ 刻画激发态的能量. 最终我们得到 Yang-Yang 方程, 即 Lieb-Liniger 玻色气体满足的 TBA 方程:

$$
\varepsilon(k) = k^2 - \mu - \frac{Tc}{\pi}\int_{-\infty}^{+\infty} \frac{\mathrm{d}k'}{c^2+(k-k')^2}\ln\left(1+\mathrm{e}^{-\frac{\varepsilon(k')}{T}}\right). \tag{2.7.17}
$$

相应的压强为

$$
\begin{aligned}
p &= -\left(\frac{\partial G}{\partial L}\right)_{T,\mu} \\
&= \int_{-\infty}^{+\infty} \{(-k^2+\mu)\rho(k) + T[(\rho(k)+\rho^h(k))\ln(\rho(k)+\rho^h(k)) - \rho\ln\rho - \rho^h\ln\rho^h]\}\mathrm{d}k \\
&= T\int_{-\infty}^{+\infty} \left\{\left[\rho(k)+\rho^h(k) - \frac{c}{\pi}\int \frac{\rho(k')\mathrm{d}k'}{c^2+(k-k')^2}\right] \times \ln\left[1+\frac{\rho(k)}{\rho^h(k)}\right]\right\}\mathrm{d}k \\
&= \frac{T}{2\pi}\int_{-\infty}^{+\infty} \ln\left(1+\mathrm{e}^{-\frac{\varepsilon(k)}{T}}\right)\mathrm{d}k,
\end{aligned}
$$

即

$$
p = \frac{T}{2\pi}\int_{-\infty}^{+\infty} \ln\left(1+\mathrm{e}^{-\frac{\varepsilon(k)}{T}}\right)\mathrm{d}k. \tag{2.7.18}
$$

根据压强便可导出体系的其他热力学量, 如粒子数密度、熵、压缩率、比热等.

$$
n = \frac{\partial p}{\partial \mu}\bigg|_{c,T}, \quad s = \frac{\partial p}{\partial T}\bigg|_{c,\mu}, \quad \kappa = \frac{\partial n}{\partial \mu}\bigg|_{c,T}, \quad \frac{C_v}{T} = \frac{\partial^2 p}{\partial T^2}\bigg|_{c,\mu}. \tag{2.7.19}
$$

2.8 Lieb-Liniger 模型中的量子统计

目前为止, 我们已经推导出系统满足的 Yang-Yang 方程 (2.7.17) [3], 得到了缀饰能 (dressed energy) 的表达式. 数学上没有求解非线性积分方程的标准方法. 我们一般利用数值的方法求解整个温度范围的 TBA 方程. 此外, 在特定条件下, 我们能够解析地求解 TBA 方程, 并由此得到普适的低温物理性质、临界现象、量子统计及量子关联特性.

我们分别在零温、高温、强相互作用、弱相互作用的条件下讨论 TBA 方程 (2.7.17) 的解.

1. 弱相互作用极限 $(c \to 0)$—— 玻色统计

在弱相互作用极限下, 系统的粒子可以看成自由的玻色子. 现在我们考虑 $c \to 0$ 的极限情况, 则

$$\lim_{c \to 0} \frac{c/\pi}{c^2 + (k-q)^2} = \lim_{c \to 0} \frac{1}{\pi} \frac{1}{c + \frac{(k-q)^2}{c}} = \delta(k-q). \qquad (2.8.1)$$

因此, TBA 方程可写为

$$\varepsilon(k) = k^2 - \mu - T \ln\left[1 + e^{-\frac{\varepsilon(k)}{T}}\right]. \qquad (2.8.2)$$

利用式 (2.7.4) 和式 (2.8.1),

$$\rho(k) + \rho^h(k) = \frac{1}{2\pi} + \frac{c}{\pi} \int_{-\infty}^{+\infty} \frac{\rho(k')\mathrm{d}k'}{c^2 + (k-k')^2} = \frac{1}{2\pi} + \rho(k), \qquad (2.8.3)$$

可解得 $\rho^h(k) = 1/2\pi$. 又由定义式 (2.7.16), 我们得到 $\rho^h/\rho = e^{\varepsilon(k)/T}$ 及 $2\pi\rho(k) = e^{-\varepsilon(k)/T}$. 由 TBA 方程可解出:

$$e^{-\varepsilon(k)/T} = \frac{1}{e^{(k^2-\mu)/T} - 1}.$$

于是

$$\rho(k) = e^{-\varepsilon(k)/T}/2\pi = \frac{1}{2\pi[e^{(k^2-\mu)/T} - 1]}. \qquad (2.8.4)$$

由此, 我们利用分部积分, 得到压强:

$$p = \frac{T}{2\pi} \int_{-\infty}^{+\infty} \ln\left(1 + e^{-\frac{\varepsilon(k)}{T}}\right) \mathrm{d}k$$

$$= \frac{T}{2\pi} \int_{-\infty}^{+\infty} k \frac{e^{-(k^2-\mu)/T}\left(-\frac{2k}{T}\right)}{1 - e^{-(k^2-\mu)/T}} \mathrm{d}k$$

$$= \frac{1}{\pi} \int_{-\infty}^{+\infty} \frac{\sqrt{\varepsilon} d\varepsilon}{e^{(\varepsilon-\mu)/T} - 1}.$$

所以，在弱相互作用极限下，压强正是自由玻色气体状态方程的生成函数：

$$p = \frac{1}{\pi} \int_{-\infty}^{+\infty} \frac{\sqrt{\varepsilon} d\varepsilon}{e^{(\varepsilon-\mu)/T} - 1}. \tag{2.8.5}$$

由 Polylog 函数的定义 $Li_s(z) = \frac{1}{\Gamma(s)} \int_0^{+\infty} \frac{t^{s-1}}{e^t/z - 1} dt$，我们可以得到压强

$$p = \frac{1}{2\sqrt{\pi}} T^{\frac{3}{2}} Li_{\frac{3}{2}}(e^{\mu/T}). \tag{2.8.6}$$

2. 高温 —— 玻尔兹曼统计

对 TBA 方程 (2.7.17) 做维里展开可得

$$\begin{aligned} e^{-\frac{\varepsilon(k)}{T}} &= \mathcal{Z} e^{-\frac{k^2}{T}} e^{\int_{-\infty}^{+\infty} dq a_2(k-q) \ln\left(1+\mathcal{Z}e^{-\frac{q^2}{T}}\right)} \\ &= \mathcal{Z} e^{-\frac{k^2}{T}} \left[1 + \mathcal{Z} \int_{-\infty}^{+\infty} dq a_2(k-q) e^{-\frac{q^2}{T}} \right], \end{aligned}$$

其中，$\mathcal{Z} = e^{\mu/T}$ 是逸度，$a_2(x) = c/\pi/(c^2 + x^2)$，利用此表达式，我们得到高温条件下的压强：

$$\begin{aligned} p &= \frac{T}{2\pi} \int_{-\infty}^{+\infty} dk \ln\left[1 + \mathcal{Z} e^{-\frac{k^2}{T}} + \mathcal{Z}^2 e^{-\frac{k^2}{T}} \int_{-\infty}^{+\infty} dq a_2(k-q) e^{-\frac{q^2}{T}} \right] \\ &= -\frac{T}{2\pi} \int_{-\infty}^{+\infty} dk \ln\left(1 - \mathcal{Z} e^{-\frac{k^2}{T}} \right) + \frac{T}{2\pi} \int_{-\infty}^{+\infty} dk \\ &\quad \times \ln\left\{ \left[1 + \mathcal{Z} e^{-\frac{k^2}{T}} + \mathcal{Z}^2 e^{-\frac{k^2}{T}} \int_{-\infty}^{+\infty} dq a_2(k-q) e^{-\frac{q^2}{T}} \right] \times \left(1 - \mathcal{Z} e^{-\frac{k^2}{T}} \right) \right\} \\ &= p_0 + \frac{T}{2\pi} \int_{-\infty}^{+\infty} dk \ln\left[1 + \mathcal{Z}^2 e^{-\frac{k^2}{T}} \int_{-\infty}^{+\infty} dq a_2(k-q) e^{-\frac{q^2}{T}} - \mathcal{Z}^2 e^{-\frac{2k^2}{T}} \right], \end{aligned}$$

这里，$\int_{-\infty}^{+\infty} e^{-\frac{k^2}{T}} dk = \sqrt{\pi T}$。我们令 $k = k' + q', q = k' - q'$，则有 $dk dq = 2 dk' dq'$。于是

$$\begin{aligned} p &= p_0 + \frac{2T}{2\pi} \int_{-\infty}^{+\infty} dk' \mathcal{Z}^2 e^{-\frac{(k'+q')^2}{T}} \int_{-\infty}^{+\infty} dq' a_2(2q') e^{-\frac{(k'-q')^2}{T}} - \frac{T}{2\pi} \int_{-\infty}^{+\infty} \mathcal{Z}^2 e^{-\frac{2k^2}{T}} dk \\ &= p_0 + \frac{T}{\pi} \sqrt{\frac{T\pi}{2}} \mathcal{Z}^2 \int_{-\infty}^{+\infty} dq' a_2(2q') e^{-\frac{2q'^2}{T}} - \frac{T}{2\pi} \sqrt{\frac{T\pi}{2}} \mathcal{Z}^2, \end{aligned}$$

即

$$p = p_0 + \frac{T^{\frac{3}{2}}}{\sqrt{2\pi}} \mathcal{Z}^2 p_2, \tag{2.8.7}$$

2.8 Lieb-Liniger 模型中的量子统计

其中

$$p_0 = -\frac{T}{2\pi} \int_{-\infty}^{+\infty} dk \ln\left(1 - \mathcal{Z}e^{-\frac{k^2}{T}}\right), \tag{2.8.8}$$

$$p_2 = -\frac{1}{2} + \int_{-\infty}^{+\infty} dq' a_2(2q') e^{-\frac{2q'^2}{T}}, \tag{2.8.9}$$

分别表示自由玻色子的压强及两体相互作用项. 高温下, 我们可以直接得到玻尔兹曼统计.

3. 温度趋于零的情况 ($T \to 0$)

在温度趋于零的情况下, 我们可以按 $\frac{1}{c}$ 做展开 [11]:

$$\varepsilon(k) = k^2 - \mu - \frac{Tc}{\pi} \int \frac{dq}{c^2 + (k-q)^2} \left[-\frac{\varepsilon(q)}{T}\right] \tag{2.8.10}$$

$$= k^2 - \mu + \frac{c}{\pi} \int_{-Q}^{Q} \frac{\varepsilon(q) dq}{c^2 + (k-q)^2}. \tag{2.8.11}$$

其中, $\pm Q$ 为 $\mathcal{E}(\pm Q) = 0$ 的动量. 考虑到 $c^2 + k^2 \gg -2kq + q^2$, 我们对上式中的积分核做泰勒展开, 如:

$$\frac{c}{\pi} \frac{1}{c^2 + (k-q)^2} = \frac{c}{\pi} \frac{1}{c^2 + k^2 + q^2 - 2kq}$$

$$\approx \frac{c}{\pi} \frac{1}{c^2 + k^2} \left[1 - \frac{q^2 - 2kq}{c^2 + k^2} + \left(\frac{q^2 - 2kq}{c^2 + k^2}\right)^2 + O\left(\frac{1}{c^6}\right)\right].$$

于是

$$\varepsilon(k) = k^2 - \mu + \frac{c}{\pi} \frac{1}{c^2 + k^2} \int_{-Q}^{Q} dq\, \varepsilon(q)$$

$$- \frac{c}{\pi} \int_{-Q}^{Q} dq \frac{\varepsilon(q)}{(c^2 + k^2)^2} (-2kq + q^2)$$

$$+ \frac{c}{\pi} \int_{-Q}^{Q} dq \frac{\varepsilon(q)}{(c^2 + k^2)^3} (-2kq + q^2)^2 + O\left(\frac{1}{c^6}\right). \tag{2.8.12}$$

注意到 $\varepsilon(k) = \varepsilon(-k)$ 及 $\varepsilon(\pm Q) = 0$, 我们可以得到

$$\varepsilon(k) \approx \varepsilon_0 - \mu - \frac{2pc}{c^2 + k^2} + \frac{4\mu^{\frac{5}{2}}}{15\pi|c|^3}, \tag{2.8.13}$$

其中 $\varepsilon_0 = k^2$. 可以直接计算得到基态的能量:

$$E \approx \frac{1}{3} n^3 \pi^2 \left(1 - \frac{4}{\gamma} + \frac{12}{\gamma^2} - \frac{32}{\gamma^3} + \frac{32\pi^2}{15\gamma^3}\right), \tag{2.8.14}$$

这与前面得到的基态能量式 (2.5.6) 是一致的.

4. 强相互作用极限和非零温 ($c \to +\infty$ 和 $T \ll E_F$) —— 费米统计

在 $c \to \infty$ 时，Yang - Yang 方程可近似展开为

$$\varepsilon(k) = \varepsilon_0 - \mu - \frac{Tc}{\pi} \int \frac{1}{c^2 + k^2 + q^2 - 2kq} dq \ln\left[1 + e^{-\frac{\varepsilon(q)}{T}}\right]$$

$$\approx \varepsilon_0 - \mu - \frac{2c}{c^2 + k^2} \frac{T}{2\pi} \int \ln\left[1 + e^{-\frac{\varepsilon(q)}{T}}\right] dq + \frac{Tc}{\pi} \int \frac{q^2}{(c^2+k^2)^2}$$

$$\times \ln\left[1 + e^{-\frac{\varepsilon(q)}{T}}\right] dq.$$

其中

$$\frac{Tc}{\pi} \int \frac{q^2}{(c^2+k^2)^2} \ln\left[1 + e^{-\frac{\varepsilon(q)}{T}}\right] dq \approx \frac{2q^2 c}{(c^2+k^2)^2} \frac{T}{2\pi} \int \ln\left[1 + e^{-\frac{\varepsilon(q)}{T}}\right] dq$$

$$\approx \frac{2q^2 p}{c^3} = \frac{2p}{c^3} \varepsilon_0.$$

利用分部积分，我们得到

$$\varepsilon(k) \approx \varepsilon_0 - \mu - \frac{2c}{c^2+k^2} p - \frac{T}{2\pi} \frac{2c}{(c^2+k^2)^2} \frac{1}{3} \int_{-\infty}^{+\infty} q^3 \frac{1}{1 + e^{\frac{\varepsilon(q)}{T}}} \left(-\frac{1}{T}\right)$$

$$\times \varepsilon'(q) dq$$

$$\approx \varepsilon_0 - \mu - \frac{2cp}{c^2+k^2} + \frac{1}{2\pi} \frac{1}{c^3} \frac{2}{3} \int_{-\infty}^{+\infty} \frac{q^3 \cdot 2q dq}{1 + e^{\frac{\varepsilon(q)}{T}}}.$$

做换元：$k = \sqrt{\varepsilon_0}, dk = \frac{1}{2} \frac{d\varepsilon_0}{\sqrt{\varepsilon_0}}$,

$$\varepsilon(k) = \varepsilon_0(k) - \mu - \frac{2cp}{c^2+k^2} + \frac{4}{3\pi c^3} \int_0^{+\infty} \frac{\varepsilon_0^{\frac{3}{2}} d\varepsilon_0}{2\left[1 + e^{\frac{\varepsilon_0 - \mu - \frac{2cp}{c^2+k^2} + o\left(\frac{1}{c^3}\right)}{T}}\right]}$$

$$= \varepsilon_0(k) - \mu - \frac{2cp}{c^2+k^2} + \frac{2}{3\pi c^3} \int_0^{+\infty} \frac{\varepsilon_0^{\frac{3}{2}} d\varepsilon_0}{1 + e^{\frac{\varepsilon_0 - A_0}{T}}}, \quad (2.8.15)$$

其中，$A_0 = \mu + \frac{2p}{c} - \frac{4\mu^{\frac{5}{2}}}{15\pi|c|^3}$，又因为 $\Gamma\left(n + \frac{1}{2}\right) = \frac{\sqrt{\pi}}{2^n}(2n-1)$，由 Polylog 函数的定义可得

$$\varepsilon(k) \approx \varepsilon_0(k) - \mu - \frac{2cp}{c^2+k^2} - \frac{1}{2\sqrt{\pi} c^3} T^{\frac{5}{2}} Li_{\frac{5}{2}}\left(-e^{\frac{A_0}{T}}\right). \quad (2.8.16)$$

再计算压强

$$p = \frac{T}{2\pi} \int_{-\infty}^{+\infty} dk \ln\left[1 + e^{-\frac{\varepsilon(k)}{T}}\right]$$

2.8 Lieb-Liniger 模型中的量子统计

$$= -\frac{T}{2\pi} \int_{-\infty}^{+\infty} k \frac{e^{-\frac{\varepsilon(k)}{T}} \cdot \left(-\frac{1}{T}\right) \varepsilon'(k) dk}{1 + e^{-\frac{\varepsilon(k)}{T}}}, \tag{2.8.17}$$

其中, $\varepsilon'(k)$ 为 k 的偏导数, 得

$$\varepsilon'(k) = 2k + \frac{2cp \cdot 2k}{(c^2 + k^2)^2} = 2k\left(1 + \frac{2p}{c^3}\right), \tag{2.8.18}$$

代入上面压强公式可得

$$\begin{aligned}
p &= \frac{2}{2\pi} \int_0^{+\infty} \frac{k\varepsilon'(k) dk}{1 + e^{\frac{\varepsilon(k)}{T}}} \\
&= \frac{1}{\pi} \int_0^{+\infty} \frac{2\varepsilon_0 \left(1 + \frac{2p}{c^3}\right)\left(\frac{1}{2}(\varepsilon_0)^{-\frac{1}{2}} d\varepsilon_0\right)}{1 + e^{\frac{\varepsilon(k)}{T}}} \\
&= \frac{1}{\pi} \int_0^{+\infty} \frac{\sqrt{\varepsilon_0}\left(1 + \frac{2p}{c^3}\right) d\varepsilon_0}{\left[1 + e^{\frac{\varepsilon_0\left(1 + \frac{2p}{c^3}\right) - A}{T}}\right]},
\end{aligned} \tag{2.8.19}$$

其中, $A = \mu + \frac{2p}{c} + \frac{1}{2\sqrt{\pi}c^3} T^{\frac{5}{2}} Li_{\frac{5}{2}}\left(-e^{\frac{A_0}{T}}\right)$, $A_0 = \mu + \frac{2p}{c} - \frac{4\mu^{\frac{5}{2}}}{15\pi|c|^3}$. 基于这些表达式, 我们最终可得

$$\begin{aligned}
p &= \frac{1}{\sqrt{\pi^2}} \int_0^{+\infty} \frac{\sqrt{\varepsilon_0'} d\varepsilon_0'}{\left(1 + e^{\frac{\varepsilon_0' - A}{T}}\right) \left(1 + \frac{2p}{c^3}\right)^{\frac{1}{2}}} \\
&= -\frac{1}{2} \frac{1}{\sqrt{\pi^2}} \cdot T^{\frac{3}{2}} Li_{\frac{3}{2}}\left(-e^{\frac{A}{T}}\right) \frac{1}{\left(1 + \frac{2p}{c^3}\right)^{\frac{1}{2}}} \\
&= -\frac{1}{2}\sqrt{\frac{1}{\pi}} T^{\frac{3}{2}} Li_{\frac{3}{2}}\left(-e^{\frac{A}{T}}\right) \left[1 + \frac{1}{2c^3\sqrt{\pi}} T^{\frac{3}{2}} Li_{\frac{3}{2}}\left(-e^{\frac{A}{T}}\right)\right].
\end{aligned} \tag{2.8.20}$$

其中, $\varepsilon_0' = \varepsilon_0\left(1 + \frac{2p}{c^3}\right)$. 最后我们可以写出压强的解析结果:

$$\begin{cases}
p = -\frac{1}{2}\sqrt{\frac{1}{\pi}} T^{\frac{3}{2}} Li_{\frac{3}{2}}\left(-e^{\frac{A}{T}}\right) \left[1 + \frac{1}{2c^3\sqrt{\pi}} T^{\frac{3}{2}} Li_{\frac{3}{2}}\left(-e^{\frac{A}{T}}\right)\right] \\
A = \mu + \frac{2p}{c} + \frac{1}{2\sqrt{\pi}c^3} T^{\frac{5}{2}} Li_{\frac{5}{2}}\left(-e^{\frac{A}{T}}\right) \\
A_0 = \mu + \frac{2p}{c} - \frac{4\mu^{\frac{5}{2}}}{15\pi|c|^3}
\end{cases} \tag{2.8.21}$$

这个结果极其重要,给出低温情况下一维玻色气体的状态方程. 我们在后面将利用此表达式分析量子临界现象,并发现压强的首项恰好是自由费米气体的状态方程. 另外,在强相互作用的限制下,由 Bethe Ansatz 方程 (2.7.4), 我们可以直接看到费米统计:

$$\rho(k) + \rho^h(k) = \frac{1}{2\pi}, \tag{2.8.22}$$

密度分布为

$$2\pi\rho(x) = \frac{1}{1+e^{\frac{\varepsilon(k)}{T}}} = \frac{1}{1+e^{\frac{k^2-\mu}{T}}}. \tag{2.8.23}$$

这就是费米-狄拉克统计!

总结:

玻尔兹曼统计 $\quad p = \frac{1}{2}\sqrt{\frac{1}{\pi}}T^{\frac{3}{2}}e^{\frac{\mu}{T}}, \quad T \to \infty,$

费米统计 $\quad p = -\frac{1}{2}\sqrt{\frac{1}{\pi}}T^{\frac{3}{2}}Li_{\frac{3}{2}}\left(-e^{\frac{A}{T}}\right), \quad c \to \infty,$

玻色统计 $\quad p = \frac{1}{2}\sqrt{\frac{1}{\pi}}T^{\frac{3}{2}}Li_{\frac{3}{2}}\left(-e^{\frac{\mu}{T}}\right), \quad c \to 0.$

2.9 普适的热力学行为

考虑非零温强相互作用极限, 利用下列方程:

$$p = -\frac{1}{2}\sqrt{\frac{1}{\pi}}T^{\frac{3}{2}}Li_{\frac{3}{2}}\left(-e^{\frac{A}{T}}\right)\left[1 + \frac{1}{2c^3\sqrt{\pi}}T^{\frac{3}{2}}Li_{\frac{3}{2}}\left(-e^{\frac{A}{T}}\right)\right] \tag{2.9.1}$$

以及热力学关系求出体系粒子数密度、压缩率和比热. 其中 Polylog 函数的定义为 $Li_s(z)\frac{1}{\Gamma(s)}\int_0^{+\infty}\frac{t^{s-1}}{e^t/z-1}dt.$

1. 粒子数密度

$$n = \frac{\partial p}{\partial \mu}\bigg|_{c,T}$$
$$= -\frac{1}{2}\sqrt{\frac{1}{\pi}}T^{\frac{1}{2}}Li_{\frac{1}{2}}\left(-e^{\frac{A}{T}}\right)\left[1 + \frac{1}{c^3\sqrt{\pi}}T^{\frac{3}{2}}Li_{\frac{3}{2}}\left(-e^{\frac{A}{T}}\right)\right]\frac{\partial A}{\partial \mu},$$

其中

$$A = \mu + \frac{2p}{c} + \frac{1}{2\sqrt{\pi}c^3}T^{\frac{5}{2}}Li_{\frac{5}{2}}\left(-e^{\frac{A}{T}}\right), \tag{2.9.2}$$

对其求导数
$$\frac{\partial A}{\partial \mu}=1+\frac{2}{c}\frac{\partial p}{\partial \mu}+\frac{1}{2\sqrt{\pi}c^3}T^{\frac{3}{2}}Li_{\frac{3}{2}}\left(-e^{\frac{A_0}{T}}\right),$$

由此可以得
$$\left.\frac{\partial p}{\partial \mu}\right|_{c,T}=-\frac{1}{2}\sqrt{\frac{1}{\pi}}T^{\frac{1}{2}}Li_{\frac{1}{2}}\left(-e^{\frac{A}{T}}\right)\left[1+\frac{1}{c^3\sqrt{\pi}}T^{\frac{3}{2}}Li_{\frac{3}{2}}\left(-e^{\frac{A}{T}}\right)\right]$$
$$\times\left[1+\frac{2}{c}\frac{\partial p}{\partial \mu}+\frac{1}{2\sqrt{\pi}c^3}T^{\frac{3}{2}}Li_{\frac{3}{2}}\left(-e^{\frac{A_0}{T}}\right)\right].$$

通过迭代方法, 计算得
$$\frac{\partial p}{\partial \mu}=\frac{-\frac{1}{2}\sqrt{\frac{1}{\pi}}T^{\frac{1}{2}}Li_{\frac{1}{2}}\left(-e^{\frac{A}{T}}\right)\left[1+\frac{1}{c^3\sqrt{\pi}}T^{\frac{3}{2}}Li_{\frac{3}{2}}\left(-e^{\frac{A}{T}}\right)\right]}{1+\frac{1}{c}\sqrt{\frac{1}{\pi}}T^{\frac{1}{2}}Li_{\frac{1}{2}}\left(-e^{\frac{A}{T}}\right)}\left[1+\frac{1}{2\sqrt{\pi}c^3}T^{\frac{3}{2}}Li_{\frac{3}{2}}\left(-e^{\frac{A}{T}}\right)\right]$$
$$\approx-\frac{1}{2}\sqrt{\frac{1}{\pi}}T^{\frac{1}{2}}Li_{\frac{1}{2}}\left(-e^{\frac{A}{T}}\right)\left[1+\frac{1}{c^3\sqrt{\pi}}T^{\frac{3}{2}}Li_{\frac{3}{2}}\left(-e^{\frac{A}{T}}\right)\right]\left[1+\frac{1}{2\sqrt{\pi}c^3}T^{\frac{3}{2}}Li_{\frac{3}{2}}\left(-e^{\frac{A}{T}}\right)\right]$$
$$\times\left[1-\frac{1}{c\sqrt{\pi}}T^{\frac{1}{2}}Li_{\frac{1}{2}}\left(-e^{\frac{A}{T}}\right)+\frac{1}{c^2\pi}TLi_{\frac{1}{2}}\left(-e^{\frac{A}{T}}\right)^2-\frac{1}{c^3\pi^{\frac{3}{2}}}T^{\frac{3}{2}}Li_{\frac{1}{2}}\left(-e^{\frac{A}{T}}\right)^3\right],$$

经整理可得
$$\frac{\partial p}{\partial \mu}=-\frac{1}{2\sqrt{\pi}}T^{\frac{1}{2}}Li_{\frac{1}{2}}\left(-e^{\frac{A}{T}}\right)\left[1-\frac{1}{c\sqrt{\pi}}T^{\frac{1}{2}}Li_{\frac{1}{2}}\left(-e^{\frac{A}{T}}\right)+\frac{T}{\pi c^2}Li_{\frac{1}{2}}\left(-e^{\frac{A}{T}}\right)^2\right.$$
$$\left.+\frac{3}{2\sqrt{\pi}c^3}T^{\frac{3}{2}}Li_{\frac{3}{2}}\left(-e^{\frac{A}{T}}\right)-\frac{T^{\frac{3}{2}}}{c^3\pi^{\frac{3}{2}}}Li_{\frac{1}{2}}\left(-e^{\frac{A}{T}}\right)^3\right],$$

其中, $f_n(x)=Li_n(-e^{x/T}), Li_n(x)=\sum_{k=1}^{\infty}\frac{x^k}{k^n}.$

最后得到密度的具体表达式:
$$n=-\frac{1}{2\sqrt{\pi}}T^{\frac{1}{2}}f_{\frac{1}{2}}\left[1-\frac{1}{c\sqrt{\pi}}T^{\frac{1}{2}}f_{\frac{1}{2}}+\frac{T}{\pi c^2}f_{\frac{1}{2}}^2+\frac{1}{\sqrt{\pi}c^3}T^{\frac{3}{2}}\left(\frac{3}{2}f_{\frac{3}{2}}-\frac{1}{\pi}f_{\frac{1}{2}}^3\right)\right]. \quad (2.9.3)$$

2. 压缩率 κ

$$\kappa=\left.\frac{\partial n}{\partial \mu}\right|_{c,T}=-\frac{1}{2\sqrt{\pi}}T^{\frac{1}{2}}f_{-\frac{1}{2}}\cdot\frac{1}{T}\frac{\partial A}{\partial \mu}\left[1-\frac{1}{c\sqrt{\pi}}T^{\frac{1}{2}}f_{\frac{1}{2}}+\frac{T}{\pi c^2}f_{\frac{1}{2}}^2+\frac{3}{2\sqrt{\pi}c^3}T^{\frac{3}{2}}f_{\frac{3}{2}}\right.$$
$$\left.-\frac{T^{\frac{3}{2}}}{c^3\pi^{\frac{3}{2}}}f_{\frac{1}{2}}^3\right]-\frac{1}{2\sqrt{\pi}}T^{\frac{1}{2}}f_{\frac{1}{2}}\times\left[-\frac{1}{c\sqrt{\pi}}T^{\frac{1}{2}}f_{-\frac{1}{2}}\cdot\frac{1}{T}\frac{\partial A}{\partial \mu}+\frac{T}{\pi c^2}2\cdot f_{\frac{1}{2}}f_{-\frac{1}{2}}\frac{\partial A}{\partial \mu}\right.$$

$$+\frac{3}{2\sqrt{\pi}c^3}T^{\frac{3}{2}}f_{\frac{1}{2}}\cdot\frac{1}{T}\frac{\partial A}{\partial \mu}-\frac{T^{\frac{3}{2}}}{c^3\pi^{\frac{3}{2}}}3\cdot f_{\frac{1}{2}}^2 f_{-\frac{1}{2}}\cdot\frac{1}{T}\frac{\partial A}{\partial \mu}\Bigg],$$

$$\frac{\partial n}{\partial \mu}\Big/\frac{\partial A}{\partial \mu}=-\frac{1}{2\sqrt{\pi}}T^{-\frac{1}{2}}f_{-\frac{1}{2}}+\frac{1}{c\pi}f_{\frac{1}{2}}f_{-\frac{1}{2}}-\frac{3}{2c^2\pi^{\frac{3}{2}}}T^{\frac{1}{2}}f_{\frac{1}{2}}^2 f_{-\frac{1}{2}}+\frac{2}{c^3\pi^2}Tf_{\frac{1}{2}}^3 f_{-\frac{1}{2}}$$
$$-\frac{3}{4\pi c^3}Tf_{\frac{1}{2}}^2-\frac{3}{4\pi c^3}Tf_{\frac{3}{2}}f_{-\frac{1}{2}},$$

对函数 $A(\mu,T)$ 的导数取近似：

$$\frac{\partial A}{\partial \mu}\approx 1-\frac{1}{c\sqrt{\pi}}T^{\frac{1}{2}}f_{\frac{1}{2}}+\frac{1}{c^2\pi}Tf_{\frac{1}{2}}^2-\frac{1}{c^3\pi^{\frac{3}{2}}}T^{\frac{3}{2}}f_{\frac{1}{2}}^3-\frac{3}{4\pi c^3}T^2 f_{\frac{1}{2}}f_{\frac{3}{2}}+\frac{1}{2\sqrt{\pi}c^3}T^{\frac{3}{2}}f_{\frac{3}{2}},$$

最终得到压缩率：

$$\kappa=\frac{\partial n}{\partial \mu}\bigg|_{c,T}=-\frac{1}{2\sqrt{\pi}}T^{-\frac{1}{2}}f_{-\frac{1}{2}}+\frac{3}{2c\pi}f_{\frac{1}{2}}f_{-\frac{1}{2}}-\frac{3}{c^2\pi^{\frac{3}{2}}}T^{\frac{1}{2}}f_{\frac{1}{2}}^2 f_{-\frac{1}{2}}+\frac{5}{c^3\pi^2}Tf_{\frac{1}{2}}^3 f_{-\frac{1}{2}}$$
$$-\frac{3}{4\pi c^3}Tf_{\frac{1}{2}}^2-\frac{1}{\pi c^3}Tf_{\frac{3}{2}}f_{-\frac{1}{2}}. \tag{2.9.4}$$

3. 比热 C_v

为了便于计算，我们用前面的标记符号 $f_n(x)$ 重新改写压强：

$$p=-\frac{1}{2\sqrt{\pi}}T^{\frac{3}{2}}f_{\frac{3}{2}}\left(1+\frac{1}{2c^3\sqrt{\pi}}T^{\frac{3}{2}}f_{\frac{3}{2}}\right), \tag{2.9.5}$$

$$\frac{\partial p}{\partial T}=-\frac{1}{2\sqrt{\pi}}\left(T^{\frac{3}{2}}f_{\frac{3}{2}}\right)'\left(1+\frac{1}{2c^3\sqrt{\pi}}T^{\frac{3}{2}}f_{\frac{3}{2}}\right)-\frac{1}{2\sqrt{\pi}}T^{\frac{3}{2}}f_{\frac{3}{2}}\left[\frac{1}{2c^3\sqrt{\pi}}\left(T^{\frac{3}{2}}f_{\frac{3}{2}}\right)'\right],$$

其中

$$\left(T^{\frac{3}{2}}f_{\frac{3}{2}}\right)'=\frac{3}{2}T^{\frac{1}{2}}f_{\frac{3}{2}}+T^{\frac{1}{2}}f_{\frac{1}{2}}\frac{\partial A}{\partial T}-T^{-\frac{1}{2}}f_{\frac{1}{2}}A,$$

$$\frac{\partial p}{\partial T}=-\frac{1}{2\sqrt{\pi}}\left(\frac{3}{2}T^{\frac{1}{2}}f_{\frac{3}{2}}+T^{\frac{1}{2}}f_{\frac{1}{2}}\frac{\partial A}{\partial T}-T^{-\frac{1}{2}}f_{\frac{1}{2}}A\right)\left(1+\frac{1}{2c^3\sqrt{\pi}}T^{\frac{3}{2}}f_{\frac{3}{2}}\right)$$
$$-\frac{1}{2\sqrt{\pi}}T^{\frac{3}{2}}f_{\frac{3}{2}}\frac{1}{2c^3\sqrt{\pi}}\left(\frac{3}{2}T^{\frac{1}{2}}f_{\frac{3}{2}}+T^{\frac{1}{2}}f_{\frac{1}{2}}\frac{\partial A}{\partial T}-T^{-\frac{1}{2}}f_{\frac{1}{2}}A\right),$$

A 对 T 求偏导可得

$$\frac{\partial A}{\partial T}=\left(\frac{2}{c}\frac{\partial p}{\partial T}+\frac{5}{4\sqrt{\pi}c^3}T^{\frac{3}{2}}f_{\frac{5}{2}}-\frac{1}{2\sqrt{\pi}c^3}T^{\frac{1}{2}}f_{\frac{3}{2}}A\right)\Big/\left(1-\frac{1}{2\sqrt{\pi}c^3}T^{\frac{3}{2}}f_{\frac{3}{2}}\right)$$

2.9 普适的热力学行为

$$\Rightarrow \frac{\partial p}{\partial T} = -\frac{3}{4\sqrt{\pi}}T^{\frac{1}{2}}f_{\frac{3}{2}} - \frac{1}{c\sqrt{\pi}}T^{\frac{1}{2}}f_{\frac{1}{2}}\left(1 - \frac{1}{2\sqrt{\pi}c^3}T^{\frac{3}{2}}f_{\frac{3}{2}}\right)^{-1}\frac{\partial p}{\partial T}$$

$$-\frac{1}{2\sqrt{\pi}}T^{\frac{1}{2}}f_{\frac{1}{2}}\frac{\dfrac{5}{4\sqrt{\pi}c^3}T^{\frac{3}{2}}f_{\frac{5}{2}} - \dfrac{1}{2\sqrt{\pi}c^3}T^{\frac{1}{2}}f_{\frac{3}{2}}A}{1 - \dfrac{1}{2\sqrt{\pi}c^3}T^{\frac{3}{2}}f_{\frac{3}{2}}}$$

$$+\frac{1}{2\sqrt{\pi}}T^{-\frac{1}{2}}f_{\frac{1}{2}}A - \frac{3}{4c^3\pi}T^2f_{\frac{3}{2}}^2$$

$$-\frac{1}{2c^3\pi}T^2f_{\frac{1}{2}}f_{\frac{3}{2}}\frac{\dfrac{5}{4\sqrt{\pi}c^3}T^{\frac{3}{2}}f_{\frac{5}{2}} - \dfrac{1}{2\sqrt{\pi}c^3}T^{\frac{1}{2}}f_{\frac{3}{2}}A}{1 - \dfrac{1}{2\sqrt{\pi}c^3}T^{\frac{3}{2}}f_{\frac{3}{2}}} + \frac{1}{2c^3\pi}Tf_{\frac{1}{2}}f_{\frac{3}{2}}A.$$

通过复杂的迭代：

$$\Rightarrow \frac{\partial p}{\partial T} = \left(-\frac{3}{4\sqrt{\pi}}T^{\frac{1}{2}}f_{\frac{3}{2}} - \frac{1}{2\sqrt{\pi}}T^{\frac{1}{2}}f_{\frac{1}{2}}\frac{\dfrac{5}{4\sqrt{\pi}c^3}T^{\frac{3}{2}}f_{\frac{5}{2}} - \dfrac{1}{2\sqrt{\pi}c^3}T^{\frac{1}{2}}f_{\frac{3}{2}}A}{1 - \dfrac{1}{2\sqrt{\pi}c^3}T^{\frac{3}{2}}f_{\frac{3}{2}}}\right.$$

$$+\frac{1}{2\sqrt{\pi}}T^{-\frac{1}{2}}f_{\frac{1}{2}}A - \frac{3}{4c^3\pi}T^2f_{\frac{3}{2}}^2 + \frac{1}{2c^3\pi}Tf_{\frac{1}{2}}f_{\frac{3}{2}}A$$

$$\left.-\frac{1}{2c^3\pi}T^2f_{\frac{1}{2}}f_{\frac{3}{2}}\frac{\dfrac{5}{4\sqrt{\pi}c^3}T^{\frac{3}{2}}f_{\frac{5}{2}} - \dfrac{1}{2\sqrt{\pi}c^3}T^{\frac{1}{2}}f_{\frac{3}{2}}A}{1 - \dfrac{1}{2\sqrt{\pi}c^3}T^{\frac{3}{2}}f_{\frac{3}{2}}}\right)$$

$$\times \left[1 + \frac{1}{c\sqrt{\pi}}T^{\frac{1}{2}}f_{\frac{1}{2}}\left(1 - \frac{1}{2\sqrt{\pi}c^3}T^{\frac{3}{2}}f_{\frac{3}{2}}\right)^{-1}\right]^{-1}$$

$$= -\frac{3}{4\sqrt{\pi}}T^{\frac{1}{2}}f_{\frac{3}{2}} + \frac{3}{4c\pi}Tf_{\frac{1}{2}}f_{\frac{3}{2}} - \frac{3}{4c^2\pi^{\frac{3}{2}}}T^{\frac{3}{2}}f_{\frac{3}{2}}f_{\frac{1}{2}}^2$$

$$-\frac{5}{8c^3\pi}T^2f_{\frac{1}{2}}f_{\frac{5}{2}} + \frac{1}{4\pi c^3}Tf_{\frac{1}{2}}f_{\frac{3}{2}}A + \frac{1}{2\sqrt{\pi}}T^{-\frac{1}{2}}f_{\frac{1}{2}}A - \frac{1}{2c\pi}f_{\frac{1}{2}}^2A$$

$$+\frac{1}{2c^2\pi^{\frac{3}{2}}}T^{\frac{1}{2}}f_{\frac{1}{2}}^3A - \frac{3}{4c^3\pi}T^2f_{\frac{3}{2}}^2 + \frac{1}{2c^3\pi}Tf_{\frac{1}{2}}f_{\frac{3}{2}}A.$$

因此，得到压强对温度求二阶偏导：

$$\Rightarrow \frac{\partial^2 p}{\partial T^2} = -\frac{3}{8\sqrt{\pi}}T^{-\frac{1}{2}}f_{\frac{3}{2}} - \frac{3}{4\sqrt{\pi}}T^{-\frac{1}{2}}f_{\frac{1}{2}}\frac{\partial A}{\partial T}$$

$$+\frac{3}{4\sqrt{\pi}}T^{-\frac{3}{2}}f_{\frac{1}{2}}A - \frac{1}{4\sqrt{\pi}}T^{-\frac{3}{2}}f_{\frac{1}{2}}A$$

$$+\frac{1}{2\sqrt{\pi}}T^{-\frac{3}{2}}f_{-\frac{1}{2}}\frac{\partial A}{\partial T}A - \frac{1}{2\sqrt{\pi}}T^{-\frac{5}{2}}f_{-\frac{1}{2}}A^2$$

$$+ \frac{1}{2\sqrt{\pi}} T^{-\frac{1}{2}} f_{\frac{1}{2}} \frac{\partial A}{\partial T} + o(1/c)$$
$$\approx -\frac{3}{8\sqrt{\pi}} T^{-\frac{1}{2}} f_{\frac{3}{2}} + \frac{1}{2\sqrt{\pi}} T^{-\frac{3}{2}} f_{\frac{1}{2}} A - \frac{1}{2\sqrt{\pi}} T^{-\frac{5}{2}} f_{-\frac{1}{2}} A^2 + o(1/c).$$

由此可得比热具体表达式:

$$\frac{C_\mathrm{v}}{T} = \frac{\partial^2 p}{\partial T^2}$$
$$= -\frac{3}{8\sqrt{\pi}} T^{-\frac{1}{2}} f_{\frac{3}{2}} + \frac{1}{2\sqrt{\pi}} T^{-\frac{3}{2}} f_{\frac{1}{2}} A - \frac{1}{2\sqrt{\pi}} T^{-\frac{5}{2}} f_{-\frac{1}{2}} A^2 + o(1/c). \quad (2.9.6)$$

比热不但可以给出量子临界现象中重要的标度理论, 也可以刻画在临界点附近不同区域的量子涨落.

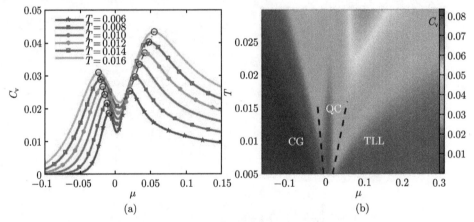

图 2.4 (a) 低温比热 C_v 与化学势 μ 的关系. 左边的峰标志着从经典气体 (CG) 区到量子临界 (QC) 区的临界温度, 右边的峰标志从量子临界 (QC) 区到 Luttinger 液体 (TLL) 区的临界温度. (b) 在无量纲的 T-μ 平面, 熵的等高线图. 右边是一个在 T-μ 平面有限温的相图. 左边的两个虚线由峰的位置决定 (扫描封底二维码可看彩图)

2.10 Luttinger 液体理论

在一维 Lieb-Liniger 模型中, 其低能物理可以用一简单场论模型来描述. 不同动量区间的激发谱有不同的形式, 在低动量, 如 $p \to 0$, 激发谱由 Luttinger 液体哈密顿量决定, 能量 $\epsilon_p = v_s p$ 是由声子线性谱决定的, 其中 v_s 是声速. 由于低能激发是玻色集体运动模式, 可用玻色化方法处理低能物理问题 [12]. 体系的密度可以用连续的玻色场 $\phi(x)$ 来表示 $\rho(x) = \left[\rho_0 - \frac{1}{\pi} \nabla \phi(x) \right] \sum_p \mathrm{e}^{\mathrm{i}p[2\pi\rho_0 x - 2\phi(x)]}$, ρ_0 是基态的密

2.10 Luttinger 液体理论

度. Lieb-Liniger 模型的玻色场算符可写为 $\Psi(x) = [\rho(x)]^{1/2}e^{i\theta(x)}$, 其中 $\theta(x)$ 是相位, 在长波极限下, 玻色气体的动能部分可表示为

$$\int dx \, (\nabla \Psi^\dagger \nabla \Psi) = \int dx \left\{ \left[\frac{\nabla \rho(x)}{2\rho(x)^{1/2}}\right]^2 + \rho(x)(\nabla \theta)^2 \right\}$$
$$\approx \int dx \left\{ \frac{1}{4\pi^2}\frac{1}{\rho_0}\left[\nabla^2 \phi(x)\right]^2 + \rho_0(\nabla \theta)^2 \right\}. \tag{2.10.1}$$

相互作用项表示为

$$g \int dx \left[|\Psi(x)|^2\right]^2 = g \int dx \, [\rho(x) - \rho_0]^2 = g \int dx \frac{1}{\pi^2}[\nabla \phi(x)]^2, \tag{2.10.2}$$

其中, $c = g/2$. 由此得到低能有效哈密顿量

$$H = \int dx \left\{ \frac{1}{4\pi^2}\frac{1}{\rho_0}\left[\nabla^2 \phi(x)\right]^2 + \rho_0(\nabla \theta)^2 \right\} + \int dx \frac{1}{\pi^2}[\nabla \phi(x)]^2$$
$$\approx \int dx \left\{ \rho_0(\nabla \theta)^2 + \frac{g}{\pi^2}[\nabla \phi(x)]^2 \right\}. \tag{2.10.3}$$

在长波极限下, 即 $p \to 0$, 有效哈密顿量通常表示为

$$H = \int dx \left[\frac{\pi v_s K}{2}\Pi^2 + \frac{v_s}{2\pi K}(\partial_x \phi)^2\right], \tag{2.10.4}$$

这里, 引入正则动量 $\Pi = \nabla \theta$ 与场 $\phi(x)$ 满足对易关系 $[\phi(x), \Pi(y)] = i\delta(x-y)$; v_s 是声速; K 是 Luttinger 参量; $\frac{v_s}{2\pi K}$ 是粒子密度变化带来的能量变化. Luttinger 参量及声速决定了体系的低能普适行为, 例如, 单粒子关联函数由 Luttinger 参量确定 $\langle \psi^\dagger(x)\psi(0)\rangle \sim 1/x^{1/2K}$. 声速和刚度定义为

$$v_s = \sqrt{\frac{L}{mn}\frac{\partial^2 E}{\partial L^2}}, \tag{2.10.5}$$

$$v_N = \frac{L}{\pi \hbar}\left(\frac{\partial^2 E}{\partial N^2}\right). \tag{2.10.6}$$

通常 Luttinger 参量可以表示为声速 v_s 与刚度 v_N 之比, 也就是

$$K = \frac{v_s}{v_N}. \tag{2.10.7}$$

Wilson 比定义为 [13]

$$R_W = \frac{k_B^2 \pi^2}{3}\frac{\kappa}{C_v/T}. \tag{2.10.8}$$

由能量的定义我们知道

$$E = n^3 e(\gamma). \tag{2.10.9}$$

因此, 推导出声速 v_s 和刚度 v_N 在基态的表达式:

$$v_s = \frac{\hbar n}{m}\sqrt{3e - 2\gamma\frac{de}{d\gamma} + \frac{1}{2}\gamma^2\frac{d^2e}{d\gamma^2}}, \tag{2.10.10}$$

$$v_N = \frac{\hbar n}{\pi m}\left(3e - 2\gamma\frac{de}{d\gamma} + \frac{1}{2}\gamma^2\frac{d^2e}{d\gamma^2}\right), \tag{2.10.11}$$

由此可以得出 Luttinger 参量:

$$K = \frac{\pi}{\sqrt{3e - 2\gamma\frac{de}{d\gamma} + \frac{1}{2}\gamma^2\frac{d^2e}{d\gamma^2}}}. \tag{2.10.12}$$

显然, 我们可以得到 Luttinger 参量与声速之间简单的关系:

$$K = \frac{\pi\hbar n}{m v_s}. \tag{2.10.13}$$

因此提供了一种直接测量 Luttinger 参数的方法.

另外, 对于一个伽利略不变系统, 刚度 v_N 和声速 v_s 与压缩率 κ 和比热 C_v 有直接的联系, 即

$$\kappa = \frac{1}{\hbar\pi v_N} \tag{2.10.14}$$

$$C_v = \frac{\pi k_B^2 T}{3}\frac{1}{\hbar v_s} \tag{2.10.15}$$

利用上式中压缩率 κ 和比热 C_v 的关系可以整理得 Wilson 比表达式, 将其与 Luttinger 参量比较可得

$$R_w = K. \tag{2.10.16}$$

在量子流体中, Wilson 比是一个可测量的无量纲的常数, 它提供了一种测量一维多体系统的 Luttinger 参量的重要方法, 见图 2.5.

另外, 当动量大时, 激发谱中需要考虑高阶修正 [8], 弱相互作用下

$$\begin{cases} \text{低动量} & \epsilon_p = v_s p\sqrt{1 + \frac{p^2}{4m^2 v_c^2}}, \\ \text{高动量} & \epsilon_p = \frac{p^2}{2m} + \gamma\frac{\hbar^2 n^2}{m}. \end{cases} \tag{2.10.17}$$

强相互作用下, 高动量为

$$\epsilon_p = \frac{p^2}{2m} + 2\gamma\frac{\hbar^2 n^2}{m} - \frac{\pi^2\hbar^2 n^2}{2m} + o\left(\frac{1}{p^2}\right), \tag{2.10.18}$$

2.11 量子临界性

强相互作用下, 低动量为

$$\epsilon_\mathrm{p} = v_\mathrm{s} p + \frac{p^2}{2m^*} + \left(\frac{8\pi}{3\gamma^3} - \frac{80\pi}{3\gamma^4}\right)\frac{p^3}{\hbar n m}$$
$$+ \left(\frac{2}{3\gamma^3} - \frac{20}{3\gamma^4}\right)\frac{p^4}{\hbar^2 n^2 m} + o(\gamma^{-5}). \tag{2.10.19}$$

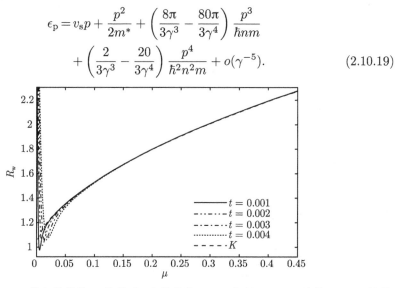

图 2.5 低温 Wilson 比与化学势 μ 的关系: 实线是从 TBA 方程 (2.7.17) 计算 Wilson 比的数值结果, 虚线是从普适关系 (2.10.12) 得到的结果. 在临界点附近两者明显不等, 它们的一致性表明 Luttinger 液体 (TLL) 的存在

2.11 量子临界性

临界现象在自然界中随处可见, 包括从热涨落引起的经典相变到量子涨落引起的量子相变. 量子临界现象描述在低温相变点附近量子多体系统热力学量的普适标度律. 通常, 量子临界点附近量子涨落伴随着热涨落. 当热能 $k_\mathrm{B}T$(其中 k_B 是玻尔兹曼常数) 小于能隙 Δ 时, 量子涨落占主导, 当 $k_\mathrm{B}T$ 大于能隙 Δ 时, 热涨落占主导, 决定序参量涨落, 控制临界行为. 在研究量子临界性的问题中, 我们主要关注的是临界指数和普适的标度函数, 它们可以描述在临界点附近两个稳定相的热力学性质. 至今在凝聚态物理中理解临界性仍然面临很大的挑战.

对于一个二级相变, 在临界点附近的临界行为是由发散的关联长度 $\xi \sim |g - g_\mathrm{c}|^{-\nu}$ 和能隙 Δ 来表征的, 它们之间有这样的关系 $\Delta \sim \xi^{-z} \sim |g - g_\mathrm{c}|^{z\nu}$. 其中动力学临界指数 z 和关联长度指数 ν 都是普适的, g 是驱动参数, 如化学势、磁场、相互作用强度等. 动力学临界指数 z 是描述能隙趋向于零的特征时间尺度, 如 $\tau_\mathrm{c} \sim \xi^z$, 将会导致在量子临界区的相位相干性. 在这个临界区, 人们认为有一个普适的标度不变量来描述这个系统的热力学性质. 序参量的奇异性, 展示了一个特殊的临界现象[15]. 就这一点而言, 量子相变的临界温度在 $T_\mathrm{c} = 0$.

在零温时, 在化学势的驱动下, Lieb - Liniger 玻色气体在临界点 $\mu_c = 0$ 处可以发生从真空态到 Luttinger 液体 (TLL) 态的量子相变. 当温度趋向零温时, 在相变点附近量子涨落及热涨落的竞争诱导了三个不同的区域, 见图 2.6: ①德布罗意波长 $\lambda_T^{-1} = \sqrt{mk_BT/2\pi\hbar^2}$ 远远小于粒子平均距离, 称为经典的区域; ②热力学量服从普适的标度不变性, 称为量子临界区 (QC); ③Luttinger 流体相 (TLL), 其自由能可以表示为 [15]

$$F(T) \approx E_0 - \frac{\pi C(k_B T)^2}{6\hbar v_s}. \tag{2.11.1}$$

v_s 是声速, 前面已给出.

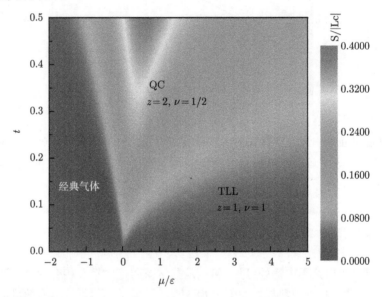

图 2.6 有限温的 Lieb-Liniger 玻色气体的量子相图: 在无量纲的 t-μ 平面, 熵的等高线图 [14]. 其中纵坐标温度重新定义为 $t = k_BT/\varepsilon$. 临界点在 $\mu_c = 0$ 处. 在低温 $T \ll |\mu - \mu_c|$ 且 $\mu < 0$ 区域, 体系是稀薄的经典气体. 在温度 $T \ll |\mu - \mu_c|$ 且 $\mu > 0$ 区域, 体系是一个 LL 的低能物理系统, 其中动力学指数 $z = 1$, 关联长度指数 $\nu = 1$. 在靠近临界点 $\mu = 0$, $T \gg |\mu - \mu_c|$ 区域, 体系是一个普适性的量子临界区 (QC), 其中动力学指数 $z = 2$, 关联长度指数 $\nu = 1/2$. 也可以参考文献 [11], 此图从文献 [14], [15] 中摘录 (扫描封底二维码可看彩图)

通过非重整化群理论, 人们可以得到在低温临界区普适标度特性. 在 2011 年, Guan 和 Batchelor 首次从精确解研究了玻色气体的量子临界性. 他们发现了量子临界性的普适标度律. 这里我们可以利用粒子数方程 (2.9.3) 推出密度标度理论 [16]:

$$n(T,\mu) \approx n_0 + T^{\frac{d}{z}+1-\frac{1}{\nu z}} \mathcal{F}\left(\frac{\mu - \mu_c}{T^{\frac{1}{\nu z}}}\right), \; \xi \sim |\mu - \mu_c|^{-2}, \Delta \sim \xi^{-z} \tag{2.11.2}$$

2.11 量子临界性

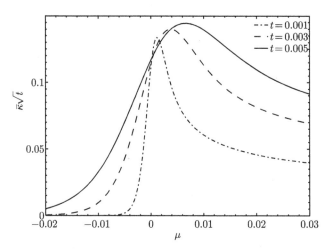

图 2.7 均匀 Lieb-Liniger 气体在量子临界处关于压缩率的标度律. 其中无量纲温度压缩率 $\bar{\kappa}\sqrt{t}$ 标出动力学指数 $z=2$ 和关联长度指数 $\nu=1/2$

其中, 背景密度 $n_0 = 0$. 由式 (2.9.3) 我们可以得到标度函数 $\mathcal{F}(x) = -\frac{1}{2\sqrt{\pi}}Li_{\frac{1}{2}}(-e^x)$, 并且临界指数满足:

$$\frac{d}{z} + 1 - \frac{1}{\nu z} = \frac{1}{2}$$

$$\frac{1}{\nu z} = 1$$

考虑一维 $d=1$ 的情况解上面方程组, 可得动力学临界指数 $z=2$, 关联长度指数 $\nu = \frac{1}{2}$. 一维 Lieb - Liniger 气体在有限温情况下, 可以根据普适标度理论定出临界指数. 压缩率的普适标度律形式为

$$\kappa^* = \kappa_0 + T^{\frac{d}{z}+1-\frac{2}{\nu z}}\mathcal{K}\left(\frac{\mu-\mu_\mathrm{c}}{T^{\frac{1}{\nu z}}}\right), \tag{2.11.3}$$

其中, $\kappa_0 = 0, \mathcal{K}(x) = -\frac{1}{2\sqrt{\pi}}Li_{-\frac{1}{2}}(-e^x)$. 可以根据同样的方法, 得到动力学临界指数 $z=2$, 关联长度指数 $\nu = \frac{1}{2}$. 压强、熵及比热的普适标度律形式为

$$p = T^{\frac{d}{z}+1}\mathcal{P}\left(\frac{\mu-\mu_\mathrm{c}}{T^{\frac{1}{\nu z}}}\right), \tag{2.11.4}$$

$$s = T^{\frac{d}{z}}\mathcal{S}\left(\frac{\mu-\mu_\mathrm{c}}{T^{\frac{1}{\nu z}}}\right), \tag{2.11.5}$$

$$C_\mathrm{v}/T = T^{\frac{d}{z}-1}\mathcal{Q}\left(\frac{\mu-\mu_\mathrm{c}}{T^{\frac{1}{\nu z}}}\right). \tag{2.11.6}$$

2.12 关联函数

关联函数是一个很重要的物理量, 它提供了很多有关量子多体波函数的信息. 因此, 超冷原子气体中两体和 M 阶高阶关联性是一个很重要的研究方向. 较早 Hanbury Brown 和 Twiss 利用高阶关联性测量遥远双星的距离大小[17]. 其后, 很多小组研究零温和有限温的一维玻色气体的局域和非局域的关联函数. 最近, 有些小组利用高阶关联性研究三维强相互作用下玻色气体的动力学问题[18,19]. 可见关联函数是一个很重要的物理量.

在本章中, 我们只考虑在 Lieb-Liniger 模型中, 一维强排斥作用下玻色气体的 M 阶关联函数 g_M [20]. 在强排斥相互作用极限下, 一维玻色气体相当于一维无相互作用的费米气体, 其统计因子满足 $\alpha = 1 - \dfrac{2}{\gamma}$, 其中 γ 表示相互作用强度. 在这里, 我们利用 Bethe Ansatz 方法讨论一维强相互作用玻色气体的高阶关联函数.

在 Lieb-Liniger 模型中, 假设波函数为

$$\psi(x_1, x_2, \cdots, x_N) = \sum_{\mathcal{P}} A(k_{\mathcal{P}_1}, k_{\mathcal{P}_2}, \cdots, k_{\mathcal{P}_N}) e^{i\sum_j k_{\mathcal{P}_j} x_j}, \tag{2.12.1}$$

由周期性边界条件 $\psi(0, x_2, \cdots, x_N) = \psi(x_2, \cdots, x_N, L)$, 我们得到能量本征函数

$$\psi(x_1, x_2, \cdots, x_N) = \sum_{\mathcal{P}} (-1)^{\mathcal{P}} \left[\prod_{1 \leqslant i < j \leqslant N} \left(1 + \frac{ik_{\mathcal{P}_j} - ik_{\mathcal{P}_i}}{c} \right) \right] \exp\left(\sum_{j=1}^N ik_{\mathcal{P}_j} x_j \right)$$

$$\approx \left[\prod_{1 \leqslant i < j \leqslant N} \left(1 + \frac{\partial_{x_j} - \partial_{x_i}}{c} \right) \right] \phi^{(0)}(x_1, x_2, \cdots, x_N), \tag{2.12.2}$$

其中

$$\phi^{(0)}(x_1, x_2, \cdots, x_N) = \sum_{\mathcal{P}} (-1)^{\mathcal{P}} \exp\left(\sum_{j=1}^N ik_{\mathcal{P}_j} x_j \right) \tag{2.12.3}$$

是全反对称函数; $k_1, \cdots, k_N$ 是准动量. M 阶局域关联函数定义为

$$g_M = \frac{\langle \psi | (\Psi^+(0))^M (\Psi(0))^M | \psi \rangle}{\langle \psi | \psi \rangle},$$

其中, Ψ^+ 和 Ψ 分别表示玻色子的产生与湮没算符. 通过分析与推导, 我们可以得出 M 阶局域关联函数的表达式为

$$g_M = \frac{\langle \psi | (\Psi^+(0))^M (\Psi(0))^M | \psi \rangle}{\langle \psi | \psi \rangle} = \langle (\Psi^+(0)^M)(\Psi(0))^M \rangle$$

2.12 关联函数

$$= \frac{N!}{(N-M)!} \frac{\int |\psi(0,\cdots,0,x_{M+1},\cdots,x_N)|^2 \mathrm{d}x_{M+1}\cdots \mathrm{d}x_N}{\int |\psi(x_1,\cdots,x_N)|^2 \mathrm{d}x_1\cdots \mathrm{d}x_N}. \qquad (2.12.4)$$

为了计算出 M 阶局域关联函数，需要将其分子与分母按 $1/c$ 不同阶展开，接下来我们分别讨论上式中的分子与分母部分.

在 $0 \leqslant x_{M+1} \leqslant x_{M+2} \leqslant \cdots \leqslant x_N \leqslant L$ 区域，我们可以将波函数 ψ 写成

$$\begin{aligned}
&\psi(0,\cdots,x_{M+1},\cdots,x_N) \\
&= \prod_{1\leqslant i<j\leqslant M} \frac{\partial_{x_j}-\partial_{x_i}}{c} \prod_{1\leqslant i\leqslant M; M+1\leqslant j\leqslant N} \left(1+\frac{\partial_{x_j}-\partial_{x_i}}{c}\right) \\
&\quad \prod_{1\leqslant i\leqslant M; M+1\leqslant j\leqslant M} \left(1+\frac{\partial_{x_j}-\partial_{x_i}}{c}\right) \phi^{(0)}(x_1,\cdots,x_N)|_{x_1=\cdots=x_M=0}. \qquad (2.12.5)
\end{aligned}$$

在强相互作用极限下，$c\to\infty$，可以将上式展开为

$$\begin{aligned}
\psi(0,\cdots,x_{M+1},\cdots,x_N) &= \chi^{(0)}(x_{M+1},\cdots,x_N) \\
&\quad + \chi^{(1)}(x_{M+1},\cdots,x_N) + o(c^{-M(M-1)/2-2}),
\end{aligned}$$

其中

$$\begin{aligned}
\chi^{(0)}(x_{M+1},\cdots,x_N) &= c^{-M(M-1)/2} \\
&\quad \times \left[\prod_{1\leqslant i<j\leqslant M} \frac{\partial_{x_j}-\partial_{x_i}}{c}\right] \phi^{(0)}(x_1,\cdots,x_N)|_{x_1=\cdots=x_M=0}.
\end{aligned}$$

$$\begin{aligned}
\chi^{(1)}(x_{M+1},\cdots,x_N) &= c^{-M(M-1)/2-1}\left[-(N-M)\sum_{l=1}^M \partial_{x_l} + \sum_{l=M+1}^N (2l-N-1)\partial_{x_l}\right] \\
&\quad \times \left[\prod_{1\leqslant i<j\leqslant M} \partial_{x_j}-\partial_{x_i}\right] \phi^{(0)}(x_1,\cdots,x_N)|_{x_1=\cdots=x_M=0}. \qquad (2.12.6)
\end{aligned}$$

我们定义：

$$\phi^{(\Delta)}(\epsilon) = \left[\prod_{1\leqslant i<j\leqslant M} \partial_{x_j}-\partial_{x_i}\right] \phi^{(0)}(x_1,\cdots,x_N)|_{x_1=\cdots=x_M=\epsilon},$$

通过对上式的变换得

$$\int_0^L |\psi(0,\cdots,x_{M+1},\cdots,x_N)|^2 \mathrm{d}x_{M+1}\cdots \mathrm{d}x_N$$

$$= c^{-M(M-1)} \times \int_0^L |\phi^{(\Delta)}(0,\cdots,x_{M+1},\cdots,x_N)|^2 \mathrm{d}x_{M+1}\cdots \mathrm{d}x_N + o(c^{-M(M-1)/2-2}). \tag{2.12.7}$$

考虑体系的基态, 系统总的动量 $k_1 + k_2 + \cdots + k_N = 0$. 因为在强相互作用下, 一维玻色气体相当于一维无相互作用的费米气体. 我们有 $k_i = k_i^F \alpha + o(c^{-2})$, 其中 $k_i^F = \dfrac{2\pi m_i}{L}$, m_i 是整数. 在强相互作用极限下, $\gamma \gg 1$, 作一个标度变换 $x_i = \dfrac{x_i^F}{\alpha} (i=1,2,\cdots,N)$. 所以, 分子积分部分可写为

$$\int_0^L |\psi(0,\cdots,0,x_{M+1},\cdots,x_N)|^2 \mathrm{d}x_{M+1}\cdots \mathrm{d}x_N$$

$$= c^{-M(M-1)} \int_0^L \mathrm{d}x_{M+1} \cdots \mathrm{d}x_N$$

$$\times \left| \left[\prod_{1\leqslant i<j\leqslant M} (\partial_{x_j} - \partial_{x_i}) \right] \sum_{\mathcal{P}} (-1)^{\mathcal{P}} \exp\left(\sum_{j=1}^N \mathrm{i} k_{\mathcal{P}_j} x_j \right) \right|^2_{x_1=x_2=\cdots=x_M=0}$$

$$+ o(c^{-M(M-1)-2})$$

$$= c^{-M(M-1)} \int_0^{\alpha L} \alpha^{M-N} \mathrm{d}x^F_{M+1} \cdots \mathrm{d}x^F_N$$

$$\times \left| \left[\prod_{1\leqslant i<j\leqslant M} \alpha(\partial_{x_j^F} - \partial_{x_i^F}) \right] \sum_{\mathcal{P}} (-1)^{\mathcal{P}} \exp\left(\sum_{j=1}^N \mathrm{i} k^F_{\mathcal{P}_j} x_j^F \right) \right|^2_{x_1^F=x_2^F=\cdots=x_M^F=0}$$

$$+ o(c^{-M(M-1)-2})$$

$$= \frac{\alpha^{M^2-N}}{c^{M(M-1)}} \int_0^{\alpha L} |\phi^{(\Delta F)}(0, x^F_{M+1},\cdots,x^F_N)|^2 \mathrm{d}x^F_{M+1}\cdots \mathrm{d}x^F_N$$

$$+ o(c^{-M(M-1)-2}), \tag{2.12.8}$$

其中

$$\phi^{(\Delta F)}(0, x^F_{M+1}, \cdots, x^F_N)$$
$$= \left[\prod_{1\leqslant i<j\leqslant M} (\partial_{x_j^F} - \partial_{x_i^F}) \right] \phi^F(x_1^F, x_2^F, \cdots, x_N^F) \Bigg|_{x_1^F=x_2^F=\cdots=x_M^F=0}, \tag{2.12.9}$$

ϕ^F 为理想费米子的波函数:

$$\phi^F(x_1^F, x_2^F, \cdots, x_N^F) = \sum_{\mathcal{P}} (-1)^{\mathcal{P}} \exp\left(\sum_{j=1}^N \mathrm{i} k^F_{\mathcal{P}_j} x_j^F \right). \tag{2.12.10}$$

2.12 关联函数

另外, 考虑到 $\phi^{(\Delta F)}(0, x_{M+1}^F, \cdots, x_N^F)$ 是周期函数:

$$\phi^{(\Delta F)}(0, x_{M+1}^F, \cdots, x_{i-1}^F, x_i^F + L, x_{i+1}^F, \cdots, x_N^F)$$
$$= \phi^{(\Delta F)}(0, x_{M+1}^F, \cdots, x_{i-1}^F, x_i^F, x_{i+1}^F, \cdots, x_N^F),$$

因此, 在 $M+1 \leqslant i \leqslant N$ 时 $\alpha L < x_i^F < L$, 函数 $\phi^{(\Delta F)}(0, x_{M+1}^F, \cdots, x_N^F)$ 有 c^{-M} 阶.

$$\int_0^{\alpha L} |\phi^{(\Delta F)}(0, x_{M+1}^F, \cdots, x_N^F)|^2 dx_{M+1}^F \cdots dx_N^F$$
$$= \int_0^L |\phi^{(\Delta F)}(0, x_{M+1}^F, \cdots, x_N^F)|^2 dx_{M+1}^F \cdots dx_N^F + o(c^{-2M}). \quad (2.12.11)$$

假定 $M \geqslant 1$, 可以得出分子部分为

$$\int_0^L |\psi(0, \cdots, 0, x_{M+1}, \cdots, x_N)|^2 dx_{M+1} \cdots dx_N$$
$$= \frac{\alpha^{M^2-N}}{c^{M(M-1)}} \int_0^L |\phi^{(\Delta F)}(0, x_{M+1}^F, \cdots, x_N^F)|^2 dx_{M+1}^F \cdots dx_N^F + o(c^{-M(M-1)-2}). \quad (2.12.12)$$

在强相互作用极限 $c \to \infty$, $0 \leqslant x_1 \leqslant x_2 \leqslant \cdots \leqslant x_N \leqslant L$ 区域中, 波函数 $\psi(x_1, x_2, \cdots, x_N)$ 可展开为

$$\psi(x_1, x_2, \cdots, x_N) = \psi^{(0)}(x_1, x_2, \cdots, x_N) + \psi^{(1)}(x_1, x_2, \cdots, x_N) + o(c^{-2}),$$

其中

$$\psi^{(0)}(x_1, x_2, \cdots, x_N) = \sum_{\mathcal{P}} (-1)^{\mathcal{P}} \exp\left(\sum_{j=1}^N \mathrm{i} k_{\mathcal{P}_j} x_j\right)$$

$$\psi^{(1)}(x_1, x_2, \cdots, x_N) = \sum_{1 \leqslant i < j \leqslant N} \frac{\partial_{x_j} - \partial_{x_i}}{c} \psi^{(0)}(x_1, x_2, \cdots, x_N).$$

又由计算得 $\int_0^L |\phi^{(0)}(x_1, x_2, \cdots, x_N)|^2 dx_1 \cdots dx_N = N! L^N \det(S)$, 其中 S 为 $N \times N$ 矩阵, 矩阵元为 $S_{ij} = \operatorname{sinc} \dfrac{(k_i - k_j)L}{2}$, $\operatorname{sinc}(\xi) = \begin{cases} \dfrac{\sin \xi}{\xi}, & \xi \neq 0 \\ 1, & \xi = 0 \end{cases}$, 我们可以计算得到 $\det(S) = 1 + o(c^{-2})$. 因此

$$\int_0^L |\phi^{(0)}(x_1, x_2, \cdots, x_N)|^2 dx_1 \cdots dx_N = [1 + o(c^{-2})] N! L^N$$
$$= [1 + o(c^{-2})] \int_0^L |\phi^F(x_1^F, x_2^F, \cdots, x_N^F)|^2 dx_1^F dx_2^F \cdots dx_N^F. \quad (2.12.13)$$

由此我们可以得到分母部分 (更详细推导, 请参考文献 [20] 附录) 为

$$\int |\psi(x_1,\cdots,x_N)|^2 \mathrm{d}x_1\cdots \mathrm{d}x_N$$
$$= [\alpha^{1-N} + o(c^{-2})] \int_0^L |\phi^F(x_1^F,x_2^F,\cdots,x_N^F)|^2 \mathrm{d}x_1^F \mathrm{d}x_2^F \cdots \mathrm{d}x_N^F.$$

最后将分子与分母部分代入关联函数的表达式, 可得

$$g_M = \frac{\alpha^{M^2-1}}{c^{M(M-1)}} \frac{N!}{(N-M)!} \times \frac{\int_0^L |\phi^{(\Delta F)}(0,x_{M+1},\cdots,x_N)|^2 \mathrm{d}x_{M+1}\cdots \mathrm{d}x_N}{\int_0^L |\phi^F(x_1,x_2,\cdots,x_N)|^2 \mathrm{d}x_1 \mathrm{d}x_2 \cdots \mathrm{d}x_N}$$
$$+ o(c^{-M(M-1)-2}). \tag{2.12.14}$$

从 $\phi^{(\Delta F)}$ 的定义, 我们发现

$$\frac{N!}{(N-M)!} \frac{\int_0^L |\phi^{(\Delta F)}(0,x_{M+1},\cdots,x_N)|^2 \mathrm{d}x_{M+1}\cdots \mathrm{d}x_N}{\int_0^L |\phi^F(x_1,x_2,\cdots,x_N)|^2 \mathrm{d}x_1 \mathrm{d}x_2 \cdots \mathrm{d}x_N}$$

$$= \Delta_M(\partial_x)\Delta_M(\partial_y) \frac{N!}{(N-M)!}$$
$$\times \left. \frac{\int_0^L \phi^{F*}(x_1,\cdots,x_N)\phi^F(y_1,\cdots,y_M,x_{M+1},\cdots,x_N)\mathrm{d}x_{M+1}\cdots \mathrm{d}x_N}{\int_0^L |\phi^F(z_1,\cdots,z_N)|^2 \mathrm{d}z_1 \cdots \mathrm{d}z_N} \right|_{\substack{x_1=\cdots=x_M=\\y_1=\cdots=y_M=0}}$$
$$\tag{2.12.15}$$

其中, $\Delta_M(\partial_x) \equiv \Delta_M(\partial_{x_1},\cdots,\partial_{x_M})$,

$$\Delta_M(\xi_1,\cdots,\xi_M) \equiv \prod_{1\leqslant i<j \leqslant M}(\xi_j - \xi_i). \tag{2.12.16}$$

利用维克定理:

$$\frac{N! \int_0^L \phi^{F*}(x_1,\cdots,x_N)\phi^F(y_1,\cdots,y_M,x_{M+1},\cdots,x_N)\mathrm{d}x_{M+1}\cdots \mathrm{d}x_N}{(N-M)!\int_0^L |\phi^F(z_1,\cdots,z_N)|^2 \mathrm{d}z_1\cdots \mathrm{d}z_N}$$
$$\equiv \sum_q (-1)^q G(x_1,y_{q_1})G(x_2,y_{q_2})\cdots G(x_M,y_{q_M}), \tag{2.12.17}$$

2.12 关联函数

在 $x_1 = \cdots = x_M = 0$ 情况下,多体的关联函数为

$$G(x,y) = \frac{N\int_0^L \phi^{F*}(x_1,x_2,\cdots,x_N)\phi^F(y,x_{M+1},\cdots,x_N)\mathrm{d}x_2\cdots\mathrm{d}x_N}{\int_0^L |\phi^F(z_1,\cdots,z_N)|^2\mathrm{d}z_1\cdots\mathrm{d}z_N}, \quad (2.12.18)$$

又因为

$$G(x,y) = \frac{1}{L}\sum_{i=1}^N \exp[-\mathrm{i}k_i^F(x-y)], \quad (2.12.19)$$

$$\Rightarrow g_M = \frac{\alpha^{M^2-1}M!}{c^{M(M-1)}L^M}\sum_{i_1=1}^N\cdots\sum_{i_M=1}^N \Delta_M^2(k_{i_1}^F\cdots k_{i_M}^F) + o(c^{-M(M-1)-2}), \quad (2.12.20)$$

在热力学限制下,费米动量 $k_1^F,\cdots,k_N^F$ 遵从费米分布 $f(k) = 1/\left(1+\mathrm{e}^{\frac{k^2-\mu}{k_\mathrm{B}T}}\right)$,所以

$$\begin{aligned}g_M = &\frac{M!\alpha^{M^2-1}}{(2\pi)^M c^{M(M-1)}} \times \int_{-\infty}^{+\infty}\mathrm{d}p_1\cdots\mathrm{d}p_M f(p_1)\cdots f(p_M)\Delta_M^2(p_1,\cdots,p_M)\\ &+o(c^{-M(M-1)-2})\end{aligned} \quad (2.12.21)$$

及

$$N = \int_{-\infty}^{+\infty}\frac{L\mathrm{d}k}{2\pi}f(k). \quad (2.12.22)$$

做变量代换 $k = 2\pi n z$,得

$$\begin{aligned}\frac{g_M}{n^M} = &M!\left(\frac{2\pi}{r}\right)^{M(M-1)}\alpha^{M^2-1} \times \int_{-\infty}^{+\infty}\mathrm{d}z_1\cdots\mathrm{d}z_M N(z_1)\cdots N(z_M)\Delta_M^2(z_1,\cdots,z_M)\\&+o(r^{-M(M-1)-2}).\end{aligned} \quad (2.12.23)$$

由式 (2.12.22),我们可得 $\int_{-\infty}^{+\infty} N(z)\mathrm{d}z = 1$. 通过利用随机矩阵理论进行正交多项式的计算,式 (2.12.23) 等号右边可写为

$$\int_{-\infty}^{+\infty}\mathrm{d}z_1\cdots\mathrm{d}z_M N(z_1)\cdots N(z_M)\Delta_M^2(z_1,\cdots,z_M) = M!\prod_{j=0}^{M-1}h_j,$$

$$\int_{-\infty}^{+\infty}P_i(z)P_j(z)N(z)\mathrm{d}z = h_j\delta_{ij}. \quad (2.12.24)$$

利用 Gram-schmidt 方法,

$$P_j(z) = z^j - \sum_{i=0}^{j-1}\frac{\langle z^j, P_i(z)\rangle}{\langle P_i(z), P_i(z)\rangle}P_i(z), \quad (2.12.25)$$

其中, $P_j(z)$ 为首一正交多项式 (monic orthogonal polynomials), 最后我们可得多体关联函数:

$$\frac{g_M}{n^M} = (M!)^2 \left(\frac{2\pi}{\gamma}\right)^{M(M-1)} \alpha^{M^2-1} \prod_{j=0}^{M-1} h_j + o(r^{-M(M-1)-2}). \qquad (2.12.26)$$

上面的方法也可以用来计算在不同温度的 M 体局域关联函数[20].

2.13 关于 Lieb-Liniger 玻色气体的实验发展

在过去的几十年里, 囚禁与冷却超冷原子气体的实验很好地揭示了量子世界的一些奇特的物理性质. 尤其是, 最近在囚禁一维冷原子技术上的突破, 使我们对量子多体系统中的量子统计和强关联效应有了更精确的理解. 一维冷原子气体通常可以通过将冷原子气体囚禁在横向上强束缚而在纵向上弱束缚的光晶格中得到. 因为横向上的束缚势非常强, 横向上的零点能 $\hbar\omega_\perp$ 远大于粒子的化学势 μ 及热能 $k_B T$, 所以粒子在横向上的运动被冻结到基态上, 这样就形成一个准一维的系统, 而有效的一维散射长度为

$$a_{1D} = -\frac{a_\perp^2}{2a_s}\left(1 - 1.4603\frac{a_s}{a_\perp}\right) \qquad (2.13.1)$$

这里的 a_s 为原子的 s 波散射长度, $a_\perp = \sqrt{2\hbar/(m\omega_\perp)}$.

实验上较早研究 Lieb-Liniger 气体的是 Bloch 及 Weiss 的小组[21,22]. 早在 2004 年, 他们就分别在实验上实现了一维 Tonks-Girardeau 气体, 并且分别测量了该系统的动量分布及基态能量. 实验上对于基态能量与相互作用的关系的测量结果, 有助于我们深刻理解在强排斥极限时玻色气体的"费米化"行为. 而在 2009 年, Haller[23] 小组又在实验上测量出高激发的亚稳态–超 Tonks-Girardeau 气体. 他们通过囚禁诱导共振得到玻色 Cs 原子强吸引区域中的稳定高激发态, 取得了一个新的实验突破. 这个独特的态第一次是由 Astrakharchik[24] 和其他人用 Monte Carlo 方法和 ANU 小组从吸引相互作用的玻色气体中通过理论预测到的. 这个模型改变了我们对量子统计学和多体物理中的相互作用效应动力学的理解. 原来高激发的类气相气体在相互作用下从强排斥到强吸引相互作用的时候, 由于类费米压的存在而不会塌缩, 并形成一个亚稳态. 这个现象在理论上引起了许多关注[15]. 除此之外还有很多关于 Lieb-Liniger 气体的实验研究, 如测量关联函数[25-30]、热力学[31,32], 涨落[33-35] 等.

原位 (in situ) 测量是现在冷原子实验中常用的一种测量密度分布的手段. 而体系的其他热力学量, 如压强、压缩率、熵、比热等都可以通过密度分布得到[36].

2.13 关于 Lieb-Liniger 玻色气体的实验发展

在 2008 年,就有实验小组通过原位测量得到密度分布,并且证实与 Yang-Yang 热力学预测的结果非常吻合 [37]。由于当时实验条件的制约,所以很难从低精度的密度分布中得到如熵、比热等热力学量。近些年,由于实验技术的成熟及高精度 CCD 的发展,精确测量密度分布成为可能,最近就有实验小组完成了这方面的工作 [38]。他们通过精确测量一维导管中囚禁的超冷 ^{87}Rb 原子气体的密度分布,观测到量子临界现象和 Tomonaga-Luttinger 液体行为。他们还通过密度分布精确测量普适标度律,并且利用比热标定了临界温度。实验观察的结果很好地证实了 Yang-Yang 巨正则理论,并让我们对前几章所讨论的量子临界行为、集体激发及 Luttinger 液体有了更直观、更深刻的认识。

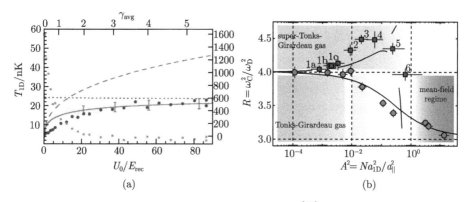

图 2.8 (a) 基态能量 T_{1D} 随横向势阱深度 U_0 的变化关系 [22];(b) 压缩模频率 ω_C 与束缚势频率 ω_D 的比率随相互作用参数 A 的变化关系 [23](扫描封底二维码可看彩图)

如上所述,该小组通过在 y 及 z 方向上施加强的束缚势 $\omega_\perp = \sqrt{\omega_y \omega_z} = 2\pi \times 7.99\text{kHz}$,而在 x 方向上施加弱的束缚势 $\omega_x = 2\pi \times 22.2\text{Hz}$,得到了一维的冷原子气体簇,如图 2.9 所示。而探测光经过原子气体后被吸收,再通过 CCD 成像就能得到势阱中的原子气体密度。从图中可以看到,他们测量得到的原子气体密度与 Yang-Yang 方程的预测结果非常吻合。由前几章节的讨论我们知道,在化学势 $\mu = 0$ 附近是量子临界区,这里密度具有普适的标度律。他们通过局域密度近似 (LDA),得到了密度随化学势的变化,然后再通过对不同温度的密度分布做拟合,从而标定了临界指数 $z = 2.3^{+0.6}_{-0.3}, \nu = 0.56^{+0.07}_{-0.08}$,这与理论预言符合得很好。他们还通过对密度分部积得到局域压强随化学势的变化,即 $p(\mu,T) = \int_{-\infty}^{\mu} n(\mu',T)\mathrm{d}\mu'$,并且验证了压强也具有普适的标度律,参见图 2.10。

我们知道局域的压强扮演着物态方程的角色,其他热力学量都可以由压强的偏导得到。他们通过对不同温度的压强做一次差分及二次差分,分别得到了熵和比热。前面我们已经讨论过通过比热的两个极值可以给出从经典气体到临界区再到

Luttinger 液体区的临界温度. 而他们的实验结果证实了我们前面的理论. 他们从实验结果清晰地观测到了比热的双峰结构, 并且由比热的双峰定出了临界温度. 为了进一步证实比热的极值给出的临界温度能够提供一个很好的判据, 他们通过测量动量分布证实了在势阱中心附近是 Luttinger 液体.

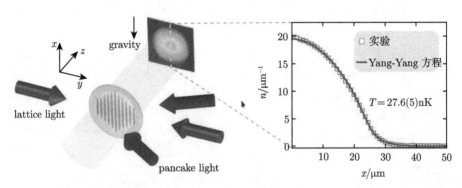

图 2.9 实验原理图: 通过吸收成像方法得到一维导管中的原子气体密度. 实验结果与 Yang-Yang 方程预测结果非常吻合. 此图取自文献 [38] (扫描封底二维码可看彩图)

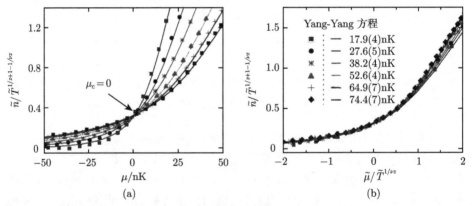

图 2.10 密度标度律: 不同温度的密度交于相变点 $\mu_c = 0$(a), 不同温度的密度塌缩到同一曲线 (b). 此图取自文献 [38]

前面我们还提到 Wilson 系数与 Luttinger 参数的关系, 在这个实验中也得到了较好的证实. Wilson 系数由压缩率及比热的比值得到, 而压缩率可以由密度分布对坐标做差分很容易得到. 在 Luttinger 液体区域, 体系的低能集体激发是声子的激发. 他们首先通过激光将势阱中心的原子打掉一小部分产生一个密度缺陷. 由于 x 方向的束缚较弱, 密度变化较慢, 所以在 Luttinger 液体区缺陷会近似以恒定速度传播, 这个速度就是声速. 有了声速, 通过公式 (2.10.13), 就得到了 Luttinger 参数. 他们的实验结果显示, 在 Luttinger 液体区域 Wilson 系数与 Luttinger 参数已经符

2.13 关于 Lieb-Liniger 玻色气体的实验发展

合得较好, 但是仍然有些差距, 这主要是因为势阱中心的 Luttinger 液体区域很小, 受到量子临界区的影响. 如果降低温度, 那么随着 Luttinger 液体区域的扩大, 可以预期这两个值的差距将会越来越小.

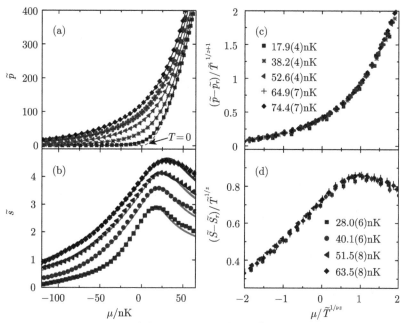

图 2.11 无量纲的压强 $\tilde{p} = p/[\hbar^2 c^3/(2m)]$ 及无量纲熵 $\tilde{S} = S/(k_B c)$ 的标度律: 阴影实线是 Yang-Yang 方程的理论结果, 其线宽度对应着温度不确定性. (a) 实验与理论的压强比较; (b) 实验与理论的熵值比较; (c) 和 (d) 是无量纲的压强和熵的标度函数. 这里由于温度相对值高, 它们的背景值不可忽略. 塌缩的行为定出的临界指数与图 2.10 给出的类似. 此图取自文献 [38]

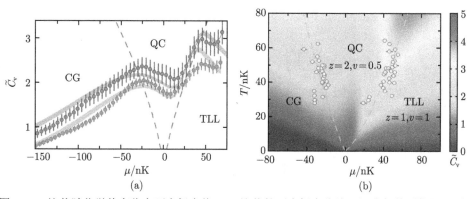

图 2.12 比热随化学势变化有两个极大值 (a); 比热的两个极大值给出经典气体到临界区, 临界区到 Luttinger 液体区的临界温度 (b). 此图取自文献 [38] (扫描封底二维码可看彩图)

近二十年来简并冷原子气体的实验制备和调控技术的发展为研究量子多体现象提供了具有里程碑意义的平台．一维实验的发展推动了理论物理和数学物理方法的研究，从而量子多体系统中的许多基本物理问题得以细致的研究，这包括冷原体系及强相互作用电子体系中的量子临界现象、多体关联、非平衡态动力学、临界动力学、量子多体的普适规律、量子磁学及普适规律 [39].

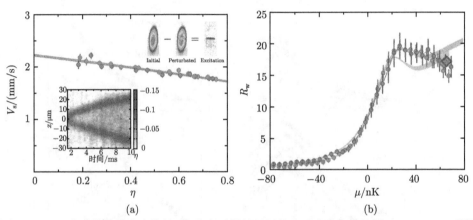

图 2.13　(a) 声速随缺陷大小的关系，其中子图显示了缺陷以声速向外传播；(b) Wilson 系数随化学的关系，其中圆圈为 Wilson 系数的实验结果，方块为 Luttinger 参数的实验结果. 此图取自文献 [38] (扫描封底二维码可看彩图)

致谢：感谢杨文力教授邀请本文作者之一在"第九期理论物理前沿暑期讲习班"授课，感谢何丰、贺文斌、陈洋洋、E. K. P. Nandani 对准备此讲义的帮助.

参 考 文 献

[1] Bethe H. Zeitschrift für Physik, 1931, 71: 205.
[2] Lieb E H, Liniger W. Phys. Rev., 1963, 130: 1605.
[3] Yang C N, Yang C P. Journal of Mathematical physics, 1969, 10: 1115.
[4] Chen S, Guan L M, Yin X G, et al. Phys. Rev. A, 2010, 81: 031609.
[5] Lieb E H. Phys. Rev., 1963, 130: 1616.
[6] Gaudin M. Physical Review A, 1971, 4: 386.
[7] Jiang Y Z, Chen Y Y, Guan X W. Chinese Physics B, 2015, 24: 050311.
[8] Ristivojevic Z. Phys. Rev. Lett., 2014, 113: 015301.
[9] Sato J, Kanamoto R, Kaminishi E, et al. New Journal of Physics, 2016, 18: 075008.
[10] Yang C N. Selected Papers, 1945-1980, with Commentary, World Scientific, 2005, 1.
[11] Guan X W, Batchelor M T. Journal of Physics A: Mathematical and Theoretical, 2011, 44: 102001.

[12] Giamarchi T. Quantum Physics in One Dimension. Oxford University Press, 2004.
[13] Yu Y C, Chen Y Y, Lin H Q, et al. Physical Review B, 2016, 94: 195129.
[14] Wang M S, Huang J H, Lee C H, et al. Phys. Rev. A, 2013, 87: 043634.
[15] Guan X W. International Journal of Modern Physics B, 2014, 28: 1430015.
[16] Fisher M E, Ma S K, Nickel B. Physical Review Letters, 1972, 29: 917.
[17] Li Y, Qiao N S. J. At. Mol. Sci., 2013, 4: 155.
[18] Kira M. Nature Communications, 2015, 6: 031609.
[19] Makotyn P, Klauss C E, Goldberger D L, et al. Nature Physics, 2014, 10: 116.
[20] Nandani E, Römer R A, Tan S, et al. New Journal of Physics, 2016, 18: 055014.
[21] Paredes B, Widera A, Murg V, et al. Nature, 2004, 429: 277.
[22] Kinoshita T, Wenger T, Weiss D S. Science, 2004, 305: 1125.
[23] Haller E, Gustavsson M, Mark M J, et al. Science, 2009, 325: 1224.
[24] Astrakharchik G, Boronat J, Casulleras J, et al. Physical Review Letters, 2005, 95: 190407.
[25] Kinoshita T, Wenger T, Weiss D S. Physical Review Letters, 2005, 95: 190406.
[26] Gando A, Gando Y, Hanakago H, et al. Physical Review C, 2012, 86: 021601.
[27] Armijo J, Jacqmin T, Kheruntsyan K, et al. Physical Review Letters, 2010, 105: 230402.
[28] Tolra B L, O'hara K, Huckans J, et al. Physical Review Letters, 2004, 92: 190401.
[29] Haller E, Rabie M, Mark M J, et al. Physical Review Letters, 2011, 107: 230404.
[30] Kuhnert M, Geiger R, Langen T, et al. Physical Review Letters, 2013, 110: 090405.
[31] Amerongen A V, Es J V, Wicke P, et al. Physical Review Letters, 2008, 100: 090402.
[32] Vogler A, Labouvie R, Stubenrauch F, et al. Physical Review A, 2013, 88: 031603.
[33] Esteve J, Trebbia J B, Schumm T, et al. Physical Review Letters, 2006, 96: 130403.
[34] Armijo J. Physical Review Letters, 2012, 108: 225306.
[35] Jacqmin T, Armijo J, Berrada T, et al. Phys. Rev. Lett., 2011, 106: 230405.
[36] Ho T L, Zhou Q. Nature Physics, 2010, 6: 131.
[37] van Amerongen A H, van Es J J P, Wicke P, et al. Phys. Rev. Lett., 2008, 100: 090402.
[38] Yang B, Chen Y Y, Zheng Y G, et al. arXiv preprint arXiv:1611.00426, 2016.
[39] Guan X W, Batchelor M T, Lee C. Rev. Mod. Phys., 2013, 85: 1633.

第 3 章 共形场论入门

丁祥茂

3.1 共形变换

如同牛顿力学反映宏观物质的运动规律, 量子力学反映微观粒子运动规律, 是人类在 20 世纪探索自然方面最重要的发现之一. 量子力学有所谓的薛定谔图景, 即所谓波动力学以及海森伯图景, 即矩阵力学. 量子力学有哈密顿表示和拉格朗日表示. 哈密顿表示的优点是有哈密顿量和相空间, 存在经典泊松括号, 与量子对易括号对应等; 拉格朗日表示的优点是存在作用量原理, 由极值条件可得到欧拉–拉格朗日运动方程, 有路径积分和费曼图微扰展开等. 但量子力学中不能解决粒子产生或湮没等问题, 这就需要引入量子场论. 量子场论可以看成是无限维的量子力学. 量子场论是物理学理论的集大成者, 也是处理许多物理问题强有力的工具. 如今, 量子场论的影响已不限于物理学, 而是渗透到自然科学的诸多方面, 尤其是现代数学科学.

始于外尔 (Weyl) 和维格纳 (Wigner) 开创性的工作, 对称性在自然科学中的重要性被逐渐揭示出来. 现在, 对称性原则已经成为理论物理学的基石之一, 在物理学中的作用亦日趋显著. 对称性有助于简化物理问题; 在一些情形, 对称甚至完全确定相关的理论值. 对任何给定的系统, 最简单的对称性是在有限数目的反射变换下系统不变; 或在有限数量的方向上系统是位置平移不变的. 反映这种对称性自然的数学理论, 分别是离散群和有限维李群. 另外, 随着研究的深入, 许多新的定理被发现, 用以解构自然规律, 如诺特定理. 在当下的物理学研究中, 关于对称性原则的研究是物理理论的核心问题之一. 可以说, 对称性的形式决定了物理内容, 如狭义相对论, 是庞加莱群变换下不变的; $U(1)$ 不变规范理论, 就是麦克斯韦电磁理论; 而麦克斯韦理论的非阿贝尔推广, 是著名的杨–米尔斯规范理论 [28, 36].

所谓共形场论, 就是共形变换下不变的场论. 最初研究共形场的动机之一是分析统计模型在临界点的二级相变行为 [2]. 后继的研究表明, 二维共形场论可以视为世界表 (worldsheet) 上的场论, 反映无质量粒子的规律, 这样的观点在弦理论研究中有广泛应用 [20, 29]; 共形场论也是严格的数学理论, 诸如顶点算子理论, 就是具体的事例 [13, 16]. 共形变换又名保角变换, 即保持两矢量间夹角不变的变换.

3.1 共形变换

本讲义先考虑一般任意维的共形变换, 继之以二维的情形.

3.1.1 d 维共形变换

设流形是欧几里得空间 R^d, 具有平坦度规 $g_{\mu\nu} = \eta_{\mu\nu}$, 设号差为 (p, q). 线元的定义为

$$\mathrm{d}s^2 = g_{\mu\nu}\mathrm{d}x^\mu \mathrm{d}x^\nu, \tag{3.1.1}$$

如果局域坐标有变换 $x \to x'$, 度规有相应的变化:

$$g_{\mu\nu}(x) \to g'_{\mu\nu}(x') = \frac{x^\alpha}{x'^\mu}\frac{x^\beta}{x'^\nu}g_{\alpha\beta}(x).$$

如果在此坐标变换下允许度规间有任意函数因子的差别, 且该函数因子依赖于局域坐标 x,

$$g_{\mu\nu}(x) \to g'_{\mu\nu} = \Omega(x)g_{\mu\nu}(x), \tag{3.1.2}$$

容易验证, 对于任意两矢量 $\boldsymbol{u}, \boldsymbol{v}$, 此变换保持二者间的夹角 $\boldsymbol{u}\cdot\boldsymbol{v}/(\boldsymbol{u}^2\boldsymbol{v}^2)^{1/2}$ 不变. 共形变换群是坐标变换群的子群, 在共形变换下, 允许度规有依赖局域坐标因子的差别. 研究群的性质, 只需考虑其独立生成元, 以及它们间的关系. 共形变换群的生成元可由无穷小共形变换得到. 若作坐标变换 $x^\mu \to x^\mu + \epsilon^\mu$, 则线元有相应的改变:

$$\mathrm{d}s^2 \to \mathrm{d}s^2 + (\partial_\mu\epsilon_\nu + \partial_\nu\epsilon_\mu)\mathrm{d}x^\mu \mathrm{d}x^\nu. \tag{3.1.3}$$

如果变换是共形的, 要求式 (3.1.3) 右侧第二项正比于度规 $g_{\mu\nu} = \eta_{\mu\nu}$, 即

$$(\partial_\mu\epsilon_\nu + \partial_\nu\epsilon_\mu) = \lambda(x)\eta_{\mu\nu},$$

等式两边同乘以度规 $\eta^{\mu\lambda}$, 有等式:

$$\eta^{\mu\lambda}(\partial_\mu\epsilon_\nu + \partial_\nu\epsilon_\mu) = \lambda(x)\eta^{\mu\lambda}\eta_{\mu\nu},$$

利用度规的定义并求迹, 可以得到

$$2\partial\cdot\epsilon = \lambda(x)d \Rightarrow \lambda(x) = \frac{2}{d}\partial\cdot\epsilon,$$

式中, $\partial\cdot\epsilon \equiv \partial_\mu\epsilon^\mu$. 重新表述上式得到

$$(\partial_\mu\epsilon_\nu + \partial_\nu\epsilon_\mu) = \frac{2}{d}(\partial\cdot\epsilon)\eta_{\mu\nu}, \tag{3.1.4}$$

用微分算子 ∂^ν 作用于上式两端, 有等式

$$\partial_\mu(\partial\cdot\epsilon) + \Box\epsilon_\mu = \frac{2}{d}\partial_\mu(\partial\cdot\epsilon), \tag{3.1.5}$$

式中, 达朗贝尔算子 $\Box \equiv \partial_\mu \partial^\mu$. 再用微分算子 ∂_ν 作用于等式 (3.1.5) 两端, 易得

$$\partial_\mu \partial_\nu (\partial \cdot \epsilon) + \Box \partial_\nu \epsilon_\mu = \frac{2}{d} \partial_\mu \partial_\nu (\partial \cdot \epsilon). \tag{3.1.6}$$

将上式中的下标 μ, ν 对称化, 即将式中的下标 μ, ν 互换并求和:

$$[\eta_{\mu\nu} \Box + (d-2) \partial_\mu \partial_\nu)] \partial \cdot \epsilon = 0, \tag{3.1.7}$$

如果继续对式 (3.1.7) 中的矢量指标实施收缩:

$$(d-1) \Box (\partial \cdot \epsilon) = 0, \tag{3.1.8}$$

由于式 (3.1.7) 中存在因子 $(d-2)$, 故对不同时空维度上的共形变换, 需要分开讨论. 先考虑 $d > 2$ 的情形, 由式 (3.1.8) 可知, ϵ^μ 至多是 x 二次函数, 只存在四种可能的变换, 分别为

- 平移变换 $x'^\mu = x^\mu + a^\mu$, $\Omega(x) = 1$;
- 转动变换 $x'^\mu = \omega^\mu_\nu x^\nu$, $\Omega(x) = 1$;
- 伸缩变换 $x'^\mu = \lambda x^\mu$, $\Omega(x) = \lambda^{-2}$;
- 特殊共形变换 $x'^\mu = \dfrac{x^\mu + x^2 b^\mu}{1 + 2\boldsymbol{b} \cdot \boldsymbol{x} + b^2 x^2}$, $\Omega(x) = (1 + 2\boldsymbol{b} \cdot \boldsymbol{x} + b^2 x^2)^2$;

其中, 特殊共形变换也可以写为 $\dfrac{x'^\mu}{x'^2} = \dfrac{x^\mu}{x^2} + b^\mu$, 即反演并平移变换, 如图 3.1 所示. 所谓反演变换就是将圆周 $x \cdot x = 1$ 的内、外点沿径向互换. 需要注意的是, 特殊共形变换不是全域 (global) 定义的. 容易验证, 给定非零矢量 b^μ, 存在点 $(\boldsymbol{b} \cdot \boldsymbol{b}) x^\mu = -b^\mu$, 使得 $1 + 2\boldsymbol{b} \cdot \boldsymbol{x} + b^2 x^2 = 0$, 从而使得 x^μ 被映射到无穷远. 因此, 需要考虑共形紧化的问题, 即要加上这些无穷远点. 这样, 经过紧致化的共形变换, 就是一个全域定义的、可逆的有限维变换群, 即共形群. 很明显, 庞加莱群 ($\Omega(x) = 1$), 即洛伦兹群加平移, 是共形群的子群, 即狭义相对论是共形理论的一个特例, 在一定意义上, 狭义相对论涉及的问题可以归结为研究庞加莱群的不可约表示. 共形变换群对应的有限维李代数被称为共形代数, 是无穷小共形变换的生成元构成的代数. 一般而言, 生成元构成的代数比共形代数更大, 共形代数只是它的有限维子代数. 关于这一点, 在二维欧几里得空间的情形会变得非常清楚. 局域地, 对上述共形变换, 可用无穷小共形变换的生成元来表达: $a^\mu \partial_\mu$; $\omega^\mu_\nu x^\nu \partial_\mu$; $\lambda x \cdot \partial$; 以及 $b^\mu (x^2 \partial_\mu - 2 x^\mu x \cdot \partial)$, 对应的独立生成元的数目分别为: d, $d(d-1)/2$, 1 和 d, 所以, 共形代数独立生成元数目总计有 $(d+1)(d+2)/2$. 考虑到流形度规的号差, 该共形代数与 $so(p+1, q+1)$ 同构. 一般的情形, 空间的维度 $d \neq 2$ 时, 共形群同构于转动群 $SO(p+1, d-p+1)$. 由共形代数的生成元得到共形群的线性表达式. 将共形代数生成元间的关系表述

3.1 共形变换

为与庞加莱群的生成元类似的形式:

$$\begin{cases} P_\mu = -\mathrm{i}\partial_\mu, \\ L_{\mu\nu} = \mathrm{i}(x_\mu \partial_\nu - x_\nu \partial_\mu), \\ D = -\mathrm{i}x^\mu \partial_\mu, \\ K_\mu = -\mathrm{i}(2x_\mu x^\nu \partial_\nu - x^2 \partial_\mu). \end{cases} \tag{3.1.9}$$

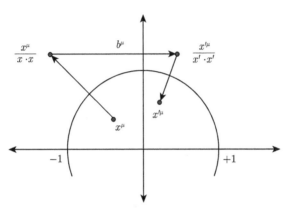

图 3.1 特殊共形变换

生成元间存在如下交换关系:

$$\begin{cases} [D, P_\mu] = \mathrm{i}P_\mu, \\ [D, K_\mu] = -\mathrm{i}K_\mu, \\ [K_\mu, P_\nu] = 2\mathrm{i}(\eta_{\mu\nu} D - L_{\mu\nu}), \\ [K_\rho, L_{\mu\nu}] = \mathrm{i}(\eta_{\rho\mu} K_\nu - \eta_{\rho\nu} K_\mu), \\ [P_\rho, L_{\mu\nu}] = \mathrm{i}(\eta_{\rho\mu} P_\nu - \eta_{\rho\nu} P_\mu), \\ [L_{\mu\nu}, L_{\rho\sigma}] = \mathrm{i}(\eta_{\nu\rho} L_{\mu\sigma} + \eta_{\mu\sigma} L_{\nu\rho} - \eta_{\mu\rho} L_{\nu\sigma} - \eta_{\nu\sigma} L_{\mu\rho}). \end{cases} \tag{3.1.10}$$

对庞加莱群的生成元进行线性组合, 定义一组新的基:

$$\begin{cases} J_{\mu,\nu} \equiv L_{\mu\nu}, \\ J_{-1,\mu} \equiv \dfrac{1}{2}(P_\mu - K_\mu), \\ J_{0,\mu} \equiv \dfrac{1}{2}(P_\mu + K_\mu), \\ J_{-1,0} \equiv D. \end{cases} \tag{3.1.11}$$

容易验证, 有如下等式:

$$[J_{mn}, J_{pq}] = \mathrm{i}\left(\eta_{mq} J_{np} + \eta_{np} J_{mq} - \eta_{mp} J_{nq} - \eta_{nq} J_{mp}\right). \tag{3.1.12}$$

可见, $d \geqslant 3$ 时的共形变换并不会给出比庞加莱群更多的信息 [9,17]. 尽管如此, 从整体共形变换条件仍可以得到一些普适的结果.

通常, 量子共形不变源自经典共形不变. 临界点的共形不变性在临界点之外就可能被破坏. 如果无穷小共形变换由 ϵ_a 来参数化, T_a 是相应的矩阵表示, 考虑它对于场量 $\phi(x)$ 的影响:

$$\phi'(x') = (1 - i\epsilon_a T_a)\phi(x). \tag{3.1.13}$$

洛伦兹群作为庞加莱群的子群, 有保持原点不变的性质, 存在矩阵表示:

$$L_{\mu\nu}\phi(0) = S_{\mu\nu}\phi(0). \tag{3.1.14}$$

用算子关系的豪斯多夫公式, 容易得到

$$e^{ix^\lambda P_\lambda} L_{\mu\nu} e^{-ix^\lambda P_\lambda} = L_{\mu\nu} - x_\mu P_\nu + x_\nu P_\mu. \tag{3.1.15}$$

让生成元 $L_{\mu\nu}$, P_μ 分别作用在场量上

$$P_\mu \phi(x) = -i\partial_\mu \phi(x),$$
$$L_{\mu\nu}\phi(x) = i(x_\mu \partial_\nu - x_\nu \partial_\mu)\phi(x) + S_{\mu\nu}\phi(x). \tag{3.1.16}$$

同样, 有等式

$$\begin{cases} e^{ix\cdot P} D e^{-ix\cdot P} = D + x^\nu P_\nu, \\ e^{ix\cdot P} K_\mu e^{-ix\cdot P} = K_\mu + 2x_\mu D - 2x^\nu L_{\mu\nu} + 2x_\mu(x^\nu P_\nu) - x^2 P_\mu. \end{cases} \tag{3.1.17}$$

这样就得到保持原点的伸缩以及特殊共形变换的矩阵表示, 分别记为 $-i\Delta$ 和 κ_μ, 容易得到这些矩阵表示的代数式:

$$\begin{cases} D\phi(x) = -i(x^\nu \partial_\nu + \Delta)\phi(x), \\ K_\mu \phi(x) = \left(\kappa_\mu - 2ix_\mu \Delta - x^\nu S_{\mu\nu} - 2ix_\mu x^\nu \partial_\nu + ix^2 \partial_\mu\right)\phi(x), \end{cases} \tag{3.1.18}$$

其中, Δ 就是场 $\phi(x)$ 的标度维. 但到目前为止, 并没有明确给出标度维的定义. 场论中, 不同的场量遵从不同的变换规律. 自然, 在共形变换下, 不同性质的场量也具有不同的变换性质. 伴随伸缩变换 $x \to x\lambda$, $\phi(x)$ 的变换性质为

$$\phi(\lambda x) = \lambda^{-\Delta} \phi(x), \tag{3.1.19}$$

其中, Δ 是由 ϕ 场在伸缩变换下给定的, 被称为标度维. 在量子场论中, 作用量是无量纲的. 例如, 考虑 d-维平坦空间零质量自由标量场

$$S = \int d^d x \ \partial_\mu \phi(x) \partial^\mu \phi(x). \tag{3.1.20}$$

3.1 共形变换

容易得到 d-维平坦空间零质量自由标量场的标量维度:

$$\Delta = \frac{d}{2} - 1. \tag{3.1.21}$$

进一步, 利用 $-\mathrm{i}\Delta$ 正比于单位矩阵这一性质, 有 $\kappa_\mu = 0$. 这样就得到 ϕ 场的变换性质:

$$\begin{cases} P_\mu \Phi(x) = -\mathrm{i}\partial_\mu \Phi(x) \\ L_{\mu\nu}\Phi(x) = \mathrm{i}(x_\mu\partial_\nu - x_\nu\partial_\mu)\Phi(x) + S_{\mu\nu}\Phi(x) \\ D\Phi(x) = -\mathrm{i}(x^\mu\partial_\mu + \Delta)\Phi(x) \\ K_\mu\Phi(x) = (-2\mathrm{i}\Delta x_\mu - x^\nu S_{\mu\nu} - 2\mathrm{i}x_\mu x^\nu\partial_\nu + \mathrm{i}x^2\partial_\mu)\Phi(x) \end{cases} \tag{3.1.22}$$

利用等式 (3.1.22), 可以得到 ϕ 场在有限共形变换下的变更. 为简单起见, 只考虑无自旋场. 对局域坐标变换, 有雅可比行列式 $\left|\frac{\partial x'}{\partial x}\right| = (\det g'_{\mu\nu})^{-1/2} = \Omega^{-d/2}$. 存在一类子集合, 使得在全域共形变换 $x \to x'$ 下, 有标度维度为 Δ_j 的场量 ϕ_j:

$$\phi_j(x) \to \left|\frac{\partial x'}{\partial x}\right|^{\Delta_j/d} \phi_j(x'), \tag{3.1.23}$$

在共形变换下协变. 具有这类性质的场量, 被称为拟元场 (quasi-primary), 或拟初场. 任何其他的场量都可以通过拟元场的线性组合或微分得到. 因此, ϕ_j 的 n-点函数为

$$\langle \phi_1(x_1)\cdots\phi_n(x_n)\rangle = \left|\frac{\partial x'}{\partial x}\right|^{\Delta_1/d}_{x=x_1}\cdots\left|\frac{\partial x'}{\partial x}\right|^{\Delta_n/d}_{x=x_n}\langle \phi_1(x'_1)\cdots\phi_n(x'_n)\rangle. \tag{3.1.24}$$

如果存在全域共形变换不变的真空 $|0\rangle$, 共形协变条件 (3.1.23) 对多点函数 (3.1.24) 给出了严苛的限制. 具体到平移、伸缩和特殊共形变换, 约束了多点函数与局域坐标间的关系. 由于存在平移不变性, 多点函数只依赖于坐标差 $(x_i - x_j)$, 再考虑到场无自旋, 具有转动不变性, 这样对坐标的依赖关系就成了 $r_{ij} \equiv |x_i - x_j|$; 由伸缩变换, 可知依赖关系变成 r_{ij}/r_{kl}; 最后, 考虑到特殊共形变换, 只有所谓的交比

$$\frac{r_{ij}r_{kl}}{r_{ik}r_{jl}} \tag{3.1.25}$$

在全部共形变换下不变的量. 对 N-点函数, 独立交比 (3.1.25) 的数目为 $N(N-3)/2$. 具体到两点函数

$$\langle \phi_1(x_1)\phi_2(x_2)\rangle = c_{12}|x_1 - x_2|^{-\Delta_1-\Delta_2}, \tag{3.1.26}$$

其中, c_{12} 是场的归一化决定的常数. 由特殊共形变换可知, 如果 $c_{12} \neq 0$, 则要求 $\Delta_1 = \Delta_2$, 因此

$$\langle \phi_1(x_1)\phi_2(x_2)\rangle = \begin{cases} c_{12}|x_1 - x_2|^{-2\Delta}, & \Delta = \Delta_1 = \Delta_2, \\ 0, & \Delta_1 \neq \Delta_2. \end{cases} \tag{3.1.27}$$

同样, 三点函数有

$$\langle \phi_1(x_1)\phi_2(x_2)\phi_3(x_3)\rangle = c_{123}\prod |x_i-x_j|^{\Delta_k-\Delta_i-\Delta_j}, \quad i,j,k=1,2,3;\ i\neq j\neq k, \tag{3.1.28}$$

式中的 c_{123} 是不依赖坐标点的常数. 四点函数

$$\langle \phi_1(x_1)\phi_2(x_2)\phi_3(x_3)\phi_4(x_4)\rangle = F\left(\frac{r_{12}r_{34}}{r_{13}r_{24}},\frac{r_{12}r_{34}}{r_{14}r_{23}}\right)\prod_{i<j}|x_i-x_j|^{\Delta/3-\Delta_i-\Delta_j}, \tag{3.1.29}$$

其中, $\Delta = \sum_{i=1}^{4}\Delta_i$; F 是两变量的未定函数. 对于四点和四点以上的函数, 整体共形并不能完全确定其表达式. 进一步的讨论可知, 由全域共形变换出发, 如果加上场的幺正性、关联函数的单值性等, 会给关联函数加上更多的约束, 多点函数成为特定微分方程的解. 通过这些特定微分方程, 任意 N- 点关联函数都可以通过一定的方法约化为两点或三点关联函数. 因此, 原则上任意 N- 点关联函数都是完全确定的.

$d=1$ 对应于时间、空间退化为一个点, 此时的理论简化为共形量子力学, 式 (3.1.8) 不能给出关于 ϵ 的任何约束, 所有光滑变换都是共形的. 在 $d=1$ 时, 任意两矢量间的夹角为零, 不存在非平凡的局域转动, 共形变换只有平移、伸缩和特殊共形三种, 共形群同构于群 $SO(2,1)$.

而当 $d=2$ 时, 共形变换的性质完全不同于其他情形, 后面给出更详细的讨论.

3.1.2 二维共形变换

当流形为实二维时, 共形场论有很多新特性, 使其完全不同于其他维. 二维共形场是反映临界点二级相变行为的理论. 在临界点, 系统的关联长度变成无限大, 系统位形空间允许各种尺度的涨落. 反映临界点行为的理论模型, 至少在局域范围内应该是共形不变的. 本节我们简述二维共形场的一些基本知识.

二维共形对称性, 是由无限维李代数 —— Virasoro 代数给出的, 它是 Witt 代数的中心扩张. Virasoro 代数的表示论具有特殊结构, 由幺正性出发, 可以得到一类二维共形场的分类, 尤其是中心荷 $c<1$ 的情形. 但是研究表明, 对于二维共形对称性的完整分类, 即包括中心荷 $c>1$ 的情形, 则有必要考虑扩展的共形对称.

将任意维度共形变换的一般表达式具体到二维情形. 设有坐标变换 $z^\mu \to w^\mu$, 相应度规的变化式为

$$g^{\mu\nu} \to \left(\frac{\partial w^\mu}{\partial z^\alpha}\right)\left(\frac{\partial w^\nu}{\partial z^\beta}\right)g^{\alpha\beta} \propto g^{\mu\nu}, \tag{3.1.30}$$

若要求变换是共形的, 则必须有等式:

$$\left(\frac{\partial w^0}{\partial z^0}\right)^2 + \left(\frac{\partial w^0}{\partial z^1}\right)^2 = \left(\frac{\partial w^1}{\partial z^0}\right)^2 + \left(\frac{\partial w^1}{\partial z^1}\right)^2, \tag{3.1.31}$$

3.1 共形变换

$$\frac{\partial w^0}{\partial z^0}\frac{\partial w^1}{\partial z^0} + \frac{\partial w^0}{\partial z^1}\frac{\partial w^1}{\partial z^1} = 0, \tag{3.1.32}$$

此方程组等价于

$$\partial_0 w_1 = \pm \partial_1 w_0, \quad \partial_0 w_0 = \mp \partial_1 w_1. \tag{3.1.33}$$

可以用复坐标 $z, \bar{z}$ 来参数化此共形变换. 先引入下面的记号:

$$\epsilon \equiv \epsilon^0 + \mathrm{i}\epsilon^1 \quad z \equiv x^0 + \mathrm{i}x^1, \quad \partial_z \equiv \frac{1}{2}(\partial_0 - \mathrm{i}\partial_1),$$

$$\bar{\epsilon} \equiv \epsilon^0 - \mathrm{i}\epsilon^1 \quad \bar{z} \equiv x^0 - \mathrm{i}x^1, \quad \partial_{\bar{z}} \equiv \frac{1}{2}(\partial_0 + \mathrm{i}\partial_1). \tag{3.1.34}$$

显然, 在复坐标下度规矩阵为

$$g_{\mu\nu} = \begin{pmatrix} 0 & \frac{1}{2} \\ \frac{1}{2} & 0 \end{pmatrix}, \tag{3.1.35}$$

而线元 $\mathrm{d}s^2 = \mathrm{d}z\mathrm{d}\bar{z}$. 这里采用的符号 $\partial = \partial_z, \bar{\partial} = \partial_{\bar{z}}$. 显而易见, 保角变换等同于柯西–黎曼方程. 由前面的讨论可知,

$$z \to f(z), \quad \bar{z} \to \bar{f}(\bar{z}), \tag{3.1.36}$$

所以当共形群变换限制在实二维时, 在复坐标下, 就等价于复平面上任意解析变换. 这样的解析变换是不可数的, 所以在二维共形变换背后隐藏着无穷维对称性. 与其他维度共形变换相比, 具有隐藏的无穷维对称性, 是二维共形变换的主要特性. 事实证明, 无穷维对称性的存在影响深远. 肇始于文献 [2] 的发表, 在此后短短数十年间, 共形场论发展成为一门全新的学科, 并影响到许多其他学科的进程. 本节中我们给出了二维共形场的一些简单概念以及更细致和系统的介绍和评述, 请参考文献 [2], [9], [16], [17], [24].

3.1.3 Witt 代数

为叙述方便, 首先介绍一些相关的定义. 给定两个拓扑空间 X 和 Y, X 和 Y 同胚, 如果它们之间存在连续映射 f 及其逆映射 f^{-1}:

$$f: X \to Y, \quad f^{-1}: Y \to X, \quad f \circ f^{-1} = f^{-1} \circ f = 1.$$

设 n- 维拓扑流形 $\mathcal{M}$ 是由坐标卡 U_α 覆盖的豪斯多夫空间, 其中每个 U_α 都是同胚到 R^n 的开集. 通常情况下, 一对坐标卡 U_α 和 U_β 之间的转换函数 $g_{\alpha\cdot\beta}$ 是同胚的. 更进一步, 如果要求转换函数 $g_{\alpha\cdot\beta}$ 是微分同胚的, 即 $g_{\alpha\cdot\beta}$ 和 $g_{\alpha\cdot\beta}^{-1}$ 都是 C^∞,

即无限可微的函数. X 的微分同胚流形构成一个群, 记为 $\mathrm{Diff}(X)$. 两个微分同胚 $f, g \in \mathrm{Diff}(X)$ 的积由下面定义的复合构成:

$$f \circ g : X \to X, \quad x \to f(g(x)).$$

由 $\mathrm{Diff}(X)$ 定义的李代数, 可以有如下的等价表述: 设流形 X 上的 (光滑) 矢量场 V, 其积分曲线为 $v(t)$. 然后, 沿 X 积分曲线走距离 t, 这样就产生一个微分同胚族, 用 f_t 标记. 将 $\mathrm{Diff}(X)$ 的李代数等价为 X 上的光滑矢量场, 记为 $\mathrm{vect}(t)$, 然后有等式:

$$\mathcal{L}(\mathrm{Diff}(X)) = \mathrm{vect}(x).$$

在局域坐标下, 可以写出

$$V = \sum_i v_i \frac{\partial}{\partial x^i}, \quad f_t = \exp(tV).$$

具体到圆周 S^1 上, 与圆周 S^1 微分同胚的李代数 $\mathrm{Diff}(S^1)$ 的生成元就是矢量 $f(\theta)\frac{\mathrm{d}}{\mathrm{d}\theta}$, 其中 θ 是角参数, 其赋值范围为 $0 \leqslant \theta \leqslant 2\pi$. 对易子 (亦称交换子) [34] 为

$$[f, g] = \left(f\frac{\mathrm{d}g}{\mathrm{d}\theta} - \frac{\mathrm{d}f}{\mathrm{d}\theta}g \right), \tag{3.1.37}$$

因此, 共形微分同胚由 $\mathrm{Diff}_c(M) = \mathrm{Diff}(S^1) \times \mathrm{Diff}(S^1)$ 给出, 其中, 在复坐标表示下, 第一因子对应于变换 $z \to f(z)$, 而第二因子对应 $\bar{z} \to \bar{f}(\bar{z})$.

为了计算二维无穷小共形变换的生成元, 即形式 (3.1.36) 的无限小变换的对易关系, 如果我们选择变换为

$$z \to z' = z + \epsilon(z),$$
$$\bar{z} \to \bar{z}' = \bar{z} + \bar{\epsilon}(\bar{z}),$$

由于 $\epsilon(z)$ 与 $\bar{\epsilon}(\bar{z})$ 是完全类似的, 可以只表一枝. 一般而言, $\epsilon(z)$ 是亚纯函数, 在开集外存在奇异点, 因此可以在 $z = 0$ 附近作洛朗 (Laurent) 展开:

$$\begin{cases} z \to z' = z + \displaystyle\sum_{n \in \mathbb{Z}} \epsilon_n(z), \\ \bar{z} \to \bar{z}' = \bar{z} + \displaystyle\sum_{n \in \mathbb{Z}} \bar{\epsilon}_n(\bar{z}), \end{cases} \tag{3.1.38}$$

其中

$$\epsilon_n(z) = -z^{n+1}, \quad \bar{\epsilon}_n(\bar{z}) = -\bar{z}^{n+1}, \tag{3.1.39}$$

3.1 共形变换

相应的无穷小共形变换的生成元分别为

$$\ell_n = -z^{n+1}\partial_z, \quad \bar{\ell}_n = -\bar{z}^{n+1}\partial_{\bar{z}}. \tag{3.1.40}$$

重要的是注意到 $n \in \mathbb{Z}$, 即无穷小共形变换独立生成元的数目为无限多. 这是二维共形场所特有的, 将会导致影响深远的结果. 下一步是计算生成元之间的对易关系, 容易验证

$$\begin{cases} [\ell_m, \ell_n] = (m-n)\ell_{m+n}, \\ [\bar{\ell}_m, \bar{\ell}_n] = (m-n)\bar{\ell}_{m+n}, \quad m, n \in \mathbb{Z}, \\ [\ell_m, \bar{\ell}_n] = 0. \end{cases} \tag{3.1.41}$$

3.1.4 共形子代数

需要特别提及, 代数 (3.1.41) 是局域定义的, 因为在黎曼球面 $S^2 = \mathbb{C} \cup \{\infty\}$ 上, 代数 (3.1.41) 不存在处处恰当的定义. 由定义显见, 当 $z \to 0$ 时, 要使 ℓ_n 非奇异, 当且仅当 $n \geqslant -1$; 若进行坐标变换 $z = w^{-1}$, 又得到 ℓ_n 非奇异, 当且仅当 $n \leqslant 1$. 即在黎曼球面上, 全纯部分的全域共形变换是由 $\ell_{-1}, \ell_0, \ell_1$ 生成的. 对 $\bar{\ell}_n$ 亦有同样的结果. 因此, 就全域而言, 黎曼球面上的整体共形变换是由 $\{\ell_{-1}, \ell_0, \ell_1\} \cup \{\bar{\ell}_{-1}, \bar{\ell}_0, \bar{\ell}_1\}$ 生成的. 如果引入复的极坐标 $z = re^{i\theta}$, 并对微分的结果进行适当的组合:

$$\ell_0 + \bar{\ell}_0 = -r\partial_r, \quad \text{以及} \quad i(\ell_0 - \bar{\ell}_0) = -\partial_\theta, \tag{3.1.42}$$

分别是径向伸缩和角向转动生成元. 不难分析, l_{-1} 产生变换 $z \to z+b$; l_0 产生变换 $z \to az$, $a \in \mathbb{C}$; l_{+1} 对应于特殊共形变换, 它对 $w = z^{-1}$ 的平移, 因为 $c l_1 z = -cz^2$ 是 $z \to \dfrac{z}{cz+1}$ 的无穷小变换, 所以它对应于 $w \to w - c$. 归纳起来, 它们的有限变换的表达式为分式线性变换:

$$z \to \frac{az+b}{cz+d}, \quad \bar{z} \to \frac{a\bar{z}+b}{c\bar{z}+d},$$

式中, $a, b, c, d \in \mathbb{C}$, 且 $ad - bc = 1$. 所以, 莫比乌斯 (Möbius) 变换 $SL(2,\mathbb{C})/\mathbb{Z}_2 \cong SO(3,1)$, 这正是所谓的射影共形变换群. 共形变换限制在实二维时, 它构成共形变换的子群. 在其他维数, 只存在全域共形变换. 由于 $\ell_n, n \in \mathbb{Z}$ 只有局域的定义, 因此后面使用代数而非群的语言.

如前文所分析, 微分算子 $\ell_0 + \bar{\ell}_0 = -r\partial_r$ 与 $i(\ell_0 - \bar{\ell}_0) = -\partial_\theta$, 分别产生伸缩和旋转变换. 假设存在微分算子 $\ell_0, \bar{\ell}_0$ 的本征态 $|h, \bar{h}\rangle$, 其特征值分别为 $h, \bar{h}$, 即 $\ell_0|h, \bar{h}\rangle = h|h, \bar{h}\rangle$, $\bar{\ell}_0|h, \bar{h}\rangle = \bar{h}|h, \bar{h}\rangle$, 则称 h 和 $\bar{h}$ 为态 $|h, \bar{h}\rangle$ 的左、右共形维, 而相应地有态的 $|h, \bar{h}\rangle$ 的标度维 $\Delta = h + \bar{h}$, 以及自旋 $s = h - \bar{h}$. 从现在开始, 我们引入

元场 (primary field), 其定义如下：

$$\phi(z,\bar{z}) \to \left(\frac{\partial f}{\partial z}\right)^h \left(\frac{\partial \bar{f}}{\partial \bar{z}}\right)^{\bar{h}} \phi(f(z),\bar{f}(\bar{z})), \tag{3.1.43}$$

其中, $(h,\bar{h})$ 分别是元场 ϕ 的左、右共形维. 当然, 不是所有共形场都具有这样的性质. 除了元场之外的场, 其余称为属场 (secondary fields).

3.1.5 元场的多点函数

考虑无穷小坐标变换 $z \to \epsilon(z)$, $\bar{z} \to \bar{\epsilon}(\bar{z})$, $\epsilon, \bar{\epsilon}$ 参数化的全纯、反全纯无穷小变换:

$$\delta_{\epsilon,\bar{\epsilon}}\phi(z,\bar{z}) = \left(h\partial_z\epsilon + \epsilon\partial_z + \bar{h}\partial_{\bar{z}}\bar{\epsilon} + \bar{\epsilon}\partial_{\bar{z}}\right)\phi(z,\bar{z}). \tag{3.1.44}$$

对于莫比乌斯变换, 则有

$$\phi(z,\bar{z}) \to (cz+d)^{-2h}(\bar{c}\bar{z}+\bar{d})^{-2\bar{h}}\phi(f(z),\bar{f}(\bar{z})).$$

如果场共形维是 $(h,0)$, 则被称为"手征左", 即不依赖 $\bar{z}$; 相反, 共形维是 $(0,\bar{h})$ 的场, 被称为"手征右", 不依赖 z. 由于多点函数要满足 (3.1.24), 具体到二维情形:

$$G^{(2)}(z_i,\bar{z}_i) = \langle \phi_1(z_1,\bar{z}_1)\phi_2(z_2,\bar{z}_2)\rangle = \frac{c_{12}\delta_{h_i,h_j}}{|z_{12}|^{2h_i}|\bar{z}_{12}|^{2\bar{h}_i}}, \tag{3.1.45}$$

其中, c_{12} 是归一常数, $z_{ij} = z_i - z_j$. 同样, 三点函数有

$$\begin{aligned}G^{(3)}(z_i,\bar{z}_i) &= \langle \phi_1(z_1,\bar{z}_1)\phi_2(z_2,\bar{z}_2)\phi_3(z_3,\bar{z}_3)\rangle \\ &= \prod C_{ijk}|z_{ij}|^{h_k-h_i-h_j}|\bar{z}_{ij}|^{\bar{h}_k-\bar{h}_i-\bar{h}_j}, \quad i,\ j,\ k=1,\ 2,\ 3;\ i\neq j\neq k,\end{aligned} \tag{3.1.46}$$

其中, C_{ijk} 是不依赖坐标点的常数, 利用共形变换的性质, 总可以选择 $\infty, 1, 0$ 三点作为参考点. 四点函数

$$\begin{aligned}G^{(4)}(z_i,\bar{z}_i) &= \langle \phi_1(z_1,\bar{z}_1)\phi_2(z_2,\bar{z}_2)\phi_3(z_3,\bar{z}_3)\phi_4(z_4,\bar{z}_4)\rangle \\ &= f(\rho,\bar{\rho})\prod_{i<j}|z_{ij}|^{h/3-h_i-h_j}|\bar{z}_{ij}|^{\bar{h}/3-\bar{h}_i-\bar{h}_j},\end{aligned} \tag{3.1.47}$$

其中, $h = \sum_{i=1}^{4} h_i$, $\bar{h} = \sum_{i=1}^{4} \bar{h}_i$, $\rho = z_{12}z_{34}/z_{13}z_{24}$, f 是关于交比的未定函数. 同样, 也许选择 $\infty, 1, z, 0$ 四点作为参考点. 原则上, 式 (3.1.45)~ 式 (3.1.47) 中出现的共形维是任意的, 实际上却不是这样, 单值条件以及后面将要讨论的酉条件, 会对多点函数加上很强的约束.

3.1.6　二维共形代数的中心扩张

微分运算 (3.1.37) 定义的李代数, 存在泛中心扩张:

$$\phi(f,g) = \frac{1}{48\pi} \int d\theta \left(f \frac{d^3 g}{d\theta^3} - \frac{d^3 f}{d\theta^3} g \right), \tag{3.1.48}$$

$\phi(f,g)$ 是上闭链 (cocycle)[34], 为记号方便, 已经采用了归一化. 这样, 中心扩张的 $\text{Diff}(S^1)$ 生成元就通过二元组 (a,f) 来表示, 其中 a 为待定实数, 则中心扩张的李代数的对易子为

$$[(a,f),(b,g)] = \left(c\phi(f,g), f\frac{dg}{d\theta} - \frac{df}{d\theta}g \right), \tag{3.1.49}$$

这里, c 可以是任意实数, 与所有代数生成元可交换, 被称为代数的中心. 对于 S^1 上的任意三个矢量 f, g, h, 上闭链满足雅可比恒等式:

$$\phi(f,[g,h]) + \phi(g,[h,f]) + \phi(h,[f,g]) = 0, \tag{3.1.50}$$

如果用代数的生成元写出, 有如下结果 [11]:

$$[L_m, L_n] = (m-n)L_{m+n} + c\alpha(m,n), \quad [L_n, c] = 0,$$
$$[\bar{L}_m, \bar{L}_n] = (m-n)\bar{L}_{m+n} + \bar{c}\alpha(m,n), \quad [\bar{L}_n, \bar{c}] = 0, \tag{3.1.51}$$
$$[L_m, \bar{L}_n] = 0.$$

式中, $c, \bar{c}$ 是代数的中心荷; 系数 $\alpha(m,n) \in \mathbb{C}, m, n \in \mathbb{Z}$. 从前面的讨论知, 式 (3.1.51) 是式 (3.1.41) 的中心扩张. 由于对易子的反对称性, 以及雅可比恒等式的约束, 系数 $\alpha(m,n)$ 不可能是关于 m, n 的任意函数. 注意到 $[L_0, L_n] = -nL_n + c\alpha(0,n)$, 可以重新定义代数的基, 即 $L_0' = L_0, L_n' = L_n - c\alpha(0,n)/n\ (n \neq 0)$. 显然, 这样重新定义的基 L_n', 与 L_m 之间只差一个变换, 从代数意义讲, 二者并无差别. 因此, 为简化符号, 下面将省去 L_n' 的上标. 利用雅可比等式:

$$[L_0, [L_m, L_n]] = -(m+n)[L_m, L_n] + c\alpha(m,n), \quad [L_n, c] = 0, \tag{3.1.52}$$

将式 (3.1.51) 代入上式, 有 $(m+n)\alpha(m,n)c = 0$. 这表示 $\alpha(m,n) = \alpha(m)\delta_{m+n,0}$, 考虑到对易子的反对称性, 有 $\alpha(-m) = -\alpha(m)$. 因此, $\alpha(m)$ 是关于 m 的奇函数. 计算任意三生成元 L_m, L_n, L_l 的雅可比关系, 限定在 $m+n+l=0$ 情形, 得到等式:

$$(m-n)\alpha(m+n) - (m+2n)\alpha(m) + (2m+n)\alpha(n) = 0. \tag{3.1.53}$$

这是一个线性递归关系, 因此, 其解空间最多是二维的, 也就是说, 最多只存在两个线性独立的解. 显然, $\alpha(m) = m$ 和 $\alpha(m) = m^3$ 都是式 (3.1.53) 的解, 所以它的通

解式可以写成 $\alpha(m) = \gamma m + \beta m^3$, 其中 γ, β 是与 m 无关的任意常数. 从前面的讨论可知, 如果取 $\beta = 0$, 则可以通过重新定义代数基, 而回到无中心扩张的代数. 若要求存在非平凡的中心扩张, $\beta \neq 0$, 而 γ 的取值存在任意性, 一种方便的选择是 $\beta = -\gamma = 1/12$. 这一结果与上链定义式 (3.1.48) 相同, 于是就得到所谓的 Virasoro 代数 [2,17]:

$$\begin{cases} [L_m, L_n] = (m-n)L_{m+n} + \dfrac{c}{12}(m^3 - m)\delta_{m+n,0}, \\ [\bar{L}_m, \bar{L}_n] = (m-n)\bar{L}_{m+n} + \dfrac{\bar{c}}{12}(m^3 - m)\delta_{m+n,0}, \\ [L_m, \bar{L}_n] = 0. \end{cases} \tag{3.1.54}$$

这是二维共形场最重要的代数关系. 从一定意义讲, 二维共形场研究就是分析此无限维代数的各类表示. 由于全纯生成元与反全纯生成元对易, 共形变换分化为两同构子代数的直和 $\mathcal{A} \oplus \bar{\mathcal{A}}$, 通常将 z 和 $\bar{z}$ 视为互为独立的、而不是彼此共轭的变量. 单纯为简单计, 在不引起混淆的情况下, 常常忽略反全纯的部分.

3.1.7 守恒量与能-动张量

作用量变分原理是协变场论的核心, 它将场的运动方程、连续性条件和对称性置于同等地位. 可以将作用量视为各种动力学量的生成泛函. 所谓的诺特定理, 就具体揭示了连续对称性与守恒量之间的关系. 如果物理系统的作用量可由拉格朗日函数的积分得到:

$$S = \int L \mathrm{d}t = \int \mathrm{d}^d x \mathcal{L}(\phi, \partial_\mu \phi),$$

通过最小作用原理 $\delta S = 0$, 得到欧拉-拉格朗日方程:

$$\frac{\partial \mathcal{L}}{\partial \phi} - \partial_\mu \left(\frac{\partial \mathcal{L}}{\partial (\partial_\mu \phi)} \right) = 0.$$

如果场量作无穷小的改变

$$\phi(x) \to \phi'(x) = \phi(x) + \alpha \Delta \phi(x),$$

α 为无穷小参数, $\Delta \phi(x)$ 是场位形空间的变更. 如果拉格朗日作用量在此无穷小变更下保持不变, 则欧拉-拉格朗日运动方程保持不变. 这样的变换就是该拉格朗日系统的一种对称性, 即存在一确定的 $\mathcal{J}^\mu$, 使得拉氏量

$$\mathcal{L}(x) \longrightarrow \mathcal{L}(x) + \alpha \partial_\mu \mathcal{J}^\mu(x)$$

保持不变, 因此有

$$\alpha \Delta \mathcal{L} = \alpha \partial_\mu \left(\frac{\partial \mathcal{L}}{\partial (\partial_\mu \phi)} \Delta \phi \right) + \alpha \left[\frac{\partial \mathcal{L}}{\partial \phi} - \partial_\mu \left(\frac{\partial \mathcal{L}}{\partial (\partial_\mu \phi)} \right) \right] \Delta \phi.$$

3.1 共形变换

如果定义
$$j^\mu(x) = \frac{\partial \mathcal{L}}{\partial(\partial_\mu \phi)} \Delta\phi - \mathcal{J}^\mu,$$
则有 $\partial_\mu j^\mu(x) = 0$, 相应于守恒流 $\partial_\mu j^\mu(x)$ 的守恒荷为
$$Q = \int \mathrm{d}^{d-1}x\, j^0(x).$$
如果定义
$$T^{\mu\nu} \equiv \frac{\partial \mathcal{L}}{\partial(\partial_\mu \phi)} \partial^\nu \phi - \mathcal{L}\eta^{\mu\nu},$$
这就是所谓的能–动张量. 显而易见, 能量守恒是该系统时间平移不变导致的; 而动量守恒则是空间平移不变的结果. 如果将时间、空间的平移写成统一的形式, 则在无穷小变换下, 作用量的变分为
$$\delta S = -\int \mathrm{d}^d x\, j_a^\mu \partial_\mu \epsilon_a, \tag{3.1.55}$$
即能–动张量是时间、空间平移变换共同的生成元.

如果存在依赖局域坐标的平移变更 $x^\mu \to x^\mu + \epsilon^\mu(x)$, 而能–动张量是相应的守恒量, 则作用量的变分为
$$\delta S = -\frac{1}{2}\int \mathrm{d}^d x\, T^{\mu\nu}(\partial_\mu \epsilon_\nu + \partial_\nu \epsilon_\mu). \tag{3.1.56}$$
同时, 该微分同胚变换又会诱导度规改变
$$\delta g_{\mu\nu} = -(\partial_\mu \epsilon_\nu + \partial_\nu \epsilon_\mu).$$
因此, 作用量总变分为
$$\delta S = -\frac{1}{2}\int \mathrm{d}^d x\, \left(T^{\mu\nu} + 2\frac{\delta S}{\delta g_{\mu\nu}}\right)(\partial_\mu \epsilon_\nu + \partial_\nu \epsilon_\mu). \tag{3.1.57}$$
若要求作用量在共形变换下是不变的, 则有
$$T^{\mu\nu} = -2\frac{\delta S}{\delta g_{\mu\nu}}. \tag{3.1.58}$$
这可以看成是能–动张量的另一种定义.

通常情况下, 对任意给定的经典作用量, 能–动张量不一定是对称的, 但通过一定的构造, 在不破坏经典守恒量的情况下能得到对称且无散的能–动张量. 共形变换要求 $\partial^\mu \cdot j_\mu = 0$, 对于伸缩变换, 有 $j_\mu = T_{\mu\nu}\epsilon^\nu$, 且 ϵ^ν 是方程 (3.1.4) 的解, 这等价于
$$\partial \cdot j = \frac{1}{2}T^\mu_\mu(\partial \cdot \epsilon) = 0,$$

即能-动张量是无迹的, $T^\mu_\mu = 0$.

为了更具体落实共形 z- 平面上的守恒荷, 需要作一些复分析, 因此使用复坐标 $z, \bar{z}$ 更为方便, 其定义为式 (3.1.36). 仍用符号 $\mathrm{d}s^2 = \mathrm{d}z\mathrm{d}\bar{z}$, 则

$$\begin{cases} T_{z\bar{z}} = T_{\bar{z}z} = \dfrac{1}{4}T^\mu_\mu = 0, \\ T_{zz} = \dfrac{1}{4}(T_{00} - 2\mathrm{i}T_{10} - T_{11}) = \dfrac{1}{2}(T_{00} - \mathrm{i}T_{10}), \\ T_{\bar{z}\bar{z}} = \dfrac{1}{4}(T_{00} + 2\mathrm{i}T_{10} - T_{11}) = \dfrac{1}{2}(T_{00} + \mathrm{i}T_{10}). \end{cases}$$

不难看出, 能-动张量仅具有两个独立的非零分量, 一个分量是纯全纯, 用 $T(z)$ 表示, 而另一个分量为反全纯的, 记为 $\bar{T}(\bar{z})$, 即

$$T(z) \equiv T_{zz}(z), \quad \bar{T}(\bar{z}) \equiv \bar{T}_{\bar{z}\bar{z}}(\bar{z}), \tag{3.1.59}$$

且

$$\partial_{\bar{z}}T(z) = 0, \quad \partial_z \bar{T}(\bar{z}) = 0. \tag{3.1.60}$$

由于伴随共形变换的流 $j_\mu = T_{\mu\nu}\epsilon^\nu$ 是守恒的, 因此可定义相应的守恒荷为

$$Q = \int j_0(x)\,\mathrm{d}x. \tag{3.1.61}$$

另一方面, 此守恒荷又是对称变换的生成元, 即

$$\delta H = [Q, H]. \tag{3.1.62}$$

此对易子是等时演化的, 在径向量子化时对应于 $|z|$ 为常值, 即常值半径. 径向量子化的定义在 3.2 节给出. 因此, 有 $\int j_0(x)\,\mathrm{d}x \to \int j_r(\theta)\,\mathrm{d}\theta$. 类似于式 (3.1.61), 更一般地有守恒荷:

$$Q = \frac{1}{2\pi\mathrm{i}} \oint_C \left[\mathrm{d}z T(z)\epsilon(z) + \mathrm{d}\bar{z}\bar{T}(\bar{z})\bar{\epsilon}(\bar{z})\right]. \tag{3.1.63}$$

积分围道的半径取定值, 且积分方向约定为沿逆时针. 根据诺特定理, 任何场量 ϕ 的变分都可以表示为守恒荷 (3.1.63) 的等时对易子:

$$\delta_{\epsilon,\bar{\epsilon}}\phi(w,\bar{w}) = \frac{1}{2\pi\mathrm{i}}\oint_C \mathrm{d}z\,[T(z)\epsilon(z), \phi(w,\bar{w})] + \frac{1}{2\pi\mathrm{i}}\oint_C \mathrm{d}\bar{z}\,[\bar{T}(\bar{z})\bar{\epsilon}(\bar{z}), \phi(w,\bar{w})]. \tag{3.1.64}$$

3.2 几个常用概念

在本节中, 简单介绍几个在共形场论中经常用到的概念, 如径向积 (radial ordered product, ROD)、算子积展开 (operator product expansion, OPE) 和正规积 (normal ordered product, NOD) 等.

3.2 几个常用概念

3.2.1 径向积

二维平坦欧几里得空间 E, 记其时间、空间坐标分别为 σ^0, σ^1, 对应的闵可夫斯基空间 M 的光锥坐标分别为 $\sigma\pm\sigma^1$. 将闵可夫斯基空间威克 (Wick) 转动到欧几里得空间, 则光锥坐标分别为 $\zeta, \bar{\zeta} = \sigma \pm \mathrm{i}\sigma^1$. 基于这样的坐标, 二维闵可夫斯基空间左、右运动的零质量场就分别对应于二维欧几里得空间的全纯、反全纯场. 因此, 我们也将全纯、反全纯场分别称为左移、右移场, 或者叫做手征场. 在场论中, 为了消除红外发散, 通常需要将空间坐标紧化, 即 $\sigma^1 \equiv \sigma^1 + 2\pi$. 这样, 就在圆柱上定义了坐标 σ^0, σ^1. 用柱面上的坐标, 可以定义共形映射 $\zeta \to z = \exp\zeta = \exp(\sigma + \mathrm{i}\sigma^1)$, 将圆柱映射为由 z 坐标的复平面 (图 3.2). 于是, 柱面上的过去和未来无穷远是 $\sigma^0 = \mp\infty$, 分别对应于复平面上的原点和无穷大; 而 σ^0 为常数所定义的等时曲面, 成为复平面上常值半径的圆周. 这样, 柱面上的时间平移变换等价于复平面上的伸缩映射. 因此, 复平面上的伸缩算子, 可以看成是系统的哈密顿量; 而系统的希尔伯特空间, 由常值半径的曲面张成. 在复平面上, 用这样的方式定义量子场, 被称为径向量子化. 基于量子场论的基础知识, 如果要在欧几里得空间恰当地定义场算符的关联函数, 则关联函数中的场算符必须是编时排序的[28,36]. 因此, 在欧几里得空间定义关联函数的泛函积分也必须是编时排序的. 需要留意的是, 如果径向图景有 $|z| > |w|$, 则表明时间 z 是晚于 w 的. 在欧几里得空间定义的场算符 $A(z)$, $B(w)$ 的关联函数, 是由其算子积 $A(z)B(w)$ 定义的, 且该定义在 $|z| > |w|$ 时才有意义, 此即所谓的径向积 (ROP). 二维欧几里得空间的编时积映射到复平面上, 就体现为径向积. 因此, 可如下定义场算符径向排序算子 R:

$$R(A(z)B(w)) = \begin{cases} +A(z)B(w), & |z| > |w|, \\ \pm B(w)A(z), & |z| < |w|, \end{cases} \tag{3.2.1}$$

其中, $\pm$ 分别对应于玻色子场和费米子场. 根据场算符 $A(z)$, $B(w)$ 是玻色子或费

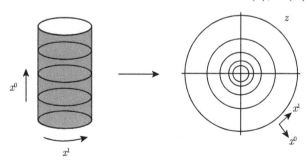

图 3.2　柱面与平面间映射

米子, 分别考虑其对易子或反对易子:

$$\frac{1}{2\pi i}\oint [A(z),B(w)]\,dz = \frac{1}{2\pi i}\oint_{|z|>|w|} A(z)B(w)\,dz - \frac{1}{2\pi i}\oint_{|z|<|w|} B(w)A(z)\,dz$$
$$= \frac{1}{2\pi i}\oint_{\mathcal{C}(w)} dz\, RA(z)B(w), \tag{3.2.2}$$

此处的围道积分是绕点 w 的. 具体到元场

$$\delta_{\epsilon,\bar{\epsilon}}\phi(w,\bar{w}) = \frac{1}{2\pi i}\oint_{\mathcal{C}} dz\,[T(z)\epsilon(z),\phi(w,\bar{w})] + \frac{1}{2\pi i}\oint_{\mathcal{C}} d\bar{z}\,[\bar{T}(\bar{z})\bar{\epsilon}(\bar{z}),\phi(w,\bar{w})]$$
$$= \left(h\partial_z\epsilon + \epsilon\partial_z + \bar{h}\partial_{\bar{z}}\bar{\epsilon} + \bar{\epsilon}\partial_{\bar{z}}\right)\phi(w,\bar{w}), \tag{3.2.3}$$

上式适用任意的元场, 因此对于由元场构成的 n- 点关联函数 (3.1.24), 在共形变换下有等式 (图 3.3)

$$\langle \phi_1(z_1,\bar{z}_1)\cdots\phi_n(z_n,\bar{z}_n)\rangle$$
$$= \prod_{j=1}^{n}(\partial f(z_j))^{h_j}(\bar{\partial}\bar{f}(\bar{z}_j))^{\bar{h}_j}\langle \phi_1(w_1,\bar{w}_1))\cdots\phi_n(w_n,\bar{w}_n)\rangle, \tag{3.2.4}$$

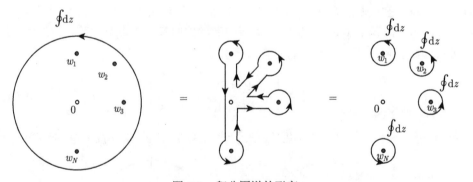

图 3.3 积分围道的形变

具体到由 $\epsilon(z)T(z)$ 导致的局域共形变换:

$$\left\langle \oint \frac{dz}{2\pi i}\epsilon(z)T(z)\phi_1(w_1,\bar{w}_1)\cdots\phi_n(w_n,\bar{w}_n)\right\rangle$$
$$= \sum_{i=1}^{n}\left\langle \phi_1(w)\cdots\left(\oint \frac{dz}{2\pi i}\epsilon(z)T(z)\phi_i(w_i,\bar{w}_i)\right)\cdots\phi_n(w_n,\bar{w}_n)\right\rangle$$
$$= \sum_{i=1}^{n}\langle \phi_1(w_1,\bar{w}_1)\cdots\delta_\epsilon\phi_i(w_i,\bar{w}_i)\cdots\phi_n(w_n,\bar{w}_n)\rangle, \tag{3.2.5}$$

对应式 (3.2.3) 给定的任意 $\delta_\epsilon \phi$:

$$0 = \oint \frac{\mathrm{d}z}{2\pi\mathrm{i}} \epsilon(z) \Big\{ \langle T(z)\phi_1(w_1, \bar{w}_1) \cdots \phi_n(w_n, \bar{w}_n) \rangle$$
$$- \sum_{i=1}^n \left[\frac{h_i}{(z-w_i)^2} + \frac{1}{z-w_i} \partial_{w_i} \right] \langle \phi_1(w_1, \bar{w}_1) \cdots \phi_n(w_n, \bar{w}_n) \rangle \Big\}. \quad (3.2.6)$$

注意到 $\oint \frac{1}{2\pi\mathrm{i}} \mathrm{d}\bar{z} T(z) = 0$, 即得到所谓的共形瓦德 (Ward) 恒等式:

$$\langle T(z)\phi_1(w_1, \bar{w}_1) \cdots \phi_n(w_n, \bar{w}_n) \rangle$$
$$= \sum_{i=1}^n \left[\frac{h_i}{(z-w_i)^2} + \frac{1}{z-w_i} \partial_{w_i} \right] \langle \phi_1(w_1, \bar{w}_1) \cdots \phi_n(w_n, \bar{w}_n) \rangle. \quad (3.2.7)$$

3.2.2 算子积展开

不失一般性, 考虑局域算子期望值

$$\langle \mathcal{O}_1(z) \mathcal{O}_2(w) \psi_1(z_1) \cdots \psi_n(z_n) \rangle, \quad (3.2.8)$$

其中, $\mathcal{O}_i(z_i)$ 是局域算子集的基. 所以, 当算子 $\mathcal{O}_1$ 趋近算子 $\mathcal{O}_2$ 时, 上面定义算子期望值一般是奇异的, 如何反映两算子趋近所导致的奇异行为! 算子积展开就是处理这类奇异行为的工具. 对任意两个局域算子的积, 可以在任意精度上表示成局域算子的和:

$$\mathcal{O}_i(z)\mathcal{O}_j(w) = \sum_{k=1} c_{ij}^k(z-w) \mathcal{O}_k(w), \quad (3.2.9)$$

其中, 数值系数 $c_{ij}^k(z-w)$ 是奇异的, 依赖于指标 i, j, k, 但不依赖期望值中其他算子.

$$\langle \mathcal{O}_1(z)\mathcal{O}_2(w)\psi_1(z_1)\cdots\psi_n(z_n) \rangle = \sum_{k=1} c_{ij}^k(z-w) \langle \mathcal{O}_k(w)\psi_1(z_1)\cdots\psi_n(z_n) \rangle, \quad (3.2.10)$$

这就是所谓的算子乘积展开. 在共形场中, 可以得到算子积更具体的表述. 假设场算子 $A(z)$ 和 $B(w)$ 为全纯的, 其共形维分别是 h_A 和 h_B. 由这些场具有 $\bar{z}$ 平移不变性的事实, 很容易证明, $[A(z), B(w)]$ 为零, 除非 $z = w$. 另外, 等时对易子可以表示为 $\delta(z-w)$ 及其各层导数的线性组合, 即

$$[A(z), B(w)] = \sum_{n=1}^{h_A+h_B} [AB]_n(w) \frac{(-\partial)^{n-1}}{(n-1)!} \delta(z-w), \quad (3.2.11)$$

式中, $[AB]_n(z)$ 是场代数的元素. 该求和至多到 $h_A + h_B$ 的原因是, 如果对上述等式进行伸缩变换, 可以发现场 $[AB]_n$ 具有共形维 $h_A + h_B - n$. 由于不存在共形维

为负的场, 所以上式的求和公式在共形维等于 $h_A + h_B$ 时截止. 等时对易子 (3.2.11) 的左侧可以写成

$$[A(z), B(w)] = \left(\lim_{|z|>|w|} - \lim_{|z|<|w|}\right) R(A(z)B(w)), \qquad (3.2.12)$$

δ 函数存在类似表述

$$\delta(z-w) = \left(\lim_{|z|>|w|} - \lim_{|z|<|w|}\right) \frac{1}{z-w}. \qquad (3.2.13)$$

从上述可知, 函数 $A(z)B(w) - \sum_n (z-w)^{-n}[AB]_n(w)$ 和 $B(w)A(z) - \sum_n (z-w)^{-n}[AB]_n(w)$, 分别在 $|z|>|w|$ 区域、$|z|<|w|$ 区域是全纯的. 由于它们在边界 $|z|=|w|$ 具有相同值, 不难发现, 它们对所有的 z 都是全纯的, 根据复变函数理论, 它们是相同的函数. 现在可以考虑此函数的幂级数展开, 其结果为

$$R(A(z)B(w)) = \sum_{n=-(h_A+h_B)}^{\infty} (z-w)^n [AB]_{-n}(w). \qquad (3.2.14)$$

也就是说, 两个场算子的积可以表示为 $\delta(z-w)$ 的幂级数展开. 这就是共形场中所谓算子的积展开 (OPE). 可以看一个具体的例子, ϕ_i 是具有确定共形维的算子基, 使得两点函数可写为

$$\langle \phi_i(z,\bar{z})\phi_j(w,\bar{w})\rangle = \delta_{ij} \frac{1}{(z-w)^{2h_i}} \frac{1}{(\bar{z}-\bar{w})^{2\bar{h}_i}}, \qquad (3.2.15)$$

于是, 它们的算子积展开式的系数 C_{ijk}:

$$\phi_i(z,\bar{z})\phi_j(w,\bar{w}) \sim \sum_k C_{ijk}(z-w)^{h_k-h_i-h_j}(\bar{z}-\bar{w})^{\bar{h}_k-\bar{h}_i-\bar{h}_j}\phi_k(w,\bar{w}), \qquad (3.2.16)$$

上式关于指标 i,j,k 是对称的. 选择三点函数 $\langle\phi_i\phi_j\phi_k\rangle$ 任意两点, 让它们趋近, 利用等式 (3.2.15) 可以得到式 (3.2.16) 中的结构常数 C_{ijk} 就是三点函数 (3.1.46) 中出现的结构常数.

利用算子 A, B 之间算子积展开式, 可以将等时对易子的结果用场量的模式表示出来. 定义场 $A(z)$ 模式展开为 $A(z) = \sum_{n\in\mathbb{Z}} A_n z^{-n-h_A}$, h_A 是场 $A(z)$ 的共形维, 于是有

$$A_n = \oint \mathrm{d}z\, z^{n+h_A-1} A(z), \qquad (3.2.17)$$

为简便起见, 这里采用了符号 $\oint_w \mathrm{d}z \equiv \oint_{\mathcal{C}_w} \frac{\mathrm{d}z}{2\pi\mathrm{i}}$, 后面会默认此记号. 对 B_n 也有类

3.2 几个常用概念

似的方程. 容易知道, 对易子 $[A_n, B_m]$ 意为包含原点的两重积分的差. 此两重积分的差等价于两重围道积分: 一次绕原点, 一次绕 w (图 3.4) [17].

$$[A_m, B_n] = \oint_0 dw \oint_w dz \, z^{m+h_A-1} w^{n+h_B-1} R(A(z)B(w)). \tag{3.2.18}$$

将式 (3.2.14) 的左边代入式 (3.2.18), 就可以得到 A_m 和 B_n 的对易子.

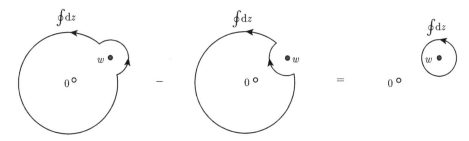

图 3.4 等时对易子积分围道变换

3.2.3 正规积

通常当 $z \to w$ 时, 局域场算子 $A(z)$, $B(w)$ 的真空期望值 $\langle A(z)B(w) \rangle$ 是发散的. 为了得到有限的结果, 需要定义如下的正规积 (NOP): 场算子的编时积, 具体到二维共形场, 正是径向积. 它等同于正规算子积, 以及这些场算子的各种可能的收缩的和, 这就是所谓的威克定理 [28,36]. 因此, 正规积就是在作算子积时将湮没算子置于右侧. 设局域场 $A(z)$, $B(w)$ 共形维分别为 h_A, h_B (我们只考虑同胚叶, 即左手扇区), 则场 $A(z)$, $B(w)$ 的正规积 $:AB:(z)$ 可定义为

$$:AB:(z) = A_-(z)B(z) + B(z)A(z)_+,$$

其中 $A_- = \sum\limits_{n \leqslant -h_A} A_n z^{-n-h_A}$, $A_+ = A - A_-$. 由算子积展开的定义可知,

$$:AB:(z) = [AB]_0(z), \tag{3.2.19}$$

用算子乘积展开重写能–动张量与元场 $\phi(z)$ 径向积:

$$\begin{cases} RT(z)\phi_h(w,\bar{w}) = \dfrac{h}{(z-w)^2}\phi_h(w,\bar{w}) + \dfrac{1}{z-w}\partial_w \phi_h(w,\bar{w}) + \cdots, \\ R\bar{T}(\bar{z})\phi_{\bar{h}}(w,\bar{w}) = \dfrac{\bar{h}}{(\bar{z}-\bar{w})^2}\phi_{\bar{h}}(w,\bar{w}) + \dfrac{1}{\bar{z}-\bar{w}}\partial_{\bar{w}} \phi_{\bar{h}}(w,\bar{w}) + \cdots, \end{cases} \tag{3.2.20}$$

式中, $\cdots$ 代表正则项 (regular terms), 包括所有那些于对易子没有贡献的场量集合, 后文中将不加解释地使用这个符号. 为简化记号, 此后会将表示径向积符号 R 也略

去. 用模式重写了上面的方程:

$$[L_m, \phi_h(z)] = z^{m+1}\partial\phi_h(z) + h(m+1)z^m\phi_h(z), \tag{3.2.21}$$

$$[\bar{L}_m, \phi_{\bar{h}}(\bar{z})] = \bar{z}^{m+1}\partial\phi_{\bar{h}}(\bar{z}) + \bar{h}(m+1)\bar{z}^m\phi_{\bar{h}}(\bar{z}), \tag{3.2.22}$$

为了醒目,此处特意将共形维标出.如果在上式中还存在其他的非正则项,则被称为拟元场.若等式 (3.2.20) 两边对 w 微分,显然有

$$T(z)\partial_w\phi(w) = \frac{2h\phi(w)}{(z-w)^3} + \frac{(h+1)\partial\phi(w)}{(z-w)^2} + \frac{\partial^2\phi(w)}{z-w} + \cdots,$$

因此,对元场 $\phi(z)$ 进行微分,就成为拟元场.

3.2.4 能-动张量展开

现在分析能-动张量算子自身在共形变换下的性质.由前面的讨论,我们已经知道 Virasoro 代数的基是无穷小共形变换的生成元,故将能-动张量 $T(z)$ 表述为关于 Virasoro 代数的洛朗级数:

$$T(z) = \sum_{n\in\mathbb{Z}} L_n z^{-n-2}, \quad L_n = \oint \mathrm{d}z\, z^{n+1}T(z). \tag{3.2.23}$$

这就是能-动张量的模式展开.若要求能-动张量的模式 L_n 的对易子满足 Virasoro 代数式,这要求

$$T(z)T(w) = \frac{c/2}{(z-w)^4} + \frac{2T(w)}{(z-w)^2} + \frac{\partial_w T(w)}{z-w} + \cdots. \tag{3.2.24}$$

容易证明

$$\begin{aligned}[L_n, L_m] &= \Big(\oint \mathrm{d}z \oint \mathrm{d}w - \oint \mathrm{d}w \oint \mathrm{d}z\Big) z^{n+1}T(z)w^{m+1}T(w) \\
&= \oint_w \mathrm{d}z \oint_0 \mathrm{d}w\, z^{n+1}w^{m+1}\Big[\frac{c/2}{(z-w)^4} + \frac{2T(w)}{(z-w)^2} + \frac{\partial_w T(w)}{z-w} + \cdots\Big] \\
&= \oint_0 \mathrm{d}w\Big[\frac{c}{12}(n-1)n(n-1)w^{n-2}w^{m+1} \\
&\quad + 2(n+1)w^n w^{m+1}T(w) + w^{n+1}w^{m+1}\partial T(w)\Big],\end{aligned} \tag{3.2.25}$$

对上式最后一项作分部积分,并与第二项合并,得到项 $(n-m)w^{n+m+1}T(w)$,再对 w 积分,结果为

$$[L_m, L_n] = (m-n)L_{m+n} + \frac{c}{12}(m^3 - m)\delta_{m+n,0}.$$

能-动张量间的算子积展开式表明，在一般情况，能-动张量不是 Virasoro 代数的元场，而只是其拟元场，当且仅当中心荷 $c=0$，能-动张量是元场，但 $c=0$ 是一个平凡的场，可以证明，常值场的中心荷 $c=0$. 所以中心荷 $c \neq 0$ 又被称为共形反常. 关于共形反常项的物理意义，我们留待后面讨论. 类似地，能-动张量的无穷小共形变换为

$$\delta_\epsilon T(w) = \oint_w \mathrm{d}z \epsilon(z) T(z) T(w)$$
$$= \frac{c}{12}\partial_w^3 \epsilon(w) + 2T(w)\partial_z \epsilon(w) + \epsilon(w)\partial_w T(w). \tag{3.2.26}$$

可以证明，能-动张量的有限共形变换 [2,4,9]：

$$T(z) \to T'(z) = \left(\frac{\partial f}{\partial z}\right)^2 T(f(z)) + \frac{c}{12}S(f(z), z), \tag{3.2.27}$$

上式中的

$$S(w, z) = \frac{(\partial_z w)(\partial_z^3 w)}{(\partial_z w)^2} - \frac{3(\partial_z^2 w)^2}{2(\partial_z w)^2}, \tag{3.2.28}$$

是著名的施瓦兹导数. 施瓦兹导数是二层的，且是复平面上唯一自洽的二层导数，在 $SL(2,\mathcal{R})$ 变换下，施瓦兹导数为零，即能-动张量是 $SL(2,\mathcal{R})$ 变换的元场. 施瓦兹导数与射影联络相对应，而与一层导数相关的是仿射联络. 具体到柱面与复平面之间的映射，$\zeta \to z = e^w = \exp\zeta = \exp(\sigma + i\sigma^1)$，容易得到

$$T_{cyl}(w) = z^2 T(z) - \frac{c}{24}, \tag{3.2.29}$$

写成洛朗展开式

$$T_{cyl}(w) = \sum_{n\in\mathbb{Z}} L_n z^{-n} - \frac{c}{24} = \sum_{n\in\mathbb{Z}} \left(L_n - \frac{c}{24}\delta_{n,0}\right) e^{-nw}. \tag{3.2.30}$$

显然，Virasoro-代数其他生成元没有任何改变，只有其零模产生移动：

$$(L_{cyl})_0 = L_0 - \frac{c}{24}. \tag{3.2.31}$$

同理，在 $\bar{z}$ 与 $\bar{L}_0$ 之间也存在这样的关系. 这说明柱面的真空能为

$$E_0 = -\frac{c+\bar{c}}{24}, \tag{3.2.32}$$

这一等式的意义将稍后讨论.

3.3 共形场：举例

本节中我们将给出几个共形场的实例，并分析它们的一些特性. 例如通常的量子场论，分析这些特定的共形场，出发点还是拉氏作用量.

3.3.1 自由玻色子

考虑取值在柱面上的无质量标量场 $\phi(\sigma^0,\sigma^1)$, 取柱面的半径 $R=1$, 它的作用量为

$$\begin{aligned}S&=\frac{1}{8\pi}\int \mathrm{d}\sigma^0\mathrm{d}\sigma^1\sqrt{|h|}h^{\alpha\beta}\partial_\alpha\phi\partial_\beta\phi\\&=\frac{1}{8\pi}\int \mathrm{d}\sigma^0\mathrm{d}\sigma^1\left[(\partial_0\phi)^2+(\partial_1\phi)^2\right],\end{aligned} \quad (3.3.1)$$

其中, $h\equiv \det h_{\alpha\beta}$, $h_{\alpha\beta}=\mathrm{diag}(1,1)$. 如果引入坐标 $z=(\sigma^0+\mathrm{i}\sigma^1)$, 则积分元 $\mathrm{i}\mathrm{d}z\wedge \mathrm{d}\bar{z}=2\mathrm{d}\sigma^0\wedge \mathrm{d}\sigma^1$, 这样就将柱面映为复平面. 相应地, 作用量可重写为

$$S=\frac{1}{8\pi}\int \mathrm{d}z\mathrm{d}\bar{z}\,\partial\phi\cdot\bar{\partial}\phi, \quad (3.3.2)$$

对该作用量进行关于 ϕ 的变分:

$$\begin{aligned}0=&\delta_\phi S\\=&\frac{1}{8\pi}\int \mathrm{d}z\mathrm{d}\bar{z}\,(\partial\delta\phi\cdot\bar{\partial}\phi+\partial\phi\cdot\bar{\partial}\delta\phi)\\=&\frac{1}{8\pi}\int \mathrm{d}z\mathrm{d}\bar{z}\,[\partial(\delta\phi\cdot\bar{\partial}\phi)-\delta\phi\cdot\partial\bar{\partial}\phi+\bar{\partial}(\partial\phi\cdot\delta\phi)-\bar{\partial}\partial\phi\cdot\delta\phi]\\=&-\frac{1}{4\pi}\int \mathrm{d}z\mathrm{d}\bar{z}\,\delta\phi(\partial\bar{\partial}\phi),\end{aligned} \quad (3.3.3)$$

相应的运动方程为

$$\partial\bar{\partial}\phi(z,\bar{z})=0. \quad (3.3.4)$$

这表明, 玻色场 ϕ 可以分解为全纯和反全纯, 或者说左移和右移两部分:

$$\phi(z,\bar{z})=(x(z)+\bar{x}(\bar{z})). \quad (3.3.5)$$

显然, ∂x 和 $\bar{\partial}\bar{x}$ 分别是手征和反手征的. 对作用量进行共形变换可以证明, 如果 $\phi'(w,\bar{w})=\phi(z,\bar{z})$, 即 $\phi(z,\bar{z})$ 的共形维是零, 则式 (3.3.2) 是共形不变的. 也就是说, 如果 $x(z)$ 和 $\bar{x}(\bar{z})$ 的共形维 $h=\bar{h}=0$, 上述作用量是共形不变. 利用场论的基础知识, 此玻色场的传播子满足方程:

$$\partial\bar{\partial}G=-2\pi\delta^{(2)}(z-w). \quad (3.3.6)$$

利用等式

$$\bar{\partial}\partial\ln|z-w|^2=\bar{\partial}\left(\frac{1}{z-w}\right)=2\pi\delta^{(2)}(z-w), \quad (3.3.7)$$

3.3 共形场：举例

由此可得传播子

$$G(z,\bar{z},w,\bar{w}) = \langle \phi(z,\bar{z})\phi(w,\bar{w}) \rangle = -\log|z-w|^2, \tag{3.3.8}$$

将传播子分成全纯和反全纯部分：

$$\langle x(z)x(w) \rangle = -\log|z-w|, \quad \langle \bar{x}(\bar{z})\bar{x}(\bar{w}) \rangle = -\log|\bar{z}-\bar{w}|, \tag{3.3.9}$$

全纯分量与反全纯分量之间的关联函数为零. 由共形场的定义可知, $x(z)$ 自身不是元场, 它的传播子为

$$\langle \partial x(z)\partial x(w) \rangle = -\frac{1}{(z-w)^2}, \tag{3.3.10}$$

因此, 有短距展开式

$$\partial x(z)\partial x(w) = -\frac{1}{(z-w)^2} + \cdots, \tag{3.3.11}$$

这正是对式 (3.3.9) 二次微分的结果. 交换 $\partial x(z)$ 与 $\partial x(w)$ 径向积的顺序, 结果不变, 因此 $\partial x(z)$ 是玻色场. 由上式右侧的标度性质, $\partial x(z)$ 是共形维 $h = (1, 0)$ 元场的候选者. 下面就来具体分析它的共形性质. 为简单起见, 分析仅仅限于全纯部分. 反全纯部分的结果是类似的. 由正规积可得到能-动张量

$$\begin{aligned} T(w) &= -\frac{1}{2} : \partial x(z)\partial x(w) : \\ &\equiv -\frac{1}{2} \lim_{z \to w} [\partial x(z)\partial x(w) + \langle \partial x(z)\partial x(w) \rangle]. \end{aligned} \tag{3.3.12}$$

计算能-动张量与 $\partial x(z)$ 算子积, 其奇异部分为

$$\begin{aligned} T(z)\partial x(w) &= -\frac{1}{2} : \partial x(z)\partial x(z) : \partial x(w) \\ &= -\frac{1}{2}\partial x(z)\langle \partial x(z)\partial x(w)\rangle 2 + \cdots \\ &= \frac{\partial x(z)}{(z-w)^2} + \cdots \\ &= [\partial x(w) + (z-w)\partial^2 x(w)]\frac{1}{(z-w)^2} + \cdots, \end{aligned} \tag{3.3.13}$$

上式的最后一行利用了 $\partial x(z)$ 在 w 点附近的泰勒展式：

$$T(z)\partial x(w) = \frac{\partial x(w)}{(z-w)^2} + \frac{\partial_w^2 x(w)}{(z-w)}. \tag{3.3.14}$$

由前面关于元场的定义式 (3.2.20), 可知 $\partial x(z)$ 正是能-动张量 $T(z) = -\frac{1}{2} : (\partial x)^2(z) :$ 的元场, 并且共形维 $h = (1, 0)$, ∂x 的模式展开式为

$$\mathrm{i}\partial x(z) = \sum_{n \in \mathbb{Z}} \alpha_n z^{-n-1}, \tag{3.3.15}$$

这等价于
$$\alpha_n = \oint_0 \mathrm{d}z\, z^n \mathrm{i}\partial x(z). \tag{3.3.16}$$

进一步可算出对易子
$$\begin{aligned}[\alpha_m,\ \alpha_n] &= \mathrm{i}^2 \left[\oint \mathrm{d}z, \oint \mathrm{d}w\right] z^n \partial_z x(z) w^m \partial_w x(w) \\ &= \oint \mathrm{d}w\, w^m \oint \mathrm{d}z\, z^n \frac{1}{(z-w)^2} = \oint \mathrm{d}w\, n w^m w^{n-1} \\ &= m\delta_{m+n,0}. \end{aligned} \tag{3.3.17}$$

不难证明,Virasoro 是 α_m 的二次式:
$$\begin{aligned} L_m &= \sum_{n\in\mathbb{Z}} :\alpha_n \alpha_{m-n}: \\ &= \sum_{l\leqslant -1} \alpha_l \alpha_{m-l} + \sum_{l > -1} \alpha_{m-l}\alpha_l. \end{aligned} \tag{3.3.18}$$

用同样的方法,我们可以计算能-动张量与自身的算子积:
$$\begin{aligned} T(z)T(w) &= \left(-\frac{1}{2}\right)^2 \partial\phi(z)\partial\phi(z) :: \partial\phi(w)\partial\phi(w): \\ &= \left(-\frac{1}{2}\right)^2 \{2[\langle\partial\phi(z)\partial\phi(w)\rangle]^2 + 4\partial\phi(z)\partial\phi(w)\langle\partial\phi(z)\partial\phi(w)\rangle + \cdots\} \\ &= \frac{1/2}{(z-w)^4} + \frac{2}{(z-w)^2}\left\{-\frac{1}{2}[\partial\phi(w)]^2\right\} + \frac{1}{z-w}\partial\left\{-\frac{1}{2}[\partial\phi(w)]^2\right\} \\ &= \frac{1/2}{(z-w)^4} + \frac{2T(w)}{(z-w)^2} + \frac{\partial T(w)}{(z-w)}, \end{aligned} \tag{3.3.19}$$

即自由玻色场的中心荷为 $c=1$. 无需重复,能-动张量不是元场.

3.3.2 顶点算子

另一个与自由玻色场有密切关系的例子,是顶点算子. 早在 20 世纪 60 年代,在量子场论中,为反映传播子间的相互作用,即引入了顶点算子的概念. 此后,在 20 世纪 80 年代,顶点算子被扩展为一类新型的代数,即顶点算子代数. 顶点算子代数在研究 Moonshine 猜测中具有关键作用 [13]. 现在,顶点算子代数已经发展成为数学的重要分支学科,在贯通不同分支学科,如代数、几何、拓扑、概率论以及物理理论等方面,具有桥梁作用 [13,16]. 定义顶点算子

$$V_\alpha(z) =: \mathrm{e}^{\mathrm{i}\alpha x(z)} :, \tag{3.3.20}$$

3.3 共形场：举例

它和能-动张量 (3.3.12) 算子积：

$$T(z):\mathrm{e}^{\mathrm{i}\alpha x(w)}:= -\frac{1}{2}[\langle\partial x(z)\mathrm{i}\alpha x(w)\rangle]^2:\mathrm{e}^{\mathrm{i}\alpha x(w)}:$$

$$-\frac{1}{2}x2\partial x(z)\langle\partial x(z)\mathrm{i}\alpha x(w)\rangle:\mathrm{e}^{\mathrm{i}\alpha x(w)}:$$

$$=\frac{\alpha^2/2}{(z-w)^2}:\mathrm{e}^{\mathrm{i}\alpha x(w)}:+\frac{\mathrm{i}\alpha\partial x(z)}{z-w}:\mathrm{e}^{\mathrm{i}\alpha x(w)}:+\cdots$$

$$=\frac{\alpha^2/2}{(z-w)^2}:\mathrm{e}^{\mathrm{i}\alpha x(w)}:+\frac{1}{z-w}\partial:\mathrm{e}^{\mathrm{i}\alpha x(w)}:+\cdots \quad (3.3.21)$$

根据定义易知, $V_\alpha(z)=:\mathrm{e}^{\mathrm{i}\alpha x(z)}:$ 是元场, 并且共形维 $h=\alpha^2/2$. 由两点函数的一般定义知, 两点函数非零, 当且仅当两局域场具有相同的共形维. 显然, $:\mathrm{e}^{\mathrm{i}\alpha x(z)}:$ 与 $:\mathrm{e}^{-\mathrm{i}\alpha x(z)}:$ 具有相同的共形维. 另外, 对数个顶点算子 $:\mathrm{e}^{\mathrm{i}\alpha_m x(z)}:$, 它们的关联函数为零, 除非它们荷的和 $\sum_m \alpha_m = 0$, 这就是所谓的荷守恒条件.

$$\langle\mathrm{e}^{\mathrm{i}\alpha x(z)}\mathrm{e}^{-\mathrm{i}\alpha x(w)}\rangle = \mathrm{e}^{\alpha^2\langle x(z)x(w)\rangle} = \frac{1}{(z-w)^{\alpha^2}}. \quad (3.3.22)$$

如果对自由玻色的能-动张量 (3.3.12) 作稍许修正：

$$T(z) = -\frac{1}{2}:[\partial x(z)]^2:+\mathrm{i}\sqrt{2}\alpha_0\partial^2 x(z), \quad (3.3.23)$$

与自由玻色的能-动张量相比较, 此式包含场量的全微分项, 该场是有确切定义的. 修正后的能-动张量 $T(z)$ 仍是共形变换的生成函数, 附加的全微分项并不会影响这一点. 重复此节的过程, 并利用等式 (3.3.9), 对应于自由玻色场的能-动张量的中心荷 $c=1$, 变为

$$c = 1 - 24\alpha_0^2.$$

可以将修正后的式 (3.3.23) 理解为, 在无穷远点放置有值为 $-2\alpha_0$ 的"背景荷". 如果有顶点算子 $V_{-2\sqrt{2}\alpha_0}(z) =:\mathrm{e}^{-\mathrm{i}2\sqrt{2}\alpha_0 x(z)}:$, 于是可以定义出态：

$$\langle -2\alpha_0| = \frac{\langle 0|V_{-2\sqrt{2}\alpha_0}(\infty)}{\langle 0|V_{-2\sqrt{2}\alpha_0}(\infty)V_{2\sqrt{2}\alpha_0}(0)|0\rangle}.$$

对于给定的实数 α_0, 式 (3.3.23) 中的例外项会将中心荷 $c=1$ 变更为 $c<1$. 由于能-动张量 (3.3.23) 包含有取值为虚的部分, 对于任意给定的 α_0, 并不能得到酉理论. 但是, 对一些给定 α_0 的特殊值, 可能包含一个自洽的酉子空间. $c<1$ 的酉理论, 在研究统计物理模型中具有重要作用. 相应地, 非零两点关联函数为

$$\langle V_\beta(z) V_{2\alpha_0-\beta}(w)\rangle = \frac{1}{(z-w)^{2\beta(\beta-2\alpha_0)}},$$

其中, 荷满足约束条件 $\sum_j \beta_j = 2\alpha_0$.

3.3.3 $\widehat{su(2)}_1$ 代数

流代数, 也被称为仿射李代数 (affine Lie algebra), 或者 Kac-Moody 代数. 如果将顶点算子 V_α 的参数设定为 $\alpha^2 = 2$, 即 $\alpha = \sqrt{2}$, 顶点代数 $:\mathrm{e}^{\mathrm{i}\alpha x(z)}:$ 就给出所谓流代数 $\widehat{su(2)}_k$ 的层 (level)$k=1$ 的表示:

$$\begin{cases} J^\pm(z) =: \mathrm{e}^{\pm\mathrm{i}\sqrt{2}x(z)} :, \\ J^3(z) = \dfrac{\mathrm{i}}{\sqrt{2}}\partial x(z). \end{cases} \tag{3.3.24}$$

它们的算子积展开式为

$$\begin{cases} J^3(z)J^\pm(w) = \pm\dfrac{1}{z-w}J^\pm + \cdots, \\ J^+(z)J^-(w) = \dfrac{1}{(z-w)^2} + \dfrac{2}{z-w}J^3 + \cdots, \\ J^3(z)J^3(w) = \dfrac{1/2}{(z-w)^2} + \cdots. \end{cases} \tag{3.3.25}$$

它们的洛朗展开为 $J^a(z) = \sum\limits_{n\in\mathbb{Z}} J^a_n z^{-n-1}$, $a = 3, \pm$, 而能-动张量为

$$T(z) = \frac{1}{6}\sum_a :J^aJ^a:(z), \tag{3.3.26}$$

这样定义的能-动张量被称为菅原构造 (Sugawara construction):

$$T(z)J^a(w) = \frac{J^a(w)}{(z-w)^2} + \frac{\partial J^a(w)}{z-w}, \tag{3.3.27}$$

或者, 等价地有

$$[L_m, J^a_n] = -n J^a_{m+n}, \tag{3.3.28}$$

$$L_n = \frac{1}{6}\sum_{a=3,\pm}\sum_{m\in\mathbb{Z}} :J^a_{m+n}J^a_{-m}:(z), \tag{3.3.29}$$

3.3.4 自由费米子

类似于自由玻色场, 我们考虑二维欧几里得空间自由马约拉纳费米子, 它的量为

$$S = \frac{1}{4\pi}\int \mathrm{d}x^0 \mathrm{d}x^1 \bar{\Psi}\gamma^\alpha \partial_\alpha \Psi, \tag{3.3.30}$$

式中, $\bar{\Psi} \equiv \Psi^\dagger \gamma^0$, $\dagger$ 为厄米共轭; γ^α 是两行两列的矩阵, 且有克利福德 (Clifford) 代数关系:

$$\{\gamma^\alpha, \gamma^\beta\} = 2\eta^{\alpha\beta}\mathbf{1}.$$

3.3 共形场: 举例

但克利福德代数并不足以完全确定该 γ^α, 通常我们采用表示

$$\gamma^0 = \begin{pmatrix} 0 & 1 \\ 1 & 0 \end{pmatrix}, \quad \gamma^1 = \begin{pmatrix} 0 & -\mathrm{i} \\ \mathrm{i} & 0 \end{pmatrix}. \tag{3.3.31}$$

利用此 γ^α 表示以及复坐标 $z = (x^0 + \mathrm{i} x^1)$:

$$\gamma^0(\gamma^0 \partial_0 + \gamma^1 \partial_1) = 2\begin{pmatrix} \partial_{\bar z} & 0 \\ 0 & \partial_z \end{pmatrix}. \tag{3.3.32}$$

使用记号 $\Psi = (\psi, \bar\psi)$, 作用量可重写为

$$S = \frac{1}{8\pi}\int \mathrm{d}z\mathrm{d}\bar z\, (\bar\psi \partial \bar\psi + \psi \bar\partial \psi). \tag{3.3.33}$$

对作用量作关于 ψ 的变分:

$$\begin{aligned}
0 =& \delta_\psi S \\
=& \frac{1}{8\pi}\int \mathrm{d}z\mathrm{d}\bar z\, [\delta\psi \partial\psi + \psi \bar\partial(\delta\psi)] \\
=& \frac{1}{8\pi}\int \mathrm{d}z\mathrm{d}\bar z\, [\delta\psi \partial\psi + \bar\partial(\psi\delta\psi) - (\bar\partial\psi)\delta\psi] \\
=& \frac{1}{4\pi}\int \mathrm{d}z\mathrm{d}\bar z\, \delta\psi \partial\bar\psi.
\end{aligned} \tag{3.3.34}$$

在分部积分时, 需用到费米变量的反交换性, 对 $\bar\psi$ 重复上述过程. 这样就得到相应的运动方程, $\partial\bar\psi = \bar\partial\psi = 0$. 它们的解就分别对应于全纯的 $\psi(z)$ 和反全纯的 $\bar\psi(\bar z)$. 对作用量进行共形变换, 可以证明, 当元场 $\psi(z), \bar\psi$ 分别具有 $h = \left(\frac{1}{2}, 0\right)$, $\bar h = \left(0, \frac{1}{2}\right)$ 的变换性质时, 作用量 (3.3.33) 是共形不变的. 由传播子满足的微分方程, 得到两点函数:

$$\langle \psi(z)\psi(w) \rangle = -\frac{1}{z-w}, \quad \langle \bar\psi(\bar z)\bar\psi(\bar w) \rangle = -\frac{1}{\bar z - \bar w}, \tag{3.3.35}$$

全纯场 ψ 与反全纯场 $\bar\psi$ 之间的关联函数为零. 因此, 其短距展开的算子积为

$$\psi(z)\psi(w) = -\frac{1}{z-w} + \cdots, \quad \bar\psi(\bar z)\bar\psi(\bar w) = -\frac{1}{\bar z - \bar w} + \cdots. \tag{3.3.36}$$

回顾一下径向积的定义, 对费米场

$$R(\psi_1(z)\psi_2(w)) = \begin{cases} +\psi_1(z)\psi_2(w), & |z| > |w|, \\ -\psi_2(w)\psi_1(z), & |z| < |w|, \end{cases} \tag{3.3.37}$$

这恰好是算子积 (3.3.36) 反映的性质. 由于费米场的取值在主标架丛的二重覆盖空间上, 又名旋量丛, 所以在欧几里得空间, 只有费米场的双线性形式才是单值的. 在柱面上, 如果转动 2π, 就存在两种可能的边界条件:

$$\psi(\mathrm{e}^{2\pi\mathrm{i}}z) = +\psi(z), \quad \text{列维–施瓦兹 (Neveu-Schwarz) 叶, 周期边界,}$$
$$\psi(\mathrm{e}^{2\pi\mathrm{i}}z) = -\psi(z), \quad \text{雷蒙德 (Ramond) 叶, 反周期边界.}$$

对于共形维 $h = \frac{1}{2}$ 的费米场, 同样有洛朗级数展开:

$$\mathrm{i}\psi(z) = \sum_r \psi_r z^{-r-1/2}, \tag{3.3.38}$$

其模式为

$$\psi_r = \oint_0 \mathrm{d}z\, z^{r-1/2} \mathrm{i}\psi(z). \tag{3.3.39}$$

由于存在两种可能的边界条件, 相应地 r 的取值也存在两种可能的选择:

$$r \in \mathbb{Z} + \frac{1}{2}, \quad \text{列维–施瓦兹 (Neveu-Schwarz) 叶, 周期边界,}$$
$$r \in \mathbb{Z}, \quad \text{雷蒙德 (Ramond) 叶, 反周期边界.}$$

类似于玻色场的情形, 由费米场的短距展开式, 可得反对易关系:

$$\begin{aligned}
\{\psi_r, \psi_s\} &= \mathrm{i}^2 \oint \mathrm{d}z \oint \mathrm{d}w \{\psi(z), \psi(w)\} x(z) z^{r-1/2} w^{s-1/2} \\
&= \oint \mathrm{d}w\, w^{s-1/2} \oint \mathrm{d}z\, z^{r-1/2} \frac{1}{z-w} = \oint \mathrm{d}w\, w^{r+s-1} \\
&= \delta_{r+s,0}.
\end{aligned} \tag{3.3.40}$$

同样由作用量出发, 可导出能–动张量. 这里略去详细的过程, 仅仅给出结果. 其全纯和反全纯的能–动张量分别为

$$T(z) = \frac{1}{2} :\psi(z)\partial\psi(z):, \quad \bar{T}(\bar{z}) = \frac{1}{2} :\bar{\psi}(\bar{z})\bar{\partial}\bar{\psi}(\bar{z}):, \tag{3.3.41}$$

它与全纯费米场的算子积展式为

$$T(z)\psi(w) = \frac{1/2}{(z-w)^2}\psi(w) + \frac{\partial\psi(w)}{z-w} + \cdots, \tag{3.3.42}$$

反全纯的有类似的结果. 这再次印证了费米场 ψ 具有共形维 $h = \frac{1}{2}$. 同理可得

$$T(z)T(w) = \frac{1/4}{(z-w)^4} + \frac{2T(w)}{(z-w)^2} + \frac{\partial T(w)}{z-w} + \cdots, \tag{3.3.43}$$

3.3 共形场：举例

即自由费米场的中心荷 $c = 1/2$. 如果对能-动张量和自由费米都进行模式展开，可得到等式：

$$L_n = \sum_{r > -3/2} \left(\frac{r}{2} + \frac{1}{4}\right) \psi_{n-r}\psi_r - \sum_{r \leqslant -3/2} \left(\frac{r}{2} + \frac{1}{4}\right) \psi_r \psi_{n-r}, \quad (3.3.44)$$

容易得到

$$[L_n, \psi_r] = -(\frac{n}{2} + r)\psi_{n+r}, \quad (3.3.45)$$

尤其是对于零模 L_0，有

$$[L_0, \psi_r] = -r\psi_r. \quad (3.3.46)$$

3.3.5 鬼系统

现在我们考虑另一类理论，它的性质既不同于自由玻色场也不同于自由费米场. 它的作用量为

$$S = \frac{1}{2\pi} \int \mathrm{d}^2 x\, b_{\mu\nu} \partial^\mu c^\nu, \quad (3.3.47)$$

其中，b, c 都是费米场，即反交换的场量，且要求场量 b 是对称、无迹的. 它们不是基本的动力学场，而是在进行泛函积分时作为规范固定的雅可比场引入，它的运动方程为

$$\partial^\mu b_{\mu\nu} = 0, \quad \partial^\mu c^\nu + \partial^\nu c^\mu = 0. \quad (3.3.48)$$

变到复坐标，定义 $c = c(z),\ \bar{c} = c(\bar{z}),\ b$ 场的非零分量 $b_{\mu\nu}$ 分别为 $b = b_{zz}, \bar{b} = b_{\bar{z}\bar{z}}$. 重写运动方程为

$$\bar{\partial}b = \partial\bar{b} = 0, \quad (3.3.49)$$

$$\bar{\partial}c = \partial\bar{c} = 0, \quad (3.3.50)$$

$$\partial c = -\bar{\partial}\bar{c}. \quad (3.3.51)$$

与自由费米场情形类似，可以得到如下的传播子：

$$\langle b(z)c(w)\rangle = \frac{1}{z-w}, \quad (3.3.52)$$

相应的算子积展开式为

$$b(z)c(w) = \frac{1}{z-w} + \cdots, \quad c(z)b(w) = \frac{1}{z-w} + \cdots. \quad (3.3.53)$$

从作用量出发，可以得到正则的能-动张量，但它不是对称的. 通过代数化的构造程式，可以得到对称无迹的能-动张量：

$$T(z) =: (\partial b)c : -\lambda \partial (: bc :)$$

$$= (1-\lambda):(\partial b)c: -\lambda:b\partial(c):. \tag{3.3.54}$$

容易证明能-动张量与场量 c, b 的算子积展式:

$$\begin{cases} T(z)c(w) = \dfrac{(1-\lambda)c(w)}{(z-w)^2} + \dfrac{\partial_w c(w)}{z-w} + \cdots, \\ T(z)b(w) = \dfrac{\lambda b(w)}{(z-w)^2} + \dfrac{\partial_w b(w)}{z-w} + \cdots. \end{cases} \tag{3.3.55}$$

根据定义, 可知元场 b, c 的共形维分别为 $(h_b, h_c) = (\lambda, (1-\lambda))$. 同样可得, 能-动张量自身的算子积展式为

$$T(z)T(w) = \dfrac{[1-3(2\lambda-1)^2]/2}{(z-w)^4} + \dfrac{2T(w)}{(z-w)^2} + \dfrac{\partial T(w)}{z-w} + \cdots. \tag{3.3.56}$$

一如其前, T 的共形维 $h=2$, 而相应的 bc- 鬼系统的中心荷 $c = 1 - 3(2\lambda-1)^2$, 被称为玻色鬼系统.

如果置定 $\lambda = 2$, 则 $(h_b, h_c) = (2, -1)$, 中心荷 $c = -26$. 之所以称之为鬼系统, 是因为根据统计-自旋之间的对应关系, 自旋为半整数的场遵循费米-狄拉克统计, 而自旋为整数的场应遵循玻色-爱因斯坦统计. 在量子场论中, 就是 Faddeev-Popov 鬼.

如果取 $\lambda = 0$, 即将能-动张量的全微分项去掉, 尽管不会影响局域共形变换, 但得到另一个鬼系统:

$$\begin{cases} T(z) =: \partial cb:, \\ T(z)b(w) = \dfrac{\partial_w b(w)}{z-w} + \cdots, \\ T(z)c(w) = \dfrac{c(w)}{(z-w)^2} + \dfrac{\partial_w c(w)}{z-w} + \cdots. \end{cases} \tag{3.3.57}$$

易知元场 b, c 的共形维分别为 $(h_b, h_c) = (0, 1)$. 同样可得, 能-动张量自身的算子积展式为

$$T(z)T(w) = \dfrac{(-2)/2}{(z-w)^4} + \dfrac{2T(w)}{(z-w)^2} + \dfrac{\partial T(w)}{z-w} + \cdots. \tag{3.3.58}$$

为区别起见, 称修正后的能-动张量 (3.3.57) 所定义的共形场为素鬼 (simple ghost) 系统.

与 bc-鬼系统对应, 还存在另一类鬼系统, 即所谓的 $\beta\gamma$-鬼系统, 它们的行为与 bc- 玻色鬼系统完全相反:

$$S = \dfrac{1}{2\pi} \int \mathrm{d}^2 z \beta \bar\partial \gamma, \tag{3.3.59}$$

其中, β, γ 都是可交换的场量, 即玻色场. 同样有运动方程:

$$\bar\partial \gamma(z) = \bar\partial \beta(z) = 0, \tag{3.3.60}$$

相应的短距展开的算子积展式为

$$\beta(z)\gamma(w) = -\frac{1}{z-w} + \cdots, \quad \gamma(z)\beta(w) = \frac{1}{z-w} + \cdots, \tag{3.3.61}$$

对称无迹的能-动张量为

$$\begin{aligned}T(z) &= :(\partial\beta)\gamma: -\lambda\partial(:\beta\gamma:) \\ &= (1-\lambda):(\partial\beta)\gamma: -\lambda:\beta\partial(\gamma):.\end{aligned} \tag{3.3.62}$$

简单地重复 bc- 系统的计算, 有 β, γ 的算子积展式:

$$\begin{cases} T(z)\gamma(w) = \dfrac{(1-\lambda)\gamma(w)}{(z-w)^2} + \dfrac{\partial_w\gamma(w)}{z-w} + \cdots, \\ T(z)\beta(w) = \dfrac{\lambda\beta(w)}{(z-w)^2} + \dfrac{\partial_w\beta(w)}{z-w} + \cdots. \end{cases} \tag{3.3.63}$$

可知 β, γ 的共形维分别为 $(h_\beta, h_\gamma) = (\lambda, 1-\lambda)$, 相应的 $\beta\gamma$- 鬼系统的中心荷为 $c = (3(2\lambda-1)^2 - 1)$. 当 $\lambda = 3/2$ 时, 交换的 $\beta\gamma$-费米鬼系统, 它们的共形维分别是 $h = (3/2, -1/2)$, 相应的中心荷为 $c = 11$.

对所谓的弦理论, 其反常消除的条件是理论总的中心荷相消. 若时空维度为 d, 对玻色弦理论而言, 其玻色场的数目, 亦即中心荷为 d; 对超弦理论而言, 其玻色、费米场贡献的中心荷总计为 $\dfrac{3d}{2}$. 因此, 反常消除的条件, 玻色弦 $d = 26$, 费米弦 $d = \dfrac{2}{3}(26-11) = 10$.

3.3.6 中心荷

从前面几节的讨论可知, 从一定意义上讲, 共形理论中最重要的参数就是中心荷 c. 中心荷反映了共形理论的特点, 不同的共形场可以有相同的中心荷, 但中心荷不同的共形场一定是不同的理论. 所以, 有必要再用一点篇幅, 对中心荷的物理意义略加分析. 前面已经提到, 中心荷不为零, 对应于共形反常效应. 共形反常是对共形不变性的"软破坏", 源自于有限宏观尺度效应. 引入有限宏观尺度效应的方式, 最常见的就是加入边界条件. 为了更具体地说明这一点, 回忆柱面到复平面的映射 $z = e^w$, 同时造成能-动张量的改变, 二者间存在等式:

$$T_{\text{cyl}}(w) = z^2 T_{\text{plane}}(z) - \frac{c}{24}, \tag{3.3.64}$$

这导致零模产生移动

$$(L_{\text{cyl}})_0 = L_0 - \frac{c}{24}. \tag{3.3.65}$$

$\bar{z}$ 与 $\bar{L}_0$ 之间也存在这样的关系，即柱面的零点能为

$$E_0 = -\frac{c+\bar{c}}{24}. \tag{3.3.66}$$

式 (3.3.64) 的导出，是默认柱面圆周是单位半径的. 更一般地，如果柱面的周长是 L，

$$T_{\text{cyl}}(w) = \left(\frac{2\pi}{L}\right)^2 \left[z^2 T_{\text{plane}}(z) - \frac{c}{24}\right], \tag{3.3.67}$$

因此有

$$\langle T_{\text{cyl}}(w)\rangle = -\frac{c\pi^2}{6L^2}, \tag{3.3.68}$$

也就是说，中心荷正比于 Casimir 能量. 显然，这一真空能量密度的改变是柱面的周期边界条件导致的，是一种边界效应. 随着柱面的宏观尺度变得很大，L 趋近无穷，Casimir 能量就趋近于零. 所以说，中心荷也是真空边界效应的体现. 对一个定义在柱面上的统计系统，中心荷是单位长度上的自由能.

另一个会影响到中心荷的因素，是伸缩变换导出的能-动张量无迹条件：

$$T^\mu_\mu = 0.$$

无疑，在经典意义下，系统在伸缩变换下不变，要求能-动张量的真空期望值为零，即

$$\langle T^\mu_\mu\rangle = 0, \tag{3.3.69}$$

限定在经典意义下的复平面上，无迹条件自然成立，无迹假设得以满足. 对于完全的量子理论，能-动张量的真空期望值并不一定为零. 在弯曲几何的背景空间，会导致迹反常 (trace anomaly)，这是弯曲空间的场论必须面对的问题：

$$\langle T^\mu_\mu\rangle = -\frac{c}{12}R, \tag{3.3.70}$$

其中，R 是弯曲空间的标量曲率. 关于式 (3.3.70) 的详细讨论以及弯曲空间的场论，可参见文献 [3]. 通过比较 3.3 节的例子，表明中心荷的值又是反映迹反常的量，或者说，共形反常量也反映系统偏离平坦空间程度的量. 也可以从另一个角度讲，中心荷的值非零反映了在平坦空间情形，系统的临界点已经偏离了重整化群的不动点. 而在经典图像中，临界点应该是标度不变的.

3.4 态

在量子场论中，算子-态对应是最基本的概念之一. 先定义算子 $[A(z,\bar{z})]$ 的伴随算子：

$$[A(z,\bar{z})]^\dagger = \left[A\left(\frac{1}{\bar{z}},\frac{1}{z}\right)\right]\frac{1}{\bar{z}^{2h}}\frac{1}{z^{2\bar{h}}}, \tag{3.4.1}$$

3.4 态

这看似奇怪的定义实为共形变换的自洽性, 更详细的讨论, 请参见文献 [9], [17]. 首先定义入态

$$|A_{\text{in}}\rangle \equiv \lim_{z,\bar{z}\to 0} A(z,\bar{z})|0\rangle, \tag{3.4.2}$$

$|0\rangle$ 为真空态, 相应的出态

$$\begin{aligned}\langle A_{\text{out}}| &\equiv \lim_{w,\bar{w}\to 0}\langle 0|A\left(\frac{1}{w},\frac{1}{\bar{w}}\right)(-w^{-2})^h(-\bar{w}^{-2})^{\bar{h}} \\ &= \lim_{z,\bar{z}\to 0}\langle A_{\text{out}}|[A(z,\bar{z})]^\dagger \\ &= \lim_{z,\bar{z}\to 0}[A(z,\bar{z})|0\rangle]^\dagger = |A_{\text{in}}\rangle^\dagger,\end{aligned} \tag{3.4.3}$$

相应有能–动张量为厄米算子的条件 $L_m^\dagger = L_{-m}$. 如果要求 $z=0$, 能–动张量作用在真空态 $T(z)|0\rangle$ 正则, 要求满足条件:

$$L_n|0\rangle = 0, \quad n \geqslant -1. \tag{3.4.4}$$

由共轭条件又得到

$$\langle 0|L_n = 0, \quad n \leqslant 1. \tag{3.4.5}$$

不难看出, $\langle 0|$ 和 $|0\rangle$ 同时为 $L_{\pm 1;0}$ 所湮没, 因此说 $\langle 0|$ 和 $|0\rangle$ 是 $SL(2,\mathcal{R})$ 不变的, 即真空态在平移、伸缩和特殊共形变换下不变. 对 $\bar{L}_{\pm 1;0}$ 有同样的结论. $L_{\pm 1;0}$ 和 $\bar{L}_{\pm 1;0}$ 是 Virasoro 代数仅存的有限维子代数. 换言之, $|0\rangle \otimes |\bar{0}\rangle$ 是 $SL(2,\mathcal{C})$ 不变的真空. 共形场真空的这一性质, 也说明共形场意义下的多点函数是可能恰当定义的. 在不引起歧义的情况下, 通常将 $|0\rangle \otimes |\bar{0}\rangle$ 简记为 $|0\rangle$. 由能–动张量是厄米算子的条件

$$\langle 0|[L_2, L_{-2}]|0\rangle = \frac{c}{2} = \langle 0|[L_2, L_2^\dagger]|0\rangle \geqslant 0, \tag{3.4.6}$$

在群论中, 将保持向量空间内积的变换, 称为幺正变换表示, 由幺正变换定义的表示, 被称为酉表示. 显而易见, 如果要求 Virasoro 代数存在酉表示, 则要求

$$c \geqslant 0.$$

3.4.1 最高权态

下面重点介绍最高权态. 如果全纯元场 $\phi(z)$ 的共形维为 h, 考虑 $\phi(0)$ 作用于真空 $|0\rangle$ 产生的态:

$$|h\rangle = \phi(0)|0\rangle, \tag{3.4.7}$$

这就是算子–态的对应. 根据 $\phi(z)$ 的定义, 不难计算它与能–动张量间的对易括号:

$$[L_n,\phi(w)] = \oint_w dz\, z^{n+1}T(z)\phi(w) = h(n+1)w^n\phi(w) + w^{n+1}\partial\phi(w). \tag{3.4.8}$$

容易验证

$$[L_n, \phi(0)] = 0, \quad n > 0. \tag{3.4.9}$$

也就是说, 对态 $|h\rangle$ 存在等式:

$$L_0|h\rangle = h|h\rangle, \quad L_n|h\rangle = 0, \, n > 0, \tag{3.4.10}$$

式 (3.4.10) 所定义的态, 就是熟知的最高权态, 更准确地说, 是 Virasoro 代数的最高权态, 也就是共形场论的元场. 因为这一原因, 元场的共形维又被称为共形权, 文献中经常不加区分地使用它们. 由 Virasoro 代数生成元的厄米条件:

$$\begin{aligned}\langle h|L_{-n}^\dagger L_{-n}|h\rangle &= \langle h|[L_n, L_{-n}]|h\rangle \\ &= 2n\langle h|L_0|h\rangle + \frac{c}{12}(n^3 - n)\langle h|h\rangle \\ &= \left[2nh + \frac{c}{12}(n^3 - n)\right]\langle h|h\rangle,\end{aligned}$$

幺正条件要求上式结果非负. 当 $n = 1$ 时, 得到 $h \geqslant 0$; 当 n 的取值足够大时, 第二项起主导作用, 所以有 $c > 0$. 由后面的讨论可知, 如果 $h = 0$, 则需 $L_{-1}|h\rangle = 0$, 即要求 $|0\rangle$ 是 $SL(2, \mathcal{C})$ 不变的真空态. 如果不考虑共形场取值的李群, 具有特殊的李超代数结构, 其超维数为零的情形[12], 诸如 $SL(n|n)$, 则中心荷 $c = 0$ 的共形场不存在非平凡的表示. 不难证明, 如果 L_{-n} 的模长为零, 故可设 $L_n|0\rangle = 0$. 利用式 (3.2.23) 定义的能–动张量模式展开, 因此, 式 (3.4.8) 给出了 Virasoro 代数生成元与共形权 $(h, 0)$- 元场 $\phi(z)$ 的对易子. 若将任意共形权为 $(h, 0)$ 的元场 $\phi(z)$ 写成展开式:

$$\phi(z) = \sum_{n \in \mathcal{Z}} \phi_n z^{-n-h}, \tag{3.4.11}$$

则模式 ϕ_n 由傅里叶变换给出:

$$\phi_n = \oint_0 dz\, z^{n+h-1} \phi(z). \tag{3.4.12}$$

若要求 $\phi(z)|0\rangle$ 在 $z = 0$ 点的作用是正则的, 这等价于 $\phi_n|0\rangle = 0$ $(n \geqslant 1 - h)$. 仿前面的定义, 如果设 $|h\rangle = \phi_{-h}|0\rangle$, 由式 (3.4.8):

$$\begin{aligned}[L_n, \phi_m] &= \oint_0 dw\, w^{n+h-1}[h(n+1)w^n \phi(w) + w^{n+1}\partial\phi(w)] \\ &= \oint_0 dw\, w^{n+m+h-1}[h(n+1) - (n+m+h)]\phi(w) \\ &= [n(h-1) - m]\phi_{n+m}, \end{aligned} \tag{3.4.13}$$

即 $[L_0, \phi_m] = -m\phi_{n+m}$,这与 $L_0|h\rangle = L_0\phi_{-h}|0\rangle = h|h\rangle$ 的定义相融. 利用真空态 $\langle 0|$ 与 $|0\rangle$ 的 $SL(2,\mathcal{C})$ 不变性,因为 $U(2) \in SL(2,\mathcal{C})$,所以 $U(2)|0\rangle = |0\rangle$,对于拟元场 ϕ_i 的 n-点函数:

$$\langle 0|U^{-1}\phi_1 U \cdots U^{-1}\phi_n U|0\rangle = \langle 0|\phi_1 \cdots \phi_n|0\rangle, \tag{3.4.14}$$

具体到分式线性变换,只考虑无穷小变换:

$$\begin{aligned}0 &= \langle 0|L_k\phi_1\cdots\phi_n|0\rangle \\ &= \langle 0|[L_k,\phi_1]\cdots\phi_n|0\rangle + \cdots + \langle 0|\phi_1\cdots[L_k,\phi_n]|0\rangle,\end{aligned} \tag{3.4.15}$$

式中,$k = 0, \pm$,分别对应于平移、伸缩和特殊共形变换. 利用式 (3.4.8),可以将上式等价地写为

$$\sum_i \partial_{z_i}\langle\phi_1(z_1)\cdots\phi_n(z_n)\rangle = 0, \tag{3.4.16}$$

$$\sum_i (z_i\partial_{z_i} + h_i)\langle\phi_1(z_1)\cdots\phi_n(z_n)\rangle = 0, \tag{3.4.17}$$

$$\sum_i (z_i^2\partial_{z_i} + 2h_i z_i)\langle\phi_1(z_1)\cdots\phi_n(z_n)\rangle = 0. \tag{3.4.18}$$

重复上面的过程,鉴于左、右手征场彼此对易,由 $\phi(z,\bar{z})$ 可得 $|h,\bar{h}\rangle$,所以共形场论的希尔伯特空间为

$$\mathcal{H} = \oplus_{i,\bar{i}} N_{i,\bar{i}} \mathcal{H}_i \otimes \mathcal{H}_{\bar{i}}, \tag{3.4.19}$$

式中,$\mathcal{H}_i = L(h_i, c)$,$\mathcal{H}_{\bar{i}} = L(\bar{h}_i, \bar{c})$,$N_{i,\bar{i}}$ 是重数.

3.4.2 嗣场

前面主要介绍了最高权态是元场作用在真空上的结果 $|h\rangle = \phi(0)|0\rangle$. 下面我们考虑 Virasoro 代数的嗣场 (descendant fields),具体地说,是 $\hat{L}_{-n}$ 作用在 $|h\rangle = \phi(0)|0\rangle$ 生成的无穷序列. 由给定的最高权态,用 L_{-n_i} 反复作用于最高权态,可构造出其他的态. 诸如 $L_{-n_1}\cdots L_{-n_k}|h\rangle$ ($n_i \geqslant 0$) 类型的态,就是所谓入最高权态的嗣场 (descendant fields),基于算子–态对应,嗣场作用于真空,就产生出嗣态. 这样得到的 Virasoro 嗣场具有更高的共形权,且嗣场之间可用共形变换联系起来. 定义嗣场为

$$\phi^{(-n)}(w) \equiv (\hat{L}_{-n}\phi)(w) = \oint dz\, \frac{1}{(z-w)^{n-1}} T(z)\phi(w), \tag{3.4.20}$$

不难证明,$\hat{L}_{-n}\phi$ 的共形维是 $(h+n)$. 显然,$\phi^{(0)} = \hat{L}_0\phi = h\phi$,而 $\phi^{(-1)} = \hat{L}_{-1}\phi = \partial\phi$. 看一个简单的例子

$$\hat{L}_{-2}\mathbf{1}(w) = \oint dz\, \frac{1}{z-w} T(z)\mathbf{1} = T(w),$$

即 $\mathbf{1}^{(-2)}(w) = (\hat{L}_{-n}\mathbf{1})(w) = T(w)$. 这表明能-动张量恒为单位算子的二层嗣算子, 这就是式 (3.2.24) 有别于式 (3.2.20) 的原因. 嗣场的关联函数, 可以从元场的关联函数得到. 考虑关联函数 $\langle(\hat{L}_{-n}\phi)(w)\Phi\rangle$, 其中 $\Phi = \phi_1(w_1)\cdots\phi_N(w_N)$, 而 ϕ_i 是共形权为 h_i 的元场. 利用嗣场的定义, 以及能-动张量 T 与元场的算子积关系, 可得到 $\hat{L}_{-n}$ 作用在 $\langle\phi(w)\Phi\rangle$ 上的结果:

$$\langle(\hat{L}_{-n}\phi)(w)\Phi\rangle \equiv L_{-n}\langle\phi(w)\Phi\rangle, \quad n \geqslant 1. \tag{3.4.21}$$

重复这样的过程, 可以得到嗣场与多个 L_{-i} 算子的关系, 诸如

$$(\hat{L}_{-m}\hat{L}_{-n}\phi)(w) = \oint_w \mathrm{d}z\ (z-w)^{1-m} T(z)(L_{-n}\phi)(w).$$

同理可证, 嗣场的多点函数完全由相关的元场确定. 如法炮制, 可得到出态、或称为最低权态的嗣场 $\langle h|_{n_1}\cdots L_{n_k}\ (n_i \geqslant 0)$.

一个关键问题是, 从嗣场的构造关系是否可以得到共形理论的更多限制. 由前面的讨论可知, 酉性对共形场的中心荷、共形权都做出了限制. 模不变性和超对称性等, 都对共形场做出了非常强的限制. 除了上面提到的约束, 这里有必要提到整体性对多点函数的限制. 以四点函数为例

$$\langle \phi_1(z_1,\bar{z}_1)\phi_2(z_2,\bar{z}_2)\phi_3(z_3,\bar{z}_3)\phi_4(z_4,\bar{z}_4)\rangle, \tag{3.4.22}$$

由于共形不变性, 为了简单且不失一般性原则, 我们总可以选择其中的三点 $z = 0, 1, \infty$. 考虑在四点函数中插入元场的算子积展开:

$$\begin{aligned}&\langle V_{h_1}(z)V_{h_2}(0)V_{h_3}(\infty)V_{h_4}(1)\rangle \\ &= \sum_{h\in\Delta} C_{h_1,h_2,h}|z|^{2(h-h_1-h_2)}\left[\langle V_h(0)V_{h_3}(\infty)V_{h_4}(1)\rangle + O(z)\right] \\ &= \sum_{h\in\Delta} C_{h_1,h_2,h}C_{h,h_3,h_4}|z|^{2(h-h_1-h_2)}\left[1 + O(z)\right]. \end{aligned} \tag{3.4.23}$$

由于全纯与反全纯部分是可交换的, 所以由 $L_{n<0}$ 产生的左移嗣场与 $\bar{L}_{n<0}$ 产生的右移嗣场也是可因式化的. 显然, 对四点函数有两种不同的求解方式, 一种是 $z_1 \to z_2,\ z_3 \to z_4$, 即先对 $\phi_1\phi_2, \phi_3\phi_4$ 作算子积 (也称为收缩), 然后对所有的元场和嗣场求和, 记为 s 道, 因此上式有因式化:

$$\langle V_{h_1}(z)V_{h_2}(0)V_{h_3}(\infty)V_{h_4}(1)\rangle = \sum_{h\in\Delta_s} C_{h_1,h_2,h}C_{h,h_3,h_4}\mathcal{F}_h^{(s)}(z)\mathcal{F}_h^{(s)}(\bar{z}). \tag{3.4.24}$$

上式中的 $\mathcal{F}_h^{(s)}(z)$ 是黎曼球面上的四点共形块:

$$\mathcal{F}_h^{(s)}(z) = z^{h-h_1-h_2}\left[1 + O(z)\right]. \tag{3.4.25}$$

3.4 态

一般地, $\mathcal{F}_h^{(s)}(z)$ 是 c, h_1, h_2, h_3, h_4 的函数. 还有另一种求解方式, $z_2 \to z_3$, $z_1 \to z_4$, 即先对 $\phi_2\phi_3$, $\phi_1\phi_4$ 作算子积, 然后对所有的元场和嗣场求和, 记为 t 道.

$$\langle V_{h_1}(z)V_{h_2}(0)V_{h_3}(\infty)V_{h_4}(1)\rangle = \sum_{h\in\Delta_t} C_{h,h_1,h_4} C_{h_2,h_3,h} \mathcal{F}_h^{(t)}(z)\mathcal{F}_h^{(t)}(\bar{z}), \quad (3.4.26)$$

其中, $\mathcal{F}_h^{(t)}(z) = (z-1)^{h-h_1-h_4}\bigl[1+O(z-1)\bigr]$ 是 t 道的共形块. 数学上将 $V_h(h\in\Delta)$ 这种含三个指标的算子称为顶点算子. 原则上, 共形对称性完全决定了共形块的函数形式. 由共形场的可交换性, 可知场的算子积展式是可结合的. 因此, 四点函数的结果应该与算子展式的顺序无关, 这就是所谓的交叉对称性, 可以用下面的公式表示:

$$\sum_{h_s\in\Delta} C_{12s}C_{s34} \;\;{}^2_1\!\!\!>\!\!\!\!-\!\!s\!\!-\!\!<\!\!\!\!{}^3_4 = \sum_{h_t\in\Delta} C_{23t}C_{t41} \;\;{}^2\!\!\!|\!\!|^3 \atop {}^1\!\!\!|\!\!|^4 t. \quad (3.4.27)$$

给定谱 Δ, 交叉对称方程变成结构常数的二次方程, 如果要求交叉对称式有解, 会对系统的谱给出很强的限制.

3.4.3 Kac 行列式和酉表示

下面我们更详细地考察共形场的希尔伯特空间, 出发点依然是元场, 或最高权表示 $|h\rangle = \phi(0)|0\rangle$. 考虑能-动张量的生成元 $L_n(n<0)$, 它作用于最高权态 $|h\rangle$, 就得到其嗣场. 所有嗣场的集合 $V_{h,c}$ 构成所谓的 Verma 模, 其中的参数 c, h 分别是中心荷和共形权. 最开始的几个 Verma 模为

$$L_{-1}|h\rangle, \quad L_{-2}|h\rangle, \quad L_{-1}^2|h\rangle, \quad L_{-3}|h\rangle, \quad L_{-1}L_{-2}|h\rangle, \quad L_{-1}^3|h\rangle, \quad \cdots$$

这里不包括 $L_{-2}L_{-1}|v\rangle$, 原因有等式 $L_{-2}L_{-1} = L_{-1}L_{-2} - L_{-3}$. 更一般地, 如果考虑形如 $L_{-n_1}L_{-n_2}\cdots L_{-n_r}|h\rangle$ 嗣场, 若 $\sum_{j=1}^r = N$, 则可以按 N 的值将嗣场排列, N 称为层 (level). 以 N 取值最大至 3 为例:

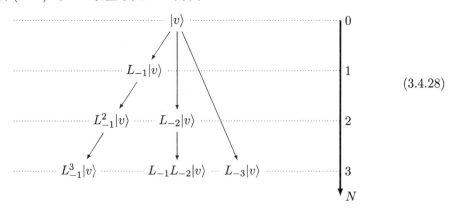

(3.4.28)

将这些 Verma 模视为相应的元场构成的态的集合. 容易看出, $N = 1, 2, 3$ 时, 嗣态的数目分别为 1, 2 和 3. 如果数目更大一些, 比如 $N = 6$, 则相应嗣场的数目为 11. 一般地, 任意给定正整数 N, 相应的数目为 $P(N)$. $P(N)$ 为正整数 N 的拆分数, 即将 N 写成不同正整数之和的方式. 拆分数可有生成函数得到

$$\prod_{n=1}^{\infty} \frac{1}{1-q^n} = (1 + q + q^2 + q^3 + q^4 + q^5 + q^6 + \cdots)(1 + q^2 + q^4 + q^6 + \cdots)$$
$$\times (1 + q^3 + q^6 + \cdots)$$
$$= 1 + q + 2q^2 + 3q^3 + 5q^4 + 7q^5 + 11q^6 + \cdots$$
$$= \sum_{N=0}^{\infty} P(N) q^N. \tag{3.4.29}$$

当然, 这不是说在层 N 的 Verma 模就存在 $P(N)$ 个独立的物理态. 如果态的模为零, 就称之为空态 (null state), 这样的态不是物理的. 要使得从 Virasoro 代数的最高权态 $|h\rangle$ 构造的 Verma 模有意义, 就必须排除所有的空态. 实际情况是, 嗣场的线性组合有可能使得其组合态的模为零. 看几个简单的例子. 显然, 在层 $N = 1$ 时,

$$L_{-1}|h\rangle = 0, \tag{3.4.30}$$

这只是表明, 此时的最高权态 $|h\rangle = |0\rangle$ 为真空, 不会给出新的约束. 在层 $N = 2$ 时, 可以假设存在常数 a, 使得

$$(L_{-2} + aL_{-1}^2)|h\rangle = 0, \tag{3.4.31}$$

显然, 其非平凡解应为参数 h, c 的函数. 如果分别将 $L_{1,2}$ 作用于上式两端, 其对易括号分别为

$$[L_1, (L_{-2} + aL_{-1}^2)]|h\rangle = [2a(2h+1) + 3]L_{-1}|h\rangle,$$
$$[L_2, (L_{-2} + aL_{-1}^2)]|h\rangle = \left[2(2+3a)h + \frac{c}{2}\right]|h\rangle,$$

这要求 $a = -3/2(2h+1)$, $c = -4(2+3a)h = 2h(5-8h)/(2h+1)$, 即

$$\left[L_{-2} - \frac{3}{2(2h+1)} L_{-1}^2\right] |h\rangle = 0. \tag{3.4.32}$$

由式 (3.2.7), 可得到多点函数满足的微分方程:

$$-a \frac{\partial^2}{\partial w_1^2} \langle \phi_1(w_1, \bar{w}_1) \cdots \phi_n(w_n, \bar{w}_n) \rangle$$
$$= \sum_{j \neq 1} \left[\frac{h_j}{(w_1 - w_j)^2} + \frac{1}{w_1 - w_j} \partial_{w_j} \right] \langle \phi_1(w_1, \bar{w}_1) \cdots \phi_n(w_n, \bar{w}_n) \rangle. \tag{3.4.33}$$

3.4 态

线性代数告诉我们,对于 N 维正定的希尔伯特空间,如果其模长为零,则其本征矢量具有零本征值,而具有零本征值的本征矢量的数目等于 N 行、N 列矩阵 M_N 的行列式

$$\det M_N = 0$$

根的重数. 由前文的讨论可知,对于嗣场的线性组合,如果出现空态,其系数一定是 Virasoro 代数的中心荷 c 及其元场的共形权 h 的函数,因为这二者是仅有的自由参数. 显然,当 $N=1$ 时,

$$\det M_1(h,c) = 2h, \tag{3.4.34}$$

所以,在层 $N=1$ 时,有空态的条件就是 $h=0$,即前面提到的 $L_{-1}|h\rangle = 0$. 当 $N=2$ 时,有如下矩阵:

$$\begin{pmatrix} \langle h|L_2 L_{-2}|h\rangle & \langle h|L_1^2 L_{-2}|h\rangle \\ \langle h|L_2 L_{-1}^2|h\rangle & \langle h|L_1^2 L_{-1}^2|h\rangle \end{pmatrix} = \begin{pmatrix} 4h+\dfrac{c}{2} & 6h \\ 6h & 4h(1+2h) \end{pmatrix},$$

相应的 Kac-行列式:

$$\begin{aligned} \det M_2(h,c) &= 32h\left(h^2 - \frac{5h}{8} + \frac{hc}{8} + \frac{c}{16}\right) \\ &= 32[h-h_{1,1}(c)][h-h_{1,2}(c)][h-h_{2,1}(c)], \end{aligned} \tag{3.4.35}$$

其中

$$\begin{cases} h_{1,1}(c) = 0, \\ h_{1,2}(c) = \dfrac{5-c}{16} - \dfrac{1}{16}\sqrt{(1-c)(25-c)}, \\ h_{2,1}(c) = \dfrac{5-c}{16} + \dfrac{1}{16}\sqrt{(1-c)(25-c)}. \end{cases}$$

而 $h_{1,1}(c) = 0$ 已经出现在层 $N=1$ 的空态中,显然 $L_{-1}(L_{-1}|h\rangle) = 0$. 更一般地,如果嗣态 $|h+n\rangle$ 在层 n 时出现空态,则层为 N 的嗣态 $L_{-n_1}L_{-n_2}\cdots L_{-n_r}|h\rangle$,存在 $P(N-n) = P(\sum\limits_{i=1}^{r} n_i)$ 个空态. 如上式,$[h-h_{1,1}(c)]^{P(2-1)} = [h-h_{1,1}(c)]^{P(1)}$. 对任意给定的 N,行列式的表述是非常复杂的,Kac 猜测有普适的表述式:

$$\begin{cases} \det M_N(h,c) = \alpha_N \prod\limits_{\substack{r,s\leqslant N \\ p,q>0}} [h-h_{r,s}(c)]^{P(N-rs)}, \\ h_{r,s}(m) = \dfrac{[(m+1)r-ms]^2-1}{4m(m+1)}, \\ m = -\dfrac{1}{2} \pm \dfrac{1}{2}\sqrt{\dfrac{25-c}{1-c}}, \end{cases} \tag{3.4.36}$$

其中, α_N 是不依赖于 h, c 的正常数; 变量 r, s 为正整数; m 可以取复值. 对于 $c<1$, 通常可选择解的支集为 $m\in(0,\infty)$. 实际上, 由于函数 $h_{r,s}$ 中的 r, s 具有可换性, 即 $r\leftrightarrow s$, 具体而言, 存在 $r\to m-r$, $s\to m+1-s$ 的对称性, 所以解的支集选择对行列式 $\det M$ 并没有影响. 如果将中心荷用 m 解出

$$c = 1 - \frac{6}{m(m+1)}.$$

参数 c, h 的值, 决定 Virasoro 代数表示的类型. 如果要求 Virasoro 代数的表示为酉, 或者说表示是不可约的, 就需要将所有正规子表示减除. 从前面的分析可以看出, Kac 行列式就具有这样的作用. 也就是说, 在任何给定的层, 只要 Kac 行列式为负, 就表明行列式存在奇数个负的本征值. 换言之, 对应于这些取值的 c, h, 包含模长为负值的态, 因此表示非酉. 在量子场论中, 幺正性与概率守恒有关; 在统计物理中, 幺正性表明转移矩阵是厄米算子. 讨论理论的幺正性, 并不是说非酉理论就没有意义, 或不重要. 在研究渗滤 (percolation) 问题、SLE 方程等时 [7], 共形场论的空态具有重要作用; 还有一个著名的例子, 李–杨格点气体模型的边缘奇点. 李–杨格点气体模型包含取虚值的场, 因此, 李–杨边缘奇点理论为非酉的共形场, 相应共形场的中心荷为 $c=-22/5$. 但这里, 我们仍主要局限于讨论酉表示. 略去这节内容的相关证明过程, 详细内容可参见文献 [9].

如果要求 Virasoro 代数存在酉表示, 则要求 c, $h\geqslant 0$. 显然, 酉表示将中心荷的取值限定在四个区间: ① $1<c<25$; ② $c\geqslant 25$; ③ $c=1$; ④ $0\leqslant c<1$. 容易证明, 如果 $c>1$, $h\geqslant 0$, Kac 行列式在任意给定的层都非零, 当 $1<c<25$ 时, 让 m 取非实值, 则 $h_{r,s}$ 存在虚部, 或者为负值; 当 $c\geqslant 25$ 时, 可以让 $-1<m<0$, 则所有 $h_{r,s}$ 都取负值, 不难证明, 如果此种情形的 Kac 行列式非零, 则其所有本征值为正; 当 $c=1$ 时, 只需让 $h=n^2/4$, $n\in\mathbb{Z}$, Kac 行列式为零, 但并不会变成负值. 从分类上说, $c=1$ 时的共形理论是单独的一类. 也就是说, 对 $c\geqslant 1$, $h\geqslant 0$ 的共形理论, Kac 行列式并不会给出新的约束, 对此种情形的共形场论, 需要借助于其他的方法. 也就是说, 现在唯一需要细致分析的共形理论, 是中心荷 $0\leqslant c<1$, $h>0$ 这种情形, 出发点依然是等式 (3.4.36), 以中心荷 c 为自变量, 重写 $h_{r,s}$ 为

$$h_{r,s}(c) = \frac{1-c}{96}\left[\left((r+s)\pm(r-s)\sqrt{\frac{25-c}{1-c}}\right)^2 - 4\right]. \tag{3.4.37}$$

在 (c,h) 定义的平面, Kac 行列式在 $h=h_{r,s}(c)$ 定义的曲线上为零, 如图 3.5 所示, 称 $h=h_{r,s}(c)$ 为 Kac 行列式的零曲线. 研究表明, 只有某些出现在 Kac 行列式零曲线上的点才有可能对应酉共形场论, 零曲线外所有的点对应非酉理论. 由于 m 的选择, 可由 $r\leftrightarrow s$ 的可换性补偿. 在中心荷 $c=1$ 曲线附近, 且设 $c=1-6\epsilon$. 考虑 ϵ

3.4 态

的主导项:

$$h_{r,s}(1-6\epsilon) = \frac{1}{4}(r-s)^2 + \frac{1}{4}(r^2-s^2)\sqrt{\epsilon}, \quad r \neq s,$$
$$h_{r,r}(1-6\epsilon) = \frac{1}{4}(r^2-1)\epsilon,$$

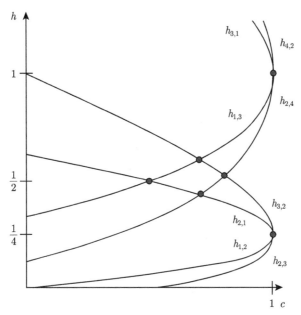

图 3.5 行列式定义的曲线

对曲线 (3.4.37) 进行分析可知, 给定层 N 的 Kac 行列式, 从零曲线外任意点出发, 都存在一条路径, 穿越唯一的 Kac 行列式的零曲线, 则路径的一端是 $0 < c < 1$, $h > 0$, 另一端是 $c > 1$, $h > 0$. 穿越零曲线时, Kac 行列式要变号, 因此在给定层 N, 一定存在负模态. 这样, 就完全排除了 Kac 行列式零曲线之外的点存在酉表示的可能, 更为详细的讨论, 可参见文献 [14]. 更细致的分析发现, Kac 行列式零曲线的点仍然存在负模态, 即不能给出酉表示. 在零曲线上存在这样的点, 某层为酉, 但在更高的层, 则可能成为负模态, 除非它们是两条零曲线的交点, 如图 3.5 所示. 因此, 这样的点是离散的, 在这些零曲线的交点, 中心荷的表示式为

$$c = 1 - \frac{6}{m(m+1)}, \quad m = 3, 4, \cdots. \tag{3.4.38}$$

$m = 2$ 时, 对应的中心荷 $c = 0$. 给定中心荷 c, h 存在 $m(m-1)/2$ 个允许的取值:

$$h_{r,s}(m) = \frac{[(m+1)r - ms]^2 - 1}{4m(m+1)}, \tag{3.4.39}$$

其中, $1 \leqslant r \leqslant m-1, 1 \leqslant s \leqslant r$. 因此, Virasoro-代数的最高权为酉的必要条件, 要么 $c \geqslant 1, h \geqslant 0$, 要么 $c=1, h=n^2/4, n \in \mathbb{Z}$, Kac 行列式为零, 要么式 (3.4.38) 和式 (3.4.39) 同时成立. 研究表明, 最后一种情形, 式 (3.4.38) 和式 (3.4.39) 也是充分条件, 这可以用陪集构造法加以证明 [18,19].

3.4.4 极小模型

代数的酉表示, 就是模具有不可分解性, 或者说表示是不可约的. 中心荷 $c \leqslant 1$ 的 Virasoro 代数不可约表示, 对所有可能二级相变的临界行为, 给出了一个完全的分类, 这几乎是不可思议的结果. 这个系列的最开始的几个, 其中心荷 $c = \frac{1}{2}, \frac{7}{10}, \frac{4}{5}, \frac{6}{7}$, 分别对应于伊辛模型 (Ising model), 二临界伊辛-模型, 3-态 Potts 模型, 以及二临界 3-态 Potts 模型. 显然, 酉表示是非常强的条件, 对一些问题而言, 是不必要的. 如果将酉表示的条件弱化, 就得到所谓的极小模型系列 [2]. 该系列的中心荷取值为

$$c(p,q) = 1 - 6\frac{(p-q)^2}{pq}, \tag{3.4.40}$$

其中, 要求 $p, q \geqslant 2, p, q$ 互质, 共形权的取值为

$$h_{r,s}(p,q) = \frac{(pr-qs)^2 - (p-q)^2}{4pq}, \tag{3.4.41}$$

其中, $1 \leqslant r \leqslant q-1, 1 \leqslant s \leqslant p-1$. 之所以称之为极小模型, 是因为在这个离散序列中, 只存在有限多个 Virasoro-代数的最高权表示, 或者说局域场, 且它们都具有较好的标度性质; 它的谱是最简单的, 不能表示为其他模型谱的乘积; 除了共形对称性外, 模型本身没有其他的对称性.

极小模型 Kac 行列式, 可采用多种参数化方式. 不同的参数化方式, 针对不同的问题, 各有擅场. 如果将中心荷写为

$$c = 1 - 24\alpha_0^2, \tag{3.4.42}$$

其中

$$\begin{cases} 2\alpha_0 = \alpha_+ + \alpha_-, \quad \alpha_+\alpha_- = -1, \\ \alpha_\pm = \alpha_0 \pm \sqrt{\alpha_0^2 + 1} = \dfrac{\sqrt{1-c} \pm \sqrt{25-c}}{\sqrt{24}}, \end{cases} \tag{3.4.43}$$

相应地

$$\begin{aligned} h_{r,s}(c) = &\frac{1}{48}[(13-c)(r^2+s^2) - 24rs - 2(1-c) \\ &+ \sqrt{(1-c)(25-c)}(r^2-s^2)] \end{aligned}$$

3.4 态

$$=h_0 + \frac{1}{4}(r\alpha_+ + s\alpha_-)^2$$
$$= -\alpha_0^2 + \frac{1}{4}(r\alpha_+ + s\alpha_-)^2, \tag{3.4.44}$$

式中的 $h_0 = \frac{(c-1)}{24}$. 另一种常见的参数化为

$$c = 13 - 6\left(t + \frac{1}{t}\right),$$
$$h_{r,s}(t) = \frac{t}{4}(r^2-1) + \frac{1}{4t}(s^2-1) - \frac{1}{2}(rs-1), \tag{3.4.45}$$

其中

$$t = 1 + \frac{1}{12}\left[1 - c \pm \sqrt{(1-c)(25-c)}\right].$$

由同样的分析可知, 解的支集选择对结果并没有影响. 不同参数化, 可相互表示, 比如:

$$\alpha_+ = \sqrt{t}, \quad \alpha_- = -\frac{1}{\sqrt{t}}.$$

同理, t 也可用 m 表示, 但方式不唯一, $t = \frac{m}{m+1}$, 或者 $t = \frac{m+1}{m}$. 在研究刘维尔理论时, 也采用如下的参数化:

$$\begin{cases} c = 1 + 6Q^2, \quad Q = \left(b + \frac{1}{b}\right), \\ \alpha_{r,s}(t) = \frac{Q}{2} - \frac{1}{2}(rb + sb^{-1}), \\ h_{r,s} = \alpha_{r,s}(Q - \alpha_{r,s}). \end{cases} \tag{3.4.46}$$

由于在极小模型中存在空态, 或者说包含奇异矢量, 所以表示是可分的. 前文提到的李–杨边缘奇点理论, 作为极小模型系列中的一个例子, 其中心荷为 $c = -22/5$, 对应于 $(p,q) = (5,2)$. 如果限定 $|q-p| = 1$, 就重新得到酉条件. 要得到不可约的表示, 就必须将各种可能的子模态分离出去. 而 $h_{r,s}(p,q)$ 的对称性能在一定程度上简化分析奇异矢量问题的难度. $h_{r,s}$ 具有周期性

$$h_{r+q,s+p} = h_{r,s}, \tag{3.4.47}$$

容易得到如下等式:

$$\begin{cases} h_{r,-s} = h_{r,s} + rs = h_{q-r,p+s}, \\ h_{r,s} + (q-r)(p-s) = h_{r,2p-s} = h_{2q-r,s}. \end{cases} \tag{3.4.48}$$

若 Verma 模 $V(c(p,q), h_{r,s}(p,q))$ 的最高权为 $h_{r,s}(p,q)$, 为简单起见, 将这样的 Verma 模标记为 $V_{r,s}$. 分析可约 Verma 模的结构, 有如下包含关系:

$$V_{q+r,p-s} \cup V_{r,2p-s} \in V_{r,s}. \tag{3.4.49}$$

直观的定义就是要减除子模的和:

$$M_{r,s} = V_{r,s}/(V_{q+r,p-s} + V_{r,2p-s}), \tag{3.4.50}$$

但不幸的是, $(V_{q+r,p-s} + V_{r,2p-s})$ 结构过于复杂, 原因是 $V_{q+r,p-s}$ 与 $V_{r,2p-s}$ 又有共同的子模:

$$V_{2q+r,s} \cup V_{r,2p+s} \in V_{q+r,p-s}. \tag{3.4.51}$$

换言之, 就是减除得太多, 又需要补偿回来:

$$V_{q-r,3p-s} \cup V_{3q-r,p-s} \in V_{r,2p-s}. \tag{3.4.52}$$

$$\vdots$$

这个过程可以一直持续下去. 因此, 不可约模为

$$M_{r,s} = V_{r,s} - (V_{q+r,p-s} \cup V_{r,2p-s}) + (V_{2q+r,s} \cup V_{r,2p+s}) - \cdots, \tag{3.4.53}$$

这是一个无穷序列. 显然, 需要借助其他的手段, 才能有效处理这一问题. 对这一问题的细致分析, 请参考文献 [9].

分析这一问题最有效且简单的方法是特征标.

3.4.5 Virasoro 特征标

设有 Verma 模 $V(c,h)$, $\chi_{(c,h)}(\tau)$ 是相应的特征标:

$$\begin{aligned}\chi_{(c,h)}(\tau) &= \text{Tr}\, q^{L_0 - c/24} \\ &= \sum_{n=0}^{\infty} \dim(h+n) q^{n+h-c/24},\end{aligned} \tag{3.4.54}$$

其中, $q \equiv \exp(2\pi i\tau)$, 要求 τ 是复变量. 因子 $q^{-c/24}$ 出现的原因, 已经在式 (3.3.65) 中给出, 量 $\dim(h+n)$ 表示层为 n 的 Verma 模上的线性独立的态的数目. 回忆拆分 $P(n)$ 的定义:

$$\sum_{n=0}^{\infty} P(n) q^n = \prod_{n=1}^{\infty} \frac{1}{1-q^n} \equiv \varphi^{-1}(\tau), \tag{3.4.55}$$

其中, $\varphi(q)$ 为欧拉函数. 由于 $\dim(h+n) \leqslant p(n)$, 如果 $|q| < 1$, 即限定在上半平面, 则欧拉函数 (3.4.55) 一致收敛. 因此, 一般 Virasoro 的特征标可以写为

$$\chi_{(c,h)}(\tau) = \frac{q^{h-c/24}}{\varphi(\tau)}. \tag{3.4.56}$$

为了方便, 引入 Dedekind-函数

$$\eta(\tau) \equiv q^{1/24}\varphi(\tau) = q^{1/24}\prod_{n=1}^{\infty}(1-q^n), \tag{3.4.57}$$

则 Virasoro 的特征标重写为

$$\chi_{(c,h)}(\tau) = \frac{q^{h+(1-c)/24}}{\eta(\tau)}. \tag{3.4.58}$$

对极小模型, 如果定义

$$K_{r,s}^{(p,q)}(\tau) = \frac{1}{\eta(\tau)}\sum_{n\in\mathbb{Z}} q^{(2pqn+pr-qs)^2/4pq}, \tag{3.4.59}$$

则极小模型的特征标为

$$\chi_{r,s}(\tau) = K_{r,s}^{(p,q)}(\tau) - K_{r,-s}^{(p,q)}(\tau). \tag{3.4.60}$$

更详细的分析, 参见文献 [9].

3.5 有理共形场

极小模型是有理共形场的一个最简单的例子. 所谓有理共形场是元场数量有限, 且中心电荷、共形权都取有理值的共形场. 除了 $0 < c < 1$ 的共形场, 也包括 $c \geqslant 1$ 的共形场.

3.5.1 伊辛模型

在极小模型的酉表示系列中, 第一个非平凡的例子就是伊辛模型. 它的 $c = 1/2$, 对应的 $m = 3$, 所以 $h_{r,s}$ 有三种可能的取值, 即 $h_{1,1} = 0$, $h_{1,2} = \frac{1}{2}$, $h_{2,1} = \frac{1}{16}$. 共形权三种的可能取值, 对应于三种算子. 显然, $h_{1,1} = 0$ 就是单位算子, 要具体分析余下的两种情况, 不妨先简单回顾一下伊辛模型的临界指数. 关于统计模型临界指数的详细讨论, 请参考文献 [25].

记伊辛模型的序参量为 σ, 二级相变就是从低温的有序相 $\langle\sigma\rangle \neq 0$, 变成高温的无序相 $\langle\sigma\rangle = 0$. 在临界点, 伊辛模型的两点函数为

$$\langle\sigma_n\sigma_0\rangle \sim \frac{1}{|n|^{d-2+\eta}}, \tag{3.5.1}$$

式中, d 是系统的维数; η 是临界指数. 另一个临界指数 ν 可以通过四点函数得到:

$$\langle \varepsilon_n \varepsilon_0 \rangle \sim \langle \sigma_n \sigma_{n+1} \sigma_0 \sigma_1 \rangle \sim \frac{1}{|n|^{2(d-1/\nu)}}, \tag{3.5.2}$$

其中, σ, ε 分别是伊辛模型的自旋、能量算子. 在统计物理中, 已经知道二维伊辛模型的临界指数, $\eta = 1/4, \nu = 1$, 所以有

$$\langle \sigma_n \sigma_0 \rangle \sim \frac{1}{|n|^{1/4}} \sim \frac{1}{|n|^{2\Delta_\sigma}}, \tag{3.5.3}$$

以及

$$\langle \varepsilon_n \varepsilon_0 \rangle \sim \frac{1}{|n|^2} \sim \frac{1}{|n|^{2\Delta_\varepsilon}}. \tag{3.5.4}$$

已经知道标度维 Δ、共形权 h 以及自旋 s 之间存在等式 $\Delta_a = h_a + \bar{h}_a$, $s_a = h_a - \bar{h}_a = 0$. 容易得到, 自旋算子与能量算子的共形权分别是 $\left(\frac{1}{16}, \frac{1}{16}\right), \left(\frac{1}{2}, \frac{1}{2}\right)$, 这与 $c = 1/2$ 共形场得到的结果完全重合.

从共形场角度, 利用 Kac 行列式, 可以分析临界二维伊辛模型算子间的关系. $N = 1$ 的 Kac 行列式的解, 就对应于 $h_{1,1} = 0$, 所以有等式

$$L_{-1}|0\rangle = 0.$$

$N = 2$ 时, 将 $h_{1,2} = \frac{1}{2}, h_{2,1} = \frac{1}{16}$ 分别代入式 (3.4.32):

$$\left(L_{-2} - \frac{3}{4}L_{-1}^2\right)\left|\frac{1}{2}\right\rangle = 0,$$
$$\left(L_{-2} - \frac{4}{3}L_{-1}^2\right)\left|\frac{1}{16}\right\rangle = 0. \tag{3.5.5}$$

因此, 考虑到 $h_{r,s}$ 的对称性, 伊辛模型中只有三个元场.

三点函数 (3.1.28) 非零, 对场的共形权作了限制, 也就是决定于三个场量的配伍关系. 注意到伊辛模型有自旋翻转 $\sigma \to -\sigma$ 和对偶性 $\varepsilon \to -\varepsilon$, 因此临界伊辛模型有融合法则 (fusion rules):

$$\begin{cases} \sigma \times \sigma = \mathbb{I} + \varepsilon, \\ \sigma \times \varepsilon = \sigma, \\ \varepsilon \times \varepsilon = \mathbb{I}. \end{cases} \tag{3.5.6}$$

以 $\phi_{2,1}$ 为例, 它与临界二维伊辛模型中的算子积被限制为

$$\phi_{2,1} \times \phi_{(r,s)} = \phi_{(r+1,s)} + \phi_{(r-1,s)}, \tag{3.5.7}$$

3.5 有理共形场

其中, $\phi_{(r,s)}$ 表示所有从元场 $\phi_{r,s}$ 得到的场的集合, 而所有其他不满足此约束的场不会对算子关系有贡献. 为了方便, 也可以写出

$$\begin{aligned}\phi_{2,1} \times \phi_{(\alpha)} &= \phi_{(\alpha-\alpha_+)} + \phi_{(\alpha+\alpha_+)}, \\ \phi_{1,2} \times \phi_{(\alpha)} &= \phi_{(\alpha-\alpha_-)} + \phi_{(\alpha+\alpha_-)},\end{aligned} \tag{3.5.8}$$

上式中关于 $\alpha_\pm$ 的定义与式 (3.4.43) 相同, $\phi_{(\alpha)}$ 代表最高权为 h_α 场的集合. 上述融合关系可以推广到极小模型:

$$\phi_{r,s} \times \phi_{(\alpha)} = \sum_{\substack{k=1-r \\ k+r=1\mathrm{mod}2}}^{k=r-1} \sum_{\substack{l=1-s \\ l+s=1\mathrm{mod}2}}^{l=s-1} \phi_{(\alpha+k\alpha_++l\alpha_-)}, \tag{3.5.9}$$

容易看出, 式中 k, r 取值的奇、偶性相反. 当然, 和伊辛模型一样, 还会有融合的截断问题. 本节相关内容的详细分析, 参见文献 [9].

3.5.2 融合代数

融合代数可以看成是 3.5.1 节内容的推广. 对任意有理共形场 ϕ_i, ϕ_j 可定义融合代数 (fusion algebra):

$$\phi_i \times \phi_j = \sum_k \mathcal{N}_{ij}^k \phi_k, \tag{3.5.10}$$

其中, 融合代数的结构常数 $\mathcal{N}_{ij}^k \in \mathbb{Z}_0^+$, $\mathcal{N}_{ij}^k \neq 0$, 可等同于存在非零的三点函数. 代数的可交换, 意味着

$$\mathcal{N}_{ij}^k = \mathcal{N}_{ji}^k, \tag{3.5.11}$$

与单位算子作乘法, 显然有

$$\mathcal{N}_{i1}^k = \delta_{ik}, \tag{3.5.12}$$

代数的结合律等价于

$$\sum_l \mathcal{N}_{kj}^l \mathcal{N}_{il}^m = \sum_l \mathcal{N}_{ij}^l \mathcal{N}_{lk}^m. \tag{3.5.13}$$

定义 N_i 为 r 行 r 列的矩阵算子, 它的矩阵元为

$$(N_i)_{j,k} = \mathcal{N}_{ij}^k, \tag{3.5.14}$$

则融合代数的结合律 (3.5.13) 可表示为

$$N_i N_k = N_k N_i. \tag{3.5.15}$$

这等价于

$$N_j N_k = \sum_l \mathcal{N}_{jk}^l N_l, \tag{3.5.16}$$

也就是说, $\mathcal{N}$ 自身就构成融合代数的一个表示.

共形场论最不可思议的结果之一, 即黎曼球面上的算子积的融合代数竟然与环形面 (亏格 $g = 1$) 上模变换的 S-矩阵有深刻的关系. 环形面是复平面 $\mathbb{C}$ 减除掉 $(1+\tau)$ 生成的格点, 即 $T^2(\tau) = \mathbb{C}/(1+\tau)\mathbb{Z}$. 将柱面上的共形场映射到环面上, 更一般地, 考虑中心荷为 c 的有理共形场, 它的元场 ϕ_i 数目有限, 配分函数变化为

$$Z(q) = \mathrm{Tr}_{\mathcal{H}}\left(q^{L_0-c/24}\bar{q}^{\bar{L}_0-\bar{c}/24}\right), \tag{3.5.17}$$

其中的 c 项 (或者 $\bar{c}$) 正是来自 Schwarzian 导数的贡献. 已经注意到, 在上式中, 左移、右移部分是同时出现的, 这其实是物理条件的体现. 作为物理系统, 在离开临界点的不同扇区存在相互作用. 它的谱在参数空间作连续形变, 因此, 即使物理系统到达共形点, 也应该保留相互作用的信息. 换句话说, 共形场的全纯、反全纯部分作为独立的存在, 其实不是物理的条件. 如同几何会导致共形反常一样, 通过几何也可以导入相互作用. 在配分函数中插入分解式 (3.4.19):

$$Z(q) = \sum_{a,\bar{a}} N_{a,\bar{a}} \chi_a(q) \bar{\chi}_a(\bar{q}), \tag{3.5.18}$$

其中, $\chi_a(q)$ 是全纯代数在 $\mathcal{H}_a$ 表示上的特征标, 也被称为弦函数:

$$\chi_a(q) = \mathrm{Tr}_{\mathcal{H}}(q^{L_0-c/24}). \tag{3.5.19}$$

为了使环面上的共形场有意义, 它应独立于从柱面得到环面的映射方式. 这意味着, 配分函数 Z 应在环面模群 $SL(2,\mathbb{Z})$ 变化下不变, 见图 3.6. 如果 S, T 为

$$T: \quad \tau \to \tau + 1, \quad S: \quad \tau \to -\frac{1}{\tau}, \tag{3.5.20}$$

且有 $S^2 = (ST)^3 = 1$ 生成模群

$$\tau \to \frac{a\tau + b}{c\tau + d}, \quad \begin{pmatrix} a & b \\ c & d \end{pmatrix} \in SL(z,\mathbb{Z}), \tag{3.5.21}$$

即 $a, b, c, d \in \mathbb{Z}$, 并且 $ad - bc = 1$. 注意到, 模群的矩阵元同时变号, 对应的是同一变换, 所以真正的模变换群是 $PL(2,\mathbb{Z}) = SL(2,\mathbb{Z})/\mathbb{Z}_2$. 换句话说, 配分函数应该是 $PL(2,\mathbb{Z})$ 变换下不变的, 即配分函数必须与环面几何相融. 在共形场的特征标空间存在模群的一个表示, 对 S-矩阵有

$$\chi_i\left(-\frac{1}{\tau}\right) = \sum_{j=0}^{N-1} S_{ij} \chi_j(\tau). \tag{3.5.22}$$

3.5 有理共形场

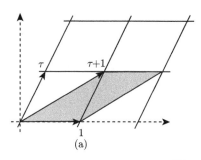
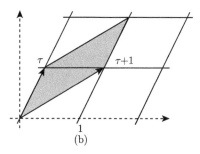

图 3.6 模变换

在已知的情形, S- 是对称酉矩阵:

$$SS^\dagger = S^\dagger S = \mathbf{1}, \quad S = S^{\mathrm{T}}. \tag{3.5.23}$$

Verlinde 公式[31]提供了一种路径, 如何用 S-矩阵的信息来计算融合代数的系数:

$$\mathcal{N}_{ij}^k = \sum_{m=0}^{N-1} \frac{S_{im} S_{jm} S_{mk}^*}{S_{0m}}. \tag{3.5.24}$$

式中, S^* 代表 S 矩阵的复共轭; 下标 0 代表单位表示. Verlinde 公式在物理和数学中都有应用, 公式的证明参见文献 [31]. 类似地, 对 T-矩阵有

$$\chi_i(\tau + 1) = \sum_{j=0}^{N-1} T_{ij} \chi_j(\tau). \tag{3.5.25}$$

略去详细推导过程, 可以选择适当基底, 使得

$$T_{ij} = \delta_{ij} \mathrm{e}^{h_i - \frac{c}{24}}, \tag{3.5.26}$$

其中, h_i 是最高权表示的共形权, 可参见特征标 $\chi_i(\tau)$ 定义 (3.4.54).

对任何给定中心荷 $c = c(m) < 1$ 的共形场, 一个显著的特征是, 只存在有限个共形权值 h, 使得相应的 Virasoro 代数存在酉表示. 给定一个共形不变的场, 通过显式计算, 人们发现, 当中心荷值契合 $c = c(m)$ 时, 如果重数 $N_{i,\bar{i}}$ 是有限的, 则直和分解 (3.4.19) 也有限. 实际上, 对于大多数可能用到的共形场, $N_{i,\bar{i}}$ 等于 0, 或者 1. Cardy 已经证明, 对于 $c < 1$ 的共形场, 只存在有限数量的元场[6]. 由环面上模群变化下不变的约束, 可以得到 $c < 1$ 共形场的约束, 即中心荷 $c \leqslant 1$ 共形场的完整的分类, 这是所谓的 ADE 分类[8].

对于中心荷 $c \geqslant 1$ 的共形场, 对任何正实值的共形权 h, 都存在 Virasoro 代数唯一、不可约的酉表示 (h, c), 但其希尔伯特空间的分解, 即使重数 $N_{i,\bar{i}}$ 有限, 但分

解式 (3.4.19) 可能是无限的. 事实上, 对于中心荷 $c \geqslant 1$ 的有理共形场, 模不变性的约束都要求共形场包含无限多个 Virasoro 代数表示.

将共形场的对称性进行扩展, 即所谓的 $\mathcal{W}$- 代数的引入, 作为一种可能的方案, 解决了这一难题. 对给定有理共形场的酉表示, 即使它相对于所给定的 Virasoro 代数表示分解是无限, 但它对于 $\mathcal{W}$-代数表示的分解仍可能有限.

3.5.3 共形块的交换关系

由局域算子三点函数的约束关系, 可以导出融合代数关系. 前面已经提到, 四点函数 (3.1.29) 存在交叉对偶关系 (3.4.27). 可以将这样的交叉对偶关系推广到任意的四点共形块, 即

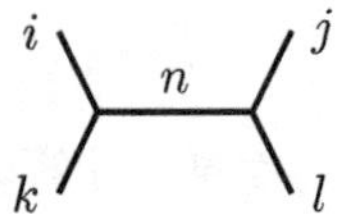

$$\begin{matrix} i & & j \\ & n & \\ k & & l \end{matrix} = \sum_m B \begin{bmatrix} i & j \\ k & l \end{bmatrix}_{n,m} \begin{matrix} j & & i \\ & m & \\ k & & l \end{matrix} . \tag{3.5.27}$$

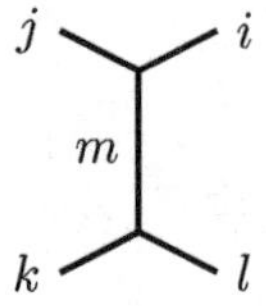

显然, B-矩阵将上面两 "腿" 的位置互换, 因此被称为 "辫子" 矩阵 (braid). 容易验证, 辫子矩阵有如图 3.7 所示的交换图, 即交换结果与交换路径无关, 这等价于矩阵等式:

$$\sum_p B \begin{bmatrix} j & k \\ i & s \end{bmatrix}_{r,p} B \begin{bmatrix} j & l \\ p & m \end{bmatrix}_{s,t} B \begin{bmatrix} k & l \\ i & t \end{bmatrix}_{p,u}$$
$$= \sum_q B \begin{bmatrix} k & l \\ r & m \end{bmatrix}_{s,q} B \begin{bmatrix} j & l \\ i & q \end{bmatrix}_{r,u} B \begin{bmatrix} j & k \\ u & m \end{bmatrix}_{q,t} . \tag{3.5.28}$$

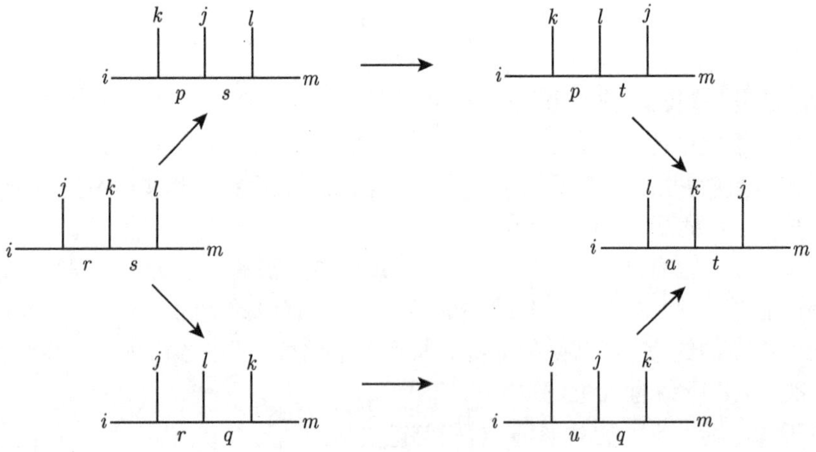

图 3.7 辫子矩阵的交换图

3.5 有理共形场

同样,还可以定义融合矩阵,即先分别将上面、下面的两"腿"粘在一起:

$$\begin{matrix} i & & j \\ & n & \\ k & & l \end{matrix} = \sum_m F\begin{bmatrix} i & j \\ k & l \end{bmatrix}_{n,m} \begin{matrix} i & & j \\ & m & \\ k & & l \end{matrix}. \tag{3.5.29}$$

它满足二次等式

$$\sum_m F\begin{bmatrix} i & j \\ k & l \end{bmatrix}_{n,m} F\begin{bmatrix} k & i \\ l & j \end{bmatrix}_{m,p} = \delta_{n,p}, \tag{3.5.30}$$

同样,融合矩阵有如图 3.8 所示的交换图,这等价于矩阵等式:

$$F\begin{bmatrix} j & k \\ i & s \end{bmatrix}_{r,t} F\begin{bmatrix} t & l \\ i & m \end{bmatrix}_{s,u} = \sum_p F\begin{bmatrix} k & l \\ r & m \end{bmatrix}_{s,p} F\begin{bmatrix} j & p \\ i & m \end{bmatrix}_{r,u} F\begin{bmatrix} j & k \\ u & l \end{bmatrix}_{p,t}, \tag{3.5.31}$$

这就是所谓的五角等式.

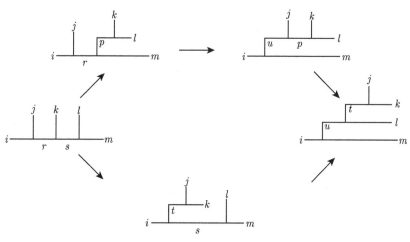

图 3.8 融合矩阵的交换图

3.5.4 流代数

一般地,对任意给定秩为 r 的李代数 $\mathcal{G}$,设它的根集为 Δ,相应的仿射李代数记为 $\hat{\mathcal{G}}$,记仿射李代数 $\hat{\mathcal{G}}$ 的生成函数为 J^a, $a \in \Delta$,这样的生成函数被称为流(current),流构成的代数就是所谓的流代数. 写成算子积的形式:

$$J^a(z)J^b(w) = \frac{k\kappa(a,b)}{(z-w)^2} + \frac{f^{abc}J^c(w)}{z-w} + \cdots, \tag{3.5.32}$$

其中, $a,b,c \in \Delta$; $\kappa(a,b)$ 是 Cartan-Killing 形式; f^{abc} 是李代数 $\mathcal{G}$ 的结构常数. 将上式写成对易子的形式:

$$[J_m^a, J_n^b] = f^{abc} J_{m+n}^c + mk\kappa(a,b)\delta_{m+n,0}, \tag{3.5.33}$$

菅原构造得到的能-动张量为

$$T(z) = \frac{1}{2(k+h^\vee)} \sum_{a=1}^{\dim \mathcal{G}} :J^a J^a:(z), \tag{3.5.34}$$

其中, $h^\vee$ 是对偶的 Coxeter 数; k 为仿射李代数 $\widehat{\mathcal{G}}$ 表示的层. 不难证明, 仿射李代数 $\widehat{\mathcal{G}}$ 是 $T(z)$ 最高权表示:

$$T(z)J^a(w) = \frac{J^a(w)}{(z-w)^2} + \frac{\partial J^a(w)}{z-w}, \tag{3.5.35}$$

或者, 等价地

$$L_n = \frac{1}{2(k+h^\vee)} \sum_{a=1}^{D} \sum_{m \in \mathbb{Z}} :J_{m+n}^a J_{-m}^a:(z), \tag{3.5.36}$$

$$[L_m, J_n^a] = -n J_{m+n}^a, \tag{3.5.37}$$

相应的 Virasoro-代数的中心荷为

$$c = \frac{k \dim \mathcal{G}}{k+h^\vee}, \tag{3.5.38}$$

其中, $\dim \mathcal{G} = r(h+1)$, h 是仿射李代数 $\widehat{\mathcal{G}}$ 的 Coxeter 数.

流代数可以看成是 Wess-Zumino-Witten (WZW) 模型的守恒流, WZW- 模型是非线性 σ- 模型的作用量, 加上拓扑项, 也就是 Wess-Zumino 项得到的 [33], 所以 WZW- 模型具体实现了流代数. 关于仿射李代数更系统的论述, 可参见文献 [22] 和 [23].

3.5.5 $\mathcal{W}$-代数

在统计物理学中, 研究二级相变的主要问题之一是对各种临界行为进行分类, 以及计算其相应的临界指数. 共形场论作为研究二级相变的工具之一, 临界指数的分类问题就对应于共形场酉表示的分类 [6,7].

二维共形场论的独特之处是其具有无限维对称性. 中心荷 $c<1$ 时, 存在无穷离散序列 $h_{r,s}$, 由 Virasoro-代数的酉表示完全分类. 对中心荷 $c>1$ 的有理共形场进行分类, 则需要更大的对称性, 也就是需要引入更多、共形权更高的元场. 这样扩大了的共形对称性被称为 $\mathcal{W}$-对称, 相应的代数为 $\mathcal{W}$-代数. $\mathcal{W}$-对称是二维共形

3.5 有理共形场

场论的重要议题, 它在理论物理和数学的不同领域有广泛应用, 尤其是近期的研究, 二维的刘维尔理论与四维 $\mathcal{N}=2$ 超对称 Yang-Mills 存在对偶关系, 在此项对偶关系中, $\mathcal{W}$-代数具有关键作用 [1]. $\mathcal{W}$-代数及其与其他领域的关系, 引起了理论物理学家和数学家的共同关注 [5], 有理共形场也是数学研究议题之一. 所以, 如何构造这样具有扩展对称性的共形场论, 也是二维共形场研究的重要问题之一. 有很多方法来构造 $\mathcal{W}$-代数, 可参见文献 [5].

我们只是列举一个 $\mathcal{W}$-代数最简单的例子, 看一看它与 Virasoro 代数有何不同. 回顾能–动张量的算子积展开:

$$T(z)T(w) = \frac{c/2}{(z-w)^4} + \frac{2T(w)}{(z-w)^2} + \frac{\partial_w T(w)}{z-w} + \cdots.$$

它的模式展开的对易子的手征代数 (全纯, 或反全纯) 就是 Virasoro 代数 (3.2.24). 从上面的算子积展式可以看出, Virasoro 代数生成元的对易子可以表示为原代数的线性组合. 这一性质与经典李代数的性质类似.

第一个非平凡 $\mathcal{W}$-代数的例子, 被称为 $\mathcal{W}_3$- 代数, 除通常的能–动张量, 还存在一个共形权为 3 的元场, 其生成函数的算子积展式为

$$\begin{aligned} W(3,z)W(3,w) = & \frac{c/3}{(z-w)^6} + \frac{2T(w)}{(z-w)^4} + \frac{\partial T(w)}{(z-w)^3} \\ & + \frac{1}{(z-w)^2}\left[2b^2\Lambda(w) + \frac{3}{10}\partial^2 T\right] \\ & + \frac{1}{z-w}[b^2\partial\Lambda(w) + \frac{1}{15}\partial^3 T] + \cdots, \end{aligned} \quad (3.5.39)$$

$$T(z)W(3,w) = \frac{3W(w)}{(z-w)^2} + \frac{\partial W(w)}{z-w} + \cdots, \quad (3.5.40)$$

其中, $\Lambda(z) = [TT]_0(z) - \frac{3}{10}\partial^2 T(z)$, $b^2 = 16/(22+5c)$ 是常数. 从 $\mathcal{W}$-代数算子积展式发现, 在 $\mathcal{W}$- 代数的对易子中, 必然出现新的场量, 它不能由原来的生成元线性地得到. 所以, 非线性是 $\mathcal{W}$-代数最显著的特性之一. $\mathcal{W}_3$-场的元场性质是由方程 (3.5.40) 确定的.

如果其中的生成函数分别有展式 $W(z) = \sum W_n z^{-n-3}$, $\Lambda = \sum \Lambda_n z^{-n-4}$, 根据定义, Λ 是共形权为 4 的场. 相应地, 生成函数展开模式之间的对易子为

$$[L_m, W_n] = (2m-n)W_{m+n}, \quad (3.5.41)$$

$$\begin{aligned}[][W_m, W_n] = & \frac{1}{30}(m-n)(2m^2 - mn + 2n^2 - 8)L_{m+n} \\ & + b^2(m-n)\Lambda_{m+n} \end{aligned}$$

$$+\frac{c}{360}m(m^2-1)(m^2-4)\delta_{m+n,0}, \qquad (3.5.42)$$

其中的中心荷为 [10,35]

$$c = 2\left[1 - \frac{12}{m(m+1)}\right], \quad m = 3, 4, \cdots. \qquad (3.5.43)$$

更一般地，可以形式地定义微分算子

$$R_N =: \prod_{m=1}^{N} \sqrt{2}\left(\mathrm{i}\alpha_0 \frac{\mathrm{d}}{\mathrm{d}z} + \frac{\boldsymbol{h}_m}{2}\partial_z\phi\right):, \qquad (3.5.44)$$

在上一个方程中出现的常数 α_0 是某些共形场论的"背景荷"。场 ϕ 和 n 向量 $\boldsymbol{h}_i(i=1,\cdots,n)$，在 $(n-1)$ 维欧几里得空间中构成超完全系统，满足条件:

$$\sum_{i}^{n} h_i = 0, \quad h_i \cdots h_j = \delta_{ij} - 1/n. \qquad (3.5.45)$$

使用微分运算符的交换规则，可以重新计算乘积 R_N:

$$R_N = \sum_{k=0}^{N} W_k(z)\left(\mathrm{i}\alpha_0\sqrt{2}\frac{\mathrm{d}}{\mathrm{d}z}\right)^{N-k}, \qquad (3.5.46)$$

由此确定场 $W_k(z)$，这是所谓的三浦 (Miura) 变换，中心荷的值为

$$c = (N-1)\left[1 - 2N(N+1)\alpha_0^2\right], \qquad (3.5.47)$$

如果限定 $N=3$，就是 $\mathcal{W}_3$ 代数; $N=2$，正是 Virasoro 代数，而中心荷表述式中的 α_0，与前文的定义差一个归一化因子。

在这简短的介绍性的文字中，除最后一笔带过，没有具体讨论环面上的共形场。几乎所有黎曼球面上共形场的内容都可以在环面上平行地讨论，只是需要留意模不变性的要求。

3.5.6 结语

尽管题目叫"共形场论入门"，但共形场作为三十多年来发展迅猛、影响广泛的一门学科，如此简短的介绍，挂一漏万，显得多少有些不够严谨。这里涉及的内容，离真正地进入这个研究领域尚有很远的距离。尚未入门，但又不能说"不入门"，我实在想不出一个恰当的题目。限于篇幅，画卷尚未展开，又不得不束之以结。

余生也晚，20 世纪 90 年代初期，在我开始攻读研究生的时候，已经是共形场论"暴涨"式发展的末期。那时网络未兴，信息闭塞，获取资料的主要方式之一还是手写卡片，向国际上几个著名的研究机构索取预印本。自己学习共形场论的方式，主要就是研读 Belavin-Polyakov-Zamolodchikov 著名的文章 [2]。托一位师兄的福，他毕业了，留下一堆故旧，我从中捡出一份 Ginsparg 讲义预印本，尽管与正式出版的稍有差

别 [17],但已经是当时最全面、权威的文献.之后,国际上出版了许多关于共形场论的优秀著作,在参考文献中,我给出了几种 [9,15,16,22-24,26].但可惜一直没有中文的介绍,这对一些人来说可能显得多余,但对刚刚接触这门学科的学生也许有一点帮助.

2016 年夏天,受国家自然科学基金委资助,西北大学承办了"第九期理论物理前沿暑期讲习班——可积模型方法及其应用".感谢杨文力教授的邀请,让我有机会和暑期班的学员一起重温"共形场论".与学员的交流,让我获益良多.

这里的内容,是根据我几次短期授课的讲义写成的.由于篇幅关系,对讲义内容进行了部分删减.尽管我在不同场合讲过几次共形场论,但都很零碎,而间隔的时间跨度又很长.2010 年年底的一次计算机故障,导致硬盘被毁,存放在计算机里的所有笔记,尽数湮没,不复可得.即使是前度"刘郎"又重来,未必"桃花"如旧栽.时过境迁,多年以后,重拾旧艺,谬误与偏颇之处,满满皆是.

由于篇幅所限,对具有非阿贝尔对称性的共形场论,本讲义几乎没有涉及,对环面上的共形场论,也只是一笔带过.实际上,这是共形场论中最优美,同时也是数学上最严格的内容之一,被广泛用于代数、几何、拓扑甚至概率论等方面.对超对称共形场论,只字未提,这其中包括 $\mathcal{N}=2$ 超共形场论,它在研究镜面对称 (mirror symmetry) 中是具有关键作用的 [21].至少还有两方面的重要内容,讲义完全没有涉及:其一,边界共形场论 [30],这是理解 D-膜的途径之一 [29],也是构造 SLE(Schramm-Loewner Evolutions) 的切入点 [32];其二,AdS_{d+1}/CFT_d 对应,即共形场论与 Anti-de Sitter 空间的边界对应.这一猜测是过去几十年来研究者在理解量子引力方面最重要的进展之一 [27].

假以时日,我希望能将眼前的讲义充实成一本像样的书,一本真正关于共形场论的入门书籍.

参 考 文 献

[1] Alday L F, Gaiotto D, Tachikawa Y. Liouville correlation functions from four-dimensional gauge theories. Lett. Math. Phys., 2010, 91(2): 167-197.

[2] Belavin A A, Polyakov A M, Zamolodchikov A B. Infinite conformal symmetry in two-dimensional quantum field theory. Nucl. Phys. B, 1984, 241: 333.

[3] Birrell N D, Davies P C W. Quantum Field in Curved Space. Cambridge University Press, 1982.

[4] Blumenhagen R, Plauschinn E. Introduction to Conformal Field Theory: With Applications to String Theory. Springer, 2009: 779.

[5] Bouwknegt P, Schoutens K. $\mathcal{W}$-Symmetry. World Scientific Publishing, 1994.

[6] Cardy J. Conformal Invariance and Statistical Mechanics//In Les Houches, session XLIX, Fields, string and critical phenomena. Elsevier Science Publishers, 1989.

[7] Cardy J. Conformal Field Theory and Statistical Mechanics//In Les Houches, session LXXXIX, Exact Methods in Low-Dimensional Statistical Mechanics and Quantum Computing. Ouvry S, Pasquier V, et al. Oxford University Press, 2010.

[8] Cappelli A, Itzykson C, Zuber J B. The A-D-E classification of minimal and conformal invariant theories. Commun. Math. Phys., 1987, 13: 1.

[9] Francesco P D, Mathieu P, Senechal D. Conformal Field Theory. Springer, 1997.

[10] Fateev V A, Zamolodchikov A B. Conformal quantum field theory models in two dimensions having $\mathbb{Z}_3$ symmetry. Nucl. Phys. B, 1987, 280: 644.

[11] Feigin B L, Fuchs D B. Skew-symmetric differential operators on the line and Verma modules over the Virasoro algebra. Funct. Anal. and Appl., 1982, 17: 114.

[12] Frappat L, Sciarrino A, Sorba P. Dictionary on Lie Algebras and Superalgebras. Academic Press, 2000.

[13] Frenkel I, Lepowsky J, Meurman A. Vertex Operators Algebras and the Monster. Academic Press, 1988.

[14] Friedan D, Qiu Z, Shenker S. Conformal invariance, unitarity and critical exponents in two dimensions. Phys. Rev. Lett., 1984, 52: 1575; Details of the non-unitarity proof for highest weight representations of the Virasoro algebra. Commun. Math. Phys., 1986, 107: 535.

[15] Frishman Y, Sonnenschein J. Non-Perturbative Field Theory from Two Dimensional Conformal Field Theory to QCD in Four Dimensions. Cambridge University Press, 2010.

[16] Gannon T. Moonshine Beyond the Monster the Bridge Connecting Algebra, Modular Forms and Physics. Cambridge University Press, 2006.

[17] Ginsparg P. Applied Conformal Field Theory//In Les Houches, session XLIX, Fields, string and critical phenomena. Elsevier, 1990.

[18] Goddard P, Kent A, Olive D. Virasoro algebras and coset space models. Phys. Lett. B, 1985, 152: 88; Unitary representations of the Virasoro and super-Virasoro algebras. Commun. Math. Phys., 1985, 103: 105.

[19] Goddard P, Olive D. Kac-Moody algebras, conformal symmetry and critical exponents, Nucl. Phys. B, 1985, 257: 226; Kac-Moody and Virasoro algebras in relation to quantum physics. Int. Journ. Mod. Phys. A, 1986, 1: 303.

[20] Green M B, Schwarz J H, Witten E. Superstring Theory. Cambridge University Press, 1987.

[21] Hori K, Katz S, Klemm A, et al. Mirror symmetry. American Mathematical Society, 2003.

[22] Kac V G. Infinite Dimensional Lie Algebras, 3rd Edition. Cambridge University, 1990.

[23] Kac V G, Raina A K, Rozhkovskaya N. Bombay Lectures on Highest Weight Representation of Infinite Dimensional Lie Algebra , 2nd Edition. World Scientific Publishing, 2013.

[24] Kaku M. String, Conformal Fields, and Topology: An Introduction, 2nd Edition. Springer, 2000.

[25] McCoy B M. Advanced Statistical Mechanics. Oxford University Press, 2010.

[26] Ketov S V. Conformal Field Theory. World Scientific Publishing, 1994.

[27] Maldacena J. The large n limit of superconformal field theories and supergravity. Adv. Theor. Math. Phys., 1998, 2: 231.

[28] Peskin M E, Schroeder D V. An Introduction to: Quantum Field Theory. Westview, 1995.

[29] Polchinski J. String Theory. Cambridge University Press, 1998.

[30] Recknagel A, Schomerus V. Boundary Conformal Field Theory and the Worldsheet Approach to D-Branes. Cambridge University Press, 2013.

[31] Verlinde E. Fusion rules and modular transformations in conformal field theory. Nucl. Phys. B, 1988, 300: 360.

[32] Tsirelson B, Werner W. Lectures on Probability Theory and Statistics. Lecture Notes in Mathematics 1840, Springer, 2004.

[33] Witten E. Nonabelian bosonization. Commun. Math. Phys., 1984, 92: 455.

[34] Witten E. Coadjoint orbits of the Virasoro group. Commun. Math. Phys., 1988, 114: 1.

[35] Zamolodchikov A B, Fateev V A. Nonlocal (parafermion) currents in two-dimensional conformal quantum field theory and self-dual critical points in $\mathbb{Z}_N$-symmetric statistical systems. Sov. Phys, JETP, 1985, 62: 215.

[36] Zee A. Quantum Field Theory in a Nutshell, 2nd Edition. Princeton University Press, 2010.

第 4 章 类非线性薛定谔可积系统中光怪波物理

刘 冲 赵立臣 杨战营

4.1 光怪波物理简介

4.1.1 怪波现象

非线性局域波动力学是非线性物理科学中的主要研究课题之一[1-3], 特别是平面波背景上局域波动力学的研究已成为当下非线性物理研究的热点和重点[4-8], 其中一个重要的原因是平面波背景上局域波的动力学特征能够很好地描述自然界中实际存在的"怪波现象"[4-8]. 那么什么是怪波现象呢? 这里我们做简单介绍. 怪波现象, 是指真实存在于自然界中的具有奇怪特征的极端波动现象, 其奇怪之处主要表现为: ①具有高的振幅能量 (一般高于背景振幅 2 倍以上); ②来无影去无踪的无法预期性和不可控性[4, 5]. 由于怪波现象最早发现于海洋并在航海历史上造成众多毁灭性的海难, 因此起初人们认为它是海怪或者某种神秘力量. 不过, 1995年科学家对怪波现象信号的一次成功探测预示着理性的科学认知的到来. 最初, 怪波现象的研究工作主要集中于水流体系统[9]. 然而, 研究之途并不平坦 —— 海洋系统本身极为复杂且不具备良好的可控性为怪波现象的研究带来极大阻碍. 不过值得指出的是, 在一批先驱科学家的不懈努力下, 怪波现象已被证实为一种由非线性效应引起的极端自然现象[9, 10]. 特别是 Zakharov 教授[10] 研究发现, 怪波现象的出现源于非线性系统中广泛存在的调制不稳定性. 至此, 非线性理论解释方案的初步建立预示着怪波现象研究春天的到来.

2007 年, 里程碑式的研究结果 ——"光怪波"的实验证实将怪波现象研究带入人们可控的非线性光学领域, 并从真正意义上开启了一个新的非线性科学研究方向 ——"光怪波物理". Solli 等[11] 创造性地类比了海洋怪波和非线性光纤中的怪波现象, 报道了光纤超连续光谱的长波长区, 光强分布的长尾直方图 ("L 形"非高斯分布)(详见图 4.1). 由图 4.1 可见, 该光学极端现象在时间–波长 (Time-Wavelength) 平面上具有很高的强度, 其出现和消失也无迹可寻, 因此该特征类似于海洋怪波. 进一步地, Solli 等[11] 通过对光怪波事件进行统计分析发现其分布特征为"L 形"长尾分布, 而非传统的高斯分布, 当增大入射脉冲的光强时亦不改变其分布特征 (图 4.1(b)). 此外, 他们通过数值模拟验证了光怪波现象的特征, 其数值结果与实验

4.1 光怪波物理简介

结果符合得很好 (图 4.1(c)). 需要注意的是, 这样的类比不是臆想的巧合, 而是严格基于以下两个物理事实: ①高度偏态分布 ("L 形"分布) 通常被认为可以定义极端怪波现象, 并预示高振幅事件虽然远离中值但依然具有不可忽略的概率可被观测; ②光纤中的超连续光谱激发源于调制不稳定性, 而后者作为能够呈指数放大光学噪声扰动的非线性过程已在之前的研究中被提议为海洋怪波的一种激发机制. 依照上述两个物理事实, 光怪波在不同光学系统中的实验验证以及相应的细致的机制分析等研究工作随之迅速扩展开来.

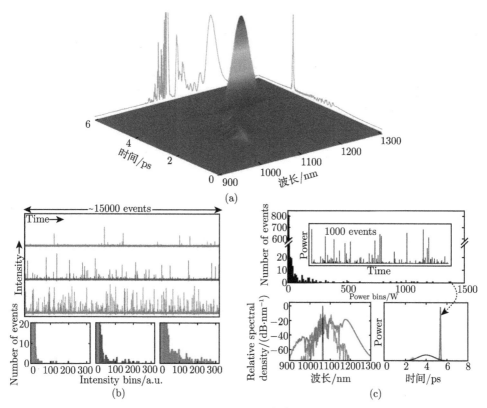

图 4.1 "光怪波现象"在光纤实验中的首次实现[11]. (a) 光怪波在时间–波长平面上的特征; (b) 光怪波特性的"L 形"长尾统计分布; (c) 光怪波数值模拟结果

(扫描封底二维码可看彩图)

需要指出的是, 上述研究主要基于统计的方法, 证实了非线性光学系统中的怪波现象, 提出了定性的产生机制、预测以及抑制方案[8]. 然而, 这种较为单一的研究方法和方式对怪波现象精确的科学认知是不够的. 幸运的是, 2008 年底, Akhmediev 等[4, 12] 开始系统地提出了怪波现象的另一种描述 —— 精确解析的解释方案. 研究表明, 非线性波动方程 (起初主要关于标准的非线性薛定谔方程) 的一系列平

面波 (非零) 背景上的局域波精确解的动力学性质, 能够很好地描述怪波现象的本质[4, 12]. 这是由于这些平面波背景上的具有呼吸特性的局域波本身就是调制不稳定性的一种精确表述形式. 鉴于标准的非线性薛定谔可广泛地运用于不同的非线性物理系统, 特别是水流体系统和非线性光纤系统, 从而怪波现象的科学研究进入了全面系统的精确研究阶段.

4.1.2 理论解释

那么, 平面波背景上的局域波具有怎样的性质以致人们坚定地认为它是严格描述怪波现象的数学原型? 此外, 众所周知的是, 非线性波动方程中也存在着一类具有稳定传输动力学的局域波解 —— "孤子解". 那么两者的区别是什么? 接下来, 我们以标准的 (1+1) 维非线性薛定谔方程为例解释上述疑问. 我们首先简单地区分标准的孤子和描述怪波现象的平面波背景上的局域波, 之后将详细阐明平面波背景上局域波作为描述怪波现象的理论原型的原因.

一般而言, 由 1+1 维非线性薛定谔方程表征的非线性物理系统中的经典孤子, 依照其振幅强度的分布不同主要分为亮孤子和暗孤子两类, 其中亮孤子为零背景 (背景强度为零) 上的稳定局域波; 而暗孤子则是出现在平面波 (非零) 背景上的稳定 "凹陷"[1]. 暗孤子与描述怪波现象的局域波都存在于平面波背景上, 不同之处在于, 前者的存在性不满足调制不稳定性条件, 而后者的产生精确表征了调制不稳定性. 这里需要注意的是, 由于非线性薛定谔方程的普适性, 相关的不同非线性物理系统中的局域波可以进行类比研究. 但我们也不能过分地简化或夸大这种类比, 因为不同物理系统中的局域波描述的物理过程是截然不同的. 如图 4.2 所示, 在非线性光纤光学中, 标准的非线性薛定谔方程描述了一个光脉冲包络演化调制的电场, 而在海洋深水系统中则表征了一群包络演化调制的表面波. 特别是, 水域流体系统中的非线性薛定谔模型描述了整个调制包络的形状, 而非单个波动周期的形状. 因此, 从物理的角度来看, 相应的具体的局域包络解无法被看成单一的非线性局域波; 在薛定谔方程的窄带近似下, 局域波分布的底层总会有许多表面波存在.

接下来, 我们将具体阐明平面波背景上的局域波作为解释怪波现象的有效原型的原因. 如上文所述, 1+1 维非线性薛定谔方程存在着一系列的局域在平面波背景上的局域波解, 最著名的有 Kuznetsov-Ma 呼吸子解[13]、Peregrine 怪波解[14]、Akhmediev 呼吸子解[15] 以及相应的非线性叠加态[4, 12] (高阶局域波解) (图 4.3). 从研究的历史发展角度来看, 最早的一类平面波背景上的局域波精确解的给出可追溯到 20 世纪 70 年代末, Kuznetsov 教授和 Ma 教授在研究标准的 1+1 维非线性薛定谔方程的相关非线性激发单元时, 分别给出了横向分布局域纵向传输周期性呼吸的局域波解[13]. 这类解的特征与之前人们广泛报道的 "孤子" 截然不同, 它与平面波背景之间具有稳定的周期性能量交换, 其峰值高度以指数放大又

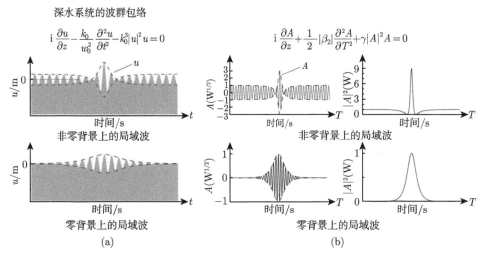

图 4.2 由标准的 1+1 维非线性薛定谔模型描述的不同物理系统中的局域波演化 [8]. (a) 深水系统的波群包络 u, (b) 光纤中群速度反常色散区的光脉冲包络 A. 该图举例阐明了平面波 (非零) 背景上的局域波 (上) 和零背景上的局域波 (下). 对于水域系统需要注意的是, $u(z,t)$ 描述的表面波下总存在深水波. (a)k_0 表示波数, ω_0 为载波频率; (b)$\beta_2 < 0$ 描述反常群速度色散效应, γ 表征非线性效应强弱

以指数衰减, 如此周而复始. 因此人们称之为 "Kuznetsov-Ma 呼吸子". 随后的 1983 年, Peregrine 教授 [14] 将该呼吸子的纵向周期增大至无穷时发现了一类时空双重局域的 "单振幅波". 这个特殊的结构就是近期被人们广泛接受的描述 "怪波现象" 的最基本原型 [16]——"Peregrine 怪波解". 该解以其特有的有理分式著名, 描述了基于调制不稳定性的单峰弱信号被指数放大的不稳定过程. 遗憾的是, 即便如此, 起初的很长时间该解未受到人们的重视, 以致解的特性分析以及实验验证等科学工作直到近几年才被系统清晰地揭示. 之后 Akhmediev 教授 [15] 发现了一类与 Kuznetsov-Ma 呼吸子特征恰好相反的呼吸子, 即纵向局域而横向呼吸的 Akhmediev 呼吸子. 有趣的是, 上述三种典型的平面波背景上的局域波并不是孤立存在的, 当上述呼吸子的周期增至无穷大时, Akhmediev 呼吸子与 Kuznetsov-Ma 呼吸子将退化为 Peregrine 怪波. 相应的一系列局域波的非线性叠加态是由 Akhmediev 教授在 2008 年率先精确给出的 [4, 12]. 研究发现呼吸子碰撞的中心位置可以形成一个振幅更大的波峰. 有趣的是, 他们精确地证实了该波峰的解析表达式是一类更高次的有理分式, 描述着若干 Peregrine 怪波的非线性叠加态. 因此, 目前人们将非线性模型中的高阶有理分式解也叫 "高阶怪波解". 最近的研究表明, 高阶怪波能够表现出结构的多样性. 这里需要指出的是, 求解相关物理模型的高阶怪波解已成为精确认知怪波现象的重要方式之一. 不过目前穷尽高阶怪波的结构类型揭示其形

成规律还是一个极具挑战性的公开命题. 从物理的角度上来看, 这些呈现出不稳定的振幅演化过程的局域波解描述了平面波背景上不同形式扰动的调制不稳定性特征 [17], 并与著名的 Fermi-Pasta-Ulam 循环 [18, 19] 紧密相关. 其中, Peregrine 怪波解和 Akhmediev 呼吸子解分别解析地描述了单峰和周期小振幅扰动的调制不稳定性; Kuznetsov-Ma 呼吸子解表征了强调制的平面波的不稳定性; 相应的局域波非线性叠加态对应于"高阶调制不稳定性" [20]. 基于这些平面波背景上局域波的特征及其描述的调制不稳定性的物理本质, 人们才将这样的一组解析解当成描述怪波现象的有效的理论原型.

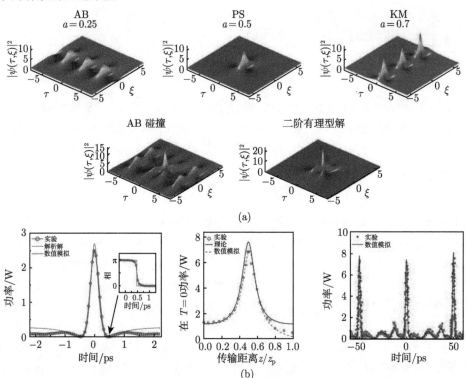

图 4.3 标准的非线性薛定谔模型中的几种典型的非零背景上的非线性激发以及相应的光纤实验中的实现 [8]. (a) 不同特征的精确解, 从左至右依次为: Akhmediev 呼吸子, Peregrine 怪波, Kuznetsov-Ma 呼吸子, Akhmediev 呼吸子碰撞和单峰二阶怪波; (b) 相应的实验验证, 从左到右分别为: Peregrine 怪波, Kuznetsov-Ma 呼吸子, Akhmediev 呼吸子碰撞 (扫描封底二维码可看彩图)

4.1.3 研究进展

上述的这些精确的局域波解为实验上的实现提供了严格的可控的初态激发条件 (振幅和相位初态). 对目前的非线性物理系统而言, 非线性光纤是实验科学上发

4.1 光怪波物理简介

展成熟的非线性实验平台. 原因是人们可以十分方便地设置光纤系统中的色散和非线性参数以使其匹配可用的光源. 基于此, 人们可以有效地设计满足标准的非线性薛定谔模型的非线性局域波传输方案, 从而实现解析结果的实验观测. 令人兴奋的是, 自 2010 年始, 上述平面波背景上的局域波在单模光纤中得到了完美地证实 (图 4.3). 实验的最主要环节是涉及了一个多频的光场在光纤的注入过程. 首先, Kibler 等 [21] 利用相应的频域分辨光开关技术 (frequency-resolved optical gating) 证实了 Peregrine 怪波的存在性, 经过对比发现, 实验观测的 Peregrine 怪波的强度和相位信息与数值模拟以及解析结果的预期几近完美地吻合. 需要注意的是, 上述实验结果是人们通过设置实验参数逼近 Akhmediev 呼吸子的周期极限得到的 [22]. Dudley 等 [23] 实现了 Akhmediev 呼吸子的特征观测, 揭示了 Akhmediev 呼吸子增长和衰减的谱演化规律. Kuznetsov-Ma 呼吸子在传输方向上激发也得到实验证实 [24], 其光强增大和衰减的规律与解析结果一致. 此外, 人们利用光频梳的光谱整形合成的初始条件激发了呼吸子的相互碰撞 [25]. 这个重要的实验结果证实了呼吸子的碰撞是如何激发更高的光强输出的. 实质上, 呼吸子碰撞的动力学是高阶调制不稳定性的表现 [20], 即在调制不稳定频宽范围内多重不稳定模的同时激发导致了 Akhmediev 呼吸子的非线性叠加.

上述实验所用的调制入射场表明不同的初态对应不同的不稳定动力学 (Akhmediev 呼吸子的初态: Akhmediev 呼吸子精确解的有限泰勒级数展开; Peregrine 怪波的初态: 一个远离其最大强度位置的初始点; Kuznetsov-Ma 呼吸子的初态: Kuznetsov-Ma 呼吸子周期演化中某一个强度最小位置). 这些重要的实验进展揭示了合适的初态制备会激发理想解析情形下的非零背景上的局域波. 这些结果在"光怪波"物理研究中的重要性体现在: ①验证了解析解的可靠性和重要性, 精确揭示了"光怪波"现象的物理本质; ②最新的研究结果表明利用由噪声诱发的调制不稳定性的混沌场得到的实验结果与利用解析解得到的结果完全相符 [8, 26]. 如图 4.4 所示, Dudley 等 [8, 26] 基于标准的非线性薛定谔模型, 利用宽频噪声背景数值研究了随机初态的演化动力学过程. 数值结果表明, Kuznetsov-Ma 呼吸子、Akhmediev 呼吸子、Peregrine 怪波甚至是相应的非线性叠加态的动力学可以通过随机噪声初态诱发的调制不稳定性成功"映射"在混沌场中. 这些结果证明了这一系列的解析解代表着一类普遍存在的具有重要物理意义的非线性激发单元. 此外, 值得注意的是, 人们基于标准非线性薛定谔模型的可积湍流理论可以对混沌场中的怪波现象给出很好的理论阐释 [27].

基于这些解析解得到的可控的非线性光学激发单元, 人们也进一步开展了"光怪波物理"应用方面的研究. 譬如, Fatome 等 [28] 利用 Akhmediev 呼吸子的特性, 在单模光纤的反常色散区设计了具有高质量和高重复率的脉冲序列激发器. Bludov 等 [29] 利用 Peregrine 怪波在非线性空间波导阵列中理论设计了能量集中器. 杨

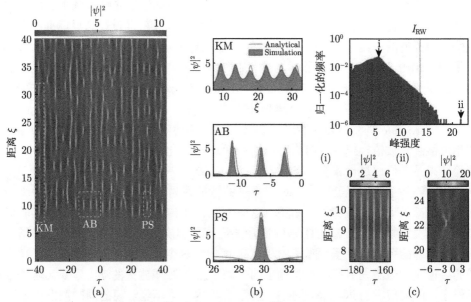

图 4.4　宽频噪声背景上的随机扰动激发 [8](扫描封底二维码可看彩图)

光眸和李禄等 [30, 31] 利用 Peregrine 怪波成功获取稳定传输的高功率脉冲以及能够稳定传输的具有呼吸特征的孤子. 张贻齐等 [32] 利用 Akhmediev 呼吸子预言了光学中的非线性泰伯效应. Tiofack 等 [33] 利用周期性色散管理的光纤成功得到了"Peregrine 怪波梳".

鉴于标准的非线性薛定谔模型的普适性, 怪波现象在其他由非线性物理系统中的研究也迅速扩展开来, 包括 Bose-Einstein 凝聚体 [34−36]、等离子体 [37]、超流体 [38]、大气 [39]、铁磁链 [40]、毛细管 [41]、金融系统 [42] 等.

另外, 考虑到实际非线性光学系统的复杂性, 标准的非线性薛定谔模型不能全面精确地描述不同情形下的局域波动力学. 因此, 对于标准的非线性薛定谔模型的推广以及相应局域波的动力学性质分析就成为亟待解决的问题. 目前研究已经取得了不少重要进展, 主要包含了以下几个方面.

1. 不同的非线性光学系统中平面波背景上局域波解析解的构造

鉴于标准的非线性薛定谔模型中怪波解、呼吸子解以及相应的高阶解在非线性光纤中的实验证实, 理论上构造不同非线性光学系统中的局域波解析解已成为研究怪波现象的重点之一. 目前, 不同非线性光学系统中的局域波解析解的构造主要集中于以下可积模型中: ①"非自治"(nonautonomous) 非线性薛定谔模型及其推广形式 [43−46]; ②高阶非线性薛定谔方程 [47−59]; ③多组分 (大于等于两组分) 耦合非线性模型 [60−72]. 相应的精确解构造方法主要包括: 达布 (Darboux) 变换、相似变换以及 Hirota 双线性方法等. 对于非自治非线性薛定谔系统, 人们已经通过有效的数学方法 (主要是相似变换) 精确求解了 1+1 维至 1+3 维的非线性薛定谔模型及其相应的推广形式 [43-46], 从而在理论上实现了光怪波和呼吸子的有效操控和管理, 为今后的实验实现提供了理论基础. 对于高阶非线性薛定谔方程 (标准非线性薛定谔方程考虑若干高阶扰动项), 怪波和呼吸子的研究主要集中于: Hirota 模型 [48-51]、Sasa-Satsuma 模型 [52-54]、Kundu-Eckhaus 模型 [55]、Lakshmanan-Porsezian-Daniel 模型 [56, 57]、五阶非线性薛定谔模型 [58, 59] 等. 最近的研究表明, 光怪波和呼吸子在高阶效应下表现出结构的多样性. 对于多组分耦合非线性模型, 矢量怪波和呼吸子的研究主要集中于: 多组分耦合非线性薛定谔方程 (包括自聚焦、自散焦以及混合非线性效应三种情况) [60-64]、多组分耦合高阶非线性薛定谔方程 (耦合 Hirota 和耦合 Sasa-Satsuma 模型) [65-67]、多波共振模型 [68, 69]、对遂穿模型 [70]、Maxwell-Bloch 以及非线性薛定谔与 Maxwell-Bloch 耦合模型 [71, 72] 等.

2. 可控光学系统中平面波背景上局域波的特性研究

非线性光学系统是非线性科学中具有良好可控性的实验平台之一. 其原因是: ①人们可以十分方便地设置光纤系统中的色散和非线性参数, 以使其匹配可用的光源; ②非线性平面波导中的"非线性管理"以及非线性光纤中的"色散管理"技术的成熟. 对于后者, 人们发现"非自治"非线性薛定谔模型及其推广形式可以很好地描述相应系统中的局域波动力学 [43-46]. 非自治怪波的精确解最早由闫振亚等 [43] 利用相似变换的方法给出, 其研究表明, 非自治怪波动力学具有较好的可控性. 之后, 戴朝卿等 [44] 对光怪波进行了系统地操控, 实现了光怪波的维持、湮没和快速激发. 需要注意的是, 目前多数的怪波性质研究都基于无限宽的平面波背景, 而实际情况中, 无限宽的平面波背景是不存在的, 实验上光怪波的实现往往利用超宽包络背景来逼近理想的平面波背景 [11]. 因此, 从怪波实验和应用的角度来看, 研究怪波在局域背景上的激发具有重要的科学意义. 我们利用相似变换和达布变换在平面波导中构造了描述光怪波在高斯背景光束上激发的精确解, 并给出了实验上制备激发初态的密度调制和相位调制的解析表达式 [46].

3. 高阶效应下平面波背景上局域波的性质研究

考虑光纤中传输超短光脉冲时, 高阶效应 (高阶色散、自陡峭、自频移、非线性延迟响应等) 对局域波动力学性质具有显著影响. Akhmediev 等 [47] 首先证实了光怪波在高阶效应中的存在性. 有趣的是, 之后的解析结果表明高阶效应能够诱发怪波和呼吸子表现出结构的多样性. Bandelow 等 [52] 发现怪波在三阶色散、自陡峭以及非线性色散效应下能够表现出倾斜或者双峰的结构特性. Soto-Crespo 等 [53] 在混沌波场中数值验证了这些结构出现的可能性, 并对这些新颖的怪波结构的谱特征进行了系统的研究. 贺劲松等 [56] 考虑四阶色散等高阶效应, 发现怪波结构的压缩效应. 另外, 人们研究发现考虑高阶效应的情况下用来描述怪波现象的平面波背景上的局域波解也有可能描述其他一些有趣的孤子态. Mahnke 等 [73] 首先数值研究了考虑拉曼自频移或者三阶色散效应情况下 Akhmediev 呼吸子转换为孤子的可能性. 贺劲松等 [74] 发现考虑自陡峭效应有理分式解在特定参数条件下可以描述孤子和孤子对结构. 赵立臣等 [54] 发现考虑三阶色散、自陡峭以及自频移效应有理分式解在特定背景频率范围内描述了 W 型孤子特征. 这些结果预示着怪波与孤子之间有可能存在态转换. 因此, 如何严格证实这种人们预期存在的态转换并严格描述态转换过程就成了一个有趣的课题. 我们通过解析解和调制不稳定性相结合的理论研究方法, 揭示了怪波与 W 型孤子的态转换特征. 该结果首次建立了解析的态转换与定性的增长率之间的严格关系 [49]. 此外, 我们最近的研究表明呼吸子与孤子以及周期波之间也存在着有趣的态转换 [51]. 之后王雷等 [57] 和 Chowdury 等 [59] 发现更高阶的色散效应也能诱发怪波与孤子间的态转换.

综上所述, 随着标准非线性薛定谔方程中一系列平面波背景上局域波精确解在光学实验中的证实, 理论上构造了不同非线性光学系统中的局域波精确解, 并通过相应解的非平庸性质理解"光怪波现象"越来越受人们重视. 需要指出的是, "光怪波物理" 源于 "怪波现象" 而不仅仅局限于此. 它激发了人们对 "老问题" —— 调制不稳定性非线性阶段以及可积湍流的进一步研究, 同时也刺激了人们建立新的理论 (Superregular 呼吸子以及渐近分析等 [17,75-77]). 目前, 不同光学系统中光怪波和呼吸子的存在性已经被较好地证明. 然而鉴于实际非线性系统的复杂性, 标准的非线性薛定谔系统已不再适用, 对怪模型的推广以及相应的研究就变得极具意义. 因此, 就引出了 "类非线性薛定谔可积模型" 的研究. 接下来, 我们将详细讲述求解类非线性薛定谔可积模型的方法, 基于此, 我们将解决两个重要的物理问题: ①光怪波能否在局域背景上激发; ②怪波和呼吸子在什么条件下能够转换为孤子或其他类型的非线性波.

4.2 类非线性薛定谔可积模型方法

局域波动力学的精确研究一般依赖于对相应的非线性模型的精确求解. 因此, 如何通过合适的数学物理方法得到局域波精确解就成为非线性科学理论研究中的重点和难点. 目前精确求解的方法主要包括: 反散射法、达布变换、Hirota 双线性、Bäcklund 变换、Painlevé分析, 以及相似变换等 [78]. 需要指出的是每种方法在精确求解中都具有其特定的优势. 最近的研究表明, 达布变换、Hirota 双线性和相似变换这三种方法能够较为系统且简洁地得到平面波背景上的局域波精确解. 本节主要运用达布变换和相似变换的精确求解方法, 精确求解了描述不同的非线性光学系统的理论模型. 下文中我们将介绍达布变换方法和相似变换的主要步骤. 另外, 由于平面波背景上的局域波动力学与调制不稳定性密切相关, 因此, 我们也将在本章节最后一部分介绍线性稳定性分析的主要步骤以及最新研究进展. 需要指出的是, 调制不稳定性分析虽然是一种定性的描述局域波性质的方法, 但能够从另一个角度较为宏观地预期平面波背景上局域波的性质. 因此, 通过精确解和相应的调制不稳定性分析结果相结合的研究方法, 人们可以更为全面地揭示平面波背景上局域波动力学的特性.

4.2.1 达布变换

达布变换, 是人们以法国数学家 Darboux 的名字命名的精确求解非线性方程的方法. 这种方法起初是 Darboux 教授在研究线性的 Sturm-Liouville 问题时所采用的变换. 在人们将 Korteweg-de Vries 模型初值问题和 1+1 维线性薛定谔方程的反散射问题建立起联系之后, 达布变换才引起非线性局域波领域 (最初主要是孤子) 的重视和关注. 目前, 达布变换广泛用于求解 Lax 可积的非线性系统. 实质上, 该方法是基于 Lax 可积将非线性问题的求解通过达布变换转化为线性问题求解. 达布变换的基本步骤是利用平庸或简单的种子解 (零解、平面波解以及最近报道的椭圆函数种子解 [79, 80]) 得到精确的单局域波和多局域波解. 其优势在于自身无限次的迭代性质能够得到相应非线性系统的无穷阶数的局域波解. 下面我们以标准的 1+1 维非线性薛定谔方程为例, 简述达布变换在求解平面波背景上局域波精确解的过程.

考虑皮秒脉冲在单模光纤中的传输, 标准的 1+1 维非线性薛定谔方程如下:

$$\mathrm{i}\frac{\partial u}{\partial z} + \frac{1}{2}\frac{\partial^2 u}{\partial t^2} + |u|^2 u = 0, \tag{4.2.1}$$

其中, $u(z,t)$ 表示脉冲的慢变振幅; z 和 t 分别为传输距离和延迟时间. 方程 (4.2.1) 中的第二项和第三项分别描述了群速度色散效应和自相位调制效应 (非线性项). 群速度色散项的正负分别表征了光纤的反常和正常色散区. 需要注意的是, 这里取群速度色散项为正, 是因为在标准的非线性薛定谔框架下, 本文所研究的平面波背景上的局域波 (怪波和呼吸子等) 存在于反常色散区. 上式被证明是可积的, 根据 Ablowitz-Kaup-Newell-Segur 系统的标准步骤 [81], 易得相应的 Lax pair:

$$\begin{cases} \boldsymbol{\Phi}_t = \boldsymbol{U}\boldsymbol{\Phi}, & (4.2.2\mathrm{a}) \\ \boldsymbol{\Phi}_z = \boldsymbol{V}\boldsymbol{\Phi}, & (4.2.2\mathrm{b}) \end{cases}$$

其中, $\boldsymbol{\Phi} = (\Phi_1, \Phi_2)^{\mathrm{T}}$, 矩阵 $\boldsymbol{U}$ 和 $\boldsymbol{V}$ 为

$$\boldsymbol{U} = \lambda \begin{pmatrix} -\mathrm{i} & 0 \\ 0 & \mathrm{i} \end{pmatrix} + \begin{pmatrix} 0 & u \\ -u^* & 0 \end{pmatrix}, \tag{4.2.3}$$

$$V = \lambda^2 \begin{pmatrix} -\mathrm{i} & 0 \\ 0 & \mathrm{i} \end{pmatrix} + \lambda \begin{pmatrix} 0 & u \\ -u^* & 0 \end{pmatrix} + \frac{1}{2} \begin{pmatrix} \mathrm{i}|u|^2 & \mathrm{i}u_t \\ \mathrm{i}u_t^* & -\mathrm{i}|u|^2 \end{pmatrix}, \quad (4.2.4)$$

其中, λ 为谱参量 (一般为复常数, 因此可设为 $\lambda = \mathrm{i}a_1 - q_1/2$, 其中 q_1 为任意实数, a_1 为非零实数). 上述 Lax pair(4.2.2) 将非线性方程的求解问题直接转化为线性方程组求解本征值问题. 由零曲率方程 (可积条件)

$$U_z - V_t + [U, V] = 0, \quad (4.2.5)$$

我们可以直接导出方程 (4.2.1). 其具体的达布变换形式为

$$u(z,t) = u_0(z,t) - \frac{2\mathrm{i}(\lambda - \lambda^*)\Phi_1(z,t)\Phi_2^*(z,t)}{|\Phi_1(z,t)|^2 + |\Phi_2(z,t)|^2}, \quad (4.2.6)$$

这里, $\Phi_1(z,t)$ 和 $\Phi_2(z,t)$ 为相应的 Lax pair(4.2.2) 在 $u(z,t) = u_0(z,t)$ 时的解, 其中 $u_0(z,t)$ 即为所谓的初始 "种子解". 因此, 接下来的主要问题就是, 选取适当形式的 "种子解", 求解 $\Phi_1(z,t)$ 和 $\Phi_2(z,t)$ 的解析表达式.

首先, 我们考虑最简单的平庸种子解 ($u_0(z,t) = 0$) 来构造非线性薛定谔方程局域波解. 当 $u_0(z,t) = 0$ 时, 相应的 Lax pair 写为

$$\begin{cases} \boldsymbol{\Phi}_t = \boldsymbol{U}\boldsymbol{\Phi}, & (4.2.7\mathrm{a}) \\ \boldsymbol{\Phi}_z = \boldsymbol{V}\boldsymbol{\Phi}, & (4.2.7\mathrm{b}) \end{cases}$$

其中, $\boldsymbol{\Phi} = (\Phi_1, \Phi_2)^\mathrm{T}$,

$$U = \lambda \begin{pmatrix} -\mathrm{i} & 0 \\ 0 & \mathrm{i} \end{pmatrix}, \quad V = \lambda^2 \begin{pmatrix} -\mathrm{i} & 0 \\ 0 & \mathrm{i} \end{pmatrix}, \quad (4.2.8)$$

这里, $\lambda = \mathrm{i}a_1 - q_1/2$, 其中 q_1 为任意实数, a_1 为非零实数. 求解上式易得

$$\Phi_1(z,t) = \exp[-\mathrm{i}(\lambda t + \lambda^2 z)], \quad (4.2.9)$$

$$\Phi_2(z,t) = \exp[-\mathrm{i}(\lambda t + \lambda^2 z)]. \quad (4.2.10)$$

将上式代入达布变换表达式 (4.2.6), 我们得零背景上局域波解:

$$u(z,t) = 2a_1 \mathrm{sech}[2a_1(t - q_1 z)] \exp\left[\mathrm{i}\left(q_1 t + 2a_1^2 z - \frac{1}{2}q_1^2 z\right)\right], \quad (4.2.11)$$

其中, a_1 为局域波振幅参数; q_1 表示局域波速度. 由图 4.5 可见, 该解描述了标准的亮孤子动力学, 其最大振幅为 $|u|_{\max} = 2a_1$.

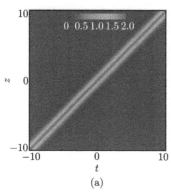 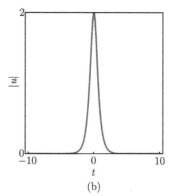

图 4.5　(a) 标准的非线性薛定谔系统中的亮孤子 $|u|$, 精确解 (4.2.11); (b) 亮孤子剖面图. 参数设置为: $a_1 = 1$, $q_1 = 1$(扫描封底二维码可看彩图)

接下来我们构造平面波背景上局域波解. 为此我们取如下一般形式的平面波种子解:

$$u_0(z,t) = a\exp[\mathrm{i}\theta(z,t)], \tag{4.2.12}$$

其中

$$\theta(z,t) = qt + (a^2 - q^2/2)z, \tag{4.2.13}$$

a 和 q 分别为平面波背景振幅和频率. 当 $a = 0$ 时, 平面波种子解 (4.2.12) 退化为平庸解 ($u_0 = 0$). 此时, 我们求解相应的 Lax pair(4.2.2) 且利用达布变换 (4.2.6) 就可以得到标准的亮孤子解. 这里, 我们从平面波种子解 (4.2.12) 出发, 构造平面波背景上的局域波解. 其难点在于, 在平面波种子解 (4.2.12) 的条件下, 如何得到 $\Phi_1(z,t)$ 和 $\Phi_2(z,t)$ 的解析表达式. 接下来, 我们首先将相应的 Lax pair 变换为常数形式, 通过对常数 Lax pair 的解的构造, 继而得到原始偏微分方程的解. 具体步骤如下:

首先, 我们引入矩阵 $\boldsymbol{P}$ 将矩阵 $\boldsymbol{U}$ 和 $\boldsymbol{V}$ 转化为常数矩阵 $\tilde{\boldsymbol{U}}$ 和 $\tilde{\boldsymbol{V}}$, 变换后的 Lax pair 具有如下形式:

$$\begin{cases} (\boldsymbol{P\Phi})_t = \tilde{\boldsymbol{U}}(\boldsymbol{P\Phi}), & (4.2.14\mathrm{a}) \\ (\boldsymbol{P\Phi})_z = \tilde{\boldsymbol{V}}(\boldsymbol{P\Phi}), & (4.2.14\mathrm{b}) \end{cases}$$

其中, 常数矩阵 $\tilde{\boldsymbol{U}}$ 和 $\tilde{\boldsymbol{V}}$ 为

$$\tilde{\boldsymbol{U}} = \boldsymbol{PUP}^{-1} + \boldsymbol{P}_t \boldsymbol{P}^{-1}, \tag{4.2.15}$$

$$\tilde{\boldsymbol{V}} = \boldsymbol{PVP}^{-1} + \boldsymbol{P}_z \boldsymbol{P}^{-1}. \tag{4.2.16}$$

这里, 我们选取矩阵 $\boldsymbol{P}$ 为

$$\boldsymbol{P} = \begin{pmatrix} \mathrm{e}^{-\frac{\mathrm{i}}{2}\theta} & 0 \\ 0 & \mathrm{e}^{\frac{\mathrm{i}}{2}\theta} \end{pmatrix}, \tag{4.2.17}$$

因此，$\tilde{U}$ 和 $\tilde{V}$ 的具体表达式分别为

$$\tilde{U} = \begin{pmatrix} -\mathrm{i}\lambda - \dfrac{\mathrm{i}}{2}q & a \\ -a & \mathrm{i}\lambda + \dfrac{\mathrm{i}}{2}q \end{pmatrix}, \tag{4.2.18}$$

$$\tilde{V} = \begin{pmatrix} -\mathrm{i}\lambda^2 + \dfrac{\mathrm{i}}{4}q^2 & -\dfrac{1}{2}aq + a\lambda \\ \dfrac{1}{2}aq - a\lambda & \mathrm{i}\lambda^2 - \dfrac{\mathrm{i}}{4}q^2 \end{pmatrix}. \tag{4.2.19}$$

这里，我们可以方便验证 $\tilde{U}$ 和 $\tilde{V}$ 满足可积条件 $[\tilde{U}, \tilde{V}] = 0$。

接下来，我们将求解变换后的常系数偏微分方程组 (4.2.14)。一般而言，传统的方式是对矩阵 $\tilde{U}$ 和 $\tilde{V}$ 进行对角化，得到对角矩阵继而求解。不过这里需要指出的是，考虑矩阵 $\tilde{U}$ 本征值方程是否具有重根，我们需要将 $\tilde{U}$ 变换为相应的对角矩阵或约当矩阵。这里，常数矩阵 $\tilde{U}$ 的本征值方程可直接由 $\mathrm{Det}[\tilde{U} - \tau \mathrm{I}] = 0$ 给出，其具体形式如下：

$$\tau^2 + a^2 + (\lambda + q/2)^2 = 0. \tag{4.2.20}$$

以上本征值方程具有两个根：$\tau_{1,2} = \pm \mathrm{i}\sqrt{a^2 + (\lambda + q/2)^2}$。因此，我们需要考虑两种不同的情况：① $\tau_1 \neq \tau_2$；② $\tau_1 = \tau_2$，分别对应着将 $\tilde{U}$ 转化为对角矩阵和约当矩阵的条件。接下来我们将介绍情况①，情况②与①本质上相同，只是在 $\tau_1 = \tau_2$ 情况下需要将 $\tilde{U}$ 转化为约当矩阵。

考虑 $\tau_1 \neq \tau_2$，我们引入变换矩阵 D，将常数矩阵 $\tilde{U}$ 和 $\tilde{V}$ 分别转化为对角矩阵 $\tilde{U}_\mathrm{d}$ 和 $\tilde{V}_\mathrm{d}$：

$$D^{-1}\tilde{U}D = \tilde{U}_\mathrm{d}, \tag{4.2.21}$$

$$D^{-1}\tilde{V}D = \tilde{V}_\mathrm{d}. \tag{4.2.22}$$

相应的变换后的 Lax pair 为

$$\begin{cases} \boldsymbol{\Phi}_{0t} = \tilde{U}_\mathrm{d}\boldsymbol{\Phi}_0, & (4.2.23\mathrm{a}) \\ \boldsymbol{\Phi}_{0z} = \tilde{V}_\mathrm{d}\boldsymbol{\Phi}_0. & (4.2.23\mathrm{b}) \end{cases}$$

对角矩阵 $\tilde{U}_\mathrm{d}$ 和 $\tilde{V}_\mathrm{d}$ 分别为

$$\tilde{U}_\mathrm{d} = \begin{pmatrix} \tau_1 & 0 \\ 0 & \tau_2 \end{pmatrix}, \tag{4.2.24}$$

$$\tilde{V}_\mathrm{d} = \begin{pmatrix} \mathrm{i}\tau_1^2 + b_1\tau_1 + b_0 & 0 \\ 0 & \mathrm{i}\tau_2^2 + b_1\tau_2 + b_0 \end{pmatrix}, \tag{4.2.25}$$

其中, $b_1 = \lambda - \frac{1}{2}q$, $b_0 = \mathrm{i}a^2 + \mathrm{i}\left(\lambda + \frac{1}{2}q\right)^2$. 需要指出的是, 变换矩阵 D 的选取不是唯一的, 为了使局域波的中心位置恰好处于 $(z,t) = (0,0)$ 点处, 我们选取变换矩阵 D 的具体表达式为

$$D = \begin{pmatrix} \sqrt{-\mathrm{i}(\lambda + q/2) + \tau_1} & \sqrt{-\mathrm{i}(\lambda + q/2) + \tau_2} \\ -\sqrt{-\mathrm{i}(\lambda + q/2) - \tau_1} & -\sqrt{-\mathrm{i}(\lambda + q/2) - \tau_2} \end{pmatrix}. \quad (4.2.26)$$

通过求解偏微分方程组 (4.2.23), 我们易得矩阵 Φ_0 的矩阵元 Φ_{01} 和 Φ_{02} 的精确表达式:

$$\Phi_{01}(z,t) = A_1 \exp\left(\tau_1 t + \mathrm{i}\tau_1^2 z + b_1 \tau_1 z + b_0 \cdot z\right), \quad (4.2.27)$$

$$\Phi_{02}(z,t) = A_2 \exp\left(\tau_2 t + \mathrm{i}\tau_2^2 z + b_1 \tau_2 z + b_0 \cdot z\right). \quad (4.2.28)$$

考虑上述变换 (4.2.14), 我们就得到原始 Lax pair(4.2.2) 的解:

$$\Phi_1(z,t) = \left\{-\left[\lambda + q/2 + \sqrt{a^2 + (\lambda + q/2)^2}\right]\Phi_{01}(z,t) + \mathrm{i}a\Phi_{02}(z,t)\right\}$$
$$\times \exp[\mathrm{i}\theta(z,t)/2], \quad (4.2.29)$$

$$\Phi_2(z,t) = \left\{-\left[\lambda + q/2 + \sqrt{a^2 + (\lambda + q/2)^2}\right]\Phi_{02}(z,t) + \mathrm{i}a\Phi_{01}(z,t)\right\}$$
$$\times \exp[-\mathrm{i}\theta(z,t)/2]. \quad (4.2.30)$$

将式 (4.2.29) 和式 (4.2.30) 代入式 (4.2.6) 经过细致的化简, 我们就得到平面波背景式 (4.2.12) 上具有一般形式的一阶局域波精确解的解析表达式, 如下:

$$u(z,t) = \left[2a_1\frac{\chi\cos\phi - \varsigma_2\cosh\varphi - \mathrm{i}(\chi - 2a^2)\sin\phi + \mathrm{i}\varsigma_1\sinh\varphi}{\chi\cosh\varphi - \varsigma_2\cos\phi} + a\right]\mathrm{e}^{\mathrm{i}\theta}, \quad (4.2.31)$$

这里

$$\varphi = 2\eta_i(t + V_1 z), \quad \phi = 2\eta_r(t + V_2 z), \quad (4.2.32)$$

$$\chi = (\chi_1^2 + \chi_2^2 + a^2), \quad \varsigma_2 = 2\chi_2 a, \quad \varsigma_1 = 2\chi_1 a, \quad (4.2.33)$$

其中

$$V_1 = v_1 + a_1\eta_r/\eta_i, \quad V_2 = v_1 - a_1\eta_i/\eta_r, \quad v_1 = -(q_1 + q)/2, \quad (4.2.34)$$

$$\chi_1 = \eta_r + (q - q_1)/2, \quad \chi_2 = \eta_i + a_1. \quad (4.2.35)$$

$$\eta_r + \mathrm{i}\eta_i = [a^2 - a_1^2 + (q - q_1)^2/4 + \mathrm{i}a_1(q - q_1)]^{1/2}. \quad (4.2.36)$$

一般而言, 以上求得的局域波解 (4.2.31) 很好地描述了呼吸子特征. 对于怪波解我们则需考虑 $\tau_1 = \tau_2$ 的简并情况, 即 $\tau_1 = \tau_2 = 0$ (意味着 $\lambda = \mathrm{i}a - q/2$), 继而重新求解 Lax pair. 然而有趣的是, 由于怪波本身为呼吸子的周期无穷大的极限情况, 因此当我们考虑 $\lambda \to \mathrm{i}a - q/2$ 的情况时, 呼吸子的结构特征就无限趋近于理想的怪波结构. 这样的数学极限过程已经在单模光纤中被实验所证实, Kibler 小组[21]成功地利用周期趋于无穷大的 Akhmediev 呼吸子得到理想的 Peregrine 怪波. 事实上, 利用泰勒展开, 上述呼吸子解在极限情形 $\lambda = \mathrm{i}a - q/2$ 可直接退化为有理形式的怪波解. 这种方法与将矩阵 $\tilde{U}$ 转化为约当矩阵的求解方法是完全自洽的. 图 4.6 中, 我们展示了运用上述达布变换方法得到的几种典型的局域波结构特征.

图 4.6(a) 呈现了标准的 Kuznetsov-Ma 呼吸子, 其特征为: 在分布方向 t 局域, 在传输方向 z 呈现出周期性, 相应的条件 $q_1 = q$, $a_1 > a$; 反之 $a_1 < a$, 我们得到标准的 Akhmediev 呼吸子, 其分布特征与 Kuznetsov-Ma 呼吸子恰好相反, 即其在分布方向 t 上呈现周期性, 在传输方向 z 呈现出局域性 (图 4.6(b)). 当 $a_1 \to a$ 时, Akhmediev 呼吸子的周期趋于无穷, $a_1 = a$ 时, 退化为 Peregrine 怪波 (标准的一阶怪波)(图 4.6(c)). 除上述条件外, 该一阶局域波解描述了"一般呼吸子"的特征 (图 4.6(d)), 其中, 一般呼吸子是指沿某一方向局域, 但在传输方向 z 以及分布方向 t 上皆有周期性的呼吸子. 历史上, 关于 Akhmediev 呼吸子、Kuznetsov-Ma 呼吸子以及 Peregrine 怪波的研究颇多, 相较而言, 一般呼吸子的研究偏少. 不过, 近期的研究取得重要进展, 首先是 Zakharov 和 Gelash[17] 在理论上揭示了一般呼吸子碰撞性质可以用来描述调制不稳定性的非线性演化阶段. 之后 Kibler 和 Chabchoub[75] 分别在光纤和水槽系统对理论预言进行了完美的实验验证.

另外需要指出的是, 高阶怪波解的求解也是怪波研究中的重点. 而幸运的是, 目前已有较为一般的方法来求解高阶怪波解[82, 83]. 这里, 我们可以方便地将式 (4.2.29) 和式 (4.2.30) 在 $\lambda = \mathrm{i}a - q/2$ 情况下作泰勒展开, 相应的不同阶数的系数的非线性叠加就构成对应的高阶怪波解. 图 4.6(e) 和 (f) 描绘了两种具有不同结构的典型的二阶怪波密度分布, 分别为单峰结构和"三胞胎"结构[84]. 可以看到, 二阶怪波实际上是三个基本一阶怪波的非线性叠加, 这些叠加由于相位参数的不同表现出不同的结构. 就目前的研究现状而言, 高阶怪波具有丰富的结构特征, 想要穷尽其结构类型并详细分类依然是该方向的公开命题. 不过, 目前人们已经初步归类了部分较低阶的怪波结构[82, 83], 并揭示了不同阶数的怪波表现出相似的轨迹特征[85], 其中的部分高阶解已经在实验上得到证实. 关于光纤系统, 人们已经在单模光纤系统中证实了二阶怪波的存在[25]. 相较于光纤系统, 人们利用水槽在实验上验证了二至五阶的怪波结构[86]. 这些重要的实验结果将极大地丰富我们对自然界真实怪波现象的理解, 并且给予理论学家强大的信心, 用以继续探寻丰富的高阶怪波结构.

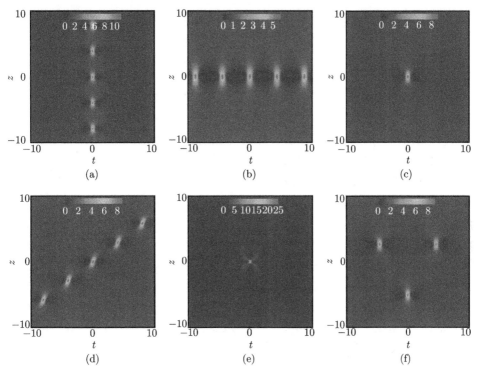

图 4.6　几种典型的平面波背景上局域波的密度分布图 $|u(z,t)|^2$. (a)Kuznetsov-Ma 呼吸子, 参数设置为: $a=1$, $q=0$, $q_1=0$, $a_1=1.2$; (b)Akhmediev 呼吸子, 参数设置为: $a=1$, $q=0$, $q_1=0$, $a_1=0.7$; (c)Peregrine 怪波, 参数设置为: $a=1$, $q=0$, $q_1=0$, $a_1=1$; (d) 一般呼吸子, 参数设置为: $a=1$, $q=0$, $q_1=0.5$, $a_1=1$; (e) 单峰二阶怪波; (f) "三胞胎" 二阶怪波 (扫描封底二维码可看彩图)

此外, 我们注意到多组分耦合系统中的矢量局域波性质研究成为该领域的另一个热点课题. 单是考虑一阶怪波情形, 科学家已经发现了怪波和呼吸子在矢量系统中更加丰富的结构以及动力学. 最著名的例子是多组分耦合系统中发现的 "暗怪波"、"四花瓣怪波"、"多怪波" 以及相应的呼吸子的有趣结构[62-64]. 这些结构不仅存在于多组分耦合非线性薛定谔框架下, 也存在于多波共振系统和多组分耦合高阶非线性薛定谔系统等. 研究表明, 这些结构不存在于人们已知的标量系统. 目前, 实验科学上对于矢量系统中的怪波和呼吸子的新结构的验证处于探究阶段, 这是由于多组分耦合系统较标量系统更为复杂, 实验参数更多, 对实验操控手段更为敏感. 不过, 最近据报道, 人们在双模光纤中已经成功观测到 "暗怪波" 的存在[87]. 更令人意想不到是, 人们还发现不同种类的局域波可以在多组分耦合系统中共存且表现出非平庸的相互作用性质. 例如, 怪波与亮暗孤子、怪波与呼吸子的共存以及相互吸引[62-64]. 上述结果表明, 理论上得到的平面波上多种不同类型的局域波解是极其关键的,它不仅以严格精确的方式发展了非线性局域波理论,也为实际非线性物理系统中局域波的实现提供了不可替代的参考.

4.2.2 相似变换

随着现代非线性科学的不断发展,相应的非线性物理实验中的操作手段也越来越成熟. 在非线性光学中,影响局域波动力学性质的主要物理参量,如色散、Kerr 非线性 (自相位调制) 以及增益 (或损耗) 等均能在实验中得到精确操控[88]. 特别地,人们将色散和非线性的有效操控称为色散和非线性管理[88]. 此外, Bose-Einstein 凝聚体中的散射长度也可以在实验上由 Feshbach-resonance 技术进行精确操控[3]. 针对以上实验事实,相应的非线性局域波的动力学需要由推广的"非自治"模型[89]来描述.

非自治非线性模型最早由 Serkin 等[89] 在研究孤子管理问题时提出. 他们发现非自治非线性模型可以很好地描述物理参数可变的实际物理系统中的局域波动力学,从而实现了相应系统中局域波动力学的精确操控. 自此,非自治局域波管理的研究已经成为非线性局域波研究中必不可少的课题之一. 非自治局域波的研究的难点在于一般的非自治系统是不可积的,因此,如何得到方程的可积性条件就成了首要问题. Serkin 等[89] 首先利用反散射方法给出了非自治非线性薛定谔方程的可积条件 (参数约束条件),实现了孤子宽度、振幅、相位的有效操控. Agrawal 等[90] 研究了平面波导中非自治孤子的存在性和双孤子的碰撞性质. Belmonte-Beitia 等[91] 利用相似变换研究了时空调制非线性效应对局域波动力学的影响. 国内的研究中,刘伍明等[92] 系统研究了 Bose-Einstein 凝聚体中非自治亮孤子的动力学特征. 罗洪刚和赵敦等[93] 发现了一种有效的变换,将非自治非线性模型转换为传统的自治的非线性模型,从而大大简化了直接求非自治局域波解的难度. 闫振亚等[94] 研究了 Bose-Einstein 凝聚体中不同结构的非自治物质波. 李禄等[95] 利用达布变换等方法研究了色散管理光纤中的时域光孤子的动力学演化性质. 张解放等[96] 利用相似变换研究了平面波导中非自治空间光孤子性质. 钟卫平等[97] 利用相似变换研究了不同维度的非自治光孤子的动力学性质. 戴朝卿等[98] 利用相似变换研究了考虑 parity-time 对称势的 1+3 维光孤子. 我们[99] 重新构造了非自治薛定谔模型的 Lax pair,继而利用达布变换研究了不同物理系统中的非自治局域波性质.

接下来我们将介绍局域波精确解构造的另一种常用方法——相似变换方法. 对于非自治模型描述的物理系统,其相应的物理参数不再是常量,而是可以随时空变化的变量 (如色散、非线性、增益以及外势等参量依赖于纵向传输变量 z,或横向分布变量 t,或两者兼有). 因此从理论角度而言,该方法对实验中的局域波操控和管理有着较为重要的意义. 接下来,我们从推广的 1+1 维非自治非线性薛定谔方

程出发, 简述通过相似变换得到局域波精确解的主要过程, 相应的非自治模型具有如下形式:

$$\mathrm{i}\frac{\partial \boldsymbol{u}}{\partial z} + D(z,t)\frac{\partial^2 \boldsymbol{u}}{\partial t^2} + R(z,t)|\boldsymbol{u}|^2\boldsymbol{u} + V(z,t)\boldsymbol{u} + \mathrm{i}G(z,t)\boldsymbol{u} = 0, \qquad (4.2.37)$$

其中, $\boldsymbol{u}(z,t)$ 表示慢变场; z 和 t 分别表示纵向传输方向和横向分布方向; $D(z,t)$ 是色散或衍射系数; $R(z,t)$ 为非线性系数; $V(z,t)$ 表示非线性物理系统的外势调制; $G(z,t)$ 为系统的增益或损耗. 首先, 需要指出的是, 这里的调制系数是关于 z 和 t 的函数, 因此模型 (4.2.37) 描述了光场在 z 和 t 方向都具有可控性的动力学演化性质. 再者, 该模型具有一般普适性, 这意味着可以描述多种不同的实际非线性物理系统中的局域波动力学. 具体如下:

(1) 若 $D(z,t) = D(z)$, $R(z,t) = R(z)$, $G(z,t) = G(z)$, $V(z,t) = M(z)t^2$, 模型 (4.2.37) 描述了光脉冲在非均匀单模光纤中的传输 [95]. 相应的调制系数 $D(z)$ 和 $R(z)$ 分别为色散和非线性管理项, $V(z,t) = M(z)t^2$ 表示非均匀的自相位调制.

(2) 若将变量 $t \to x$, 且 $D(z,x) = 1/2$, $R(z,x) = R(z)$, $G(z,x) = G(z)$, $V(z,x) = F(z)x^2$, 则此时方程 (4.2.37) 描述了光束在非线性平面波导中的动力学演化 [3], 对应的折射率为 $n = n_0 + n_1 F(z)x^2 + n_2 R(z)I(z,x)$, 其中 $R(z)$ 为非均匀克尔非线性项, $F(z)$ 为线性折射率横向维度衍射强弱, 其值正负分别对应于梯度折射率作为自聚焦和自散焦效应的线性透镜. 另外, 若 $D(z,x) = 1/2$, $R(z,x) = R(x)$, $V(z,x) = V(x)$, $G(z,x) = 0$, 则方程 (4.2.37) 描述了光束在具有横向周期格子的非线性平面光波导的动力学 [3], 其中 $R(x)$ 和 $V(x)$ 皆为周期函数.

(3) 若将变量 $t \to x$, $z \to t$, 且 $D(t,x) = 1/2$, 则系统 (4.2.37) 退化为 1+1 维的 Gross-Pitaevskii 方程, 用以描述雪茄型 Bose-Einstein 凝聚体的动力学 [3]. 其中非线性项 $R(t,x)$ 表示由原子碰撞所决定的散射长度, 其正负分别描述吸引和排斥的相互作用. 实验上散射长度可以由 Feshbach-resonance 管理技术进行精确操控 [3].

一般而言, 方程 (4.2.37) 是不可积的. 为了精确地研究其中的局域波动力学, 我们将利用相似变换的方法获得可积性 (约束) 条件, 从而构造出一般模型 (4.2.37) 的精确解. 其主要步骤如下:

首先我们引入如下相似变换形式, 即假设方程 (4.2.37) 有如下形式解:

$$\boldsymbol{u}(z,t) = A(z,t)u[Z(z), T(z,t)]\mathrm{e}^{\mathrm{i}\Theta(z,t)}, \qquad (4.2.38)$$

其中, $A(z,t)$, $\Theta(z,t)$, $Z(z)$, $T(z,t)$ 为待定函数; $A(z,t)$ 和 $\Theta(z,t)$ 分别为振幅和相位函数; $Z(z)$ 和 $T(z,t)$ 是我们引入的两个待定的相似变量. $u[Z(z), T(z,t)]$ 满足标准的非线性薛定谔方程:

$$\mathrm{i}\frac{\partial u}{\partial Z} + \frac{1}{2}\frac{\partial^2 u}{\partial T^2} + |u|^2 u = 0. \qquad (4.2.39)$$

上文中，我们已经给出了方程 (4.2.39) 平面波背景上多种局域波解. 为了研究这些局域波在非自治光学系统中的动力学及其相关特性，我们将式 (4.2.38) 代入式 (4.2.37) 导出标准的非线性薛定谔模型 (4.2.39). 此时，$A(z,t)$, $\Theta(z,t)$, $Z(z)$, $T(t,z)$ 满足一系列偏微分方程. 为了方便求解，我们这里设: $T(z,t) = \int_0^\varsigma F[\varsigma(z,t)]d\varsigma$, $\varsigma(z,t) = \alpha(z)t$, $R(z,t) = \alpha(z)F[\varsigma(z,t)]$. 基于此我们易得如下表达式:

$$\rho(z,t) = \frac{\kappa(z)}{\alpha(z)F(z,t)}, \tag{4.2.40}$$

$$Z(z) = \int_0^z \kappa(z)\mathrm{d}z + Z_0, \tag{4.2.41}$$

$$\Theta(z,t) = \frac{\alpha(z)\alpha_z(z)}{\kappa(z)} \int_0^t F^2[\varsigma(z,t)]t\mathrm{d}t, \tag{4.2.42}$$

$$V(z,t) = \eta_1(\varsigma,z)t^2 + \int_0^t [\eta_2(\varsigma,z)t - \eta_3(\varsigma,z)t^2]\mathrm{d}t + \eta_4(\varsigma,z), \tag{4.2.43}$$

其中

$$\eta_1(\varsigma,z) = \frac{\alpha_z^2(z)F^2(\varsigma)}{2\kappa(z)}, \tag{4.2.44}$$

$$\eta_2(\varsigma,z) = \frac{F^2(\varsigma)}{\kappa^2(z)}[\alpha(z)\alpha_z(z)\kappa_z(z) - \kappa(z)\alpha_z^2(z) - \kappa(z)\alpha(z)\alpha_{zz}(z)], \tag{4.2.45}$$

$$\eta_3(\varsigma,z) = \frac{2}{\kappa(z)}F(\varsigma)F_\varsigma(\varsigma)\alpha(z)\alpha_z^2(z), \tag{4.2.46}$$

$$\eta_4(\varsigma,z) = \frac{\kappa(z)}{8F^4(\varsigma)}[F(\varsigma)F_{\varsigma\varsigma}(\varsigma) - 3F_\varsigma^2(\varsigma)]. \tag{4.2.47}$$

此时，我们注意到方程 (4.2.37) 的调制系数是不能任意选取的. 这些调制系数需要满足如下约束关系:

$$G(z,t) = \frac{\alpha_z(z)}{\alpha(z)}\left[\frac{R_t(z,t)}{R(z,t)}t + 1\right] - \frac{D_z(z,t)}{2D(z,t)} - \frac{R_z(z,t)}{R(z,t)}, \tag{4.2.48}$$

其中

$$D(z,t) = \frac{\kappa(z)}{2\alpha^2(z)F^2[\varsigma(z,t)]}. \tag{4.2.49}$$

因此，方程 (4.2.48) 可被认为是非自治方程 (4.2.37) 存在精确解的可积性条件. 另外，需要注意的是，我们可以通过对调制函数 $\alpha(z)$ 和 $F[\varsigma(z,t)]$ 的选取来实现实验上多种不同的非线性管理方案 $R(z,t) = \alpha(z)F[\varsigma(z,t)]$. 下文中我们研究的局域背景上光怪波的激发就是基于高斯函数的非线性管理, 并且需要注意的是, 高斯型非线性管理已经在非线性实验中得到了实现[88].

需要注意的是, 相似变换已经广泛运用于各种不同的非自治非线性物理系统中, 主要包括: ①考虑附加的实际物理效应 (高阶效应等) 模型; ②1+2 维和 1+3 维模型. 相似变换方法的限制是需要知道约化后目标方程的局域波解. 不过, 考虑到该变换本质上是一种非线性变换, 因此研究所得局域波动力学亦有其非平庸的性质.

4.2.3 调制不稳定性

调制不稳定性是普遍存在于多种非线性物理系统中的不稳定现象, 包括流体、等离子体、Bose-Einstein 凝聚体、非线性光学等. 调制不稳定性的产生机制是基于色散或衍射效应和非线性效应间的相互作用[100]. 一般而言, 在调制不稳定性的初始演化阶段, 与不稳定性关联的频谱边带在损耗泵浦波能量的前提下呈指数放大, 不过随之而来的动力学演化就变得更加复杂并且表现出多重谱模式能量的循环交换.

通常, 单模光纤系统中的调制不稳定性发生在反常色散区, 其典型的表现为: 将连续波 (或准连续波) 分裂成一系列的短脉冲. 最近, 人们将这些局域的脉冲结构和具有相似结构特征的海洋怪波进行比较, 类比定义了光怪波. 此后, 光怪波在不同非线性光学系统中的研究逐渐成为研究的热点. 不过最近的研究表明, 在交叉相位调制和高阶效应存在的条件下可以突破上述约束 (反常色散约束) 而使光纤在正常色散区同样表现出调制不稳定性[101].

另外, 时域上的调制不稳定性描述了平面波 (或连续波) 在小振幅扰动情况下的不稳定性质, 与上述不稳定的局域波 (怪波和呼吸子) 动力学性质相似. 的确, 最近的研究表明怪波呼吸子以及更为复杂的非线性叠加形式的产生机制就是调制不稳定性[8]. 因此, 在平面波背景上局域波性质的研究中, 相应的调制不稳定性特征的研究是重要且必要的. 通常, 人们采取线性稳定性分析方法对调制不稳定性进行研究.

接下来, 我们首先简要介绍标准的线性稳定性分析方法的步骤, 之后将概述调制不稳定性与平面波背景上局域波研究的最新进展. 以标准的线性薛定谔框架式 (4.2.1) 为例, 线性稳定性分析方法的步骤如下:

考虑方程 (4.2.1) 的一个小振幅扰动的平面波解:

$$u_p(z,t) = [a + p(z,t)] \exp[\mathrm{i}\theta(z,t)], \quad (4.2.50)$$

其中, $p(z,t)$ 是满足特定线性方程的微扰. 扰动平面波解 (4.2.50) 为标准平面波背

景 (4.2.12) 加上扰动项 $p(z,t)$. 线性化的主要思想就是: 由于扰动项 $p(z,t)$ 为微扰, 因此关于 $p(z,t)$ 的非线性项可以直接略去, 从而直接将非线性问题转换为易处理的线性问题.

接下来, 我们将扰动平面波解 (4.2.50) 代入方程 (4.2.1) 得如下关系:

$$2a^2 p + 2a^2 p^* + 2\mathrm{i}p_z + 2\mathrm{i}qp_t + p_{tt} + 2ap^2 + 4app^* + 2p^2 p^* = 0, \quad (4.2.51)$$

其中, p^* 为 p 的复共轭, 下脚标为相应扰动函数的偏导. 略去关于 $p(z,t)$ 的非线性项, 易得如下线性关系:

$$2a^2 p + 2a^2 p^* + 2\mathrm{i}p_z + 2\mathrm{i}qp_t + p_{tt} = 0, \quad (4.2.52)$$

通常情况下, 扰动项 $p(z,t)$ 可做如下展开:

$$p(z,t) = \eta_{1,s}(z)\mathrm{e}^{\mathrm{i}Qt} + \eta_{1,a}(z)\mathrm{e}^{-\mathrm{i}Qt}. \quad (4.2.53)$$

由上式可知, $p(z,t)$ 是在横向分布方向 t 上具有周期性且频率为 Q 的复函数. 将式 (4.2.53) 代入线性关系式 (4.2.52), 并分离 $\mathrm{e}^{\mathrm{i}Qt}$ 和 $\mathrm{e}^{-\mathrm{i}Qt}$ 项, 可得

$$\begin{cases} (2a^2 + 2qQ - Q^2)\eta_{1,a} + 2a^2 \eta_{1,s}^* + 2\mathrm{i}\eta_{1,a}' = 0, & (4.2.54\mathrm{a}) \\ (2a^2 - 2qQ - Q^2)\eta_{1,s} + 2a^2 \eta_{1,a}^* + 2\mathrm{i}\eta_{1,s}' = 0. & (4.2.54\mathrm{b}) \end{cases}$$

其中, "′" 表示相应函数关于 z 的导数. 简单起见, 我们将上式表示为

$$2\boldsymbol{\eta}' = \mathrm{i}\boldsymbol{M}\boldsymbol{\eta}, \quad (4.2.55)$$

其中, $\boldsymbol{\eta} = (\eta_{1,s}, \eta_{1,a}^*)^{\mathrm{T}}$; 矩阵 $\boldsymbol{M}$ 为

$$\boldsymbol{M} = \begin{pmatrix} 2a^2 - 2qQ - Q^2 & 2a^2 \\ -2a^2 & -2a^2 - 2qQ + Q^2 \end{pmatrix}. \quad (4.2.56)$$

对任意实频率 Q, 扰动项 $\eta(z)$ 为指数 $\exp(\mathrm{i}\omega_j z)$ 的线性组合, 其中 ω_j $(j=1,2)$ 为矩阵 $\boldsymbol{M}$ 的两个本征值. 换言之, ω_j 为矩阵 $\boldsymbol{M}$ 特征多项式的根, 相应的特征多项式如下:

$$B(\omega) = \omega^2 + 4qQ\omega + 4a^2 Q^2 + 4q^2 Q^2 - Q^4. \quad (4.2.57)$$

此时, 我们易得色散关系

$$\omega = -2qQ \pm \sqrt{Q^4 - 4a^2 Q^2}. \quad (4.2.58)$$

若 ω_j 存在非零的虚部, 则相应的系统具有调制不稳定性. 若 $\mathrm{Im}\{\omega\} < 0$, 那么, 初始的小扰动信号将会以指数形式 $\exp(Gz)$ 增长放大. 由表达式 (4.2.58) 可知, 调制

不稳定性存在于 $-2a < Q < 2a$ 的扰动频率区域. 为了描述调制不稳定性的强弱, 我们定义调制不稳定性增长率为 $G = |\text{Im}\{\omega\}|$.

这里需要指出的是, 作为一种经典的调制不稳定性分析手段, 线性稳定性分析能够较为直观地预言调制不稳定性的频谱范围和扰动增益, 但也有其自身局限性. 例如, 线性稳定性分析只给出微扰的初始指数放大情况, 而这种振幅放大的过程是不能一直持续下去的, 因此线性稳定分析在描述调制不稳定性的非线性演化阶段就会失效. 不过, 对于描述怪波现象而言, 线性稳定性分析能够精确地证实怪波的存在性. 特别是 Baronio 小组 [61] 揭示了 Peregrine 怪波的产生的确源于调制不稳定性, 但只存在于调制不稳定区域中的一个特殊子区域——"零频扰动区域". 紧接着, 最近的研究表明 Peregrine 怪波的激发源于平面波背景和扰动的零频共振 [102].

为此, 接下来我们将关注调制不稳定性在零频扰动子区域的性质. 其相应的数学处理为: 考虑调制不稳定性在 $Q \to 0$ 时的特征, 我们将特征多项式 (4.2.57) 写为 [61]

$$B(Q\omega) = Q^2 b(\omega), \tag{4.2.59}$$

其中, $b(\omega)$ 表示为

$$b(\omega) = \omega^2 + 4q\omega + 4a^2 + 4q^2 - Q^2. \tag{4.2.60}$$

此时, 我们得到色散关系

$$\omega = -2q \pm \sqrt{Q^2 - 4a^2}. \tag{4.2.61}$$

经过变换 (4.2.59) 后的色散关系 (4.2.61) 没有改变调制不稳定区域的分布范围, 不过却显著改变了零频扰动区域 ($Q = 0$) 的不稳定性质. 为了方便理解, 我们通过绘图来展示两个不同的色散关系在不同参数空间下的调制不稳定性特征.

由图 4.7 可清晰看到, 无论是在 (Q, q) 平面还是在 (Q, a) 平面, 色散关系 (4.2.58) 下的零频扰动区的增长率都为零 (图 4.7(a) 和 (c)); 而色散关系 (4.2.61) 下的零频扰动区的增长率不为零 (图 4.7(b) 和 (d)). 由于怪波的产生源于调制不稳定性, 因此怪波的存在区域的不稳定性增长率必不为零.

另外, 由方程所得怪波解可知, Peregrine 怪波的最大振幅与背景波振幅的关系为 $|u(z,t)|_{\max} = 3a$, 而在色散关系 (4.2.61) 得到的零频调制不稳定性增长率 $G_0 = 2a$. 因此, 怪波最大振幅和零频不稳定性增长率皆与背景振幅大小成正比, 意味着 a 值越大, 增长率 G_0 越大, 对应的怪波峰值 $|u(z,t)|_{\max}$ 越大; 反之, 增长率 G_0 越小, 对应的怪波峰值 $|u(z,t)|_{\max}$ 越小. 而在色散关系 (4.2.58) 中, 我们得不到这样的对应关系. 因此, 我们可以论断, 上述变换 (4.2.59) 对怪波存在于零频调制不稳定区是有效且必要的.

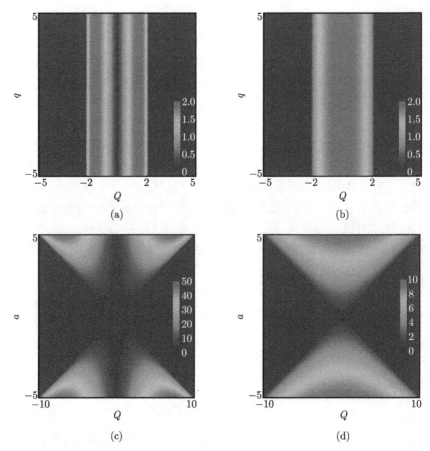

图 4.7 标准的非线性薛定谔系统中不同色散关系 (4.2.58) 和 (4.2.61) 下的调制不稳定性特征, (a) 和 (b) 分别为色散关系 (4.2.58) 和 (4.2.61) 的增长率 G 在 (Q,q) 平面上的分布特征, 参数设置为: $a=1$; (c) 和 (d) 分别为色散关系 (4.2.58) 和 (4.2.61) 的增长率 G 在 (Q,a) 平面上的分布特征, 参数设置为: $q=0$(扫描封底二维码可看彩图)

最近的研究进展 [61, 101] 发现, 在许多不同的非线性物理框架下 (描述各种各样不同的非线性物理系统), 怪波的存在条件与 $Q \to 0$ 情况下的零频调制不稳定区域 (需考虑变换 (4.2.59)) 严格一致, 包括 Fokas-Lenells 系统、矢量多组分耦合非线性薛定谔系统、长短波共振系统以及三波共振系统. 此外, 对于标准的非线性薛定谔系统, 人们也建立了调制不稳定性与多种已知的非线性激发单元 (Peregrine 怪波、Kuznetsov-Ma 呼吸子、Akhmediev 呼吸子和标准的亮孤子) 的定量关系 [102]. 因此, 下文中对不同物理模型进行的调制不稳定性分析都采取了色散关系 (4.2.61) 的处理方式.

然而需要指出的是, 上述研究只关注到零频调制不稳定性与怪波的存在性的对应关系, 未涉及怪波在零频扰动区的性质分析, 特别是零频增长率表现出新特征的情况下, 怪波会有怎样的新性质是没有研究的. 此外, 除怪波外, 呼吸子也被证明源于调制不稳定性, 而呼吸子在相应扰动区域的性质分析也鲜有报道. 对于上述这些问题, 我们将在下文做详细探讨.

4.3 高斯背景上光怪波的激发

如绪论所述, 描述怪波的局域波解一般为局域在平面波背景上具有不稳定性质的非线性结构. 然而, 在实际中无法制备无限宽的平面波背景. 目前实验上激发怪波或呼吸子是利用超宽包络背景逼近理想的平面波背景, 运用相应的密度调制和相位调制获得初始激发, 从而实现怪波和呼吸子的实验观测. 本章将从实际物理情况出发, 探究有限宽背景上怪波的激发及其性质. 利用相似变换方法, 解析地论证了怪波能够在局域的高斯背景上激发并给出相应的初始激发条件, 并发现经典的平面波背景上的怪波特征 (高峰值和双重局域性) 在高斯背景上都得到很好的保持.

绪论中我们介绍了怪波的基本性质和研究进展. 目前, 理论上对怪波的描述主要基于平面波背景上局域波的动力学特征. 这种局域波首先得有高于两倍背景波振幅以上的高幅值; 其次这种局域波需要具备鲜明的时空局域性 (演化和分布方向的双重局域性) [4, 5]. 然而, 理论上看似完美的描述在实际物理系统中却很难找到对应, 这是因为理想的平面波是无限宽的, 对应着无穷大的背景能量. 特别是在光学系统中, 理想的平面波是无法制备的. 最近, Solli 等[11] 利用准平面波来替代理想的平面波, 在非线性光纤实验平台上实现了光学怪波的激发和观测. 这里, 准平面波背景可以理解为宽度相当大, 可以逼近平面波的包络背景, 也就意味着背景波的宽度比光学怪波的局域尺度要大得多. 所以, 从怪波的应用角度来看, 怪波在有限宽背景上的研究是具有极大的实际意义的. 这里需要指出, 在光学实验中, 鉴于高斯背景光束/脉冲的易制备性, 高斯背景光束/脉冲已经作为暗孤子的激发背景被广泛使用[103]. 然而, 作为一类重要的有限宽局域背景, 高斯背景上光学怪波的激发却一直鲜有人研究和关注[5-8]. 我们将从理论上设计一种光怪波在局域的高斯背景上激发的理论方案. 利用相似变换的方法, 我们给出了一组精确描述光怪波在高斯背景上激发的怪波解的解析表达式. 我们发现, 怪波的典型特征 (高峰值和双重局域性) 在高斯背景上 (甚至是宽度较窄的高斯背景上) 都得到很好的保持. 这些结果有可能对今后光怪波在局域背景上的实现提供相应的理论支持.

4.3.1 高斯背景上光怪波精确解

如上文所述, 有限宽局域背景上的怪波性质研究目前仍处于起始阶段. 不过, 怪波在不同结构背景波上激发的研究已经得到了人们越来越多的重视. 最近的研究主要包括周期波背景上怪波的激发和管理 [79]、孤子背景上怪波的操控 [79, 104].下文中, 我们研究平面波导中光怪波在高斯背景光束上的激发以及相应的动力学性质.

我们首先考虑光束在平面波导管中的传输, 相应的折射率分布满足如下关系:

$$n = n_0 + n_1(\zeta,\chi) + R(\zeta,\chi)I(\zeta,\chi), \qquad (4.3.1)$$

其中, $I(\zeta,\chi)$ 为光强, ζ 和 χ 分别为纵向传播距离和横向分布坐标. 这里, 第一项 $n_0 + n_1(\zeta,\chi)$ 表示折射率的线性部分, 而第二项 $R(\zeta,\chi)I(\zeta,\chi)$ 为 Kerr 非线性项. Kerr 系数 $R(\zeta,\chi)$ 可取正亦可取负, 分别对应着非线性自聚焦和自散焦介质. 非线性光束在这样一个波导中传输的动力学可由如下非线性方程来描述:

$$\mathrm{i}\frac{\partial \boldsymbol{u}}{\partial \zeta} + \frac{1}{2k_0}D(\zeta,\chi)\frac{\partial^2 \boldsymbol{u}}{\partial \chi^2} + \frac{k_0}{n_0}n_1(\zeta,\chi)\boldsymbol{u} + \frac{k_0}{n_0}R(\zeta,\chi)|\boldsymbol{u}|^2\boldsymbol{u} + \mathrm{i}\frac{k_0}{n_0}G(\zeta,\chi)\boldsymbol{u} = 0, (4.3.2)$$

这里, $\boldsymbol{u}(\zeta,\chi)$ 为电场复包络; $k_0 = 2\pi n_0/\lambda_0$ 为波数 (λ_0 表示入射波的波长); $D(\zeta,\chi)$ 表示波导中的衍射效应; $G(\zeta,\chi)$ 为能量的增益或者损耗. 引入变量代换 $z = \dfrac{k_0}{n_0}\zeta$ 和 $x = \dfrac{\sqrt{2k_0}}{\sqrt{n_0}}\chi$, 上述非线性波动方程可化简为

$$\mathrm{i}\frac{\partial \boldsymbol{u}}{\partial z} + D(x,z)\frac{\partial^2 \boldsymbol{u}}{\partial x^2} + R(x,z)|\boldsymbol{u}|^2\boldsymbol{u} + n_1(x,z)\boldsymbol{u} + \mathrm{i}G(x,z)\boldsymbol{u} = 0. \qquad (4.3.3)$$

首先, 我们不考虑增益损耗项, 即 $G(x,z) = 0$. 原因是增益损耗一般只影响波峰强度的大小, 即只会造成局域波峰值的衰减或者放大, 对局域波的性质没有本质影响. 再次, 我们忽略关于传输方向 z 的函数调制, 这是因为调制函数对 z 方向的操控实质上改变了怪波的演化性质, 而无法使怪波在局域的背景上激发. 因此, 这些实际可调控物理参量都是关于空间分布变量 x 的函数, 上式变化为

$$\mathrm{i}\frac{\partial \boldsymbol{u}}{\partial z} + D(x)\frac{\partial^2 \boldsymbol{u}}{\partial x^2} + R(x)|\boldsymbol{u}|^2\boldsymbol{u} + n_1(x)\boldsymbol{u} = 0. \qquad (4.3.4)$$

为了得到模型 (4.3.4) 中高斯背景光束上的怪波解的解析表达式, 我们首先引入具有如下形式的高斯波包作为激发怪波的有限宽的局域背景:

$$\boldsymbol{u}_0(z,x) = \exp\left[-\frac{x^2}{2b^2} + \mathrm{i}z\right], \qquad (4.3.5)$$

其中, b 为非零实参数, 可用来调节高斯背景光束的宽度. b 值越大, 高斯背景光束就越宽; b 值越小, 高斯背景光束就越窄.

接下来, 利用绪论中相似变换的基本步骤, 我们将式 (4.3.5) 代入式 (4.3.4) 得到标准非线性薛定谔方程, 再结合上述达布变换方法求得标准非线性薛定谔方程的

4.3 高斯背景上光怪波的激发

平面波背景上的局域波解, 最终得到了高斯背景光束上多种局域波精确解. 为了简单起见, 这里只给出高斯背景 (4.3.5) 上的一阶和二阶怪波精确解 ($u_{1,2}$), 其表达式如下:

$$u_j(z,x) = [1+\kappa_j(z,x)]\exp\left[-\frac{x^2}{2b^2}+\mathrm{i}z\right], \quad j=1,2, \tag{4.3.6}$$

其中, $j=1,2$ 分别表示一阶和二阶怪波解; $\kappa_j(z,x)$ 为变换后的有理函数, 对于不同阶数的怪波解相应的表达式也不相同, 这里对于一阶怪波有如下表达式:

$$\kappa_1(z,x) = -\frac{4+8\mathrm{i}}{1+4X^2+4z^2}, \tag{4.3.7}$$

对于二阶怪波有如下表达式:

$$\kappa_2(z,x) = \frac{4\eta_1(z,x)\eta_2^*(z,x)}{|\eta_1(z,x)|^2+|\eta_2(z,x)|^2}, \tag{4.3.8}$$

其中

$$\begin{aligned}\eta_1 =\ & 6(1+2X+2\mathrm{i}z) - A(1+2X+2\mathrm{i}z)\\ & \times(1-2X+2\mathrm{i}z)\{-6(4b_1+4\mathrm{i}b_2+X+5\mathrm{i}z)\\ & -8(X+\mathrm{i}z)^3+3[-1+4(X+\mathrm{i}z)^2]\}\\ & +[1-A(1+2X+2\mathrm{i}z)(1+2X-2\mathrm{i}z)]\\ & \times\{6(4b_1+4\mathrm{i}b_2+X+5\mathrm{i}z)+8(X+\mathrm{i}z)^3\\ & +3[-1+4(X+\mathrm{i}z)^2]\},\end{aligned} \tag{4.3.9}$$

$$\begin{aligned}\eta_2 =\ & 6(1-2X-2\mathrm{i}z)+[1-A(1-2X-2\mathrm{i}z)\\ & \times(1-2X+2\mathrm{i}z)]\{-6(4b_1+4\mathrm{i}b_2+X+5\mathrm{i}z)\\ & -8(X+\mathrm{i}z)^3+3[-1+4(X+\mathrm{i}z)^2]\}\\ & -A(1-2X-2\mathrm{i}z)(1+2X-2\mathrm{i}z)\\ & \times\{6(4b_1+4\mathrm{i}b_2+X+5\mathrm{i}z)+8(X+\mathrm{i}z)^3\\ & +3[-1+4(X+\mathrm{i}z)^2]\}.\end{aligned} \tag{4.3.10}$$

这里, $A(x,z) = 1/[2+8X^2+8z^2]$; $X = \frac{\sqrt{\pi}b}{2}\mathrm{Erfi}\left[\frac{x}{b}\right]$ $\left(\mathrm{Erfi}(s) = \frac{2}{\sqrt{\pi}}\int_0^s \mathrm{e}^{\tau^2}\mathrm{d}\tau\right)$. b_1 和 b_2 为实常数, 决定了二阶怪波的不同构型. 需要指出的是, 怪波解的阶数越高, 引入的实常数就越多, 对应的高阶怪波结构就越丰富. 当 $|x|\to\infty$ 时, 相应的光学振幅衰减为零, 即 $|u(z,x)|=0$, 上述解就表示局域的高斯型波包. 另外, 通过下文

中对光怪波振幅分布特征的分析 (详见图 4.8~ 图 4.12), 我们可以将方程 (4.3.6) 看成表述怪波在高斯背景上激发的一组精确解.

有趣的是, 与之前研究报道的有理形式的怪波解比较 [4, 12], 这组解包含了变换了的多项式以及高斯函数. 同时需要指出的是, 由上述相似变换方法的推导可知, 这里的系统可调节参数 $D(x)$, $R(x)$ 和 $V(x)$ 是非任意选取的, 其相应的具体表达式如下:

$$D(x) = \frac{1}{2}\exp\left(-2\frac{x^2}{b^2}\right), \tag{4.3.11}$$

$$R(x) = \exp\left(\frac{x^2}{b^2}\right), \tag{4.3.12}$$

$$n_1(x) = \left(\frac{1}{2b^2} - \frac{x^2}{2b^4}\right)\exp\left(-2\frac{x^2}{b^2}\right). \tag{4.3.13}$$

通过上述怪波解的解析表达式, 我们易得光学振幅 $I = |u_j(z,x)|$ 的分布特征. 接下我们将研究 Peregrine 怪波以及高阶怪波在高斯背景上的激发性质以及相应的激发条件.

4.3.2 怪波激发性质

图 4.8 呈现了一阶 (基本的 Peregrine) 光怪波在宽度参数 $b = 6$ 的高斯背景上的振幅分布特征. 我们可以清楚地看到这个非线性波是双重 (分布和演化方向) 局域在一个有限宽背景上. 并且它的最大光学振幅是最大背景值的 3 倍. 也就是说, 经典的 Peregrine 怪波的特征在有限宽的高斯背景上得到了很好的保持. 最近的研究表明, 标准非线性薛定谔方程的有理分式解能够较好地描述怪波的本性, 并且重要的是, 这些数学上的精确解的有效性已经在众多实际物理系统中被观测证实. 这表明, 标准非线性薛定谔方程的有理分式解可以被认为是一类由调制不稳定性诱导的基本的怪波激发元. 更有趣的是, Kibler 等发现, 当怪波的初态不满足理想的数学表达式时也可以在光纤实验平台上实现光怪波的激发 [21]. 这个观点在文献 [48] 中得到了很好的证实. 因此, 通过以上分析, 我们给出的高斯背景上的怪波解不仅是非线性偏微分方程的一个特解, 也代表着一类在局域的非零背景上的怪波激发元.

众所周知的是, 平面波背景上合适的小的幅值调制会诱发怪波现象. 因此, 通过对高斯背景的适当的调制也很有可能激发怪波. 人们可以通过不同的方法对这类激发怪波的方式进行细致的研究. 例如, 人们可以通过数值模拟的方法, 类比平面波上激发怪波的初态, 对高斯背景上激发怪波的初始激励进行分析研究. 接下来, 我们将通过精确的方法对这一问题进行研究. 通过对精确解 (4.3.6) 的分析, 怪波激

4.3 高斯背景上光怪波的激发

发的初始激励可以通过如下表达式给出：

$$u_1 = \frac{\sqrt{(1+4X^2+396)^2+6400}}{1+4X^2+400}\exp\left[-\frac{x^2}{2b^2}+\mathrm{i}\Theta-\mathrm{i}10\right], \quad (4.3.14)$$

其中相对相位 Θ 表示为

$$\Theta = \mathrm{Arccos}\left\{\frac{1+4X^2+396}{\sqrt{(1+4X^2+396)^2+6400}}\right\}. \quad (4.3.15)$$

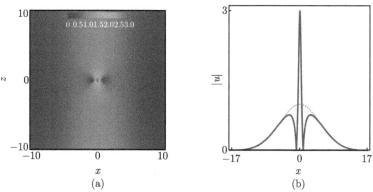

图 4.8 (a) 高斯背景上一阶怪波 (精确解 (4.3.6)) 的光学振幅 $|u_1|$ 分布特征; (b) $|u_1|$ 在中心位置 $z=0$ 的剖面图 (实线) 以及高斯背景振幅 (虚线). 由图可见, 怪波依然双重局域于高斯背景上, 且相应的一阶怪波的最大振幅是高斯背景最大振幅的 3 倍. 这表明怪波的基本性质 (双重局域性以及高峰值性) 在高斯背景上得到很好的保持. 参数选取为: $b=6$(扫描封底二维码可看彩图)

图 4.9 展示了初始激发相应的相位调制和振幅调制的曲线特征. 这里需要指出的是, 这些初态调制参量可以通过文献 [21] 中的相关密度和相位调制装置在实际光学实验中实现. 在图 4.9(a) 中, 实线表示调制的密度信号, 而虚线代表未被调制的高斯背景. 我们可以清楚地看到, 与高斯背景的振幅分布比较, 初始激励的振幅调制是非常小的 (只有在高斯背景较大幅值附近有小的调制). 在图 4.9(b) 中, 与 π 相位比较起来, 我们发现对于相对相位的调制也是比较弱的. 这样, 我们通过上述密度和相位的调制就得到了在高斯背景上激发怪波的初始激发元.

接下来, 我们再来考虑高阶怪波在高斯背景上的激发. 高阶怪波的激发已在不同非线性物理系统中被广泛研究. 特别地, 高阶怪波的实验验证工作取得了重要突破. Frisquet 等 [25] 在单模光纤中证实了二阶怪波的存在. Chabchoub 等 [86] 在水箱中成功激发了二阶到五阶的怪波. 此类高阶怪波可称为 "超级怪波", 意味着具有更高的振幅以及更为复杂的时空分布结构. 这类解在实际非线性物理系统中的实现表

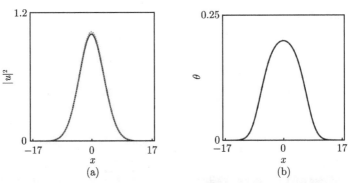

图 4.9 (a) 宽度参数 $b=6$ 的高斯背景上激发光怪波的密度调制曲线 (由方程 (4.3.14) 描述). 其中实线表示高斯背景的密度分布, 虚线代表弱调制的高斯背景的密度分布. (b) 相应的相位调制形式 (由方程 (4.3.15) 描述)

明: 非线性波动方程中的一系列具有特殊性质的非线性激发元本质上表征了实际存在的非线性自然现象. 理论上, 相应的求解高阶怪波解的解析手段主要包括上述达布变换和 Hirota 双线性方法. 目前的研究表明高阶怪波为多个基本 Peregrine 怪波的非线性叠加. 根据相位参数的不同选取, 高阶怪波往往表现出极其丰富的结构. 为简单起见, 我们以二阶怪波解 $u_2(z,x)$ 为例给出相应的光学振幅分布 $I = |u_2(z,x)|$. 如图 4.10 所示, 我们可以清楚地看到二阶的单峰怪波具有更高的光学振幅, 并且它的最大光学振幅可以达到高斯背景振幅最大值的 5 倍. 这个结果和平面波背景上的二阶怪波相一致. 不同的是, 我们的结果提供了另一种激发高阶怪波的方式, 即在实际物理系统中通过制备合适的激发条件将有可能实现怪波在高斯背景上的激发.

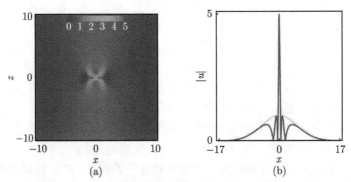

图 4.10 (a) 高斯背景上二阶单峰怪波 (精确解 (4.3.6)) 的光学振幅 $|u_2|$ 分布特征. (b) $|u_2|$ 在中心位置 $z=0$ 的剖面图 (实线) 以及高斯背景振幅 (虚线). 由图可见怪波依然双重局域于高斯背景上, 且相应的二阶怪波的最大振幅是高斯背景最大振幅的 5 倍. 这表明高阶怪波的基本性质 (双重局域性以及高峰值性) 在高斯背景上得到很好的保持. 参数选取为: $b=6$, $b_1=0$, $b_2=0$(扫描封底二维码可看彩图)

4.3 高斯背景上光怪波的激发

图 4.11 中展示了二阶的"三胞胎"怪波 (三个一阶怪波) 在高斯背景上的激发. 由图可见, "三胞胎"怪波仍然可以很好地存在于高斯背景上, 其中位于高斯背景最大振幅位置 $x=0$ 的一阶怪波的幅值可以达到最大背景振幅的 3 倍, 这与标准的一阶怪波情形是完全一致的; 其余两个一阶怪波处于 $x \neq 0$, 相应的最大振幅要小一些. 不过, 有趣的是怪波的最大幅值等于所在位置高斯背景最大振幅的 3 倍. 因此, 该结果表明 "三胞胎" 二阶怪波能够在高斯背景上激发, 且相应的幅值和局域性保持得很好.

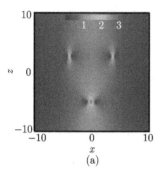 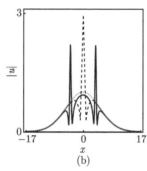

图 4.11 (a) 高斯背景上二阶"三胞胎"怪波 (精确解 (4.3.6)) 的光学振幅 $|u_2|$ 分布特征. (b) $|u_2|$ 在 $z=-5$ 的剖面图 (细实线), $z=2.5$ 的剖面图 (粗实线) 以及高斯背景振幅 (虚线). 由图可见二阶"三胞子"怪波依然存在于高斯背景上, 其中每个怪波的振幅仍然是所在高斯背景位置振幅的 3 倍. 参数选取为: $b=6$, $b_1=0$, $b_2=100$ (扫描封底二维码可看彩图)

接下来, 我们考察在高斯背景的局域性质变化的情况下怪波的激发会有怎样的性质. 有趣的是, 我们发现窄宽高斯背景上怪波仍然能够激发且保持双重局域, 且其峰值依然是高斯背景最大峰值的 3 倍 (对于一阶怪波) 或 5 倍 (对于二阶怪波). 我们以一阶怪波为例, 通过减小参数 b 得到窄宽高斯背景, 从此过程中我们观察一阶怪波的最大峰值变化. 如图 4.12 所示, 高斯背景的半波宽度是略小于怪波的局

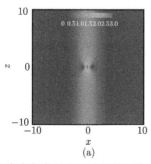 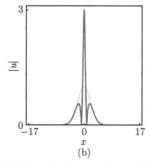

图 4.12 (a) 窄宽高斯背景上一阶怪波 (精确解 (4.3.6)) 的光学振幅 $|u_1|$ 分布特征. (b) $|u_1|$ 在中心位置 $z=0$ 的剖面图 (实线) 以及高斯背景振幅 (虚线). 由图可见怪波在窄宽高斯背景上依然保持其基本性质. 参数选取为: $b=2$ (扫描封底二维码可看彩图)

域尺寸, 但一阶怪波的最大峰值仍然是背景最大值的 3 倍. 怪波在局域的高斯背景上的这种特性为从窄宽背景上获取高强度脉冲或光束提供了一种方式.

4.4 高阶效应诱发光学局域波态转换

本章我们将系统研究高阶效应下光怪波和呼吸子的激发以及态转换问题. 通常情况下, 超短脉冲在光纤中传输需要考虑高阶效应的影响, 这些高阶效应主要包括三阶色散、自陡峭以及延迟的非线性响应. 此时, 高阶效应显著地影响局域波的性质. 因此, 我们就要回答两个问题: ①高阶效应下光怪波和呼吸子是否能够激发; ②若能激发, 会有怎样新的性质. 对于前者, Akhmediev 等[47]已经证实了光怪波和呼吸子在高阶效应下的存在性. 对于后者, 之前的研究表明高阶效应下光怪波和呼吸子表现出结构的多样性[47−59]. 本节将基于局域波精确解和调制不稳定性相结合的研究方法探究高阶效应下光怪波和光呼吸子的特性. 研究表明, 怪波和呼吸子在高阶效应下可以与其他种类的非线性波进行态转换. 我们详细地分析了相应态转换的物理性质和机制, 并论证了态转换严格存在于考虑高阶效应的情形下, 标准的非线性薛定谔系统不存在这样的态转换.

最近, Peregrine 怪波、Akhmediev 呼吸子、Kuznetsov-Ma 呼吸子以及相应的非线性叠加态的实验观测在多种不同的非线性物理系统中取得了重要进展, 包括非线性光纤[21, 25]、水槽[86, 105]以及等离子体[106]. 目前研究表明, 这些局域在平面波背景上的非线性结构是不同扰动调制下调制不稳定性的结果[10, 17]. 其中 Peregrine 怪波和 Akhmediev 呼吸子的激发分别对应着初始的单峰和周期小振幅扰动; Kuznetsov-Ma 呼吸子对应着初始的强振幅扰动; 而相应局域波的非线性叠加态对应着高阶的调制不稳定性[20]. 特别是最近的研究表明, Peregrine 怪波只存在于调制不稳定区中的一个特殊子区域 —— 零频不稳定区[61, 101, 102]. 就这点而言, 标准的非线性薛定谔系统描述的单模光纤中光怪波具有一致的结构 (其整体结构在振幅大小不同的背景上同比例地放大或缩小, 但基本结构保持不变; 在取定背景波振幅大小的情况下, 改变背景波频率对怪波的结构没有任何影响). 究其本质原因是, 标准的非线性薛定谔方程的零频调制不稳定性增长率随背景波频率的变化始终保持为一恒定值, 见图 4.13. 然而, 在特定的物理情境下考虑一些必不可少的非线性物理效应, 如交叉相位调制效应[63]以及高阶扰动项[107], 调制不稳定性往往呈现出一些不同于标准非线性薛定谔情况的新特征. 因此, 从调制不稳定性的新特征的角度出发, 研究相应的局域波动力学性质, 将是一个非常重要的课题.

最近的研究证实了光怪波和呼吸子在高阶效应下呈现出结构的多样性[47−59], 并且这些特征是无法用标准的非线性薛定谔方程来描述的. 例如, 国际上, Akhmediev 小组[52]研究发现飞秒光怪波能够呈现出 "倾斜" 和 "劈裂" 的现象, 并通过

4.4 高阶效应诱发光学局域波态转换

数值模拟验证了倾斜怪波和双峰怪波的鲁棒性. 文献 [48] 解析和数值研究了飞秒级光怪波的形成和激发性质. 文献 [56] 报道了发现光怪波和呼吸子在高阶效应下的尺寸压缩效应. 我们之前的研究发现高阶效应会造成光学怪波相位的对称性破缺 [55]. 需要指出的是, 以上研究的高阶效应下的光怪波都处于调制不稳定区域. 据我们所知, 鲜有人通过结合相应的调制不稳定性特征对怪波和呼吸子的演化性质进行系统的分析研究. 我们将通过精确解与调制不稳定性分析相结合的研究方法来揭示光怪波和呼吸子在高阶效应下的新特性. 由调制不稳定分析我们首先发现, 当考虑以下必要的高阶效应 (三阶色散和延迟的非线性响应项) 时, 调制不稳定性的增长率在低扰动频率区域呈现出有趣的非均匀分布特征. 特别是当背景频率改变时, 调制不稳定区域中出现了一个调制稳定的子区域 (详见图 4.13). 因此, 我们期望当怪波和呼吸子由不稳定区域演化且接近这个稳定区域的时候会出现一些有趣的物理现象.

超短脉冲在光纤中传输需要考虑高阶效应的影响, 这些高阶效应主要包括三阶色散、自陡峭以及延迟的非线性响应. 考虑这些高阶效应, 描述局域波动力学的有效物理模型为高阶非线性薛定谔方程 [1]. 其无量纲形式可表示如下:

$$i\frac{\partial u}{\partial z} + \frac{1}{2}\alpha\frac{\partial^2 u}{\partial t^2} + \gamma|u|^2 u - i\beta\frac{\partial^3 u}{\partial t^3} - is\frac{\partial(|u|^2 u)}{\partial t} - i\delta u\frac{\partial |u|^2}{\partial t} = 0, \quad (4.4.1)$$

其中, $u(z,t)$ 表示电场包络, z 是传输距离, t 是延迟时间; α 和 γ 分别是二阶色散和交叉相位调制系数; β 是三阶色散系数; s 是自陡峭效应系数; δ 是延迟非线性响应参数. 具有一般参数的方程 (4.4.1) 是不可积的, 不过考虑一定的参数约束的情况下, 方程 (4.4.1) 涵盖了以下三类可积的高阶非线性薛定谔模型. 具体情况如下:

(1) 考虑 $\alpha = \gamma, s = 6\beta, s + \delta = 0$, 方程 (4.4.1) 退化为著名的 Hirota 模型:

$$i\frac{\partial u}{\partial z} + \frac{1}{2}\alpha\frac{\partial^2 u}{\partial t^2} + \alpha|u|^2 u - i\beta\left(\frac{\partial^3 u}{\partial t^3} + 6|u|^2\frac{\partial u}{\partial t}\right) = 0. \quad (4.4.2)$$

由于可积性的保证, 该模型已被广泛地用于解析研究, 主要包括经典的孤子 (亮孤子和暗孤子) 以及最近的热点, 即呼吸子和怪波 [48-51]. 研究表明, 在 Hirota 模型下高阶效应会引起局域波传输速度的改变.

(2) 考虑 $\alpha = \gamma, s = 6\beta, \delta = -3\beta$, 方程 (4.4.1) 变为著名的 Sasa-Satsuma 模型:

$$i\frac{\partial u}{\partial z} + \frac{1}{2}\alpha\frac{\partial^2 u}{\partial t^2} + \alpha|u|^2 u - i\beta\left(\frac{\partial^3 u}{\partial t^3} + 6|u|^2\frac{\partial u}{\partial t} + 3u\frac{\partial |u|^2}{\partial t}\right) = 0. \quad (4.4.3)$$

该系统中的局域波在高阶效应下往往能够表现出非平庸特征. 相较于标准结构的局域波 (经典的孤子、Peregrine 怪波以及呼吸子), 最近的研究表明 Sasa-Satsuma 模型中存在着有趣的双峰局域波结构, 如双峰亮孤子、双峰怪波以及双峰呼吸子 [52-54,108].

(3) 考虑 $\alpha = \gamma$, $\beta = \delta = 0$, 方程 (4.4.1) 为只含有自陡峭效应的高阶非线性薛定谔模型:

$$i\frac{\partial u}{\partial z} + \frac{1}{2}\alpha\frac{\partial^2 u}{\partial t^2} + \alpha|u|^2 u - is\frac{\partial(|u|^2 u)}{\partial t} = 0. \tag{4.4.4}$$

该模型中的局域波在高阶效应下性质的研究主要集中于孤子和怪波, 其中自陡峭效应对亮孤子有明显的峰值压缩作用 [109]. 对于怪波, 最近人们发现自陡峭效应可以使怪波转化为一系列孤子对 [74].

本节主要研究情形 (1) Hirota 模型中的多种不同种类的局域波的态转换, 并揭示相应的物理机制. 这里需要注意的是三阶色散对局域波的性质起着重要作用. 这是因为该效应从本质上改变了平面波背景的色散关系, 意味着我们无法通过平庸的伽利略变化将背景频率抹掉. 就这点而言, 背景波频率的改变往往能诱发局域波新的性质 (详见下文调制不稳定性分析和态转换研究).

4.4.1 调制不稳定性分析

接下来, 为了揭示光怪波和呼吸子在高阶效应下丰富有趣的物理性质, 我们首先将注意力集中于标准的线性稳定性分析. 通过线性稳定性分析, 我们将定性地初步了解高阶效应对不同扰动的非线性响应. 之后, 我们将根据调制不稳定性的分析结果, 结合解析解对光怪波和呼吸子的性质进行详细研究. 下面我们具体讨论线性稳定性分析过程. 首先, 我们取如下具有一般形式的背景波解:

$$u_0(z,t) = a\exp[i\theta(z,t)], \tag{4.4.5}$$

其中

$$\theta(z,t) = qt + [a^2 - q^2/2 + \beta(6qa^2 - q^3)]z, \tag{4.4.6}$$

a 和 q 分别表示背景波的振幅和频率. 接下来, 我们考虑如下小振幅扰动的非线性背景波:

$$u_p(z,t) = [a + p(z,t)]\exp[i\theta(z,t)], \tag{4.4.7}$$

其中, $p(z,t)$ 是满足特定线性方程的微扰. 将扰动平面波解 (4.4.7) 代入方程 (4.4.2) 进行标准的线性化, 易得如下线性关系:

$$2a^2\alpha p + 12a^2 q\beta p + 2a^2\alpha p^* + 12a^2 q\beta p^* + 2ip_z + 2iq\alpha p_t - 12ia^2\beta p_t$$
$$+6iq^2\beta p_t + (\alpha + 6q\beta)p_{tt} - 2i\beta p_{ttt} = 0, \tag{4.4.8}$$

其中, p^* 为 p 的复共轭, 下脚标为相应的偏导. 通常情况下, 扰动项 $p(z,t)$ 可做如下展开:

$$p(z,t) = \eta_{1,s}(z)e^{iQt} + \eta_{1,a}(z)e^{-iQt}. \tag{4.4.9}$$

4.4 高阶效应诱发光学局域波态转换

将式 (4.4.9) 代入线性关系式 (4.4.8) 并对 e^{iQt} 和 e^{-iQt} 项进行分离,可得

$$\begin{cases} M_1\eta_{1,a} + N\eta_{1,s}^* + 2i\eta_{1,a}' = 0, & (4.4.10a) \\ M_2\eta_{1,s} + N\eta_{1,a}^* + 2i\eta_{1,s}' = 0, & (4.4.10b) \end{cases}$$

其中

$$M_1 = (2a^2 + 2qQ - Q^2)\alpha + 12a^2\beta(q-Q) + 6qQ\beta(q-Q) + 2qQ^3\beta, \quad (4.4.11)$$

$$M_2 = (2a^2 - 2qQ - Q^2)\alpha + 12a^2\beta(q+Q) - 6qQ\beta(q+Q) - 2qQ^3\beta, \quad (4.4.12)$$

$$N = 2a^2(\alpha + 6q\beta). \quad (4.4.13)$$

这里,"$'$" 表示关于 z 的导数. 上式亦可表示为

$$2\boldsymbol{\eta}' = \mathrm{i}\boldsymbol{M}\boldsymbol{\eta}, \quad (4.4.14)$$

其中, $\eta = (\eta_{1,s}, \eta_{1,a}^*)^{\mathrm{T}}$, 矩阵 $\boldsymbol{M}$ 为

$$\boldsymbol{M} = \begin{pmatrix} M_2 & N \\ -N & M_1 \end{pmatrix}. \quad (4.4.15)$$

对任意实频率 Q, 扰动项 $\eta(z)$ 为指数 $\exp(i\omega_j z)$ 的线性组合,其中 ω_j $(j=1,2)$ 满足矩阵 $\boldsymbol{M}$ 的特征多项式,

$$B(\omega) = \omega^2 + B_1\omega + B_0, \quad (4.4.16)$$

其中

$$B_1 = 4qQ\alpha - 24a^2Q\beta + 12q^2Q\beta + 4Q^3\beta, \quad (4.4.17)$$

$$B_0 = 4a^2Q^2\alpha^2 + 4q^2Q^2\alpha^2 - Q^4\alpha^2 + 24q^3Q^2\alpha\beta - 4qQ^4\alpha\beta$$
$$+ 144a^4Q^2\beta^2 + 36q^4Q^2\beta^2 - 48a^2Q^4\beta^2 - 12q^2Q^4\beta^2 + 4Q^6\beta^2. \quad (4.4.18)$$

考虑 $Q \to 0$ 时调制不稳定性特征,我们作如下变换:

$$B(Q\omega) = Q^2 b(\omega). \quad (4.4.19)$$

相应的 $b(\omega)$ 为

$$b(\omega) = \omega^2 + b_1\omega + b_0. \quad (4.4.20)$$

其中

$$b_1 = 4q\alpha - 24a^2\beta + 12q^2\beta + 4Q^2\beta, \quad (4.4.21)$$

$$b_0 = 4a^2\alpha^2 + 4q^2\alpha^2 - Q^2\alpha^2 + 24q^3\alpha\beta - 4qQ^2\alpha\beta$$
$$+ 144a^4\beta^2 + 36q^4\beta^2 - 48a^2Q^2\beta^2 - 12q^2Q^2\beta^2 + 4Q^4\beta^2. \quad (4.4.22)$$

此时我们可得如下色散关系：

$$\omega = -2q\alpha + 12a^2\beta - 6q^2\beta - 2Q^2\beta \pm \sqrt{(Q^2-4a^2)(\alpha+6q\beta)^2}. \quad (4.4.23)$$

如果 $\text{Im}\{\omega\} < 0$，调制不稳定性存在于 $-2a < Q < 2a$ 的扰动频率区域。通常，我们定义调制不稳定性增长率为 $G = |\text{Im}\{\omega\}|$。如果 $G > 0$，那么初始的小扰动信号将会以指数形式 $\exp(Gz)$ 增长放大。

如图 4.13 所示，我们给出调制不稳定性增长率在扰动频率-背景频率 (Q,q) 平面上的分布特征。我们首先考虑不存在高阶效应的情况 $(\beta = 0)$，即标准的非线性薛定谔模型的调制不稳定特征。从图 4.13(a) 可以看出，调制不稳定性存在于区域 $-2a < Q < 2a$，当 q 变化时，增长率的大小始终保持不变。然而，当考虑高阶效应 $(\beta \neq 0)$ 时，调制不稳定性的性质发生显著改变。我们看到调制不稳定性增长率随着背景频率 q 改变呈现出不均匀的分布特征。如图 4.13(b) 所示，调制不稳定性的增长率是关于直线 $q = q_s = -\alpha/(6\beta)$ 对称分布。需要注意的是，这条直线对应着低频扰动区中增长率为零的区域。因此我们将这条直线描述的区域称为"调制稳定区域"。此外，增长率的大小与 $|q - q_s|$ 的大小密切相关；$|q - q_s|$ 值越大，增长率越大；$|q - q_s|$ 值越小，增长率越小。此时，我们可以精确地给出零频增长率的表达式，如下：

$$G_0 = 2a\left|\frac{q-q_s}{q_s}\right|. \quad (4.4.24)$$

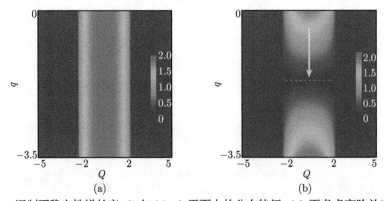

图 4.13 调制不稳定性增长率 G 在 (Q,q) 平面上的分布特征，(a) 不考虑高阶效应的情况 $(\beta = 0)$；(b) 考虑高阶效应的情况 $(\beta = 0.1)$。我们可以清楚地看到，当背景频率 q 变化时高阶效应极大地影响了调制不稳定性增长率的分布特征。这里虚线表示调制稳定区域 $-2a < Q < 2a$ 中的调制稳定区，其对应的精确描述为：$q = q_s = -\alpha/(6\beta)$。零频扰动子区域上的箭头表示怪波从调制不稳区到调制稳定区的演化过程，简单表示即为 $q \to q_s$。参数设置为：$a = 1$(扫描封底二维码可看彩图)

鉴于近期研究进展的结论，怪波只存在于零频不稳定扰动区域[61, 101, 102]，因此, 研究怪波在增长率非均匀分布的零频扰动区域会有怎样的物理性质, 将会是一个非常有趣的课题. 同样, 我们已知呼吸子存在于调制不稳定区, 那么呼吸子在 $q = q_s = -\alpha/(6\beta)$ 处会有怎样的物理性质, 也值得去研究.

4.4.2 局域波精确解构造

为研究上述问题, 我们需构造相应模型平面波背景上的局域波解. 接下来, 我们将根据绪论中介绍的达布变换方法构造 Hirota 方程 (4.4.2) 的局域波精确解. Hirota 方程相应的 Lax pair 为

$$\begin{cases} \boldsymbol{\Phi}_t = \boldsymbol{U}\boldsymbol{\Phi}, & (4.4.25\text{a}) \\ \boldsymbol{\Phi}_z = \boldsymbol{V}\boldsymbol{\Phi}. & (4.4.25\text{b}) \end{cases}$$

这里

$$\boldsymbol{U} = \lambda \boldsymbol{U}_0 + \boldsymbol{U}_1, \tag{4.4.26}$$

$$\boldsymbol{V} = \lambda^3 \boldsymbol{V}_0 + \lambda^2 \boldsymbol{V}_1 + \lambda \boldsymbol{V}_2 + \boldsymbol{V}_3, \tag{4.4.27}$$

其中矩阵 $\boldsymbol{U}_0$ 和 $\boldsymbol{U}_1$ 为

$$\boldsymbol{U}_0 = \begin{pmatrix} -\mathrm{i} & 0 \\ 0 & \mathrm{i} \end{pmatrix}, \quad \boldsymbol{U}_1 = \begin{pmatrix} 0 & u \\ -u^* & 0 \end{pmatrix}. \tag{4.4.28}$$

矩阵 $\boldsymbol{V}_0, \boldsymbol{V}_1, \boldsymbol{V}_2, \boldsymbol{V}_3$ 为

$$\boldsymbol{V}_0 = -4\beta \boldsymbol{U}_0, \quad \boldsymbol{V}_1 = \boldsymbol{U}_0 - 4\beta \boldsymbol{U}_1, \tag{4.4.29}$$

$$\boldsymbol{V}_2 = \begin{pmatrix} -2\mathrm{i}\beta|u|^2 & u - 2\mathrm{i}\beta u_t \\ -u^* - 2\mathrm{i}\beta u_t^* & 2\mathrm{i}\beta|u|^2 \end{pmatrix}, \tag{4.4.30}$$

$$\boldsymbol{V}_3 = \begin{pmatrix} \dfrac{\mathrm{i}}{2}u^2 + \beta(u^*u_t - uu_t^*) & \dfrac{\mathrm{i}}{2}u_t + \beta(u_{tt} + 2u^2 u) \\ \dfrac{\mathrm{i}}{2}u_t^* - \beta(u_{tt}^* + 2u^2 u^*) & -\dfrac{\mathrm{i}}{2}u^2 - \beta(u^*u_t - uu_t^*) \end{pmatrix}. \tag{4.4.31}$$

其中, $\lambda = \mathrm{i}a_1 - q_1/2$ 为谱参量, a_1 和 q_1 为实常数, $a_1 \neq 0$. 上述 Lax pair 将非线性求解问题转化为线性方程组的求解问题. 由零曲率方程 (可积条件) $\boldsymbol{U}_z - \boldsymbol{V}_t + [\boldsymbol{U}, \boldsymbol{V}] = 0$, 我们可以得到 Hirota 方程 (4.4.2). 具体的达布变换形式为

$$u(z,t) = u_0(z,t) - \frac{2\mathrm{i}(\lambda - \lambda^*)\Phi_1(z,t)\Phi_2^*(z,t)}{|\Phi_1(z,t)|^2 + |\Phi_2(z,t)|^2}. \tag{4.4.32}$$

这里, $\Phi_1(z,t)$ 和 $\Phi_2(z,t)$ 为相应的 Lax pair(4.4.25) 在 $u(z,t) = u_0(z,t)$ 时的解.

接下来, 我们首先将矩阵 U 转换为常数矩阵, 其相应的本征值方程为

$$\tau^2 + a^2 + (\lambda + q/2)^2 = 0. \tag{4.4.33}$$

可以看到, Hirota 系统中得到的本征值方程与标准的非线性薛定谔系统中的本征值方程是一样的. 我们在绪论中已经介绍了平面波背景上局域波的一般构造方法, 接下来将主要强调针对本章节态转换的研究在构造局域波解时需要注意的几点内容. 我们已经知道本征值方程 (4.4.33) 的两个根 $\tau_{1,2} = \pm\mathrm{i}\sqrt{a^2 + (\lambda + q/2)^2}$ 是否具有重根将决定最后的局域波类型. 具体如下:

(1) 当 $\tau_1 \neq \tau_2$ 时, 即 $\lambda \neq \mathrm{i}a - q/2$, 我们将得到具有一般形式呼吸子解, 进一步我们可以研究呼吸子与其他种类局域波的态转换.

(2) 当考虑一种特殊情况时, 即 $\tau_1 = \tau_2 = 0$(因此 $\lambda = \mathrm{i}a - q/2$), 我们最终得到了 Hirota 系统中具有一般形式的怪波解, 可以方便地研究怪波与孤子间的态转换.

我们在绪论中已经指出在求解相应 Lax pair 过程中相似矩阵 D 的选取不是唯一的. 不同形式的矩阵 D 决定了所求局域波的中心位置是否处于 $(z,t) = (0,0)$ 点处. 这里需要着重指出的是, 对于怪波解而言, 不同形式的矩阵 D 改变了怪波的中心位置, 对怪波与孤子的态转换性质却没有影响, 即中心位置分布不同的怪波能够转换为一类相同结构的孤子. 因此对于这方面的研究, 我们将取与绪论中矩阵 D 相同的矩阵表达式, 即怪波的中心位置位于 $(z,t) = (0,0)$ 处.

然而对于呼吸子解而言, 不同形式的矩阵 D 对呼吸子与其他种类非线性波的态转换具有显著影响. 因此, 我们取如下两种不同的相似矩阵 D:

$$D_1 = \begin{pmatrix} \sqrt{-\mathrm{i}(\lambda + q/2) + \tau_1} & \sqrt{-\mathrm{i}(\lambda + q/2) + \tau_2} \\ -\sqrt{-\mathrm{i}(\lambda + q/2) - \tau_1} & -\sqrt{-\mathrm{i}(\lambda + q/2) - \tau_2} \end{pmatrix}. \tag{4.4.34}$$

$$D_2 = \begin{pmatrix} 1 & 1 \\ \dfrac{1}{a}(\mathrm{i}\lambda + \tau_1) & \dfrac{1}{a}(\mathrm{i}\lambda + \tau_2) \end{pmatrix}. \tag{4.4.35}$$

其中, D_1 和 D_2 分别对应下文中的呼吸子解 u_1 和 u_2. 我们发现一系列有趣的结果:

(1) 不同 D 矩阵可诱发 Kuznetsov-Ma 呼吸子转换为 W 型孤子或反暗孤子, 而 Akhmediev 呼吸子只能转换为周期波.

(2) 对于一般的呼吸子而言, 不同 D 矩阵可诱发呼吸子至具有横向振幅对称分布和非对称分布的多峰孤子的态转换.

下文中我们将详细讨论多种不同种类局域波的态转换问题.

4.4.3 一阶怪波与孤子的态转换

本节我们讨论怪波和孤子间的态转换, 包括一阶怪波和 W 型孤子的态转换以及二阶怪波和新的孤子叠加结构的态转换. 我们首先考虑一阶怪波在增长率非均匀分布的零频扰动区域会有怎样的物理性质. 为此, 我们把注意力转移到求解平面波背景上的有理分式的局域波精确解. 利用绪论中介绍的达布变换求解怪波解方法, 一阶怪波解一般且简洁的解析表达式为

$$u_{1rw}(z,t) = u_0 \left[\frac{4 + 8\mathrm{i}a^2(1 - q/q_s)\xi}{1 + 4a^4(1 - q/q_s)^2\xi^2 + 4a^2(\tau - \upsilon\xi)^2} - 1 \right], \quad (4.4.36)$$

其中, $\upsilon = q + (2a^2 - q^2)/(2q_s)$, $\xi = z - z_0$, $\tau = t - t_0$, z_0 和 t_0 是用来决定这组解中心位置的任意实数. 当 $z_0 = t_0 = 0$ 时, 怪波的中心位置处于 $(z,t) = (0,0)$ 点.

有趣的是, 根据平面波背景频率 q 具体的取值, 这组解包含了两类动力学演化特性截然不同的局域波. 具体为, 在 $q \neq q_s$ 的情况下, 这类非线性局域波是在 (z,t) 平面上双重局域化, 并且具有一个强度极高的单峰和两个振幅零点. 这类非线性波已经被认定为解释怪波现象的理论原型 (详见图 4.14(a) 和 (b)). 的确如此, 考虑上述解的特殊参数情况, 即 $a = 1$ 和 $q = 0$, 上述解 (4.4.36) 退化为文献 [48] 中的标准的 Hirota 一阶怪波解. 然而在 $q = q_s$ 情况下, 此时的表达式表示一种具有孤子演化性质的行波 (详见图 4.14(c)). 需要注意的是, 与上述调制不稳定性的分析比较, 我们可以清楚地发现, 这类孤子刚好对应着调制稳定区. 其相应的表达式如下:

$$u_s(z,t) = a\mathrm{e}^{\mathrm{i}\theta_s} \left[\frac{4}{1 + 4a^2(\tau - \upsilon_s\xi)^2} - 1 \right], \quad (4.4.37)$$

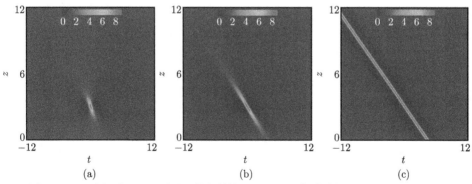

图 4.14 一阶怪波至 W 型孤子的态转换 $|u_{1rw}(z,t)|^2$(精确解 (4.4.36)): (a)$q = 0$, (b)$q = q_s/2$, (c)$q = q_s$. 其他参数设置为: $a = 1$, $\beta = 0.1$, $z_0 = 3$, $t_0 = 0$. 此态转换过程严格对应于怪波由调制不稳定区到调制稳定区的演化过程 (见图 4.13(b) 中箭头及相应描述). 这里 (a) 和 (b) 描绘了高阶效应下经典的具有倾斜结构的怪波 (具有非零速度), 而 (c) 展示了一个新型的 W 型有理分式的孤子 (扫描封底二维码可看彩图)

其中, $\theta_s = q_s\tau - q_s^2\xi/3$, $\upsilon_s = (2a^2 + q_s^2)/(2q_s)$.

需要注意的是, 方程 (4.4.37) 所描述的行波所具有的振幅结构性质与我们常见的经典的孤子有着较大的不同. 我们可以从图 4.14(c) 中看到, 这类孤子是在平面波背景上传输演化的, 它具有一个稳定演化的波峰 ($|u_s|_p = 3a$) 和两个振幅为零且同样稳定演化的波谷 ($|u_s|_v = 0$). 因此我们称之为 "W 型孤子". 值得强调的是, 由于条件 $q = q_s = -\alpha/(6\beta)$ 的约束, 这类 W 型孤子是不能存在于标准非线性薛定谔系统的. 因此, 我们可以初步判断这类波是高阶效应下所特有的局域波类型.

需要指出的是, 在高阶非线性薛定谔模型的研究中, 人们通过设置孤子解的形式 (Tanh 和 Sech 函数相结合), 精确给出具有相似结构的孤子解[110]. 与之不同的是, 我们得到的是有理分式的 W 型孤子解. 有趣的是, 人们已经在 Sasa-Satsuma 模型中得到了一组有理分式的 W 型孤子精确解[54]. 此外, 通过求解变系数的 Ginzburg-Landau 方程, 人们得到多种不同结构的孤子[111].

图 4.14 给出了非线性波由调制不稳定区到调制稳定线的态转换特征. 这个过程恰好对应于图 4.13 中描述调制不稳定增长率分布的箭头. 当 $q \to q_s$ 时, 我们可以看到怪波的分布结构变得越来越细长, 这个过程正好对应着调制不稳定性增长率的变小; 当 $q = q_s$ 时, 怪波转换为一个 W 型的孤子, 对应着增长率恰好衰减为零.

以上我们展示了一阶怪波和 W 型孤子的确存在态转换. 那么如何刻画和理解这样新颖的怪波至孤子的态转换的物理机制呢? 此外, 态转换是否稳定, 其光谱又有怎样的特征? 接下来我们将详细回答这些问题.

1. 态转换性质刻画和物理机制分析

接下来, 为了更好地理解上述一阶怪波和 W 型孤子间的态转换性质, 我们定义了局域波分布的长宽比来描述局域波转换过程的局域性变化. 从精确解 (4.4.36) 出发, 局域波长宽比表达式可以表示为

$$\frac{\Delta L}{\Delta W} = \frac{1}{\sqrt{3}a}\left|\frac{q_s}{q - q_s}\right|, \quad (4.4.38)$$

这里, ΔL 表示局域波长度, 由背景波上峰值演化的半值距离来定义; ΔW 为相应的宽度, 代表着局域波振幅零点间的距离. 如图 4.15 所示, 我们可以看到局域波长宽比的曲线是关于 $q = q_s$ 对称的. 当 $q \to q_s$ 时, 长宽比 $\frac{\Delta L}{\Delta W}$ 急剧增大; 当 $q = q_s$ 时, 其值趋于无限大. 这也意味着, W 型孤子的出现与怪波在传输方向上的退局域化过程紧密相关, 并且这种退局域化是由于高阶效应的影响.

为了揭示上述态转换的物理机制, 接下来详细分析非线性波的局域化特征和相应的调制不稳定性增长率之间的关系. 一个非常有趣的发现是, 局域波的局域化

4.4 高阶效应诱发光学局域波态转换

特征 (上文中定义的局域波长宽比 $\frac{\Delta L}{\Delta W}$) 结果是与调制不稳定性零频增长率的导数 $(1/G_0)$ 严格正相关, 即 $\frac{\Delta L}{\Delta W} \sim \frac{1}{G_0}$, 如图 4.15 所示. 值得注意的是, 怪波与零频调制不稳定性的精确关系在自散焦矢量非线性薛定谔系统中的研究取得了重要进展, Baronio 等 [61, 101] 严格地证明了怪波只存在于调制不稳定零频区域. 之后, 赵立臣等 [102] 在标准的非线性薛定谔系统中建立了多种非线性局域波与调制不稳定性的定量关系. 这里我们的结果的重要性表现在首次建立了怪波和孤子间态转换特征与相应增长率性质之间的严格关系. 这个结果极大地丰富了我们对调制不稳定性特征和怪波演化性质关系的理解; 同时, 也给出了解释怪波与孤子之间关系的一种理论方案. 此外, 这个结果也从侧面证明了怪波处在零频调制不稳定区的正确性.

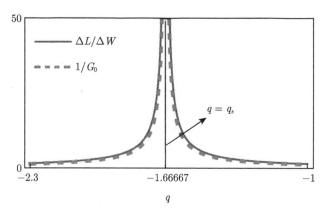

图 4.15 非线性波的局域化特征 $\Delta L/\Delta W$ 的演化分布 (实线) 和零频调制不稳定性增长率的倒数 $1/G_0$ (虚线)

2. 态转换的稳定性和光谱特征

为了检验上述态转换的鲁棒性, 我们运用经典的分步傅里叶方法对方程 (4.4.2) 进行了大量的数值模拟. 这里, 我们选取 $z=0$ 的精确解外加 0.05% 的随机白噪声扰动作为传输初态:

$$u_{in} = u_{1rw}(z_0, t)[1 + 0.005 \text{random}(t)]. \quad (4.4.39)$$

如图 4.16 所示, 在前六个传播距离中, 数值模拟结果与精确解极度吻合; 在此之后, 怪波呈现出劈裂态; 而 W 型孤子保持与初始状态相一致的稳定传输状态. 需要指出的是, 怪波的劈裂态在文献 [48] 中也被报道, 其根本机制是调制不稳定性的影响. 从数值模拟与精确解的比较结果来看, 我们报道的一阶怪波和 W 型孤子的态转换具有鲁棒性. 此外, 从中也可以看出怪波是不稳定 (演化中呈现的劈裂态), 而 W 型孤子则表现出稳定的传输特征 (稳定传输超过 12 个传播距离).

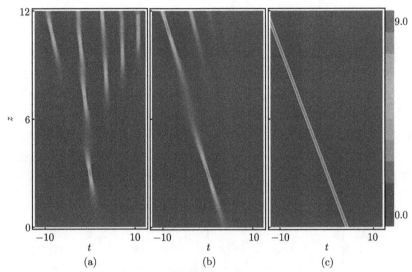

图 4.16　图 4.14 中一阶怪波和 W 型孤子态转换的数值模拟验证. 其初态为方程 (4.4.39) 加 0.05% 的随机噪声. 由数值和解析态转换的结果比较来看. 研究所得态转换具有一定抗扰性. 其中怪波在传播六个单位后出现 "劈裂" 现象是由于调制不稳定性影响；W 型孤子保持稳定是因为处在调制稳定区 (扫描封底二维码可看彩图)

众所周知, Peregrine 怪波谱具有独特的三角结构且在最大峰值处急剧展宽 (图 4.17(a) 和 (b))[21]. 这里, 我们考虑上述态转换的谱演化特征. 据上述详细分析, 可知怪波与孤子间的态转换与高阶效应下的背景波频率有密切关系, 而与背景波振幅大小无关. 因此, 为简单起见我们固定 $a = 1$, 根据标准的傅里叶变换:

$$F(\omega, z) = \frac{1}{\sqrt{2\pi}} \int_{-\infty}^{\infty} u(z,t) \exp(-\mathrm{i}\omega t) \mathrm{d}t, \tag{4.4.40}$$

局域波态转换的谱强度可由如下解析表达式描述:

$$|F(\omega, z)|^2 = 2\pi \exp\left[-|\omega'|\sqrt{1 + 4(q-q_s)^2 \xi^2 / q_s^2}\right], \tag{4.4.41}$$

其中, $\omega' = \omega + q$. 如图 4.17 所示, 在高阶效应下的怪波谱仍然具有典型的三角形结构特征, 并且在振幅强度最大值处 ($z_0 = 3$) 最大程度地展宽. 这与文献 [21] 中报道的怪波谱演化特征是完全一致的. 当 $q \to q_s$ 时, 怪波谱的最大谱宽保持不变, 但沿着传播距离方向 z 的展宽速率逐渐减小. 当态转换发生时, 怪波谱转变为行波谱. 与怪波谱比较, 此时的光谱结构具有较大谱宽且谱宽大小沿传输方向保持不变 (图 4.17 (c)).

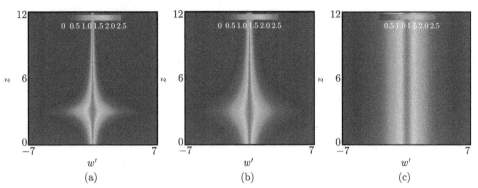

图 4.17 一阶局域波态转换 (图 4.14) 相应的谱特征 $|F(\omega,z)|$(精确解 (4.4.41)). 这里 (a) 和 (b) 呈现了典型的具有三角形特征的怪波谱. 由图可见, 怪波谱在峰值最大压缩处 ($z_0 = 3$) 极大地展宽, 而 (c) 展示了 W 型孤子稳定的宽谱特征, 其宽度与怪波最大展宽宽度一致 (扫描封底二维码可看彩图)

4.4.4 二阶怪波与孤子的态转换

接下来我们将讨论二阶怪波与孤子的态转换的特征. 我们首先根据上文中介绍的达布变换方法给出二阶怪波解. 考虑到背景振幅 a 对局域波的性质没有本质影响, 因此为了简洁起见我们取 $a=1$, 相应的二阶怪波解为

$$u_{2rw}(z,t) = u_{1rw}(z,t) + \frac{4\eta_1(z,t)\eta_2^*(z,t)}{|\eta_1(z,t)|^2 + |\eta_2(z,t)|^2} \exp[\mathrm{i}\theta(z,t)], \tag{4.4.42}$$

$$\begin{aligned}
\eta_1 = & 6[1 + 2t - 2z(-\mathrm{i} + q - 6\beta - 6\mathrm{i}q\beta + 3q^2\beta)] \\
& - A(t,z)[1 + 2t - 2z(-\mathrm{i} + q - 6\beta - 6\mathrm{i}q\beta + 3q^2\beta)] \\
& \times [1 - 2t + 2z(\mathrm{i} + q - 6\beta + 6\mathrm{i}q\beta + 3q^2\beta)] \\
& \times [-6(4b_1 + 4\mathrm{i}b_2 + t + 5\mathrm{i}z - qz + 38z\beta + 30\mathrm{i}qz\beta - 3q^2z\beta) \\
& - 8\{t + z[\mathrm{i} + q(-1 + 6\mathrm{i}\beta) + 6\beta - 3q^2\beta]\}^3 \\
& + 3(-1 + 4\{t + z[\mathrm{i} + q(-1 + 6\mathrm{i}\beta) + 6\beta - 3q^2\beta]\}^2)] \\
& + \{1 - A(t,z)[1 + 2t - 2z(-\mathrm{i} + q - 6\beta - 6\mathrm{i}q\beta + 3q^2\beta)] \\
& \times [1 + 2t - 2z(\mathrm{i} + q - 6\beta + 6\mathrm{i}q\beta + 3q^2\beta)]\} \\
& \times [6(4b_1 + 4\mathrm{i}b_2 + t + 5\mathrm{i}z - qz + 38z\beta + 30\mathrm{i}qz\beta - 3q^2z\beta) \\
& + 8\{t + z[\mathrm{i} + q(-1 + 6\mathrm{i}\beta) + 6\beta - 3q^2\beta]\}^3 \\
& + 3(-1 + 4\{t + z[\mathrm{i} + q(-1 + 6\mathrm{i}\beta) + 6\beta - 3q^2\beta]\}^2)], \tag{4.4.43}
\end{aligned}$$

$$\begin{aligned}
\eta_2 = & 6[1-2t+2z(-\mathrm{i}+q-6\beta-6\mathrm{i}q\beta+3q^2\beta)] \\
& +\{1-A(t,z)[1-2t+2z(-\mathrm{i}+q-6\beta-6\mathrm{i}q\beta+3q^2\beta)] \\
& \times[1-2t+2z(\mathrm{i}+q-6\beta+6\mathrm{i}q\beta+3q^2\beta)]\} \\
& \times[-6(4b_1+4\mathrm{i}b_2+t+5\mathrm{i}z-qz+38z\beta+30\mathrm{i}qz\beta-3q^2z\beta) \\
& -8\{t+z[\mathrm{i}+q(-1+6\mathrm{i}\beta)+6\beta-3q^2\beta]\}^3 \\
& +3(-1+4\{t+z[\mathrm{i}+q(-1+6\mathrm{i}\beta)+6\beta-3q^2\beta]\}^2)] \\
& -A(t,z)[1-2t+2z(-\mathrm{i}+q-6\beta-6\mathrm{i}q\beta+3q^2\beta)] \\
& \times[1+2t-2z(\mathrm{i}+q-6\beta+6\mathrm{i}q\beta+3q^2\beta)] \\
& \times[6(4b_1+4\mathrm{i}b_2+t+5\mathrm{i}z-qz+38z\beta+30\mathrm{i}qz\beta-3q^2z\beta) \\
& +8\{t+z[\mathrm{i}+q(-1+6\mathrm{i}\beta)+6\beta-3q^2\beta]\}^3 \\
& +3(-1+4\{t+z[\mathrm{i}+q(-1+6\mathrm{i}\beta)+6\beta-3q^2\beta]\}^2)],
\end{aligned} \qquad (4.4.44)$$

$A(z,t)$ 为关于 z 和 t 的实函数：

$$A(t,z) = 1/[2+8t^2-16tz(q-6\beta+3q^2\beta)+8z^2(1+q^2+6q^3\beta+36\beta^2+9q^4\beta^2)].$$

b_1 和 b_2 为实常数. 需要指出的是：①b_1 和 b_2 的不同取值决定了二阶怪波的不同构型；②怪波解的阶数越高, 引入的实常数就越多, 相应的高阶怪波结构就越丰富. 接下来, 我们根据 b_1 和 b_2 取值的不同分别研究两种不同结构的二阶怪波 (单峰结构以及"三胞胎"结构的二阶怪波) 的态转换特征.

1. 单峰二阶怪波的态转换特征

我们首先考虑单峰二阶怪波的态转换特征. 单峰二阶怪波的存在条件为：$b_1 = 0, b_2 = 0$. 由图 4.18 可见, 当 $b_1 = 0, b_2 = 0$ 时, 二阶怪波为标准的单峰结构, 且最大振幅为背景振幅的 5 倍. 事实上, 此时的二阶怪波为三个一阶怪波在相同位置上的非线性叠加. 这种结构与标准非线性薛定谔模型中的二阶怪波结构相似. 不同之处在于, 这里的二阶怪波由于高阶效应的影响具有非零的速度 (图 4.18(a)). 需要注意的是二阶怪波的存在条件与一阶怪波的存在条件相同, 即 $q \neq q_s$. 当 $q \to q_s$ 时, 二阶怪波的最大峰值保持不变, 而其整体的长宽比逐渐增大. 此特征与一解怪波情形完全一致. 当 $q = q_s$ 时, 我们就得到了二阶局域波在调制稳定区的非线性相互作用. 可以看到二阶怪波的双重局域性不复存在, 意味着二阶怪波的纵向局域性在调制稳定区被完全破坏, 这与一阶怪波和 W 型孤子的态转换完全一致.

4.4 高阶效应诱发光学局域波态转换

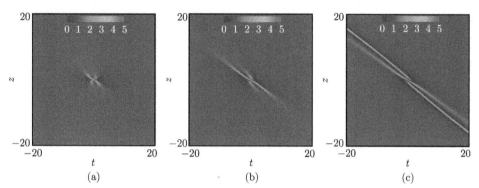

图 4.18 二阶单峰怪波至孤子的态转换 $|u_{2rw}(z,t)|$(精确解 (4.4.42)), (a) $q = 0$, (b)$q = q_s/2$, (c)$q = q_s$. 其他参数设置为: $a = 1$, $\beta = 0.1$, $z_0 = 0$, $t_0 = 0$, $b_1 = 0$, $b_2 = 0$. 这里 (a) 和 (b) 展示了调制不稳定区的单峰的二阶怪波; (c) 呈现了调制稳定区的二阶局域波 (扫描封底二维码可看彩图)

2. "三胞胎"二阶怪波的态转换特征

接下来我们考虑"三胞胎"二阶怪波的态转换特征. "三胞胎"二阶怪波的存在条件为参数 b_1 和 b_2 全不为零或其中之一不为零. 在图 4.19 中, 我们描绘了在条件 $b_1 = 0$, $b_2 = 50$ 下二阶局域波的态转换过程. 当 $q \neq q_s$ 时, 二阶怪波由三个一阶怪波组成, 被称为"三胞胎"结构. 当 $q \to q_s$ 时, "三胞胎"二阶怪波中的每个一阶怪波的最大峰值保持不变, 而其长宽比逐渐增大. 这说明随着调制不稳定性增长率的减小, 每个一阶怪波的纵向局域性都在减弱. 有趣的是, 当 $q = q_s$ 时, 三个一阶怪波并没有转换为三个 W 型孤子. 由图 4.19(c) 所示的相互作用的局域结构图来看, 调制稳定区中的局域波"似乎"呈现出两个孤子相互作用的结构.

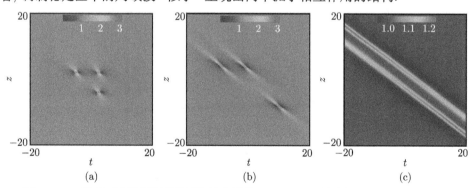

图 4.19 二阶"三胞胎"怪波至孤子的态转换 $|u_{2rw}(z,t)|$(精确解 (4.4.42)), (a)$q = 0$, (b)$q = q_s/2$, (c)$q = q_s$. $b_2 = 50$, 其他参数设置与图 4.18 相同. (a) 和 (b) 展示了调制不稳定区的"三胞胎"二阶怪波; (c) 呈现了调制稳定区的二阶局域波 (扫描封底二维码可看彩图)

由上文两种情况的讨论可知, 不同结构的二阶怪波在高阶效应下的确存在态转换, 并且相应的转换条件与一阶怪波的态转换条件相同. 然而, 调制稳定区的孤子相互作用的结构性质还没有清晰地给出. 图 4.18(c) 和图 4.19(c) 中展示了该相互作用在一定局域范围内的相互作用. 其中图 4.18(c) 中的孤子碰撞时形成一个高振幅峰值, 其大小仍然是背景振幅的 5 倍; 而图 4.19 (c) 中的孤子在碰撞时并未形成高振幅波峰. 一般而言, 对于 Hirota 系统中由达布变换构造的经典二阶孤子解 (零背景的标准亮孤子), 相应的结构一般为两个亮孤子的弹性碰撞. 人们可以利用经典的渐进分析方法得到每个孤子的性质. 不过对于上述由二阶怪波态转换得到的孤子相互作用结构而言, 我们发现经典的渐进分析失效, 原因是我们无法在无穷远处分离出两个速度不同的孤子. 因此, 接下来我们通过取不同位置剖面图的方法定性地研究孤子相互作用.

在图 4.20 中, 我们取不同的传播距离 z 做 W 型孤子相互作用结构的剖面图. 由图可见, 在 $z \to -\infty$ 处, 即 $z = -2000$, 只有一个 W 型孤子 (其大小结

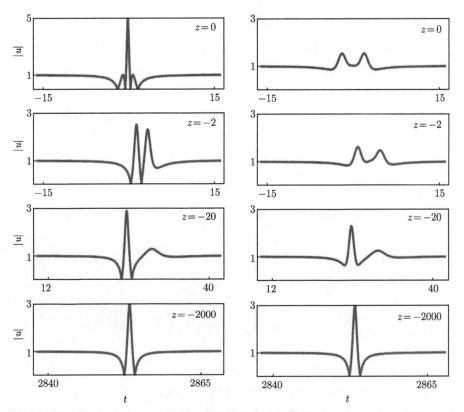

图 4.20 调制稳定区中局域波的相互作用在不同位置 ($z = -2000, -20, -2, 0$) 的振幅剖面图, 左边对应于图 4.18(c), 右边对应于图 4.19(c). 由于局域波相互作用的结构是关于中心位置 (t_0, z_0) 对称的, 因此我们略去 $z > 0$ 的情形, 只展示 $z < 0$ 的情形

构与一阶 W 型孤子一致). 一旦此 W 型孤子的振幅减小, 在其旁就会生成一个小振幅的次级波. 随着 z 增大, 主波和次级波相互作用, 主波的振幅逐渐减小, 而次级波振幅逐渐增大. 两波在中心位置 (t_0, z_0) 处发生碰撞, 并且形成一个大振幅波峰或者一个小振幅的双峰波. 碰撞后两波分开. 此后, 两个波之间依然存在能量相互交换, 其具体过程与碰撞前刚好相反. 当 $z \to \infty$ 时, 又恢复成单个 W 型孤子的初始形状. 整个相互作用过程中, 平面波背景下的能量始终相等, 易通过 $\int_{-\infty}^{+\infty} \{|u_{2rw}(z,t)|^2 - a^2\} \mathrm{d}t = 0$ 来证明.

综上所述, 我们发现具有不同结构的二阶怪波也能在调制稳定区发生态转换. 相应的态转换后的局域波相互作用结构由于在 $z \to \pm\infty$ 处为单 W 型孤子, 呈现出新颖的 "线状" 结构. 该结果与标准的孤子弹性碰撞结果不同. 因此, 经典的渐进分析不再适用. 值得注意的是, 最近 Akhmediev 小组[59]通过对可积的五阶非线性薛定谔方程的研究也发现了相似的碰撞结构. 就目前而言, 精确地分析该相互作用的性质规律仍有困难, 因此发展和改进渐进分析或采用其他分析手段来研究此类局域波相互作用的性质将是未来研究的一个有趣且具有挑战的方向.

4.4.5 呼吸子与其他非线性波的态转换

4.4.4 节中, 我们讨论了高阶效应下光怪波与 W 型孤子间的态转换. 考虑到呼吸子与怪波都产生于调制不稳定区, 那么, 呼吸子是否也存在类似的态转换? 其相应的态转换条件是否一致? 本章节将进一步讨论呼吸子与其他非线性波的态转换. 为此, 我们通过上文中的局域波精确解的构造方法, 构造了两组具有一般形式的呼吸子解. 有趣的是, 经过细致地化简整理, 这两组呼吸子解可以写成统一的形式, 其具体形式如下:

$$u_{1,2}(z,t) = \left[\frac{\Delta_{1,2}\cosh(\varphi+\delta_{1,2}) + \Xi_{1,2}\cos(\phi+\xi_{1,2})}{\Omega_{1,2}\cosh(\varphi+\omega_{1,2}) + \Gamma_{1,2}\cos(\phi+\gamma_{1,2})} + a\right]\mathrm{e}^{\mathrm{i}\theta}. \tag{4.4.45}$$

这里

$$\varphi = 2\eta_i(\tau + V_1\xi), \quad \phi = 2\eta_r(\tau + V_2\xi), \tag{4.4.46}$$

$$V_1 = v_1 + v_2\eta_r/\eta_i, \quad V_2 = v_1 - v_2\eta_i/\eta_r, \tag{4.4.47}$$

$$v_1 = \beta(2a^2 + 4a_1^2 - q'^2) - (q_1+q)(q\beta + \alpha/2), \tag{4.4.48}$$

$$v_2 = a_1[\alpha + 2\beta(q+2q_1)], \quad \eta_r + \mathrm{i}\eta_i = \sqrt{\epsilon + \mathrm{i}\epsilon'}, \tag{4.4.49}$$

$$\epsilon = a^2 - a_1^2 + (q-q_1)^2/4, \quad \epsilon' = a_1(q-q_1), \tag{4.4.50}$$

$$\Delta_1 = -4aa_1\sqrt{\rho+\rho'}, \quad \Delta_2 = -4a^2 a_1, \tag{4.4.51}$$
$$\Xi_1 = 2a_1\sqrt{\chi^2-(2a^2-\chi)^2}, \quad \Xi_2 = 4aa_1\sqrt{2(\mathrm{i}\epsilon'-\epsilon)}, \tag{4.4.52}$$
$$\Omega_1 = \rho+\rho', \quad \Omega_2 = \sqrt{\rho^2-\rho'^2}, \tag{4.4.53}$$
$$\Gamma_1 = -2a(\eta_i+a_1), \quad \Gamma_2 = \sqrt{\varrho^2+\varrho'^2}, \tag{4.4.54}$$
$$\delta_1 = \operatorname{arctanh}(-\mathrm{i}\chi_1/\chi_2), \tag{4.4.55}$$
$$\delta_2 = \operatorname{arctanh}[\mathrm{i}2(\eta_i+\mathrm{i}\eta_r)/(q-q_1-2\mathrm{i}a_1)], \tag{4.4.56}$$
$$\xi_1 = -\arctan[\mathrm{i}(2a^2-\chi)/\chi], \tag{4.4.57}$$
$$\xi_2 = -\arctan[\mathrm{i}2(\eta_i+\mathrm{i}\eta_r)/(q-q_1-2\mathrm{i}a_1)], \tag{4.4.58}$$
$$\omega_1 = 0, \quad \omega_2 = \operatorname{arctanh}(-\rho'/\rho), \tag{4.4.59}$$
$$\gamma_1 = 0, \quad \gamma_2 = -\arctan(\varrho'/\varrho), \tag{4.4.60}$$

其中

$$\rho = \epsilon + 2a_1^2 + \eta_i^2 + \eta_r^2, \quad \rho' = \eta_r(q-q_1) + 2\eta_i a_1, \tag{4.4.61}$$
$$\varrho = \epsilon + 2a_1^2 - \eta_i^2 - \eta_r^2, \quad \varrho' = \eta_i(q_1-q) + 2\eta_r a_1, \tag{4.4.62}$$
$$\chi = \chi_1^2 + \chi_2^2 + a^2, \quad \chi_1 = \eta_r + (q-q_1)/2, \tag{4.4.63}$$
$$\chi_2 = \eta_i + a_1. \tag{4.4.64}$$

上述一般解的表达式具有六个自由参量, 分别为: $a, q, a_1, q_1, \alpha, \beta$. 其中, a 和 q 分别表征平面波背景振幅和频率; α 和 β 分别描述群速度色散以及高阶效应; a_1 和 q_1 为两个实常数, 分别决定非线性波初态的形状以及速度, 不失一般性我们令 $a_1 > 0$.

我们发现, 由于上述一般解具有较多的自由物理参量, 因此在不同参数条件下精确解 (4.4.45) 描述了多种不同种类的局域波动力学特征, 其中包括: ①Akhmediev 呼吸子与周期波的态转换; ②Kuznetsov-Ma 呼吸子与非零背景上的单峰孤子的态转换; ③一般呼吸子与非零背景上多峰孤子的态转换.

这里需要指出的是, 在上述呼吸子解的构造中我们取两类非相关相似矩阵. 对于呼吸子, 这两类非相关相似矩阵只会造成呼吸子中心位置的不同; u_1 表示呼吸子的一个最大峰值在时空分布中心位置 $(z,t) = (0,0)$, 而 u_2 则描述了一类最大峰值不出现在中心位置的呼吸子. 所以, 相似矩阵的选取对呼吸子特征没有本质影响.

有趣的是, 我们发现不同的相似矩阵对呼吸子与其他种类的非线性波的态转换性质起到至关重要的作用, 包括: ①Kuznetsov-Ma 呼吸子和不同结构的单峰孤子的态转换; ②一般呼吸子与不同结构的多峰孤子的态转换.

接下来, 我们首先考虑两种特殊的呼吸子 (Akhmediev 呼吸子和 Kuznetsov-Ma 呼吸子) 的态转换特征. 随后, 我们详尽地考察一般呼吸子的态转换性质. 此外, 上

4.4 高阶效应诱发光学局域波态转换

述解 u_1 在 $a_1 = a$, $q = q_1$ 的情况下将退化为有理分式解 (4.4.36), 其中怪波与 W 孤子间的态转换已在上文中讨论.

Akhmediev 呼吸子与周期波的态转换

我们首先考察 Akhmediev 呼吸子与其他种类非线性波的态转换. 因此, 我们需要得到 Akhmediev 呼吸子解的一般表达式. 我们注意到, 上述一般解 (4.4.45) 在具体参数条件 $0 < a_1 < a$ 且 $q = q_1$ 下, 就退化到一般的 Akhmediev 呼吸子解. 经化简, Akhmediev 呼吸子解的一般表达式具有如下的简洁形式:

$$u_{1,2}(z,t) = \left[\frac{2\eta^2 \cosh(\kappa\xi) + i2\eta a_1 \sinh(\kappa\xi)}{a\cosh(\kappa\xi) - e^{i\sigma} a_1 \cos[2\eta(\tau + v_1\xi) - \mu]} - a \right] e^{i\theta}, \quad (4.4.65)$$

这里, $v_1 = \beta(2a^2 + 4a_1^2 - q^2) - 2q(q\beta + \alpha/2)$, $\kappa = 2\eta v_2$, $\eta = \pm\sqrt{a^2 - a_1^2}$, $v_2 = a_1\alpha(1 - q/q_s)$, $q_s = -\alpha/(6\beta)$, $\sigma = \sigma_{1,2} = \{0, \pi\}$, $\mu = \mu_{1,2} = \{0, \arctan(-\eta_r/a_1)\}$.

与上述 Peregrine 怪波和 W 型孤子态转换的研究方法一致, 我们将改变背景频率 $q \to q_s$, 从而使 Akhmediev 呼吸子从任意一个选定的调制不稳定区演化至调制稳定区. 如图 4.21 所示, 我们可以看到当 $q \to q_s$ 时, Akhmediev 呼吸子的传输

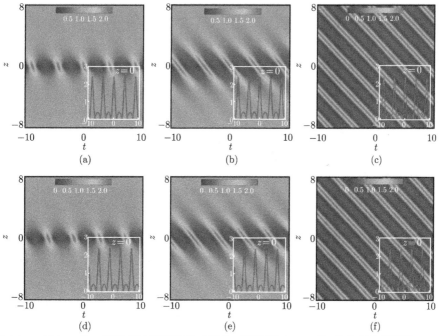

图 4.21 Akhmediev 呼吸子至周期波的态转换 (精确解 (4.4.65)), 从左至右依次取 $q = 0$, $q_s/2$, q_s. 第一行和第二行分别表示具有不同相位参数 $\sigma = 0, \pi$ 的非线性波解. 其他参数设置为: $a = 1$, $\beta = 0.1$, $a_1 = 0.7$. 由图可见, 具有不同相位参数对 Akhmediev 呼吸子至周期波的态转换没有本质影响, 只是改变了中心位置的分布 (扫描封底二维码可看彩图)

方向 z 的局域性逐渐减小, 这个变化趋势正好对应了调制不稳定性增长率的逐渐衰减. 有趣的是, 当增长率衰减至零时, Akhmediev 呼吸子的纵向局域性被彻底破坏, 意味着其在调制稳定区域 $q = q_s$ 最终转化为周期波. 此时, 我们可以精确地给出相应的周期波解的解析表达式:

$$u_{p\ 1,2}(z,t) = \left[\frac{2\eta^2}{a - e^{i\sigma}a_1 \cos[2\eta(\tau + v\xi) - \mu]} - a\right]e^{i\theta}. \quad (4.4.66)$$

需要指出的是, 上述解的不同的相位参数 $(\sigma = 0, \pi)$ 对 Akhmediev 呼吸子与周期波之间的转化没有造成本质上的影响. 此外, 上述 Akhmediev 呼吸子和转换后的周期波具有一致的横向周期, 即 $D_t = \pi/\sqrt{a^2 - a_1^2}$, 意味着高阶效应不会对局域波的横向周期分布产生影响, 而只是改变其传播方向的局域性.

4.4.6 Kuznetsov-Ma 呼吸子与单峰孤子的态转换

接下来, 我们研究 Kuznetsov-Ma 呼吸子和单峰孤子间的转换. 考虑上述一般解 (4.4.45) 的具体参数条件 $a_1 > a$ 且 $q = q_1$, Kuznetsov-Ma 呼吸子解具有如下的一般表达式:

$$u_{1,2}(z,t) = \left\{\frac{2\eta'^2 \cos(\kappa'\xi) + i2\eta' a_1 \sin(\kappa'\xi)}{e^{i\sigma}a_1 \cosh[2\eta'(\tau + v_1\xi) + \mu'] - a\cos(\kappa'\xi)} - a\right\}e^{i\theta}, \quad (4.4.67)$$

其中, $\eta' = \pm\sqrt{a_1^2 - a^2}$, $\kappa' = 2\eta' v_2$, $\sigma = \sigma_{1,2} = \{0, \pi\}$, $\mu' = \mu'_{1,2} = \{0, \text{arctanh}(-\eta_i/a_1)\}$.

由其表达式我们可得 Kuznetsov-Ma 呼吸子沿其传播方向的呼吸周期为 $D_z = \pi/[\eta' a_1 \alpha(1 - q/q_s)]$. 与上述态转换类似, 我们考虑 Kuznetsov-Ma 呼吸子从任意一个选定的调制不稳定区演化至调制稳定区, 即 $q \to q_s$. 如图 4.22 所示, 我们可以看到当 $q \to q_s$ 时, Kuznetsov-Ma 呼吸子的呼吸周期 D_z 逐渐增大, 同时每个基本单元的结构也变得细长. 这个变化趋势与 Peregrine 怪波态转换的情形是一致的.

值得注意的是, 当 $q = q_s$ 时, 具有不同相位参数的 Kuznetsov-Ma 呼吸子转化为两种结构截然不同的单峰孤子. 与标准的亮孤子或暗孤子的特征截然不同, 这两种孤子都是分布在平面波背景上局域的孤立波结构. 具体地, 当 $\sigma = 0$ 时, Kuznetsov-Ma 呼吸子转化为 W 型孤子; 当 $\sigma = \pi$ 时, Kuznetsov-Ma 呼吸子转化为反暗孤子. 有趣的是, 我们可以精确地给出描述两种不同结构的单峰孤子的统一的解析表达式, 如下:

$$u_{s\ 1,2}(z,t) = \left\{\frac{2\eta'^2}{e^{i\sigma}a_1 \cosh[2\eta'(\tau + v\xi) + \mu'] - a} - a\right\}e^{i\theta}. \quad (4.4.68)$$

综上所述, Kuznetsov-Ma 呼吸子在调制稳定区可以转换为平面波背景上的单峰孤子. 呼吸子的最大峰值是否在中心位置 $(z,t) = (0,0)$ 决定了 Kuznetsov-Ma 呼吸子与不同结构单峰孤子 (W 型孤子或反暗孤子) 的态转换.

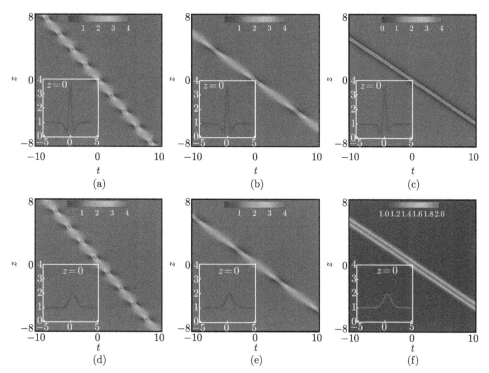

图 4.22　Kuznetsov-Ma 呼吸子至平面波背景上的单峰孤子的态转换 (精确解 (4.4.67)), 从左至右依次取 $q = 0$, $q_s/2$, q_s. 第一行和第二行分别表示具有不同相位参数 $\sigma = 0$, π 的非线性波解. 其他参数设置为: $a = 1$, $\beta = 0.1$, $a_1 = 1.5$. 由图可见, 具有不同相位参数致使 Kuznetsov-Ma 呼吸子转换为结构不同的单峰孤子 (W 型孤子和反暗孤子). 当 Kuznetsov-Ma 呼吸子的中心位置正好是最大峰值出现的位置时 $(\sigma = 0)$, 呼吸子转换为 W 型孤子. 反之, Kuznetsov-Ma 呼吸子转换为反暗孤子 (扫描封底二维码可看彩图)

4.4.7　一般呼吸子与多峰孤子的态转换

在研究了 Akhmediev 呼吸子与周期波以及 Kuznetsov-Ma 呼吸子与两种结构不同的单峰孤子之间的态转换后, 自然有一个问题: 一般呼吸子会转换成何种类型的非线性波, 相应的转换条件又是什么? 下文中, 我们将回答上述疑问. 为此, 我们先回归到一般呼吸子解, 即约束条件 $q \neq q_1$ 下的方程 (4.4.45), 并对其表达式进行细致的分析. 由精确解 (4.4.45) 可以看出, 一般呼吸子的解析表达式是由双曲函数 $\cosh\varphi$ 和三角函数 $\cos\phi$ 在平面波背景上的非线性叠加构成的. 相应的双曲函数 $\cosh\varphi$ 和三角函数 $\cos\phi$ 分别具有速度 V_1 和 V_2. 通常, 双曲函数表征非线性波的局域性, 而三角函数描述非线性波的周期性. 我们发现, 一般呼吸子与其他非线性波的态转换取决于速度差 $V_1 - V_2$ 的取值.

当 $V_1 \neq V_2$ 时, 即非线性波的局域性和周期性速度不匹配, 意味着 $v_2 \neq 0$, 非线性波表现出一般呼吸子的特征 (图 4.23). 我们可以看到这种呼吸子的结构与标准的 Akhmediev 呼吸子和 Kuznetsov-Ma 呼吸子是显著不同的, 即这类呼吸子在横向分布方向和纵向传输方向上都表现出周期性, 并沿着某一个特定方向传输. 因此, 我们可以称之为一般呼吸子. 文献 [17] 中将 $a_1 < a$ 情形下的一般呼吸子称为准 Akhmediev 呼吸子. 需要指出的是, 准 Akhmediev 呼吸子的研究最近也受到关注 [17], 特别是其相互作用表现出非平庸的调制不稳定性特征, 更重要的是这些由准 Akhmediev 呼吸子碰撞所表现出的新的调制不稳定性特征已经于近日在单模光纤系统和水流体系统中得到了实验证实 [75]. 需要强调的是, 这里给出的两类一般呼吸子的精确解具有相等大小的振幅和周期, 其唯一区别是中心位置的不同. 具体而言, u_1 的中心位置在 $(z,t) = (0,0)$ 处, 而 u_2 的中心位置不在 $(z,t) = (0,0)$ 处.

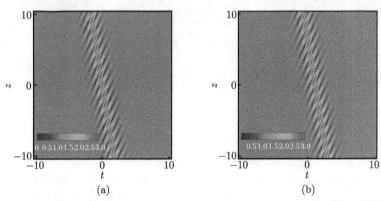

图 4.23 一般呼吸子的结构特征 $|u|$(精确解 (4.4.45)), (a)u_1, (b)u_2, 分别表示呼吸子的中心位置在 $(z,t) = (0,0)$ 处和不在 $(z,t) = (0,0)$ 处. 参数设置为: $a = 1$, $a_1 = 1$, $\beta = 0.1$, $\alpha = 1$, $q_1 = 0.9[-\alpha/(4\beta) - q/2]$, $q = 4q_s$, $q_s = -\alpha/(6\beta)$(扫描封底二维码可看彩图)

然而当 $V_1 = V_2$ 时, 意味着 $\alpha + 2\beta(q + 2q_1) = 0$, 此时非线性波的局域性和周期性速度相匹配, 因此就不再呈现出不稳定的呼吸特性, 而是呈现出一种局域在平面波背景上的"多峰孤子"结构. 图 4.24 中, 我们给出两种具有不同振幅结构的多峰孤子, 分别对应着参数条件 $V_1 = V_2$ 下的解析解 u_1 和 u_2. 值得注意的是, 这类局域在平面波背景上的多峰孤子结构与之前报道的零背景上的多峰孤子以及标准的亮暗孤子是截然不同的. 这两种多峰孤子由中心位置分布不同的两类一般呼吸子转换而来. 有趣的是, 我们发现这两类多峰孤子具有不同的振幅分布结构. 详细而言, u_1 中的多峰孤子的振幅横向分布具有对称性, 而 u_2 中的多峰孤子的振幅横向分布具有非对称性. 需要强调的是, 由于这两类多峰孤子存在条件 $\alpha + 2\beta(q + 2q_1) = 0$ 的限制, 我们得到的这种新颖的存在于非零背景上的多峰孤子只限于考虑高阶效应的情况 ($\beta \neq 0$). 因此, 标准非线性薛定谔系统中是不存在这样的局域结构的. 我们可论断,该局域波的存在反映了高阶效应对局域波性质的巨大影响.另外,由图 4.24 可见,对称和非对称多峰孤子的峰数和最大峰值强度均不同.因此,两类多峰孤子间是否存在相关联的物理量?有趣的是,我们发现,初始参数相同的条件下两类多峰孤子扣除背景后的光强大小是相等的,即

$$\int_{-\infty}^{+\infty}\left(|u_1|^2-a^2\right)\mathrm{d}t=\int_{-\infty}^{+\infty}\left(|u_2|^2-a^2\right)\mathrm{d}t. \quad (4.4.69)$$

事实上,上式也适用于相同初始参数条件下的 W 型孤子和反暗孤子.

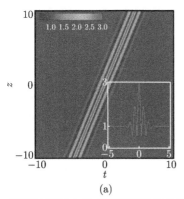
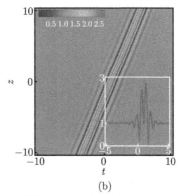

图 4.24 平面波背景上的多峰孤子的结构特征 $|u|$(精确解 (4.4.45)), (a)u_1, (b)u_2, 分别表示振幅分布对称和非对称的多峰孤子. 参数设置为: $a=1$, $a_1=1$, $\beta=0.1$, $\alpha=1$, $q_1=-\alpha/(4\beta)-q/2$, $q=4q_s$, $q_s=-\alpha/(6\beta)$(扫描封底二维码可看彩图)

为了更清晰地展示高阶效应诱发的平面波背景上多种不同种类的非线性波,我们利用表 4.1,将上述不同种类的局域波及其相应的存在条件简洁明了地呈现出来.

表 4.1 高阶效应诱发的平面波背景上多种不同种类的非线性波和相应的存在条件,其中 $q_s=-\alpha/(6\beta)$. 表格中的多峰孤子、W 型孤子、反暗孤子、周期波以及有理形式 W 型孤子的激发皆源于高阶效应的影响

非线性波的类型	精确的存在条件
呼吸子和怪波	$q_1 \neq (3q_s-q)/2$
多峰孤子	$q_1 = (3q_s-q)/2$, $q \neq q_s$
W 型孤子/反暗孤子	$q_1 = (3q_s-q)/2$, $q = q_s$, $a^2 < a_1^2$
周期波	$q_1 = (3q_s-q)/2$, $q = q_s$, $a^2 > a_1^2$
有理形式 W 型孤子	$q_1 = (3q_s-q)/2$, $q = q_s$, $a^2 = a_1^2$

参 考 文 献

[1] Akhmediev N, Ankiewicz A. Solitons: Nonlinear Pulses and Beams. Chapman & Hall, 1997; Yang J. Nonlinear Waves in Integrable and Nonintegrable Systems. SIAM, 2010;

Lou S Y, Huang F. Alice-Bob physics: Coherent solutions of nonlocal KdV systems. Scientific Reports, 2017, 7: 869.

[2] 楼森岳, 唐晓艳. 非线性数学物理方法. 北京: 科学出版社, 2006; 陈险峰, 郭旗, 佘卫龙, 等. 非线性光学研究前沿. 上海: 上海交通大学出版社, 2014; 郭柏灵, 田立新, 闫振亚, 等. 怪波及其数学理论. 杭州: 浙江科学技术出版社, 2015; 张解放, 戴朝卿, 王悦悦. 基于非线性薛定谔方程的畸形波理论及其应用. 北京: 科学出版社, 2016.

[3] Kartashov Y V, Malomed B A, Torner L. Solitons in nonlinear lattices. Reviews of Modern Physics, 2011, 83(1): 247.

[4] Akhmediev N, Soto-Crespo J M, Ankiewicz A. Extreme waves that appear from nowhere: on the nature of rogue waves. Physics Letters A, 2009, 373(25): 2137-2145; Akhmediev N, Ankiewicz A, Taki M. Waves that appear from nowhere and disappear without a trace. Physics Letters A, 2009, 373(6): 675-678.

[5] Akhmediev N, Pelinovsky E. Editorial-introductory remarks on "discussion & debate: rogue waves-towards a unifying concept?". The European Physical Journal-Special Topics, 2010, 185(1): 1-4.

[6] Onorato M, Residori S, Bortolozzo U, et al. Rogue waves and their generating mechanisms in different physical contexts. Physics Reports, 2013, 528(2): 47-89.

[7] Akhmediev N, Dudley J M, Solli D R, et al. Recent progress in investigating optical rogue waves. Journal of Optics, 2013, 15(6): 060201; Akhmediev N, Kibler B, Baronio F, et al. Roadmap on optical rogue waves and extreme events. Journal of Optics, 2016, 18(6): 063001.

[8] Dudley J M, Dias F, Erkintalo M, et al. Instabilities, breathers and rogue waves in optics. Nature Photonics, 2014, 8(10): 755-764.

[9] Kharif C, Pelinovsky E. Physical mechanisms of the rogue wave phenomenon. European Journal of Mechanics-B/Fluids, 2003, 22(6): 603-634.

[10] Zakharov V, Gelash A. Freak waves as a result of modulation instability. Procedia IUTAM, 2013, 9: 165-175.

[11] Solli D R, Ropers C, Koonath P, et al. Optical rogue waves. Nature, 2007, 450(7172): 1054-1057.

[12] Akhmediev N, Soto-Crespo J M, Ankiewicz A. How to excite a rogue wave. Physical Review A, 2009, 80(4): 043818; Akhmediev N, Ankiewicz A, Soto-Crespo J M. Rogue waves and rational solutions of the nonlinear Schrödinger equation. Physical Review E, 2009, 80(2): 026601.

[13] Kuznetsov E A. Solitons in a parametrically unstable plasma. Akademiia Nauk SSSR Doklady, 1977, 236: 575-577; Ma Y C. The perturbed plane-wave solutions of the cubic schrödinger equation. Studies in Applied Mathematics, 1979, 60(1): 43-58.

[14] Peregrine D H. Water waves, nonlinear Schrödinger equations and their solutions. The Journal of the Australian Mathematical Society. Series B. Applied Mathematics, 1983,

25(01): 16-43.
[15] Akhmediev N N, Korneev V I. Modulation instability and periodic solutions of the nonlinear Schrödinger equation. Theoretical and Mathematical Physics, 1986, 69(2): 1089-1093.
[16] Shrira V I, Geogjaev V V. What makes the Peregrine soliton so special as a prototype of freak waves? Journal of Engineering Mathematics, 2010, 67(1-2): 11-22.
[17] Zakharov V E, Gelash A A. Nonlinear stage of modulation instability. Physical Review Letters, 2013, 111(5): 054101.
[18] Akhmediev N. Nonlinear physics: Déjàvu in optics. Nature, 2001, 413(6853): 267-268.
[19] Mussot A, Kudlinski A, Droques M, et al. Fermi-Pasta-Ulam recurrence in nonlinear fiber optics: the role of reversible and irreversible losses. Physical Review X, 2014, 4(1): 011054.
[20] Erkintalo M, Hammani K, Kibler B, et al. Higher-order modulation instability in nonlinear fiber optics. Physical Review Letters, 2011, 107(25): 253901.
[21] Kibler B, Fatome J, Finot C, et al. The Peregrine soliton in nonlinear fibre optics. Nature Physics, 2010, 6(10): 790-795.
[22] Hammani K, Kibler B, Finot C, et al. Peregrine soliton generation and breakup in standard telecommunications fiber. Optics Letters, 2011, 36(2): 112-114.
[23] Dudley J M, Genty G, Dias F, et al. Modulation instability, akhmediev breathers and continuous wave supercontinuum generation. Optics Express, 2009, 17(24): 21497-21508.
[24] Kibler B, Fatome J, Finot C, et al. Observation of Kuznetsov-Ma soliton dynamics in optical fibre. Scientific Reports, 2012, 2: 463.
[25] Frisquet B, Kibler B, Millot G. Collision of Akhmediev breathers in nonlinear fiber optics. Physical Review X, 2013, 3(4): 041032.
[26] Toenger S, Godin T, Billet C, et al. Emergent rogue wave structures and statistics in spontaneous modulation instability. Scientific Reports, 2015, 5: 10380.
[27] Walczak P, Randoux S, Suret P. Optical rogue waves in integrable turbulence. Physical Review Letters, 2015, 114(14): 143903; Soto-Crespo J M, Devine N, Akhmediev N. Integrable turbulence and rogue waves: breathers or solitons? Physical Review Letters, 2016, 116(10): 103901.
[28] Fatome J, Kibler B, Finot C. High-quality optical pulse train generator based on solitons on finite background. Optics Letters, 2013, 38(10): 1663-1665.
[29] Bludov Y V, Konotop V V, Akhmediev N. Rogue waves as spatial energy concentrators in arrays of nonlinear waveguides. Optics Letters, 2009, 34(19): 3015-3017.
[30] Yang G, Li L, Jia S, et al. High power pulses extracted from the Peregrine rogue wave. Romanian Reports in Physics, 2013, 65(2): 391-400.
[31] Yang G, Wang Y, Qin Z, et al. Breatherlike solitons extracted from the Peregrine

rogue wave. Physical Review E, 2014, 90(6): 062909.

[32] Zhang Y, Belić M R, Zheng H, et al. Nonlinear Talbot effect of rogue waves. Physical Review E, 2014, 89(3): 032902.

[33] Tiofack C G L, Coulibaly S, Taki M, et al. Comb generation using multiple compression points of Peregrine rogue waves in periodically modulated nonlinear Schrödinger equations. Physical Review A, 2015, 92(4): 043837.

[34] Bludov Y V, Konotop V V, Akhmediev N. Matter rogue waves. Physical Review A, 2009, 80(3): 033610.

[35] Wen L, Li L, Li Z D, et al. Matter rogue wave in Bose-Einstein condensates with attractive atomic interaction. The European Physical Journal D, 2011, 64(2-3): 473-478; Zhao L C. Dynamics of nonautonomous rogue waves in Bose-Einstein condensate. Annals of Physics, 2013, 329: 73-79; He J S, Charalampidis E G, Kevrekidis P G, et al. Rogue waves in nonlinear Schrödinger models with variable coefficients: Application to Bose-Einstein condensates. Physics Letters A, 2014, 378(5): 577-583.

[36] Liu C, Yang Z Y, Zhao L C, et al. Long-lived rogue waves and inelastic interaction in binary mixtures of bose-einstein condensates. Chinese Physics Letters, 2013, 30(4): 040304; Zhao L C, Ling L, Yang Z Y, et al. Pair-tunneling induced localized waves in a vector nonlinear Schrödinger equation. Communications in Nonlinear Science and Numerical Simulation, 2015, 23(1): 21-27; Qin Z, Mu G. Matter rogue waves in an $F=1$ spinor Bose-Einstein condensate. Physical Review E, 2012, 86(3): 036601.

[37] Moslem W M, Shukla P K, Eliasson B. Surface plasma rogue waves. Europhysics Letters, 2011, 96(2): 25002.

[38] Ganshin A N, Efimov V B, Kolmakov G V, et al. Observation of an inverse energy cascade in developed acoustic turbulence in superfluid helium. Physical Review Letters, 2008, 101(6): 065303.

[39] Stenflo L, Marklund M. Rogue waves in the atmosphere. Journal of Plasma Physics, 2010, 76(3-4): 293-295.

[40] Zhao F, Li Z D, Li Q Y, et al. Magnetic rogue wave in a perpendicular anisotropic ferromagnetic nanowire with spin-transfer torque. Annals of Physics, 2012, 327(9): 2085-2095.

[41] Shats M, Punzmann H, Xia H. Capillary rogue waves. Physical Review Letters, 2010, 104(10): 104503.

[42] Yan Z. Financial rogue waves. Communications in Theoretical Physics, 2010, 54(5): 947; Yan Z. Vector financial rogue waves. Physics Letters A, 2011, 375(48): 4274-4279; Wu Y, Zhao L C, Lei X K. The effects of background fields on vector financial rogue wave pattern. The European Physical Journal B, 2015, 88(11): 1-5.

[43] Yan Z. Nonautonomous "rogons" in the inhomogeneous nonlinear Schrödinger equation with variable coefficients. Physics Letters A, 2010, 374(4): 672-679.

[44] Dai C Q, Zhou G Q, Zhang J F. Controllable optical rogue waves in the femtosecond regime. Physical Review E, 2012, 85(1): 016603.

[45] Wang L, Feng X Q, Zhao L C. Dynamics and trajectory of nonautonomous rogue wave in a graded-index planar waveguide with oscillating refractive index. Optics Communications, 2014, 329: 135-139.

[46] Liu C, Yang Z Y, Zhao L C, et al. Optical rogue waves generated on Gaussian background beam. Optics Letters, 2014, 39(4): 1057-1060.

[47] Ankiewicz A, Devine N, Akhmediev N. Are rogue waves robust against perturbations? Physics Letters A, 2009, 373(43): 3997-4000.

[48] Ankiewicz A, Soto-Crespo J M, Akhmediev N. Rogue waves and rational solutions of the Hirota equation. Physical Review E, 2010, 81(4): 046602; Yang G, Li L, Jia S. Peregrine rogue waves induced by the interaction between a continuous wave and a soliton. Physical Review E, 2012, 85(4): 046608; Tao Y, He J. Multisolitons, breathers, and rogue waves for the Hirota equation generated by the Darboux transformation. Physical Review E, 2012, 85(2): 026601.

[49] Liu C, Yang Z Y, Zhao L C, et al. State transition induced by higher-order effects and background frequency. Physical Review E, 2015, 91(2): 022904.

[50] Chowdury A, Ankiewicz A, Akhmediev N. Moving breathers and breather-to-soliton conversions for the Hirota equation//Proc. R. Soc. A. The Royal Society, 2015, 471(2180): 20150130.

[51] Liu C, Yang Z Y, Zhao L C, et al. Symmetric and asymmetric optical multi-peak solitons on a continuous wave background in the femtosecond regime. Physical Review E, 2016, 94(4): 042221.

[52] Bandelow U, Akhmediev N. Sasa-Satsuma equation: Soliton on a background and its limiting cases. Physical Review E, 2012, 86(2): 026606.

[53] Soto-Crespo J M, Devine N, Hoffmann N P, et al. Rogue waves of the Sasa-Satsuma equation in a chaotic wave field. Physical Review E, 2014, 90(3): 032902.

[54] Zhao L C, Li S C, Ling L. Rational W-shaped solitons on a continuous-wave background in the Sasa-Satsuma equation. Physical Review E, 2014, 89(2): 023210.

[55] Zhao L C, Liu C, Yang Z Y. The rogue waves with quintic nonlinearity and nonlinear dispersion effects in nonlinear optical fibers. Communications in Nonlinear Science and Numerical Simulation, 2015, 20(1): 9-13; Wang X, Yang B, Chen Y, et al. Higher-order rogue wave solutions of the Kundu-Eckhaus equation. Physica Scripta, 2014, 89(9): 095210.

[56] Wang L H, Porsezian K, He J S. Breather and rogue wave solutions of a generalized nonlinear Schrödinger equation. Physical Review E, 2013, 87(5): 053202.

[57] Wang L, Zhang J H, Wang Z Q, et al. Breather-to-soliton transitions, nonlinear wave interactions, and modulational instability in a higher-order generalized nonlinear

Schrödinger equation. Physical Review E, 2016, 93(1): 012214.

[58] Ankiewicz A, Wang Y, Wabnitz S, et al. Extended nonlinear Schrödinger equation with higher-order odd and even terms and its rogue wave solutions. Physical Review E, 2014, 89(1): 012907; Chowdury A, Kedziora D J, Ankiewicz A, et al. Breather solutions of the integrable quintic nonlinear Schrödinger equation and their interactions. Physical Review E, 2015, 91(2): 022919; Yang Y, Yan Z, Malomed B A. Rogue waves, rational solitons, and modulational instability in an integrable fifth-order nonlinear Schrödinger equation. Chaos: An Interdisciplinary Journal of Nonlinear Science, 2015, 25(10): 103112.

[59] Chowdury A, Kedziora D J, Ankiewicz A, et al. Breather-to-soliton conversions described by the quintic equation of the nonlinear Schrödinger hierarchy. Physical Review E, 2015, 91(3): 032928.

[60] Guo B L, Ling L M. Rogue wave, breathers and bright-dark-rogue solutions for the coupled Schrödinger equations. Chinese Physics Letters, 2011, 28(11): 110202; Baronio F, Degasperis A, Conforti M, et al. Solutions of the vector nonlinear Schrödinger equations: evidence for deterministic rogue waves. Physical Review Letters, 2012, 109(4): 044102.

[61] Baronio F, Conforti M, Degasperis A, et al. Vector rogue waves and baseband modulation instability in the defocusing regime. Physical Review Letters, 2014, 113(3): 034101.

[62] Zhao L C, Liu J. Localized nonlinear waves in a two-mode nonlinear fiber. JOSA B, 2012, 29(11): 3119-3127; Zhao L C, Liu J. Rogue-wave solutions of a three-component coupled nonlinear Schrödinger equation. Physical Review E, 2013, 87(1): 013201.

[63] Zhao L C, Xin G G, Yang Z Y. Rogue-wave pattern transition induced by relative frequency. Physical Review E, 2014, 90(2): 022918; Ling L, Guo B, Zhao L C. High-order rogue waves in vector nonlinear Schrödinger equations. Physical Review E, 2014, 89(4): 041201.

[64] Liu C, Yang Z Y, Zhao L C, et al. Vector breathers and the inelastic interaction in a three-mode nonlinear optical fiber. Physical Review A, 2014, 89(5): 055803.

[65] Chen S, Song L Y. Rogue waves in coupled Hirota systems. Physical Review E, 2013, 87(3): 032910.

[66] Liu C, Yang Z Y, Zhao L C, et al. Transition, coexistence, and interaction of vector localized waves arising from higher-order effects. Annals of Physics, 2015, 362: 130-138.

[67] Zhao L C, Yang Z Y, Ling L. Localized waves on continuous wave background in a two-mode nonlinear fiber with high-order effects. Journal of the Physical Society of Japan, 2014, 83(10): 104401.

[68] Baronio F, Conforti M, Degasperis A, et al. Rogue waves emerging from the resonant

interaction of three waves. Physical Review Letters, 2013, 111(11): 114101.

[69] Chen J, Chen Y, Feng B F, et al. Rational solutions to two-and one-dimensional multicomponent Yajima-Oikawa systems. Physics Letters A, 2015, 379(24): 1510-1519.

[70] Zhao L C, Ling L, Yang Z Y, et al. Pair-tunneling induced localized waves in a vector nonlinear Schrödinger equation. Communications in Nonlinear Science and Numerical Simulation, 2015, 23(1): 21-27; Ling L, Zhao L C. Integrable pair-transition-coupled nonlinear Schrodinger equations. Physical Review E, 2015, 92(2): 022924.

[71] He J, Xu S, Porsezian K. N-order bright and dark rogue waves in a resonant erbium-doped fiber system. Physical Review E, 2012, 86(6): 066603.

[72] Ren Y, Yang Z Y, Liu C, et al. Different types of nonlinear localized and periodic waves in an erbium-doped fiber system. Physics Letters A, 2015, 379(45): 2991-2994; Ren Y, Yang Z Y, Liu C, et al. Characteristics of optical multi-peak solitons induced by higher-order effects in an erbium-doped fiber system. The European Physical Journal D, 2016, 70(9): 187.

[73] Mahnke C, Mitschke F. Possibility of an Akhmediev breather decaying into solitons. Physical Review A, 2012, 85(3): 033808.

[74] He J S, Xu S W, Ruderman M S, et al. State transition induced by self-steepening and self phase-modulation. Chinese Physics Letters, 2014, 31(1): 010502.

[75] Kibler B, Chabchoub A, Gelash A, et al. Superregular breathers in optics and hydrodynamics: Omnipresent modulation instability beyond simple periodicity. Physical Review X, 2015, 5(4): 041026.

[76] Liu C, Ren Y, Yang Z Y, et al. Superregular breathers in a complex modified Korteweg-de Vries system. Chaos: An Interdisciplinary Journal of Nonlinear Science, 2017, 27(8): 083120; Liu C, Wang L, Yang Z Y, et al. Femtosecond Optical Superregular Breathers. arXiv preprint, 2017, arXiv:1708.03781.

[77] Zhang J H, Wang L, Liu C. Superregular breathers, characteristics of nonlinear stage of modulation instability induced by higher-order effects. Proc. R. Soc. A, 2017, 473: 20160681.

[78] 郭柏灵, 庞小峰. 孤立子. 北京: 科学出版社, 1987; 李翊神. 孤子与可积系统. 上海: 上海科技教育出版社, 1999; 谷超豪, 胡和生, 周子翔. 孤立子理论中的达布变换及其几何应用. 上海: 上海科学技术出版社, 2005.

[79] Kedziora D J, Ankiewicz A, Akhmediev N. Rogue waves and solitons on a cnoidal background. The European Physical Journal Special Topics, 2014, 223(1): 43-62.

[80] Cheng X P, Lou S Y, Chen C, et al. Interactions between solitons and other nonlinear Schrödinger waves. Physical Review E, 2014, 89(4): 043202.

[81] Ablowitz M J, Kaup D J, Newell A C, et al. The inverse scattering transform-fourier analysis for nonlinear problems. Studies in Applied Mathematics, 1974, 53(4): 249-

315.

[82] Guo B, Ling L, Liu Q P. Nonlinear Schrödinger equation: generalized Darboux transformation and rogue wave solutions. Physical Review E, 2012, 85(2): 026607.

[83] Ohta Y, Yang J. General high-order rogue waves and their dynamics in the nonlinear Schrödinger equation//Proceedings of The Royal Society of London A: Mathematical, Physical and Engineering Sciences. The Royal Society, 2012, 468(2142): 1716-1740.

[84] Ankiewicz A, Kedziora D J, Akhmediev N. Rogue wave triplets. Physics Letters A, 2011, 375(28): 2782-2785.

[85] Ling L, Zhao L C. Simple determinant representation for rogue waves of the nonlinear Schrödinger equation. Physical Review E, 2013, 88(4): 043201.

[86] Chabchoub A, Hoffmann N, Onorato M, et al. Super rogue waves: observation of a higher-order breather in water waves. Physical Review X, 2012, 2(1): 011015.

[87] Frisquet B, Kibler B, Morin P, et al. Optical dark rogue wave. Scientific Reports, 2016, 6: 20785.

[88] Turitsyn S K, Bale B G, Fedoruk M P. Dispersion-managed solitons in fibre systems and lasers. Physics Reports, 2012, 521(4): 135-203.

[89] Serkin V N, Hasegawa A, Belyaeva T L. Nonautonomous solitons in external potentials. Physical Review Letters, 2007, 98(7): 074102.

[90] Ponomarenko S A, Agrawal G P. Do solitonlike self-similar waves exist in nonlinear optical media? Physical Review Letters, 2006, 97(1): 013901.

[91] Belmonte-Beitia J, Pérez-García V M, Vekslerchik V, et al. Localized nonlinear waves in systems with time-and space-modulated nonlinearities. Physical Review Letters, 2008, 100(16): 164102.

[92] Liang Z X, Zhang Z D, Liu W M. Dynamics of a bright soliton in Bose-Einstein condensates with time-dependent atomic scattering length in an expulsive parabolic potential. Physical Review Letters, 2005, 94(5): 050402.

[93] He X G, Zhao D, Li L, et al. Engineering integrable nonautonomous nonlinear Schrödinger equations. Physical Review E, 2009, 79(5): 056610.

[94] Yan Z, Zhang X F, Liu W M. Nonautonomous matter waves in a waveguide. Physical Review A, 2011, 84(2): 023627.

[95] Li L, Li Z, Li S, et al. Modulation instability and solitons on a cw background in inhomogeneous optical fiber media. Optics Communications, 2004, 234(1): 169-176.

[96] Zhang J F, Wu L, Li L. Self-similar parabolic pulses in optical fiber amplifiers with gain dispersion and gain saturation. Physical Review A, 2008, 78(5): 055801.

[97] Zhong W P, Xie R H, Belić M, et al. Exact spatial soliton solutions of the two-dimensional generalized nonlinear Schrödinger equation with distributed coefficients. Physical Review A, 2008, 78(2): 023821.

[98] Dai C Q, Zhu S Q, Wang L L, et al. Exact spatial similaritons for the generalized (2+1)-

dimensional nonlinear Schrödinger equation with distributed coefficients. Europhysics Letters, 2010, 92(2): 24005.

[99] Yang Z Y, Zhao L C, Zhang T, et al. Snakelike nonautonomous solitons in a graded-index grating waveguide. Physical Review A, 2010, 81(4): 043826; Yang Z Y, Zhao L C, Zhang T, et al. Dynamics of a nonautonomous soliton in a generalized nonlinear Schrödinger equation. Physical Review E, 2011, 83(6): 066602; Liu C, Yang Z Y, Yang W L, et al. Nonautonomous dark solitons and rogue waves in a graded-index grating waveguide. Communications in Theoretical Physics, 2013, 59(3): 311; Liu C, Yang Z Y, Zhang M, et al. Dynamics of nonautonomous dark solitons. Communications in Theoretical Physics, 2013, 59(6): 703; Ren Y, Yang Z Y, Liu C, et al. Controllable optical superregular breathers in the femtosecond regime. Chin. Phys. B, 2018, 27: 010504.

[100] Zakharov V E, Ostrovsky L A. Modulation instability: The beginning. Physica D: Nonlinear Phenomena, 2009, 238(5): 540-548.

[101] Baronio F, Chen S, Grelu P, et al. Baseband modulation instability as the origin of rogue waves. Physical Review A, 2015, 91(3): 033804.

[102] Zhao L C, Ling L. Quantitative relation between modulational instability and several well-known nonlinear excitations. JOSA B, 2016, 33(5): 850-856.

[103] Kivshar Y S, Luther-Davies B. Dark optical solitons: physics and applications. Physics Reports, 1998, 298(2): 81-197.

[104] Duan L, Yang Z Y, Liu C, et al. Optical rogue wave excitation and modulation on a bright soliton background. Chinese Physics Letters, 2016, 33(1): 010501; Goyal A, Raju T S, Kumar C N, et al. The effect of different background beams on the optical rogue waves generated in a graded-index waveguide. Optics Communications, 2016, 364: 177-180.

[105] Chabchoub A, Hoffmann N P, Akhmediev N. Rogue wave observation in a water wave tank. Physical Review Letters, 2011, 106(20): 204502; Chabchoub A, Kibler B, Dudley J M, et al. Hydrodynamics of periodic breathers. Philosophical Transactions, 2014, 372(2027): 4152-4160.

[106] Bailung H, Sharma S K, Nakamura Y. Observation of Peregrine solitons in a multi-component plasma with negative ions. Physical Review Letters, 2011, 107(25): 255005; Pathak P, Sharma S K, Nakamura Y, et al. Observation of second order ion acoustic Peregrine breather in multicomponent plasma with negative ions. Physics of Plasmas, 2016, 23(2): 790.

[107] Wright O C. Sasa-Satsuma equation, unstable plane waves and heteroclinic connections. Chaos Solitons & Fractals, 2007, 33(2): 374-387.

[108] Xu T, Li M, Li L. Anti-dark and Mexican-hat solitons in the Sasa-Satsuma equation on the continuous wave background. Europhysics Letters, 2015, 109(3): 30006.

[109] Anderson D, Lisak M. Nonlinear asymmetric self-phase modulation and self-steepening of pulses in long optical waveguides. Physical Review A, 1983, 27(3): 1393.

[110] Li Z, Li L, Tian H, et al. New types of solitary wave solutions for the higher order nonlinear Schrödinger equation. Physical Review Letters, 2000, 84(18): 4096-4099.

[111] Huang L G, Pang L H, Wong P, et al. Analytic soliton solutions of cubic-quintic Ginzburg-Landau equation with variable nonlinearity and spectral filtering in fiber lasers. Annalen der Physik, 2016, 528(6): 493-503.

第5章 Introduction to Exactly Solvable Quantum Many-body Systems
(精确可解量子多体系统导论)

Ryu Sasaki

This is a note for a series of lectures for graduate students on elementary introduction to quantum exactly solvable many-body systems. Starting from the well-known classical Liouville theorem and its incomplete quantum counterpart, the exactly solvable 1-d quantum mechanical systems with differential as well as difference Schrödinger equations including those of the Wilson and Askey-Wilson polynomials are briefly reviewed. Various aspects of exactly solvable quantum multi-body systems governed by differential Schrödinger equations, that is the Calogero-Sutherland-Moser systems, are discussed in some detail. They are the properties of the eigenvalues and eigenfunctions, the four types of interaction potentials (rational, trigonometric, hyperbolic and elliptic), the dependence on the root systems including the exceptional ones (E_6, E_7, E_8, F_4 and G_2), various types of Lax pairs and the conserved quantities, the quantum-classical correspondence and the dynamics near the classical equilibrium points, etc. The eigenpolynomials of the Sutherland systems based on A type root systems are known as the Jack polynomials. It is interesting to see how various combinatorial concepts and methods related to partitions and Young diagrams are useful for this purely dynamical problem. The exactly solvable quantum multi-body systems governed by difference Schrödinger equations, that is the Ruijsenaars-Schneider-van Diejen-Macdonald-Koornwinder systems, are discussed in a parallel fashion. After brief introduction to the subject, the Macdonald polynomials are discussed in some detail. They are the eigenpolynomials of the quantum trigonometric Ruijsenaars-Schneider systems based on A type root systems and are a one parameter deformation of the Jack polynomials. Again combinatorial methods are useful for understanding the eigenfunctions of this dynamical problem. Since this series of lectures is meant as an introductory course, and the number of hours is quite limited, I try not to give technical details but to state the outline of the problems, general situations and the future prospects in simple languages to stimulate the thoughts. Classical cases are also included as the limiting cases of $\hbar \to 0$. In the same spirits, I would put emphasis on the unsolved problems, rather than detailing the established results. The rudimentary knowledge of classical and quantum mechanics is assumed.

5.1 Introduction (引言)

Since Isaac Newton's discovery of Equations of Motion, people tried very hard to understand Nature through their solutions. Here we consider the dynamical systems of finite degrees of freedom N, in the Hamiltonian formulation.

The generalised coordinates are: $x = (x_1, \cdots, x_N)$.

The conjugate momenta are denoted as $p = (p_1, \cdots, p_N)$. The Hamiltonian is $\mathcal{H}(p, x)$.

Throughout this lecture the rudimentary knowledge of classical [1] and quantum mechanics[2, 3] is assumed. We consider the case that the Hamiltonian $\mathcal{H}(p, x)$ does not have an explicit time-dependence. In other words, we consider autonomous systems only. The Poisson brackets (the symplectic structure) are

$$\{x_j, p_k\} = \delta_{jk}, \quad \{x_j, x_k\} = \{p_j, p_k\} = 0. \tag{5.1.1}$$

The canonical equations of motion for the canonical coordinates are:

$$\frac{\mathrm{d}}{\mathrm{d}t} x_j = \{x_j, \mathcal{H}\}, \quad \frac{\mathrm{d}}{\mathrm{d}t} p_j = \{p_j, \mathcal{H}\}, \quad j = 1, \cdots, N. \tag{5.1.2}$$

For an arbitrary quantity $A(p, x)$, the Hamiltonian equation reads

$$\frac{\mathrm{d}}{\mathrm{d}t} A(p, x) = \{A(p, x), \mathcal{H}\}. \tag{5.1.3}$$

In the corresponding quantum system, the coordinates and momenta are linear operators obeying the canonical commutation relations:

$$[x_j, p_k] = \mathrm{i}\hbar \delta_{jk}, \quad [x_j, q_k] = [p_j, p_k] = 0, \quad \mathrm{i} \stackrel{\mathrm{def}}{=} \sqrt{-1}. \tag{5.1.4}$$

The Hamiltonian $\mathcal{H}(p, x)$ is a self-adjoint linear operator governing the time translation of other operators through commutation relations called the Heisenberg equations of motion:

$$\mathrm{i}\hbar \frac{\mathrm{d}}{\mathrm{d}t} x_j = [x_j, \mathcal{H}], \quad \mathrm{i}\hbar \frac{\mathrm{d}}{\mathrm{d}t} p_j = [p_j, \mathcal{H}], \quad j = 1, \cdots, N. \tag{5.1.5}$$

For an arbitrary observable $A(p, x)$, the Heisenberg equation reads

$$\mathrm{i}\hbar \frac{\mathrm{d}}{\mathrm{d}t} A(p, x) = [A(p, x), \mathcal{H}]. \tag{5.1.6}$$

5.1 Introduction

The time-dependent Schrödinger equation for the wavefunction $\psi(x;t)$ reads

$$i\hbar \frac{\partial}{\partial t}\psi(x;t) = \mathcal{H}\psi(x;t), \qquad (5.1.7)$$

in which the momentum operator p_j in the Hamiltonian $\mathcal{H}(p,x)$ acts as a differential operator $p_j \to -i\hbar \frac{\partial}{\partial x_j}$ on the wavefunction $\psi(x;t)$. The above equation can be reduced to the eigenvalue problem for the given Hamiltonian $\mathcal{H}$:

$$\mathcal{H}\psi_n(x) = \mathcal{E}_n\psi_n(x), \quad n = 0, 1, \cdots, \qquad (5.1.8)$$

The eigenfunctions must satisfy certain boundary conditions and they have a finite norm,

$$(\psi_n, \psi_n) \stackrel{\text{def}}{=} \int |\psi_n(x)|^2 d^N x < \infty. \qquad (5.1.9)$$

The eigenfunctions belonging to distinct eigenvalues are orthogonal:

$$\mathcal{E}_n \neq \mathcal{E}_m \Rightarrow (\psi_n, \psi_m) = \int \psi_n^*(x)\psi_m(x) d^N x = 0. \qquad (5.1.10)$$

If the Hamiltonian $\mathcal{H}(p,x)$ is a polynomial in the momentum operators, for example, the Newtonian form $\mathcal{H}(p,x) = \sum_{j=1}^{N} p_j^2 + U(x)$, the eigenvalue problem (5.1.8) is a partial differential equation

$$\left[-\hbar^2 \sum_{j=1}^{N} \frac{\partial^2}{\partial x_j^2} + U(x) \right] \psi_n(x) = \mathcal{E}_n\psi_n(x). \qquad (5.1.11)$$

If the Hamiltonian $\mathcal{H}(p,x)$ contains the momentum operators in exponentiated forms, for example, $\mathcal{H}(p,x) = \sum_{j=1}^{N} \left[f_j(x)e^{\beta p_j} + g_j(x)e^{-\beta p_j} + \text{h.c.} \right]$, such operators act as difference operators on the wavefunction:

$$\left(e^{\pm \beta p_j}\psi \right)(x) = \psi(x_1, \cdots, x_j \mp i\beta\hbar, \cdots, x_N), \qquad (5.1.12)$$

and the eigenvalue problem (5.1.8) becomes a difference equation.

Quantum-Classical Correspondence The quantum theory is an $\hbar$-deformation of the classical theory. For arbitrary observables $A(p,x)$ and $B(p,x)$, the correspondence is

$$\hbar \to 0, \quad [A(p,x), B(p,x)] \Longrightarrow i\hbar\{A(p,x), B(p,x)\} + o(\hbar^2). \qquad (5.1.13)$$

On the left hand side, $A(p,x)$ and $B(p,x)$ are linear operators acting on the wavefunction, whereas on the right hand side $A(p,x)$ and $B(p,x)$ are functions on the symplectic manifold spanned by p_j and x_j, $j = 1, \cdots, N$.

A classical system is solved if one can find a transformation to the action and angle variables (J_i, φ_i):

$$(p_1, \cdots, p_N, x_1, \cdots, x_N) \to (J_1, \cdots, J_N, \varphi_1, \cdots, \varphi_N), \tag{5.1.14}$$

$$\{\varphi_j, J_k\} = \delta_{jk}, \quad \{\varphi_j, \varphi_k\} = \{J_j, J_k\} = 0, \tag{5.1.15}$$

$$\frac{d}{dt}J_k = 0, \quad \frac{d}{dt}\varphi_k = \omega_k, \quad \omega_k : \text{constant}, \quad i = 1, \cdots, N. \tag{5.1.16}$$

A quantum system is exactly solved in the Schrödinger picture if all the eigenvalues $\{\mathcal{E}_n\}$ and the corresponding eigenfunctions $\{\psi_n\}$ in (5.1.8) are explicitly constructed. In the Heisenberg picture (5.1.5)–(5.1.6) the exact solvability or integrability has not yet been clearly formulated except for the cases of one degree of freedom, see section 5.2.5.

Liouville Theorem: Classical & Quantum? (刘维尔定理：经典与量子？)

Classical Case In most cases it is impossible to reduce the system to the action-angle variables. The sufficient condition for the reduction was found by Liouville[4] (mid 19-th century). If independent conserved quantities $\{K_j\}$, in involution, as many as the degrees of freedom (N) are found

$$\{K_j, K_k\} = 0, \quad j,k = 1, \cdots, N, \quad K_1 \equiv \mathcal{H}, \tag{5.1.17}$$

$$\Rightarrow \frac{d}{dt}K_j = 0, \quad j = 2, \cdots, N, \tag{5.1.18}$$

then the generating function of the canonical transformation to reduce the system to the action-angle variables can be constructed in terms of $\{K_j\}$ by quadrature only.

Remarks
- the First Half: a sufficient condition for "integrability".
- the Second Half gives the method achieving the solution.
- the Theorem does not require that the results of the quadrature are known functions or elementary functions.
- the Theorem never tells how to find (construct) an integrable system.
- the conserved quantities $\{K_j\}$ become the *action* variables. Therefore the involution property is the necessary condition.

5.1 Introduction

- if M_k is compact and connected

$$M_k \equiv \{(p,x) : K_j(p,x) = k_j, j = 1, \cdots, N\}$$

$$\Longrightarrow M_k \cong T^N: \quad N\text{-dimensional Torus}$$

angle variables parametrise the angles along the Torus.
- Noether's Theorem: conserved quantity $\Leftrightarrow$ symmetry.
- only the commuting part (the Cartan sub-algebra) of the symmetry algebra matters, rank $= N$.
- abelian $u(1)^N$ or non-abelian symmetry?
- a system can have more than N independent conserved quantities, (called super-integrable) but only up to N are mutually in involution.
- a maximally super-integrable system has $2N - 1$ conserved quantities, which uniquely determine the time-trajectory of the system.
- an infinite degree system or a classical field theory needs an infinite number of conserved quantities for integrability. In such a case, a proper number of infinity is needed.
- they should have infinite rank symmetry algebra, very different from known affine or Kac-Moody Lie algebra.
- the well-known soliton equations, for example, KdV (Korteweg de Vries), mKdV (Modified Korteweg de Vries), NLS (Non-Linear Schrödinger), s-G (sine-Gordon) equations are known to have the proper infinite number of conserved quantities and explicit elementary solutions called soliton solutions.
- the method to generate the infinite number of conserved quantities and transformation to the action-angle variables for soliton equations are called linear-scattering method.
- $N = 1$ systems are always integrable since the Hamiltonian is always given and conserved. The equation of motion is always solvable by quadrature. Suppose the Hamiltonian has the standard Newtonian form: $\mathcal{H} = \frac{1}{2}p^2 + U(x)$. It can be easily solved from the conservation of energy

$$p = \dot{x}, \quad \frac{1}{2}\dot{x}^2 + U(x) = \mathcal{E}, \quad \Longrightarrow \quad t = \int^x \frac{dx'}{\sqrt{2(\mathcal{E} - U(x'))}}. \tag{5.1.19}$$

The situation is the same for Hamiltonians having quite general momentum dependence $\mathcal{H} = h(p,x)$, including the exponentiated form. From the energy conservation $h(p,x) = \mathcal{E} \stackrel{\text{def}}{=} h(p_0, x_0)$, one can solve the momentum $p = g(x, \mathcal{E})$, which is in general, not unique. By substituting this to the canonical equation for x (5.1.2), one obtains

$$\frac{dx}{dt} = G(x, \mathcal{E}), \quad \Longrightarrow \quad t = \int^x \frac{dx'}{G(x', \mathcal{E})}. \tag{5.1.20}$$

It should be stressed that the Classical Liouville Theorem works for quite general forms of the Hamiltonians.

Quantum Case The notion corresponding to the action-angle variables does not exist. Obviously, if all the eigenvalues and the corresponding eigenfunctions of a given Hamiltonian are known, the system can be said to be completely solved. As we will see shortly, complete solvability is much stronger than the quantum integrability.

Quantum Liouville Theorem? The First Half of the Quantum Liouville Theorem is straightforward to write down:

Let us call a Hamiltonian $\mathcal{H}$ is quantum integrable if independent conserved quantities $\{K_j\}$, in involution, as many as the degrees of freedom (N) are found

$$[K_j, K_k] = 0, \quad j, k = 1, \cdots, N, \quad K_1 \equiv \mathcal{H}, \tag{5.1.21}$$

$$\Rightarrow \frac{d}{dt} K_j = 0, \quad j = 2, \cdots, N. \tag{5.1.22}$$

The Second Half is not yet formulated. We still do not have an explicit method to solve a quantum system starting from the knowledge of conserved quantities, in a way the classical Liouville Theorem does. Of course we can say;

We should try to find the basis of the Hilbert space consisting of the simultaneous eigenstates of all the conserved quantities $\{K_j\}$.

This is still a statement of the goal, not a concrete way to go there.

Remarks:
- $N = 1$ systems are always quantum integrable.
- we know that the complete spectrum and eigenfunctions are not known for most of the $N = 1$ quantum systems. Quantum integrability is definitely not the exact solvability.
- it is a good challenge to formulate a full Quantum Liouville Theorem.

5.2 1-Degree of Freedom System (单自由度系统)

Let us consider one-dimensional QM defined in an interval (x_1, x_2), in which x_1 and/or x_2 can be infinite. For finite x_j, $j = 1, 2$, the potential must provide an infinite barrier $\lim_{x \to x_j} U(x) = +\infty$ at that boundary lest the particle tunnel out from (x_1, x_2). This fact provides proper boundary conditions of the wavefunctions. Hereafter we adopt the convention $\hbar = 1$ and $2m = 1$ and consider the following Hamiltonian

$$\mathcal{H} \stackrel{\text{def}}{=} -\frac{d^2}{dx^2} + U(x), \quad x_1 < x < x_2, \tag{5.2.1}$$

with a smooth potential $U(x) \in \mathbf{C}^\infty$. We also require that the Hamiltonian is bounded from below. The eigenvalue problem is to find all the discrete eigenvalues $\{\mathcal{E}_n\}$ and the corresponding eigenfunctions $\{\phi_n(x)\}$

$$\mathcal{H}\phi_n(x) = \mathcal{E}_n \phi_n(x), \quad n = 0, 1, \cdots, \tag{5.2.2}$$

of the given Hamiltonian $\mathcal{H}$ (5.2.1). In 1-d QM the discrete eigenvalues do not have degeneracy. The numbering of the eigenvalues is monotonously increasing:

$$0 = \mathcal{E}_0 < \mathcal{E}_1 < \mathcal{E}_2 < \cdots. \tag{5.2.3}$$

The eigenfunctions are mutually orthogonal

$$(\phi_n, \phi_m) \stackrel{\text{def}}{=} \int_{x_1}^{x_2} \phi_n(x)^* \phi_m(x) dx = h_n \delta_{nm}, \quad 0 < h_n < \infty, \; n, m = 0, 1, \cdots, \tag{5.2.4}$$

which is a consequence of the self-adjointness (or the hermiticity in physics) of the Hamiltonian $\mathcal{H}$, [2, 3]. Then the oscillation theorem[5] asserts that the n-th eigenfunction $\phi_n(x)$ has n simple zeros in (x_1, x_2). In particular the ground state eigenfunction $\phi_0(x)$ has no zero in (x_1, x_2), and we will choose the convention that it is positive $\phi_0(x) > 0$. We also choose all the eigenfunctions to be real, $\phi_n(x) \in \mathbf{R}$.

5.2.1 Factorised Hamiltonian (因式化哈密顿量)

Let us consider the eigenvalue problem of a given Hamiltonian (5.2.1) having a finite or semi-infinite number of discrete energy levels. The additive constant of the Hamiltonian is so chosen that the ground state energy vanishes, $\mathcal{E}_0 = 0$. That is, the Hamiltonian is positive semi-definite. It is a well known theorem in linear algebra that any positive semi-definite hermitian matrix can be factorised as a product of a certain matrix, say $\mathcal{A}$, and its hermitian conjugate $\mathcal{A}^\dagger$. As we will see shortly, the Hamiltonians we consider always have factorised forms in one-dimension as well as in higher dimensions.

The Hamiltonian we consider has a simple factorised form [6]

$$\mathcal{H} \stackrel{\text{def}}{=} \mathcal{A}^\dagger \mathcal{A} \quad \text{or} \quad \mathcal{H} \stackrel{\text{def}}{=} \sum_{j=1}^{D} \mathcal{A}_j^\dagger \mathcal{A}_j \quad \text{in } D \text{ dimensions.} \tag{5.2.5}$$

The operators $\mathcal{A}$ and $\mathcal{A}^\dagger$ in 1-d QM are:

$$\mathcal{A} \stackrel{\text{def}}{=} \frac{\mathrm{d}}{\mathrm{d}x} - \frac{\mathrm{d}w(x)}{\mathrm{d}x} = \frac{\mathrm{d}}{\mathrm{d}x} - \frac{\partial_x \phi_0(x)}{\phi_0(x)}, \quad w(x) \in \mathbf{R}, \quad \phi_0(x) = \mathrm{e}^{w(x)}, \tag{5.2.6}$$

$$\mathcal{A}^\dagger = -\frac{\mathrm{d}}{\mathrm{d}x} - \frac{\mathrm{d}w(x)}{\mathrm{d}x} = -\frac{\mathrm{d}}{\mathrm{d}x} - \frac{\partial_x \phi_0(x)}{\phi_0(x)},$$

$$\mathcal{H} = p^2 + U(x), \quad U(x) \stackrel{\text{def}}{=} \left[\partial_x w(x)\right]^2 + \partial_x^2 w(x), \tag{5.2.7}$$

in which a real function $w(x)$ is called a prepotential. The Hamiltonian of a multi-degrees of freedom system can be constructed in a similar way:

$$\mathcal{A}_j \stackrel{\text{def}}{=} \frac{\partial}{\partial x_j} - \frac{\partial w(x)}{\partial x_j}, \quad \mathcal{A}_j^\dagger = -\frac{\partial}{\partial x_j} - \frac{\partial w(x)}{\partial x_j} \quad (j=1,\cdots,D), \quad \phi_0(x) = \mathrm{e}^{w(x)}. \tag{5.2.8}$$

The prepotential approach is also useful in Calogero-Sutherland systems (5.3.4).

The Schrödinger equation (5.2.2) is a second order differential equation and the ground state wavefunction $\phi_0(x)$ is determined as a zero mode of the operator $\mathcal{A}$ ($\mathcal{A}_j$) which is a first order equation:

$$\mathcal{A}\phi_0(x) = 0 \quad (\mathcal{A}_j \phi_0(x) = 0, \ j=1,\cdots,D) \quad \Rightarrow \quad \mathcal{H}\phi_0(x) = 0. \tag{5.2.9}$$

It should be stressed that the inverse of the zero mode of $\mathcal{A}$ is the zero mode of $\mathcal{A}^\dagger$:

$$\mathcal{A}^\dagger \phi_0^{-1}(x) = 0 \quad (\mathcal{A}_j^\dagger \phi_0^{-1}(x) = 0, \ j=1,\cdots,D). \tag{5.2.10}$$

At the end of this subsection, let us emphasise that any non-vanishing (for example, positive) solution of the original Hamiltonian (5.2.1)

$$\mathcal{H}\tilde{\phi}(x) = \tilde{\mathcal{E}}\tilde{\phi}(x), \quad \tilde{\mathcal{E}} \in \mathbf{R}, \quad \tilde{\phi}(x) > 0, \quad x \in (x_1, x_2), \tag{5.2.11}$$

provides a non-singular factorisation of the original Hamiltonian:

$$\mathcal{H} = \tilde{\mathcal{A}}^\dagger \tilde{\mathcal{A}} + \tilde{\mathcal{E}}, \quad \tilde{\mathcal{A}} \stackrel{\text{def}}{=} \frac{\mathrm{d}}{\mathrm{d}x} - \frac{\partial_x \tilde{\phi}(x)}{\tilde{\phi}(x)}, \quad \tilde{\mathcal{A}}^\dagger = -\frac{\mathrm{d}}{\mathrm{d}x} - \frac{\partial_x \tilde{\phi}(x)}{\tilde{\phi}(x)}. \tag{5.2.12}$$

5.2.2 Intertwining Relations: Crum's Theorem (交互关系:Crum's 定理)

In this subsection we show the general structure of the solution space of 1-d QM. Let us denote by $\mathcal{H}^{[0]}$ the original factorised Hamiltonian $\mathcal{H}$ (5.2.1), (5.2.5) and by $\mathcal{H}^{[1]}$ its partner (associated) Hamiltonian obtained by changing the order of $\mathcal{A}^\dagger$ and $\mathcal{A}$:

$$\mathcal{H} \equiv \mathcal{H}^{[0]} \stackrel{\text{def}}{=} \mathcal{A}^\dagger \mathcal{A}, \qquad \mathcal{H}^{[1]} \stackrel{\text{def}}{=} \mathcal{A}\mathcal{A}^\dagger = \mathcal{H}^{[0]} - 2\partial_x^2 \log \phi_0(x). \qquad (5.2.13)$$

One simple and most important consequence of the factorised Hamiltonians (5.2.13) is the intertwining relations:

$$\mathcal{A}\mathcal{H}^{[0]} = \mathcal{A}\mathcal{A}^\dagger \mathcal{A} = \mathcal{H}^{[1]}\mathcal{A}, \qquad \mathcal{A}^\dagger \mathcal{H}^{[1]} = \mathcal{A}^\dagger \mathcal{A}\mathcal{A}^\dagger = \mathcal{H}^{[0]}\mathcal{A}^\dagger. \qquad (5.2.14)$$

The pair of Hamiltonians $\mathcal{H}^{[0]}$ and $\mathcal{H}^{[1]}$ are essentially *iso-spectral* and their eigenfunctions $\{\phi_n^{[0]}(x)\}$ and $\{\phi_n^{[1]}(x)\}$ are related by the Darboux-Crum transformations[7, 8]:

$$\mathcal{H}^{[0]}\phi_n^{[0]}(x) = \mathcal{E}_n \phi_n^{[0]}(x) \quad (n=0,1,\cdots), \qquad \mathcal{A}\phi_0^{[0]}(x) = 0, \qquad (5.2.15)$$

$$\mathcal{H}^{[1]}\phi_n^{[1]}(x) = \mathcal{E}_n \phi_n^{[1]}(x) \quad (n=1,2,\cdots), \qquad (5.2.16)$$

$$\phi_n^{[1]}(x) \stackrel{\text{def}}{=} \mathcal{A}\phi_n^{[0]}(x) = \frac{W[\phi_0, \phi_n](x)}{\phi_0(x)}, \quad \phi_n^{[0]}(x) = \frac{\mathcal{A}^\dagger}{\mathcal{E}_n}\phi_n^{[1]}(x) \quad (n=1,2,\cdots), \qquad (5.2.17)$$

$$(\phi_n^{[1]}, \phi_m^{[1]}) = \mathcal{E}_n(\phi_n^{[0]}, \phi_m^{[0]}) \quad (n,m = 1,2,\cdots). \qquad (5.2.18)$$

In (5.2.17) the Wroskian is defined by $W[f,g](x) \stackrel{\text{def}}{=} f(x)g'(x) - f'(x)g(x)$. The associated Hamiltonian $\mathcal{H}^{[1]}$, sometimes called the partner Hamiltonian, has the lowest eigenvalue $\mathcal{E}_1$ and the state corresponding to $\mathcal{E}_0$ is missing. In other words, the ground state corresponding to $\mathcal{E}_0$ is deleted by the transformation $\mathcal{H}^{[0]} \to \mathcal{H}^{[1]}$.

If the ground state energy $\mathcal{E}_1$ is subtracted from the partner Hamiltonian $\mathcal{H}^{[1]}$, it is again positive semi-definite and can be factorised in terms of new operators $\mathcal{A}^{[1]}$ and $\mathcal{A}^{[1]\dagger}$:

$$\mathcal{H}^{[1]} = \mathcal{A}^{[1]\dagger}\mathcal{A}^{[1]} + \mathcal{E}_1, \qquad \mathcal{A}^{[1]}\phi_1^{[1]}(x) = 0. \qquad (5.2.19)$$

By changing the orders of $\mathcal{A}^{[1]\dagger}$ and $\mathcal{A}^{[1]}$, a new Hamiltonian $\mathcal{H}^{[2]}$ is defined:

$$\mathcal{H}^{[2]} \stackrel{\text{def}}{=} \mathcal{A}^{[1]}\mathcal{A}^{[1]\dagger} + \mathcal{E}_1 = \mathcal{H}^{[1]} - 2\partial_x^2 \log |\phi_1^{[1]}(x)|. \qquad (5.2.20)$$

These two Hamiltonians, $\mathcal{H}^{[1]} - \mathcal{E}_1$ and $\mathcal{H}^{[2]} - \mathcal{E}_1$, are intertwined by $\mathcal{A}^{[1]}$ and $\mathcal{A}^{[1]\dagger}$:

$$\mathcal{A}^{[1]}(\mathcal{H}^{[1]} - \mathcal{E}_1) = \mathcal{A}^{[1]}\mathcal{A}^{[1]\dagger}\mathcal{A}^{[1]} = (\mathcal{H}^{[2]} - \mathcal{E}_1)\mathcal{A}^{[1]}, \qquad (5.2.21)$$

$$\mathcal{A}^{[1]\dagger}(\mathcal{H}^{[2]} - \mathcal{E}_1) = \mathcal{A}^{[1]\dagger}\mathcal{A}^{[1]}\mathcal{A}^{[1]\dagger} = (\mathcal{H}^{[1]} - \mathcal{E}_1)\mathcal{A}^{[1]\dagger}. \tag{5.2.22}$$

The iso-spectrality of the two Hamiltonians $\mathcal{H}^{[1]}$ and $\mathcal{H}^{[2]}$ and the relationship among their eigenfunctions follow as before:

$$\mathcal{H}^{[2]}\phi_n^{[2]}(x) = \mathcal{E}_n\phi_n^{[2]}(x) \quad (n = 2, 3, \cdots), \tag{5.2.23}$$

$$\phi_n^{[2]}(x) \stackrel{\text{def}}{=} \mathcal{A}^{[1]}\phi_n^{[1]}(x) = \frac{W[\phi_0, \phi_1, \phi_n](x)}{W[\phi_0, \phi_1](x)}, \quad \phi_n^{[1]}(x) = \frac{\mathcal{A}^{[1]\dagger}}{\mathcal{E}_n - \mathcal{E}_1}\phi_n^{[2]}(x) \quad (n = 2, 3, \cdots), \tag{5.2.24}$$

$$(\phi_n^{[2]}, \phi_m^{[2]}) = (\mathcal{E}_n - \mathcal{E}_1)(\phi_n^{[1]}, \phi_m^{[1]}) \quad (n, m = 2, 3, \cdots), \tag{5.2.25}$$

$$\mathcal{H}^{[2]} = \mathcal{A}^{[2]\dagger}\mathcal{A}^{[2]} + \mathcal{E}_2, \quad \mathcal{A}^{[2]}\phi_2^{[2]}(x) = 0. \tag{5.2.26}$$

By the transformation $\mathcal{H}^{[1]} \to \mathcal{H}^{[2]}$ the state corresponding to $\mathcal{E}_1$ is deleted. The Wronskian of n-functions $\{f_1, \cdots, f_n\}$ is defined by the formula

$$W[f_1, \cdots, f_n](x) \stackrel{\text{def}}{=} \det\left(\frac{d^{j-1}f_k(x)}{dx^{j-1}}\right)_{1 \leq j, k \leq n}. \tag{5.2.27}$$

The following properties of the Wronskians are instrumental:

$$W[gf_1, gf_2, \cdots, gf_n](x) = g(x)^n W[f_1, f_2, \cdots, f_n](x), \tag{5.2.28}$$

$$W[f_1(y), f_2(y), \cdots, f_n(y)](x)$$
$$= y'(x)^{n(n-1)/2} W[f_1, f_2, \cdots, f_n](y), \tag{5.2.29}$$

$$W\big[W[f_1, f_2, \cdots, f_n, g], W[f_1, f_2, \cdots, f_n, h]\big](x)$$
$$= W[f_1, f_2, \cdots, f_n](x) W[f_1, f_2, \cdots, f_n, g, h](x) \quad (n \geq 0). \tag{5.2.30}$$

This process can go on indefinitely by successively deleting the lowest lying energy level. The Hamiltonian and the eigenfunctions at the M-th step have succinct determinant forms in terms of the Wronskian: $(n \geq M \geq 0)$

$$\mathcal{H}^{[M]} = \mathcal{H}^{[0]} - 2\partial_x^2 \log |W[\phi_0, \phi_1, \cdots, \phi_{M-1}](x)|, \tag{5.2.31}$$

$$\mathcal{H}^{[M]}\phi_n^{[M]}(x) = \mathcal{E}_n\phi_n^{[M]}(x) \quad (n = M, M+1, \cdots), \quad \mathcal{A}^{[M]}\phi_M^{[M]}(x) = 0, \tag{5.2.32}$$

$$\phi_n^{[M]}(x) = \frac{W[\phi_0, \phi_1, \cdots, \phi_{M-1}, \phi_n](x)}{W[\phi_0, \phi_1, \cdots, \phi_{M-1}](x)}, \tag{5.2.33}$$

$$(\phi_n^{[M]}, \phi_m^{[M]}) = \prod_{j=0}^{M-1}(\mathcal{E}_n - \mathcal{E}_j) \cdot (\phi_n, \phi_m). \tag{5.2.34}$$

Exercise Demonstrate (5.2.31)~(5.2.34) by induction.

To sum up, we have the following

5.2 1-Degree of Freedom System

Theorem 1 (Crum [8]). For a given Hamiltonian system $\mathcal{H} \equiv \mathcal{H}^{[0]}$, there are associated Hamiltonian systems $\mathcal{H}^{[1]}$, $\mathcal{H}^{[2]}$, $\cdots$, as many as the total number of discrete eigenvalues of the original system $\mathcal{H}^{[0]}$. They share the same eigenvalues $\{\mathcal{E}_n\}$ of the original system and the eigenfunctions of $\mathcal{H}^{[j]}$ and $\mathcal{H}^{[j+1]}$ are related linearly by $\mathcal{A}^{[j]}$ and $\mathcal{A}^{[j]\dagger}$.

This situation of the Crum's theorem is illustrated in Fig.5.1. If the original system $\mathcal{H}^{[0]}$ is exactly solvable, then all the associated systems $\{\mathcal{H}^{[j]}\}$ are also exactly solvable.

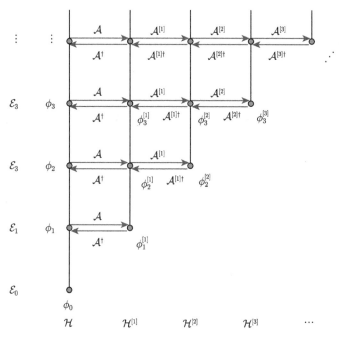

Fig. 5.1 General structure of the solution space of 1-d QM

5.2.3 Five Typical Solvable Potentials (五种典型的可解势)

Here are five elementary examples of exactly solvable potentials [6, 9, 10]. The prepotential $w(x)$ determines the potential $U(x)$ of the Hamiltonian as shown in (5.2.7):

H : $\quad w(x) = -\dfrac{1}{2}x^2, \quad \phi_0(x) = e^{-x^2/2}, \qquad\qquad -\infty < x < \infty,$

harmonic oscillator: $\quad U(x) = x^2 - 1, \qquad\qquad \eta(x) = x. \qquad (5.2.35)$

L : $\quad w(x) = -\dfrac{1}{2}x^2 + g\log x, \quad \phi_0(x; g) = e^{-x^2/2}x^g, \qquad g > \dfrac{1}{2}, \quad 0 < x < \infty,$

radial oscillator: $U(x) = x^2 + \dfrac{g(g-1)}{x^2} - 1 - 2g$, $\quad \eta(x) = x^2$. (5.2.36)

G: $w(x) = g \log \sin x$, $\phi_0(x; g, h) = (\sin x)^g$, $g > \dfrac{1}{2}$, $0 < x < \pi$,

$1/\sin^2 x$ potential: $U(x) = \dfrac{g(g-1)}{\sin^2 x} - g^2$, $\quad \eta(x) = \cos x$. (5.2.37)

J: $w(x) = g \log \sin x + h \log \cos x$, $\phi_0(x; g, h) = (\sin x)^g (\cos x)^h$,

$$g > \frac{1}{2},\ h > \frac{1}{2},\ 0 < x < \frac{\pi}{2},$$

Pöschl-Teller: $U(x) = \dfrac{g(g-1)}{\sin^2 x} + \dfrac{h(h-1)}{\cos^2 x} - (g+h)^2, \eta(x) = \cos 2x$, (5.2.38)

hJ: $w(x) = g \log \sinh x - h \log \cosh x$, $\phi_0(x; g, h) = (\sinh x)^g (\cosh x)^{-h}$,

$$h > g > \frac{1}{2},\ 0 < x < \infty,\quad \eta(x) = \cosh 2x,$$

hyperbolic Pöschl-Teller: $U(x) = \dfrac{g(g-1)}{\sinh^2 x} - \dfrac{h(h+1)}{\cosh^2 x} + (h-g)^2$. (5.2.39)

Remark: Classical Hamiltonian is obtained by $g(g - \hbar) \to g^2$, e.g. g^2/x^2, $g^2/\sin^2 x$ etc. Their eigenfunctions have a factorised form:

$$\phi_n(x) = \phi_0(x) P_n(\eta(x)), \quad \phi_0(x) = e^{w(x)}, \tag{5.2.40}$$

in which $P_n(\eta(x))$ is a degree n polynomial in the sinusoidal coordinate $\eta(x)$. $\phi_0(x)$ is the ground state eigenfunction. For the above examples, they are the three classical orthogonal polynomials; the Hermite (H), the Laguerre (L), the Gegenbauer or the ultra-spherical (G) and the Jacobi (J) polynomials (for notation, see Appendix):

H: $P_n(\eta(x)) = H_n(x) \stackrel{\text{def}}{=} (2x)^n \,_2F_0\left(\begin{array}{c} -\dfrac{n}{2}, -\dfrac{n-1}{2} \\ - \end{array}\Big| -\dfrac{1}{x^2}\right)$, (5.2.41)

L: $P_n(\eta(x)) = L_n^{(g-\frac{1}{2})}(x^2) \stackrel{\text{def}}{=} \dfrac{\left(g+\dfrac{1}{2}\right)_n}{n!} \,_1F_1\left(\begin{array}{c} -n \\ g+\dfrac{1}{2} \end{array}\Big| x^2\right)$, (5.2.42)

G: $P_n(\eta(x)) = C_n^g(\cos x) \stackrel{\text{def}}{=} \dfrac{(2g)_n}{n!} \,_2F_1\left(\begin{array}{c} -n, n+2g \\ g+\dfrac{1}{2} \end{array}\Big| \dfrac{1-\cos x}{2}\right)$, (5.2.43)

5.2 1-Degree of Freedom System

J : $P_n(\eta(x)) = P_n^{(g-\frac{1}{2},h-\frac{1}{2})}(\cos 2x)$

$$\stackrel{\text{def}}{=} \frac{\left(g+\frac{1}{2}\right)_n}{n!} {}_2F_1\left(\begin{matrix}-n, n+g+h\\ g+\frac{1}{2}\end{matrix}\bigg| \frac{1-\cos 2x}{2}\right), \quad (5.2.44)$$

hJ : $P_n(\eta(x)) = P_n^{(g-\frac{1}{2},-h-\frac{1}{2})}(\cosh 2x)$

$$\stackrel{\text{def}}{=} \frac{\left(g+\frac{1}{2}\right)_n}{n!} {}_2F_1\left(\begin{matrix}-n, n+g-h\\ g+\frac{1}{2}\end{matrix}\bigg| \frac{1-\cosh 2x}{2}\right). \quad (5.2.45)$$

The orthogonality weight function for these orthogonal polynomials are given by the square of the ground state eigenfunctions:

$$(\phi_n, \phi_m) = \int \phi_0^2(x) P_n(\eta(x)) P_m(\eta(x)) \mathrm{d}x \propto \delta_{nm}. \quad (5.2.46)$$

The similarity transformed Hamiltonian $\widetilde{\mathcal{H}}$ in terms of the ground state wavefunction $\phi_0(x)$ provides the second order equation for the polynomial $P_n(\eta(x))$:

$$\widetilde{\mathcal{H}} P_n(\eta(x)) = \mathcal{E}_n P_n(\eta(x)), \quad \widetilde{\mathcal{H}} \stackrel{\text{def}}{=} \phi_0(x)^{-1} \circ \mathcal{H} \circ \phi_0(x) = -\frac{\mathrm{d}^2}{\mathrm{d}x^2} - 2\frac{\mathrm{d}w(x)}{\mathrm{d}x}\frac{\mathrm{d}}{\mathrm{d}x}. \quad (5.2.47)$$

The exact solvability can be rephrased as the triangularity of $\widetilde{\mathcal{H}}$:

$$\widetilde{\mathcal{H}} \eta(x)^n = \mathcal{E}_n \eta(x)^n + \text{lower degree polynomials in } \eta(x), \quad (5.2.48)$$

in the special basis

$$1, \eta(x), \eta(x)^2, \cdots, \eta(x)^n, \cdots,$$

spanned by the sinusoidal coordinate $\eta(x)$. This situation is expressed as

$$\widetilde{\mathcal{H}} \mathcal{V}_n \subseteq \mathcal{V}_n, \quad \mathcal{V}_n \stackrel{\text{def}}{=} \text{Span}\left[1, \eta(x), \cdots, \eta(x)^n\right]. \quad (5.2.49)$$

Obviously, the square of the ground state wavefunction $\phi_0(x)^2 = e^{2w(x)}$ provides the positive definite orthogonality weight function for the polynomials:

$$\int_{x_1}^{x_2} \phi_0(x)^2 P_n(\eta(x)) P_n(\eta(x)) \mathrm{d}x = h_n \delta_{nm}, \quad (5.2.50)$$

$$h_n = \begin{cases} 2^n n! \sqrt{\pi} & : \text{H} \\[6pt] \dfrac{1}{2\,n!}\Gamma\left(n+g+\dfrac{1}{2}\right) & : \text{L} \\[6pt] \dfrac{2^{2g}\left[\Gamma(n+g+1/2)\right]^2}{2(n+g)n!\Gamma(n+2g)} & : \text{G} \\[6pt] \dfrac{\Gamma\left(n+g+\dfrac{1}{2}\right)\Gamma\left(n+h+\dfrac{1}{2}\right)}{2\,n!\,(2n+g+h)\Gamma(n+g+h)} & : \text{J} \\[6pt] \dfrac{\Gamma\left(n+g+\dfrac{1}{2}\right)\Gamma(h-g-n+1)}{2\,n!\,(h-g-2n)\Gamma\left(h-n+\dfrac{1}{2}\right)} & : \text{hJ} \end{cases} \qquad (5.2.51)$$

Let us emphasise that the weight function, or $\phi_0(x)$ is determined as a solution of a first order differential equation (5.2.9). In the next section we will derive these results (5.2.35)~(5.2.51) based on shape invariance.

Let us look at these examples more closely. For H, the transformed Hamiltonian $\widetilde{\mathcal{H}}$ read

$$\widetilde{\mathcal{H}} = -\frac{d^2}{dx^2} + 2x\frac{d}{dx},$$

for which the triangularity (5.2.48) in the basis $\{1, x, \cdots, x^n\}$ is obvious with $\mathcal{E}_n = 2n$. This operator governs the Hermite polynomials $\{H_n(x)\}$:

$$\widetilde{\mathcal{H}} H_n(x) = 2n H_n(x) \;\Rightarrow\; H_n''(x) - 2x H_n'(x) + 2n H_n(x) = 0. \qquad (5.2.52)$$

Solving this equation for lower n is straightforward.

For L, the transformed Hamiltonian $\widetilde{\mathcal{H}}$ read

$$\widetilde{\mathcal{H}} = -\frac{d^2}{dx^2} + 2\left(x - \frac{g}{x}\right)\frac{d}{dx}.$$

Since $\widetilde{\mathcal{H}} x = 2(x - g/x)$, the polynomial space $\{1, x, \cdots, x^n\}$ is not invariant under this $\widetilde{\mathcal{H}}$. By introducing a new independent variable $y \overset{\text{def}}{=} x^2$, this operator is rewritten as

$$\widetilde{\mathcal{H}} = -4y\frac{d^2}{dy^2} - 4(\alpha + 1 - y)\frac{d}{dy}, \qquad \alpha \overset{\text{def}}{=} g - \frac{1}{2}.$$

The triangularity (5.2.48) in the basis $\{1, y, \cdots, y^n\}$ is obvious with $\mathcal{E}_n = 4n$. This operator governs the Laguerre polynomials $\{L_n^{(\alpha)}(y)\}$:

$$\widetilde{\mathcal{H}} L_n^{(\alpha)}(y) = 4n\, L_n^{(\alpha)}(y) \;\Rightarrow\; x L_n^{(\alpha)''}(x) + (\alpha + 1 - x) L_n^{(\alpha)'}(x) + n\, L_n^{(\alpha)}(x) = 0. \quad (5.2.53)$$

5.2 1-Degree of Freedom System

For G, the transformed Hamiltonian $\tilde{\mathcal{H}}$ read

$$\tilde{\mathcal{H}} = -\frac{d^2}{dx^2} - 2g\cot x \frac{d}{dx} = -\frac{d^2}{dx^2} - 2g\frac{\cos x}{\sin x}\frac{d}{dx}.$$

By introducing an independent variable $y = \cos x$, the above operator reads

$$\tilde{\mathcal{H}} = -(1-y^2)\frac{d^2}{dy^2} + (2g+1)y\frac{d}{dy}.$$

It is rather elementary to show the triangularity (5.2.48) in the basis $\{1, \cos x, \cdots, \cos^n x\}$ with $\mathcal{E}_n = n(n+2g)$. This operator governs the Gegenbauer polynomials $\{C_n^{(g)}(y)\}$:

$$(1-x^2)C_n^{(g)''}(x) - (2g+1)xC_n^{(g)'}(x) + n(n+2g)C_n^{(g)}(x) = 0. \tag{5.2.54}$$

For J, the transformed Hamiltonian $\tilde{\mathcal{H}}$ read

$$\tilde{\mathcal{H}} = -\frac{d^2}{dx^2} - 2(g\cot x - h\tan x)\frac{d}{dx} = -\frac{d^2}{dx^2} - 2\left(g\frac{\cos x}{\sin x} - h\frac{\sin x}{\cos x}\right)\frac{d}{dx}.$$

The presence of $\sin x$ and $\cos x$ in the denominator suggests a variable $\eta(x)$ with the derivative $\eta'(x) \propto \sin x \cos x$. The simplest choice is $\eta(x) = \cos 2x$. By introducing an independent variable $y \stackrel{\text{def}}{=} \cos 2x$, this operator is rewritten as

$$\tilde{\mathcal{H}} = -4(1-y^2)\frac{d^2}{dy^2} + 4[g - h + (g+h+1)y]\frac{d}{dy},$$

$$= -4(1-y^2)\frac{d^2}{dy^2} + 4[\alpha - \beta + (\alpha+\beta+2)y]\frac{d}{dy}, \quad \alpha \stackrel{\text{def}}{=} g - \frac{1}{2}, \quad \beta \stackrel{\text{def}}{=} h - \frac{1}{2}.$$

The triangularity (5.2.48) in the basis $\{1, y, \cdots, y^n\}$ is obvious with $\mathcal{E}_n = 4n(n+g+h) = 4n(n+\alpha+\beta+1)$. This operator governs the Jacobi polynomials $\{P_n^{(\alpha,\beta)}(y)\}$:

$$\tilde{\mathcal{H}} P_n^{(\alpha,\beta)}(y) = 4n(n+\alpha+\beta+1) P_n^{(\alpha,\beta)}(y)\nu$$
$$(1-x^2)P_n^{(\alpha,\beta)''}(x) + (\beta - \alpha - (\alpha+\beta+2)x)P_n^{(\alpha,\beta)'}(x)$$
$$+ n(n+\alpha+\beta+1) P_n^{(\alpha,\beta)}(x) = 0. \tag{5.2.55}$$

Remarks on singularities Here are some elementary remarks on the regularity and singularities of linear and homogeneous second order differential equations[11] with meromorphic coefficients $f(x)$ and $g(x)$:

$$y''(x) + f(x)y'(x) + g(x)y(x) = 0. \tag{*}$$

At a holomorphic point x_0 of the coefficient functions

$$f(x) = f(x_0) + \sum_{n=1}^{\infty} A_n (x-x_0)^n, \quad g(x) = g(x_0) + \sum_{n=1}^{\infty} B_n (x-x_0)^n,$$

a general solution around x_0 is given by

$$y(x) = \sum_{n=0}^{\infty} C_n (x-x_0)^n,$$

in which C_0 and C_1 are arbitrary and the rest, $\{C_n\}$ $n = 2, \cdots$, is determined recursively by the equation. If $f(x)$ has a simple pole at x_0 and $g(x)$ up to a double pole at the same point:

$$f(x) = \frac{\alpha}{x-x_0} + \sum_{n=0}^{\infty} A_n (x-x_0)^n, \quad g(x) = \frac{\beta}{(x-x_0)^2} + \frac{\gamma}{x-x_0} + \sum_{n=0}^{\infty} B_n (x-x_0)^n,$$

such a point x_0 is called a regular singularity. At the regular singularity, the equation (∗) has two independent local solutions

$$y^{(j)}(x) = (x-x_0)^{\rho_j} \left[1 + \sum_{n=1}^{\infty} C_n^{(j)} (x-x_0)^n \right], \quad j = 1, 2, \qquad (5.2.56)$$

in which ρ_1 and ρ_2 are the roots of the following quadratic equation

$$\rho(\rho-1) + \alpha\rho + \beta = 0 \qquad (5.2.57)$$

and called the characteristic exponents. The rest of the coefficients $\{C_n^{(j)}\}$, $n = 1, \cdots$ are determined recursively by the equation. When $\rho_1 - \rho_2$ is an integer, some modification is needed. If all the singularities are regular singularities, the equation (∗) is called a *Fuchsian* differential equation.

Let us note that $x = 0$ for L and $x = 0, \pi/2$ for J are the regular singular points of second order differential equations. The monodromy at the regular singular point is determined by the characteristic exponent ρ:

$$M_\rho = e^{2\pi i \rho}. \qquad (5.2.58)$$

The corresponding exponents are expressed simply by the original parameters:

$$\rho = g, \ 1-g \text{ for } L \text{ and } \rho = g, \ 1-g \ (x=0), \quad \rho = h, \ 1-h \ (x=\pi/2) \text{ for J}.$$
$$(5.2.59)$$

It is obvious that the radial oscillator (5.2.36) and the Pöschl-Teller (5.2.38) potentials without the constant terms $(-(1+2g), -(g+h)^2)$ are invariant under the following discrete symmetry transformations:

$$\text{L: } g \leftrightarrow 1-g\,;\quad \text{J: } g \leftrightarrow 1-g \text{ and/or } h \leftrightarrow 1-h. \tag{5.2.60}$$

The same transformations also keep the above characteristic exponents (5.2.59) invariant. Likewise, the Hamiltonians of the harmonic (5.2.35) and the radial (5.2.36) oscillators (without the constant term) change sign under the discrete transformation of the coordinate, $x \to ix$:

$$\mathcal{H}_h \to -\mathcal{H}_h, \quad \mathcal{H}_r \to -\mathcal{H}_r;\quad \mathcal{H}_h \stackrel{\text{def}}{=} -\partial_x^2 + x^2,$$
$$\mathcal{H}_r \stackrel{\text{def}}{=} -\partial_x^2 + x^2 + \frac{g(g-1)}{x^2}. \tag{5.2.61}$$

It should be noted that the above symmetries (5.2.60)-(5.2.61) are not respected by the corresponding polynomial equations (5.2.52), (5.2.53) and (5.2.55), since another (singular) solution is discarded due to square integrability of the eigenfunctions.

5.2.4 Shape Invariance: Sufficient Condition of Exact Solvability (形状不变性: 完全可解性的充分条件)

Shape invariance[12] is a sufficient condition for the exact solvability in the Schrödinger picture. Combined with Crum's theorem[8], or the factorisation method[6] or the so-called supersymmetric quantum mechanics[10, 13], the totality of the discrete eigenvalues and the corresponding eigenfunctions can be easily obtained as shown in (5.2.63) and (5.2.64).

In many cases the Hamiltonian contains some parameter(s), $\boldsymbol{\lambda} = (\lambda_1, \lambda_2, \cdots)$. Here we write the parameter dependence symbolically, $\mathcal{H}(\boldsymbol{\lambda})$, $\mathcal{A}(\boldsymbol{\lambda})$, $\mathcal{E}_n(\boldsymbol{\lambda})$, $\phi_n(x;\boldsymbol{\lambda})$, $P_n(\eta(x);\boldsymbol{\lambda})$ etc, since it is the central issue. The shape invariance condition with a suitable choice of parameters is

$$\mathcal{A}(\boldsymbol{\lambda})\mathcal{A}(\boldsymbol{\lambda})^\dagger = \mathcal{A}(\boldsymbol{\lambda}+\boldsymbol{\delta})^\dagger \mathcal{A}(\boldsymbol{\lambda}+\boldsymbol{\delta}) + \mathcal{E}_1(\boldsymbol{\lambda}), \tag{5.2.62}$$

where $\boldsymbol{\delta}$ is the shift of the parameters. In other words $\mathcal{H}^{[0]}(\boldsymbol{\lambda})$ and $\mathcal{H}^{[1]}(\boldsymbol{\lambda})-\mathcal{E}_1(\boldsymbol{\lambda})$ have the same shape, only the parameters are shifted by $\boldsymbol{\delta}$. The s-th step Hamiltonian $\mathcal{H}^{[s]}$ in section 5.2.2 is $\mathcal{H}^{[s]}(\boldsymbol{\lambda}) = \mathcal{H}(\boldsymbol{\lambda}+s\boldsymbol{\delta})+\mathcal{E}(s;\boldsymbol{\lambda})$. The energy spectrum and the excited state wavefunctions are determined by the data of the ground state wavefunction

$\phi_0(x;\boldsymbol{\lambda})$ and the energy of the first excited state $\mathcal{E}_1(\boldsymbol{\lambda})$ as follows [13]:

$$\mathcal{E}_n(\boldsymbol{\lambda}) = \sum_{s=0}^{n-1} \mathcal{E}_1(\boldsymbol{\lambda}^{[s]}), \quad \boldsymbol{\lambda}^{[s]} \stackrel{\text{def}}{=} \boldsymbol{\lambda} + s\boldsymbol{\delta}, \tag{5.2.63}$$

$$\phi_n(x;\boldsymbol{\lambda}) \propto \mathcal{A}(\boldsymbol{\lambda}^{[0]})^\dagger \mathcal{A}(\boldsymbol{\lambda}^{[1]})^\dagger \mathcal{A}(\boldsymbol{\lambda}^{[2]})^\dagger \cdots \mathcal{A}(\boldsymbol{\lambda}^{[n-1]})^\dagger \phi_0(x;\boldsymbol{\lambda}^{[n]}). \tag{5.2.64}$$

The above formula for the eigenfunctions $\phi_n(x;\boldsymbol{\lambda})$ can be considered as the universal Rodrigues formula for the classical orthogonal polynomials. For the explicit form of the Rodrigues type formula for each polynomial, one only has to substitute the explicit forms of the operator $\mathcal{A}(\boldsymbol{\lambda})$ and the ground state wavefunction $\phi_0(x;\boldsymbol{\lambda})$.

In the case of a finite number of bound states, e.g. the Morse potential, the eigenvalue has a maximum at a certain level n, $\mathcal{E}_n(\boldsymbol{\lambda})$. Beyond that level the formula (5.2.63) ceases to work and the Rodrigues formula (5.2.64) does not provide square integrable eigenfunctions, although ϕ_m $(m > n)$ continues to satisfy the Schrödinger equation with $\mathcal{E}_m(\boldsymbol{\lambda})$.

The above shape invariance condition (5.2.62) is equivalent to the following condition:

$$[\partial_x w(x;\boldsymbol{\lambda})]^2 - \partial_x^2 w(x;\boldsymbol{\lambda}) = [\partial_x w(x;\boldsymbol{\lambda}+\boldsymbol{\delta})]^2 + \partial_x^2 w(x;\boldsymbol{\lambda}+\boldsymbol{\delta}) + \mathcal{E}_1(\boldsymbol{\lambda}). \tag{5.2.65}$$

It is straightforward to verify the shape invariance for the three examples (5.2.35)∼(5.2.38) in section 5.2.3 with the following data:

H: $\quad \boldsymbol{\lambda} = \phi$ (null), $\quad \boldsymbol{\delta} = \phi, \quad \mathcal{A} = \partial_x + x, \quad w(x) = -x^2/2,$ (5.2.66)

$\qquad (-x)^2 - (-1) = (-x)^2 + (-1) + \mathcal{E}(1), \quad \mathcal{E}_1 = 2 \Rightarrow \mathcal{E}_n = 2n,$

L: $\quad \boldsymbol{\lambda} = g, \quad \boldsymbol{\delta} = 1, \quad \mathcal{A}(g) = \partial_x + x - g/x, \quad w(x) = -x^2/2 + g\log x,$

(5.2.67)

$$\left(-x+\frac{g}{x}\right)^2 - \left(-1-\frac{g}{x^2}\right) = \left(-x+\frac{g+\delta}{x}\right)^2 + \left(-1-\frac{g+\delta}{x^2}\right) + \mathcal{E}_1,$$

$\qquad \delta = 1, \quad \mathcal{E}_1 = 4 \Rightarrow \mathcal{E}_n = 4n,$

J: $\quad \boldsymbol{\lambda} = (g,h), \quad \boldsymbol{\delta} = (1,1), \quad \mathcal{A}(g,h) = \partial_x - g\cot x + h\tan x,$ (5.2.68)

$$(g\cot x - h\tan x)^2 - \left(-\frac{g}{\sin^2 x} - \frac{h}{\cos^2 x}\right)$$

$$= [(g+\delta_1)\cot x - (h+\delta_2)\tan x]^2 + \left(-\frac{g+\delta_1}{\sin^2 x} - \frac{h+\delta_2}{\cos^2 x}\right) + \mathcal{E}_1,$$

$\qquad \delta_1 = \delta_2 = 1, \quad \mathcal{E}_1 = 4(g+h+1) \Rightarrow \mathcal{E}_n(g,h) = 4n(n+g+h).$

(5.2.69)

5.2 1-Degree of Freedom System

It should be stressed that the above shape invariant transformation $\lambda \to \lambda + \delta$, $\mathcal{H}^{[s]} \to \mathcal{H}^{[s+1]}$ for L and J, that is, $g \to g+1$, $h \to h+1$, preserves the monodromy (5.2.58) at the regular singularities.

The above universal Rodrigues formula (5.2.64) for the harmonic oscillator (H) reads
$$e^{-x^2/2} P_n(\eta) \propto (-\partial_x + x)^n e^{-x^2/2}.$$

By using the relation $\partial_x - x = e^{x^2/2} \circ \dfrac{d}{dx} \circ e^{-x^2/2}$, this gives $P_n(\eta) \propto (-1)^n e^{x^2} \left(\dfrac{d}{dx}\right)^n e^{-x^2}$ and the Rodrigues formula for the Hermite polynomial (5.6.8) is obtained. The universal Rodrigues formula (5.2.64) for the radial oscillator (L) reads
$$e^{-x^2/2} x^g P_n(\eta) \propto \left(-\partial_x + x - \frac{g}{x}\right) \cdots \left(-\partial_x + x - \frac{g+n-1}{x}\right) e^{-x^2/2} x^{g+n}.$$

By using the relation ($\eta = x^2$)
$$-\partial_x + x - \frac{g}{x} = -e^{x^2/2} x^{-g} \circ \frac{d}{dx} \circ e^{-x^2/2} x^g = -2 e^{\eta/2} \eta^{-(g-1)/2} \circ \frac{d}{d\eta} \circ e^{-\eta/2} \eta^{g/2},$$
the above formula gives
$$P_n(\eta) \propto (-2)^n e^\eta \eta^{-g+1/2} \left(\frac{d}{d\eta}\right)^n \left(e^{-\eta} \eta^{n+g-1/2}\right),$$
and the Rodrigues formula for the Laguerre polynomial (5.6.9) is obtained, up to an n dependent normalisation constant. For the Jacobi, we note
$$-\mathcal{A}^\dagger(\lambda) = \frac{d}{dx} + \partial_x w(x; \lambda) = \phi_0(x; \lambda)^{-1} \circ \frac{d}{dx} \circ \phi_0(x; \lambda)$$
$$= -4 \sin x \cos x \phi_0(x; \lambda)^{-1} \circ \frac{d}{d\eta} \circ \phi_0(x; \lambda)$$
$$= -4 \phi_0(x; \lambda - \delta)^{-1} \circ \frac{d}{d\eta} \circ \phi_0(x; \lambda).$$

This leads to
$$P_n(\eta; \lambda) \propto (-4)^n \phi_0(x; \lambda)^{-1} \phi_0(x; \lambda - \delta)^{-1} \left(\frac{d}{d\eta}\right)^n \phi_0(x; \lambda + (n-1)\delta) \phi_0(x; \lambda + n\delta) \nu$$
$$P_n^{(\alpha,\beta)}(\eta) \propto (-1)^n 2^{-\alpha-\beta-1} \frac{1}{(1-\eta)^\alpha (1+\eta)^\beta} \left(\frac{d}{d\eta}\right)^n [(1-\eta)^{\alpha+n}(1+\eta)^{\beta+n}]. \quad (5.2.70)$$

Exercise Derive the explicit forms of the potentials from the following prepotentials. Show the shape invariance and derive the eigenvalues of the corresponding Hamiltonians. These are all exactly solvable with a finite or infinite number of discrete eigenstates.

(1) Coulomb potential with centrifugal barrier: $w(x;\boldsymbol{\lambda}) = g\log x - \dfrac{x}{g}$, $0 < x < \infty$, $g > 1/2$.

(2) Kepler problem in spherical space: $w(x;\boldsymbol{\lambda}) = g\log\sin x - \dfrac{\mu}{g}x$, $0 < x < \pi$, $g > \dfrac{3}{2}$, $\mu > 0$.

(3) Morse potential: $w(x;\boldsymbol{\lambda}) = hx - \mu e^x$, $-\infty < x < \infty$, $h, \mu > 0$.

(4) Soliton or $1/\cosh^2 x$ potential: $w(x;\boldsymbol{\lambda}) = -h\log\cosh x$, $-\infty < x < \infty$, $h > 0$.

(5) Rosen-Morse potential: $w(x;\boldsymbol{\lambda}) = -h\log\cosh x - \dfrac{\mu}{h}x$, $-\infty < x < \infty$, $h > \sqrt{\mu} > 0$.

5.2.5 Solvability in the Heisenberg Picture (海森伯绘景下的可解性)

As is well known the Heisenberg operator formulation is central to quantum field theory. The creation/annihilation operators of the harmonic oscillators are the cornerstones of modern quantum physics. In the Heisenberg picture, observables depend on time and the state is independent of time. The Heisenberg equation of motion for an observable $\hat{\boldsymbol{A}}(t)$ reads

$$\frac{\partial}{\partial t}\hat{\boldsymbol{A}}(t) = \mathrm{i}\,[\mathcal{H}, \hat{\boldsymbol{A}}(t)].$$

By taking an arbitrary complete set of the normalised orthogonal functions $\{\hat{\phi}_n(x)\}$, $n = 0, 1, \cdots$, it is rewritten in a matrix form:

$$\frac{\partial}{\partial t}\hat{\boldsymbol{A}}_{nm}(t) = \mathrm{i}\sum_{k=0}^{\infty}\left(\mathcal{H}_{nk}\hat{\boldsymbol{A}}_{km}(t) - \hat{\boldsymbol{A}}_{nk}(t)\mathcal{H}_{km}\right),$$

$$\hat{\boldsymbol{A}}_{nm}(t) \stackrel{\text{def}}{=} \int \hat{\phi}_n(x)^*\hat{\boldsymbol{A}}(t)\hat{\phi}_m(x)\mathrm{d}x, \quad \mathcal{H}_{nm} \stackrel{\text{def}}{=} \int \hat{\phi}_n(x)^*\mathcal{H}\hat{\phi}_m(x)\mathrm{d}x.$$

A formal solution of the Heisenberg equation of motion is given by

$$\hat{\boldsymbol{A}}(t) = \mathrm{e}^{\mathrm{i}t\mathcal{H}}\hat{\boldsymbol{A}}(0)\mathrm{e}^{-\mathrm{i}t\mathcal{H}} = \sum_{n=0}^{\infty}\frac{(\mathrm{i}t)^n}{n!}(\operatorname{ad}\mathcal{H})^n\hat{\boldsymbol{A}}(0),$$

in which the definition $(\operatorname{ad}\mathcal{H})X \stackrel{\text{def}}{=} [\mathcal{H}, X]$ is used. For an explicit closed form solution, one must be able to evaluate the multiple commutators

$$[\mathcal{H}, [\mathcal{H}, \cdots, [\mathcal{H}, \hat{\boldsymbol{A}}(0)]]\cdots]$$

and to sum them up. This is why, until recently, it had been generally conceived that the Heisenberg operator solutions are intractable.

5.2 1-Degree of Freedom System

1. Closure Relations

Here we show that most of the shape invariant QM Hamiltonian systems are exactly solvable in the Heisenberg picture, too [14, 15]. To be more precise, the Heisenberg operator of the sinusoidal coordinate $\eta(x)$:

$$e^{it\mathcal{H}}\eta(x)e^{-it\mathcal{H}} \tag{5.2.71}$$

can be evaluated in a closed form. The sinusoidal coordinate $\eta(x)$ is the variable of the orthogonal polynomials constituting the eigenfunctions (5.2.40) of the Hamiltonian $\mathcal{H}$.

It is well known that any orthogonal polynomials starting at degree 0 satisfy the three term recurrence relations [16, 17]

$$\eta Q_n(\eta) = A_n Q_{n+1}(\eta) + B_n Q_n(\eta) + C_n Q_{n-1}(\eta), \quad n \geqslant 0, \tag{5.2.72}$$

with $Q_0(\eta) = $ constant, $Q_{-1}(\eta) = 0$. Here the coefficients A_n, B_n and C_n depend on the normalisation of $\{Q_n(\eta)\}$. They are real and $A_{n-1}C_n > 0$ ($n \geqslant 1$). This is a simple consequence of the orthogonality

$$(Q_n, Q_m) \stackrel{\text{def}}{=} \int W(\eta) Q_n(\eta) Q_m(\eta) d\eta = 0, \quad n > m. \tag{5.2.73}$$

Here $W(\eta)$ is the orthogonality weight function. By construction $\eta Q_n(\eta)$ is a degree $n+1$ polynomial and it is expressed by

$$\eta Q_n(\eta) = A_n Q_{n+1}(\eta) + B_n Q_n(\eta) + C_n Q_{n-1}(\eta) + D_n(\eta),$$

in which $D_n(\eta)$ is a degree $k < n-1$ polynomial in η. This means $(Q_k, D_n) \neq 0$. However, this leads to a contradiction as $(Q_k, \eta Q_n) = (\eta Q_k, Q_n) = 0$ and $(Q_k, Q_{n+1}) = (Q_k, Q_n) = (Q_k, Q_{n-1}) = 0$ and we find $D_n(\eta) \equiv 0$. The three term recurrence (5.2.72) is proved. Conversely all the polynomials starting with degree 0 and satisfying the above three term recurrence relations are orthogonal (Favard's theorem[18]).

For the factorised quantum mechanical eigenfunctions (5.2.40), these relations mean

$$\eta(x)\phi_n(x) = A_n \phi_{n+1}(x) + B_n \phi_n(x) + C_n \phi_{n-1}(x) \ (n \geqslant 0), \quad \phi_{-1}(x) = 0. \tag{5.2.74}$$

In other words, the operator $\eta(x)$ acts like a creation operator which sends the eigenstate n to $n+1$ as well as like an annihilation operator, which maps an eigenstate n to

$n-1$. This fact combined with the well known result that the annihilation/creation operators of the harmonic oscillator are the positive/negative frequency parts of the Heisenberg operator solution for the coordinate x is the starting point of this subsection. As will be shown below the sinusoidal coordinate $\eta(x)$ undergoes sinusoidal motion (5.2.77), whose frequencies depend on the energy. Thus it is not harmonic in general.

A sufficient condition for the closed form expression of the Heisenberg operator (5.2.71) is the closure relation

$$[\mathcal{H}, [\mathcal{H}, \eta(x)]] = \eta(x)\, R_0(\mathcal{H}) + [\mathcal{H}, \eta(x)]\, R_1(\mathcal{H}) + R_{-1}(\mathcal{H}). \tag{5.2.75}$$

Here the coefficients $R_i(y)$ are polynomials in y. It is easy to see that the cubic commutator $[\mathcal{H}, [\mathcal{H}, [\mathcal{H}, \eta(x)]]] \equiv (\operatorname{ad}\mathcal{H})^3 \eta(x)$ is reduced to $\eta(x)$ and $[\mathcal{H}, \eta(x)]$ with $\mathcal{H}$ depending coefficients:

$$(\operatorname{ad}\mathcal{H})^3 \eta(x) = [\mathcal{H}, \eta(x)] R_0(\mathcal{H}) + [\mathcal{H}, [\mathcal{H}, \eta(x)]]\, R_1(\mathcal{H})$$
$$= \eta(x)\, R_0(\mathcal{H}) R_1(\mathcal{H}) + [\mathcal{H}, \eta(x)]\, [R_1(\mathcal{H})^2 + R_0(\mathcal{H})] + R_{-1}(\mathcal{H}) R_1(\mathcal{H}).$$

In this notation the above closure relation (5.2.75) reads

$$(\operatorname{ad}\mathcal{H})^2 \eta(x) = \eta(x)\, R_0(\mathcal{H}) + (\operatorname{ad}\mathcal{H})\eta(x)\, R_1(\mathcal{H}) + R_{-1}(\mathcal{H}), \tag{5.2.76}$$

which can be understood as the Cayley-Hamilton equation for the operator $\operatorname{ad}\mathcal{H}$ acting on $\eta(x)$. It is trivial to see that all the higher commutators $(\operatorname{ad}\mathcal{H})^n \eta(x)$ can also be reduced to $\eta(x)$ and $[\mathcal{H}, \eta(x)]$ with $\mathcal{H}$ depending coefficients. The second order closure (5.2.75) simply reflects the Schrödinger equation, which is a second order differential equation. Thus we arrive at

$$e^{it\mathcal{H}} \eta(x) e^{-it\mathcal{H}} = \sum_{n=0}^{\infty} \frac{(it)^n}{n!} (\operatorname{ad}\mathcal{H})^n \eta(x)$$
$$= [\mathcal{H}, \eta(x)] \frac{e^{i\alpha_+(\mathcal{H})t} - e^{i\alpha_-(\mathcal{H})t}}{\alpha_+(\mathcal{H}) - \alpha_-(\mathcal{H})} - R_{-1}(\mathcal{H}) R_0(\mathcal{H})^{-1}$$
$$+ \bigl(\eta(x) + R_{-1}(\mathcal{H}) R_0(\mathcal{H})^{-1}\bigr) \frac{-\alpha_-(\mathcal{H}) e^{i\alpha_+(\mathcal{H})t} + \alpha_+(\mathcal{H}) e^{i\alpha_-(\mathcal{H})t}}{\alpha_+(\mathcal{H}) - \alpha_-(\mathcal{H})}. \tag{5.2.77}$$

This simply means that $\eta(x)$ oscillates sinusoidally with two energy-dependent "frequencies" $\alpha_\pm(\mathcal{H})$ given by

$$\alpha_\pm(\mathcal{H}) \stackrel{\text{def}}{=} \frac{1}{2}\bigl[R_1(\mathcal{H}) \pm \sqrt{R_1(\mathcal{H})^2 + 4R_0(\mathcal{H})}\,\bigr], \tag{5.2.78}$$

5.2 1-Degree of Freedom System

$$\alpha_+(\mathcal{H}) + \alpha_-(\mathcal{H}) = R_1(\mathcal{H}), \quad \alpha_+(\mathcal{H})\alpha_-(\mathcal{H}) = -R_0(\mathcal{H}). \tag{5.2.79}$$

Here is some explanation. Three term recurrence relations with constant coefficients

$$\alpha a_{n+2} + \beta a_{n+1} + \gamma a_n = 0, \quad n = 0, 1, \cdots,$$

can be solved easily by the roots of a quadratic equation:

$$\alpha x^2 + \beta x + \gamma = 0, \quad x_\pm \stackrel{\text{def}}{=} \frac{-\beta \pm \sqrt{\beta^2 - 4\alpha\gamma}}{2\alpha},$$

$$a_n = A(x_+)^n + B(x_-)^n, \quad A + B = a_0, \quad Ax_+ + Bx_- = a_1.$$

The closure relation (5.2.76) can be cast into the three term recurrence form by a shift of $\eta(x)$:

$$(\text{ad}\,\mathcal{H})^2 \tilde{\eta}(x) - (\text{ad}\,\mathcal{H})\tilde{\eta}(x) R_1(\mathcal{H}) - \tilde{\eta}(x) R_0(\mathcal{H}) = 0,$$

$$\tilde{\eta}(x) \stackrel{\text{def}}{=} \eta(x) + R_{-1}(\mathcal{H}) R_0(\mathcal{H})^{-1}. \tag{5.2.80}$$

This is solved by

$$(\text{ad}\,\mathcal{H})^n \tilde{\eta}(x) = A\alpha_+(\mathcal{H})^n + B\alpha_-(\mathcal{H})^n,$$

$$A = -\tilde{\eta}(x)\frac{\alpha_-(\mathcal{H})}{\alpha_+(\mathcal{H}) - \alpha_-(\mathcal{H})} + [\mathcal{H}, \eta(x)]\frac{1}{\alpha_+(\mathcal{H}) - \alpha_-(\mathcal{H})},$$

$$B = \tilde{\eta}(x)\frac{\alpha_+(\mathcal{H})}{\alpha_+(\mathcal{H}) - \alpha_-(\mathcal{H})} - [\mathcal{H}, \eta(x)]\frac{1}{\alpha_+(\mathcal{H}) - \alpha_-(\mathcal{H})}.$$

This leads to the above result (5.2.77).

The energy spectrum is determined by the over-determined recursion relations

$$\mathcal{E}_{n+1}(\lambda) - \mathcal{E}_n(\lambda) = \alpha_+(\mathcal{E}_n(\lambda)) = \mathcal{E}_1(\lambda + n\delta), \tag{5.2.81}$$

$$\mathcal{E}_{n-1}(\lambda) - \mathcal{E}_n(\lambda) = \alpha_-(\mathcal{E}_n(\lambda)) = -\mathcal{E}_1[\lambda + (n-1)\delta], \quad \mathcal{E}_0 = 0. \tag{5.2.82}$$

It should be stressed that for the known spectra $\{\mathcal{E}(n)\}$ determined by the shape invariance, the quantity inside the square root in the definition of $\alpha_\pm(\mathcal{H})$ (5.2.78) for each n:

$$R_1(\mathcal{E}(n))^2 + 4R_0(\mathcal{E}(n)),$$

becomes a complete square and the above two conditions are consistent. For 1-d QM, the Hamiltonians and the sinusoidal coordinates satisfying the closure relation (5.2.75) are classified and then the eigenfunctions have the factorised form (5.2.40)[14].

For the three elementary examples (5.2.35)~(5.2.38) given in section 5.2.3, the data are:

H : $\quad R_1(y) = 0, \quad R_0(y) = 4, \quad R_{-1}(y) = 0,$ \hfill (5.2.83)

$\quad\quad A_n = 1/2, \quad B_n = 0, \quad C_n = n,$ \hfill (5.2.84)

L : $\quad R_1(y) = 0, \quad R_0(y) = 16, \quad R_{-1}(y) = -8(y + 2g + 1),$ \hfill (5.2.85)

$\quad\quad A_n = -(n+1), \quad B_n = (2n + g + 1/2), \quad C_n = -(n + g - 1/2),$ \hfill (5.2.86)

J : $\quad R_1(y) = 8, \quad R_0(y) = 16[y + (g+h)^2 - 1], \quad R_{-1}(y) = 16(g - h)(g + h - 1),$ \hfill (5.2.87)

$$A_n = \frac{2(n+1)(n+g+h)}{(2n+g+h)(2n+g+h+1)}, \quad B_n = \frac{(h-g)(g+h-1)}{(2n+g+h-1)(2n+g+h+1)},$$

$$C_n = \frac{2(n+g-1/2)(n+h-1/2)}{(2n+g+h-1)(2n+g+h)}. \tag{5.2.88}$$

It is tedious but straightforward to calculate $R_j(y)$ for the three potentials. It is straight forward to verify the recursion relations (5.2.81)、(5.2.82) for the eigenvalue formulas $\mathcal{E}_n$, (5.2.66)~(5.2.69) for the three elementary examples.

In fact,

H : $\quad \alpha_\pm = \pm 2, \quad \mathcal{E}_{n+1} - \mathcal{E}_n = 2, \quad \mathcal{E}_{n-1} - \mathcal{E}_n = -2,$

L : $\quad \alpha_\pm = \pm 4, \quad \mathcal{E}_{n+1} - \mathcal{E}_n = 4, \quad \mathcal{E}_{n-1} - \mathcal{E}_n = -4,$

J : $\quad R_1(y)^2 + 4R_0(y) = 64[1 + y + (g+h)^2 - 1], \quad y \to \mathcal{E}_n = 4n(n+g+h),$

$$\Rightarrow 64\left[4n(n+g+h) + (g+h)^2\right] = 64(2n+g+h)^2,$$

$$\alpha_\pm(n) = 4 \pm 4(2n+g+h),$$

$$\mathcal{E}_{n+1} - \mathcal{E}_n = 4(n+1)(n+1+g+h) - 4n(n+g+h) = 4(2n+g+h+1),$$

$$\mathcal{E}_{n-1} - \mathcal{E}_n = 4(n-1)(n-1+g+h) - 4n(n+g+h) = -4(2n+g+h-1).$$

Exercise Solve the Hamilton equation for the $1/\sin^2 x$ potential (5.2.37) and derive

$$\cos x(t) = \cos x(0) \cdot \cos(2t\sqrt{\mathcal{E}'}) - p(0) \sin x(0) \cdot \frac{\sin(2t\sqrt{\mathcal{E}'})}{2\sqrt{\mathcal{E}'}}, \quad \mathcal{E}' \overset{\text{def}}{=} \mathcal{E} + g^2.$$

For 1-d QM, the necessary and sufficient conditions for the existence of the sinusoidal coordinate satisfying the closure relation (5.2.75) are analysed in Appendix A of [14]. It was shown that such systems constitute a sub-group of the shape invariant 1-d QM. We also mention that exact Heisenberg operator solutions for independent sinusoidal coordinates as many as the degree of freedom were derived for the Calogero systems based on any root system [20]. These are novel examples of infinitely many multi-particle Heisenberg operator solutions.

2. Annihilation and Creation Operators

The annihilation and creation operators $a^{(\pm)}$ are extracted from this exact Heisenberg operator solution:

$$e^{it\mathcal{H}}\eta(x)e^{-it\mathcal{H}} = a^{(+)}e^{i\alpha_+(\mathcal{H})t} + a^{(-)}e^{i\alpha_-(\mathcal{H})t} - R_{-1}(\mathcal{H})R_0(\mathcal{H})^{-1}, \qquad (5.2.89)$$

$$a^{(\pm)} \stackrel{\text{def}}{=} \pm\Big\{[\mathcal{H},\eta(x)] - [\eta(x) + R_{-1}(\mathcal{H})R_0(\mathcal{H})^{-1}]\alpha_{\mp}(\mathcal{H})\Big\}[\alpha_+(\mathcal{H}) - \alpha_-(\mathcal{H})]^{-1}$$

$$= \pm[\alpha_+(\mathcal{H}) - \alpha_-(\mathcal{H})]^{-1}\Big\{[\mathcal{H},\eta(x)] + \alpha_{\pm}(\mathcal{H})[\eta(x) + R_{-1}(\mathcal{H})R_0(\mathcal{H})^{-1}]\Big\},$$

$$(5.2.90)$$

$$e^{it\mathcal{H}}\eta(x)e^{-it\mathcal{H}}\phi_n(x) = e^{it(\mathcal{E}_{n+1}-\mathcal{E}_n)}A_n\phi_{n+1}(x) + B_n\phi_n(x) + e^{it(\mathcal{E}_{n-1}-\mathcal{E}_n)}C_n\phi_{n+1}(x)$$

$$= e^{it\alpha_+(\mathcal{E}_n)}a^{(+)}\phi_n(x) + e^{it\alpha_-(\mathcal{E}_n)}a^{(-)}\phi_n(x)$$

$$- R_{-1}(\mathcal{E}_n)R_0(\mathcal{E}_n)^{-1}\phi_n(x),$$

$$\Downarrow$$

$$a^{(+)\dagger} = a^{(-)}, \quad a^{(+)}\phi_n(x) = A_n\phi_{n+1}(x), \quad a^{(-)}\phi_n(x) = C_n\phi_{n-1}(x), \qquad (5.2.91)$$

$$B_n = -R_{-1}(\mathcal{E}_n)R_0(\mathcal{E}_n)^{-1}. \qquad (5.2.92)$$

5.2.6 Difference Schrödinger Equations (差分薛定谔方程)

Now we discuss 1-d QM governed by difference Schrödinger equations, which are also called 'discrete' QM (dQM). This is a process of deformation. Replacing differential operators by difference operators means introducing a parameter of the finite differences (shifts), which are either pure imaginary or real as shown shortly. In the limit of zero shifts, the original QM is obtained. This is a quite successful deformation as all the (q)-hypergeometric orthogonal polynomials belonging to the Askey scheme[16, 17, 19], e.g. the Wilson, Askey-Wilson, Racah and q-Racah polynomials, are obtained as the main part of the eigenfunctions of the corresponding exactly solvable dQM [21, 22]. The Hamiltonians contain the momentum operator in exponentiated forms $e^{\pm\beta p}$ which work as shift operators on the wavefunction

$$e^{\pm\beta p}\psi(x) = \psi(x \mp i\beta).$$

For the two choices of the parameter β, either real or pure imaginary, we have two types of dQM; with (i) pure imaginary shifts $\beta = \gamma \in \mathbf{R}_{\neq 0}$ (idQM), or (ii) real shifts $\beta = \mathrm{i}$ (rdQM), respectively. In the case of idQM, $\psi(x \mp \mathrm{i}\gamma)$, we require the wavefunctions and potential functions etc to be analytic functions of x with their domains including the real axis or a part of it on which the dynamical variable x is defined. In contrast, in the rdQM, the difference equation gives constraints on wavefunctions only on equally spaced lattice points. Then we choose, after proper rescaling, the variable x to take value of non-negative integers, with the total number either finite $(N+1)$ or infinite $(x_{\max} = N$ or $\infty)$. To sum up, the dynamical variable x of the one dimensional dQM takes continuous or discrete values:

$$\text{idQM}: x \in \mathbf{R}, x \in (x_1, x_2) \ ; \quad \text{rdQM}: x \in \mathbf{Z}_{\geqslant 0}, x \in [0, x_{\max}]. \tag{5.2.93}$$

Here x_1, x_2 may be finite, $-\infty$ or $+\infty$. Correspondingly, the inner product of the wavefunctions has the following form:

$$\text{idQM}: (f,g) = \int_{x_1}^{x_2} f(x)^* g(x) \mathrm{d}x \ ; \quad \text{rdQM}: (f,g) = \sum_{x=0}^{x_{\max}} f(x)^* g(x), \tag{5.2.94}$$

and the norm of $f(x)$ is $\|f\| = \sqrt{(f,f)}$.

As in the ordinary QM discussed above, we will consider the systems having positive semi definite spectrum. The Hamiltonian we consider has a simple factorised form (5.2.5):

$$\mathcal{H} \stackrel{\text{def}}{=} \mathcal{A}^\dagger \mathcal{A} \quad \text{or} \quad \mathcal{H} \stackrel{\text{def}}{=} \sum_{j=1}^{D} \mathcal{A}_j^\dagger \mathcal{A}_j \quad \text{in } D \text{ dimensions.}$$

The operators $\mathcal{A}$ and $\mathcal{A}^\dagger$ in one dimensional QM have the general forms:

$$\text{idQM}: \quad \mathcal{A} \stackrel{\text{def}}{=} \mathrm{i}\bigl[\mathrm{e}^{\frac{\gamma}{2}p}\sqrt{V^*(x)} - \mathrm{e}^{-\frac{\gamma}{2}p}\sqrt{V(x)}\bigr], \qquad \gamma \in \mathbf{R}_{\neq 0},$$
$$\mathcal{A}^\dagger = -\mathrm{i}\bigl[\sqrt{V(x)}\,\mathrm{e}^{\frac{\gamma}{2}p} - \sqrt{V^*(x)}\,\mathrm{e}^{-\frac{\gamma}{2}p}\bigr], \quad V(x), V^*(x) \in \mathbf{C}, \tag{5.2.95}$$
$$\mathcal{H} = \sqrt{V(x)}\mathrm{e}^{\gamma p}\sqrt{V^*(x)} + \sqrt{V^*(x)}\mathrm{e}^{-\gamma p}\sqrt{V(x)} - V(x) - V^*(x), \tag{5.2.96}$$

$$\text{rdQM}: \quad \mathcal{A} \stackrel{\text{def}}{=} \sqrt{B(x)} - \mathrm{e}^{\partial}\sqrt{D(x)}, \quad \mathcal{A}^\dagger = \sqrt{B(x)} - \sqrt{D(x)}\,\mathrm{e}^{-\partial}, \tag{5.2.97}$$
$$D(x) > 0 \ (\text{for } x > 0), \quad D(0) = 0,$$
$$B(x) > 0 \ (\text{for } x \geqslant 0), \quad B(N) = 0 \ (\text{for the finite case}), \tag{5.2.98}$$
$$\mathcal{H} = -\sqrt{B(x)}\mathrm{e}^{\partial}\sqrt{D(x)} - \sqrt{D(x)}\mathrm{e}^{-\partial}\sqrt{B(x)} + B(x) + D(x). \tag{5.2.99}$$

The function $V^*(x)$ in idQM is an analytic function of x obtained from $V(x)$ by the $*$-operation, which is defined as follows. If $f(x) = \sum_n a_n x^n$, $a_n \in \mathbf{C}$, then $f^*(x) \stackrel{\text{def}}{=}$

$\sum_n a_n^* x^n$, in which a_n^* is the complex conjugation of a_n. Obviously $f^{**}(x) = f(x)$ and $\bar{f}(x)^* = f^*(x^*)$. If a function satisfies $f^* = f$, then it takes real values on the real line. The condition $D(0) = 0$ in rdQM (5.2.98) is necessary for the term $\psi(-1)$ not to appear in $\mathcal{H}\psi(0)$. Likewise $B(N) = 0$ is necessary for the finite case.

Remark: The classical idQM Hamiltonian is

$$\mathcal{H} = 2\sqrt{|V(x)|^2}\cosh\gamma p - \text{Re}[V(x)]. \qquad (5.2.100)$$

The rdQM Hamiltonian (5.2.99) does not define classical dynamics since it does not contain the momentum operator p.

The Schrödinger equation

$$\mathcal{H}\phi_n(x) = \mathcal{E}_n\phi_n(x) \ (n = 0, 1, 2, \cdots), \quad 0 = \mathcal{E}_0 < \mathcal{E}_1 < \mathcal{E}_2 < \cdots,$$

is a second order difference equation (dQM) and the ground state wavefunction $\phi_0(x)$ is determined as the zero mode of the operator $\mathcal{A}$ ($\mathcal{A}_j$) which is a first order equation (5.2.9):

$$\mathcal{A}\phi_0(x) = 0 \ (\mathcal{A}_j\phi_0(x) = 0, j = 1, \cdots, D) \ \Rightarrow \ \mathcal{H}\phi_0(x) = 0. \qquad (5.2.101)$$

As before we look for the eigenfunctions in a factorised form:

$$\phi_n(x) = \phi_0(x)P_n(\eta(x)),$$

in which $P_n(\eta(x))$ is a degree n polynomial in the sinusoidal coordinate $\eta(x)$. The explicit forms of the eigenfunctions will be given for various examples (5.2.103)~(5.2.108). Here we require $\phi_0(x)$ to be chosen real and positive for the physical values of x. One important distinction between a differential and a difference equation is the uniqueness of the solution. For a linear differential equation, the solution is unique when the initial conditions are specified. In contrast, a solution of a difference Schrödinger equation (5.2.2) multiplied by any periodic function of a period $i\beta$ ($i\gamma$ for idQM, 1 for rdQM) is another solution. In the case of idQM, this nonuniqueness is removed by requiring the finite norm condition for the groundstate wavefunction ϕ_0, $(\phi_0, \phi_0) < \infty$ and the hermiticity (self-adjointness) condition of the Hamiltonian. For details, see [22] Appendix A. For the rdQM, the periodic ambiguity of period 1 is harmless since the values of eigenfunctions on the integer lattice points only count for the inner products, etc. It is also important to realise that the periodic ambiguity does not appear in the polynomial solutions of (5.2.47).

It is important to stress that the second order equation for $P_n(\eta(x))$ is square root free (5.2.47):
$$\tilde{\mathcal{H}} P_n(\eta(x)) = \mathcal{E}_n P_n(\eta(x)).$$

In other words, the similarity transformed Hamiltonian $\tilde{\mathcal{H}}$ in terms of the ground state wavefunction $\phi_0(x)$ has a much simpler form than the original Hamiltonian $\mathcal{H}$:

$$\tilde{\mathcal{H}} \stackrel{\text{def}}{=} \phi_0(x)^{-1} \circ \mathcal{H} \circ \phi_0(x) = \begin{cases} V(x)(e^{\gamma p} - 1) + V^*(x)(e^{-\gamma p} - 1) & : \text{idQM} \\ B(x)(1 - e^{\partial}) + D(x)(1 - e^{-\partial}) & : \text{rdQM} \end{cases}. \quad (5.2.102)$$

For all the exactly solvable examples discussed in this section, $\tilde{\mathcal{H}}$ is triangular (5.2.48)
$$\tilde{\mathcal{H}} \eta(x)^n = \mathcal{E}_n \eta(x)^n + \text{lower orders in } \eta(x).$$

Here are some explicit examples. For the idQM ($0 < q < 1$), the Meixner-Pollaczek (MP), the Wilson (W) and the Askey-Wilson (AW) polynomials:

MP : $\quad V(x) = a + ix, \quad a > 0, \quad -\infty < x < \infty, \quad \gamma = 1, \quad \eta(x) = x, \quad (5.2.103)$

W : $\quad V(x) = \dfrac{\prod_{j=1}^{4}(a_j + ix)}{2ix(2ix + 1)}, \quad \text{Re}(a_j) > 0, \quad 0 < x < \infty, \quad \gamma = 1,$

$\eta(x) = x^2, \quad \{a_1^*, a_2^*, a_3^*, a_4^*\} = \{a_1, a_2, a_3, a_4\} \text{ (as a set)}, \quad (5.2.104)$

AW : $\quad V(x) = \dfrac{\prod_{j=1}^{4}(1 - a_j e^{ix})}{(1 - e^{2ix})(1 - q e^{2ix})}, \quad |a_j| < 1, \quad 0 < x < \pi, \quad \gamma = \log q,$

$\eta(x) = 1 - \cos x, \quad \{a_1^*, a_2^*, a_3^*, a_4^*\} = \{a_1, a_2, a_3, a_4\} \text{ (as a set)}. \quad (5.2.105)$

For the rdQM ($0 < q < 1$), the Meixner (M), the Racah (R) and the q-Racah (qR) polynomials:

M : $\quad B(x) = \dfrac{c}{1 - c}(x + \beta), \quad D(x) = \dfrac{1}{1 - c}x, \quad \beta > 0, \quad 0 < c < 1,$

$\eta(x) = x, \quad x_{\max} = \infty, \quad (5.2.106)$

R : $\quad B(x) = -\dfrac{(x + a)(x + b)(x + c)(x + d)}{(2x + d)(2x + 1 + d)},$

$D(x) = -\dfrac{(x + d - a)(x + d - b)(x + d - c)x}{(2x - 1 + d)(2x + d)}, \quad \tilde{d} \stackrel{\text{def}}{=} a + b + c - d - 1,$

5.2 1-Degree of Freedom System

$$a = -N, \quad a+b > d > 0, \quad 0 < c < 1+d,$$
$$\eta(x) = x(x+d), \quad x_{\max} = N, \tag{5.2.107}$$

qR : $$B(x) = -\frac{(1-aq^x)(1-bq^x)(1-cq^x)(1-dq^x)}{(1-dq^{2x})(1-dq^{2x+1})},$$

$$D(x) = -\tilde{d}\,\frac{(1-a^{-1}dq^x)(1-b^{-1}dq^x)(1-c^{-1}dq^x)(1-q^x)}{(1-dq^{2x-1})(1-dq^{2x})}, \quad \tilde{d} \stackrel{\text{def}}{=} abcd^{-1}q^{-1},$$

$$a = q^{-N}, \quad 0 < ab < d < 1, \quad qd < c < 1,$$
$$\eta(x) = (q^{-x}-1)(1-dq^x), \quad x_{\max} = N. \tag{5.2.108}$$

The eigenpolynomials $P_n(\eta(x))$ are the following (for notation, see [19]):

idQM :

MP : $$P_n(\eta(x)) = P_n^{(a)}\!\left(x;\frac{\pi}{2}\right) \stackrel{\text{def}}{=} \frac{(2a)_n}{n!}\, i^n{}_2F_1\!\left(\begin{array}{c}-n,\,a+ix\\ 2a\end{array}\,\Big|\,2\right), \tag{5.2.109}$$

W : $$P_n(\eta(x)) = W_n(x^2;a_1,a_2,a_3,a_4) \quad (b_1 \stackrel{\text{def}}{=} a_1+a_2+a_3+a_4)$$
$$\stackrel{\text{def}}{=} (a_1+a_2)_n(a_1+a_3)_n(a_1+a_4)_n$$
$$\times {}_4F_3\!\left(\begin{array}{c}-n,\,n+b_1-1,\,a_1+ix,\,a_1-ix\\ a_1+a_2,\,a_1+a_3,\,a_1+a_4\end{array}\,\Big|\,1\right), \tag{5.2.110}$$

AW : $$P_n(\eta(x)) = p_n(\cos x;a_1,a_2,a_3,a_4|q) \quad (b_4 \stackrel{\text{def}}{=} a_1a_2a_3a_4)$$
$$\stackrel{\text{def}}{=} a_1^{-n}(a_1a_2,a_1a_3,a_1a_4;q)_n$$
$$\times {}_4\phi_3\!\left(\begin{array}{c}q^{-n},\,b_4q^{n-1},\,a_1e^{ix},\,a_1e^{-ix}\\ a_1a_2,\,a_1a_3,\,a_1a_4\end{array}\,\Big|\,q;q\right), \tag{5.2.111}$$

rdQM :

M : $$P_n(\eta(x)) = M_n(x;\beta,c) \stackrel{\text{def}}{=} {}_2F_1\!\left(\begin{array}{c}-n,\,-x\\ \beta\end{array}\,\Big|\,1-c^{-1}\right), \tag{5.2.112}$$

R : $$P_n(\eta(x)) = R_n(x(x+d);a-1,\tilde{d}-a,c-1,d-c)$$
$$\stackrel{\text{def}}{=} {}_4F_3\!\left(\begin{array}{c}-n,\,n+\tilde{d},\,-x,\,x+d\\ a,\,b,\,c\end{array}\,\Big|\,1\right), \tag{5.2.113}$$

qR : $$P_n(\eta(x)) = R_n(q^{-x}+dq^x;aq^{-1},\tilde{d}a^{-1},cq^{-1},dc^{-1}|q)$$
$$\stackrel{\text{def}}{=} {}_4\phi_3\!\left(\begin{array}{c}q^{-n},\,\tilde{d}q^n,\,q^{-x},\,dq^x\\ a,\,b,\,c\end{array}\,\Big|\,q;q\right). \tag{5.2.114}$$

Obviously, the square of the ground state wavefunction $\phi_0(x)^2$ provides the positive definite orthogonality weight function for the polynomials:

$$\text{idQM}: \quad \int_{x_1}^{x_2} \phi_0(x)^2 P_n(\eta(x)) P_n(\eta(x)) \mathrm{d}x = h_n \delta_{nm}, \qquad (5.2.115)$$

$$\text{rdQM}: \quad \sum_{x=0}^{x_{\max}} \phi_0(x)^2 P_n(\eta(x)) P_n(\eta(x)) = \frac{1}{d_n^2} \delta_{nm}. \qquad (5.2.116)$$

The explicit forms of the squared ground zstate wavefunction (weight function) $\phi_0(x)^2$ and the normalisation constants h_n for the above examples in pure imaginary shifts dQM are:

$$\phi_0(x)^2 = \begin{cases} \Gamma(a+ix)\Gamma(a-ix) & : \text{MP} \\ [\Gamma(2ix)\Gamma(-2ix)]^{-1} \prod_{j=1}^{4} \Gamma(a_j+ix)\Gamma(a_j-ix) & : \text{W} \\ (e^{2ix};q)_\infty (e^{-2ix};q)_\infty \prod_{j=1}^{4} [(a_j e^{ix};q)_\infty (a_j e^{-ix};q)_\infty]^{-1} & : \text{AW} \end{cases}, \qquad (5.2.117)$$

$$h_n = \begin{cases} 2\pi(2^{2a}n!)^{-1}\Gamma(n+2a) & : \text{MP} \\ 2\pi n!\,(n+b_1-1)_n \prod_{1\leq i<j\leq 4} \Gamma(n+a_i+a_j)\cdot \Gamma(2n+b_1)^{-1} & : \text{W} \\ 2\pi(b_4 q^{n-1};q)_n (b_4 q^{2n};q)_\infty (q^{n+1};q)_\infty^{-1} \prod_{1\leq i<j\leq 4} (a_i a_j q^n;q)_\infty^{-1} & : \text{AW} \end{cases}. \qquad (5.2.118)$$

Exercise Derive the ground state eigen function $\phi_0(x)$ for the Askey-Wilson system AW in (5.2.117) from the zero mode equation (5.2.101).

For the rdQM, the zero-mode equation $\mathcal{A}\phi_0(x) = 0$ (5.2.9) is a two term recurrence relation, which can be solved elementarily by using the boundary condition (5.2.98):

$$\phi_0(x)^2 = \prod_{y=0}^{x-1} \frac{B(y)}{D(y+1)}, \quad \phi_0(0) = 1. \qquad (5.2.119)$$

The explicit forms of $\phi_0(x)^2$ and d_n^2, which are related by duality, are:

$$\phi_0(x)^2 = \begin{cases} \dfrac{(\beta)_x\, c^x}{x!} & : \text{M} \\ \dfrac{(a,b,c,d)_x}{(1+d-a,1+d-b,1+d-c,1)_x} \dfrac{2x+d}{d} & : \text{R} \\ \dfrac{(a,b,c,d;q)_x}{(a^{-1}dq,b^{-1}dq,c^{-1}dq,q;q)_x \tilde{d}^x} \dfrac{1-dq^{2x}}{1-d} & : \text{qR} \end{cases}, \qquad (5.2.120)$$

5.2 1-Degree of Freedom System

$$d_n^2 = \begin{cases} \dfrac{(\beta)_n c^n}{n!} \times (1-c)^\beta & : \text{M} \\[2mm] \dfrac{(a,b,c,\tilde{d})_n}{(1+\tilde{d}-a, 1+\tilde{d}-b, 1+\tilde{d}-c, 1)_n} \dfrac{2n+\tilde{d}}{\tilde{d}} & \\[2mm] \quad \times \dfrac{(-1)^N(1+d-a, 1+d-b, 1+d-c)_N}{(\tilde{d}+1)_N (d+1)_{2N}} & : \text{R} \\[2mm] \dfrac{(a,b,c,\tilde{d};q)_n}{(a^{-1}\tilde{d}q, b^{-1}\tilde{d}q, c^{-1}\tilde{d}q, q;q)_n d^n} \dfrac{1-\tilde{d}q^{2n}}{1-\tilde{d}} & \\[2mm] \quad \times \dfrac{(-1)^N (a^{-1}dq, b^{-1}dq, c^{-1}dq;q)_N \tilde{d}^N q^{\frac{1}{2}N(N+1)}}{(\tilde{d}q;q)_N (dq;q)_{2N}} & : \text{qR} \end{cases} \quad . \quad (5.2.121)$$

1. Intertwining Relations: Crum's Theorem

The general structure of the intertwining relations, the consequences of the factorised Hamiltonians, works equally well for the dQM. The quantities in the s-th step are defined by those in the $(s-1)$-st step: $(s \geq 1)$

$$\text{idQM}: \quad V^{[s]}(x) \stackrel{\text{def}}{=} \sqrt{V^{[s-1]}\left(x - \mathrm{i}\frac{\gamma}{2}\right) V^{[s-1]*}\left(x - \mathrm{i}\frac{\gamma}{2}\right)} \, \frac{\phi_s^{[s]}(x - \mathrm{i}\gamma)}{\phi_s^{[s]}(x)}, \quad (5.2.122)$$

$$\mathcal{A}^{[s]} \stackrel{\text{def}}{=} \mathrm{i}\left[e^{\frac{\gamma}{2}p}\sqrt{V^{[s]*}(x)} - e^{-\frac{\gamma}{2}p}\sqrt{V^{[s]}(x)}\right],$$

$$\mathcal{A}^{[s]\dagger} = -\mathrm{i}\left[\sqrt{V^{[s]}(x)}\,e^{\frac{\gamma}{2}p} - \sqrt{V^{[s]*}(x)}\,e^{-\frac{\gamma}{2}p}\right], \quad (5.2.123)$$

$$\text{rdQM}: \quad B^{[s]}(x) \stackrel{\text{def}}{=} \sqrt{B^{[s-1]}(x+1)D^{[s-1]}(x+1)} \, \frac{\phi_s^{[s]}(x+1)}{\phi_s^{[s]}(x)}, \quad (5.2.124)$$

$$D^{[s]}(x) \stackrel{\text{def}}{=} \sqrt{B^{[s-1]}(x)D^{[s-1]}(x)} \, \frac{\phi_s^{[s]}(x-1)}{\phi_s^{[s]}(x)}, \quad (5.2.125)$$

$$\mathcal{A}^{[s]} \stackrel{\text{def}}{=} \sqrt{B^{[s]}(x)} - e^{\partial}\sqrt{D^{[s]}(x)}, \quad \mathcal{A}^{[s]\dagger} = \sqrt{B^{[s]}(x)} - \sqrt{D^{[s]}(x)}\,e^{-\partial}.$$

$$(5.2.126)$$

The eigenfunctions at the s-th step have succinct *determinant forms* in terms of the Casoratian (dQM) [8, 23, 24, 26]: $(n \geq s \geq 0)$

$$\text{idQM}: \quad \mathrm{W}_\gamma[f_1,\cdots,f_m](x) \stackrel{\text{def}}{=} \mathrm{i}^{\frac{1}{2}m(m-1)} \det\left[f_k\left(x + \mathrm{i}\frac{m+1-2j}{2}\gamma\right)\right]_{1 \leq j,k \leq m}$$

(Casoratian), $\quad (5.2.127)$

$$\phi_n^{[s]}(x) = \prod_{j=0}^{s-1}\sqrt{V^{[j]}\left(x+\mathrm{i}\frac{s-j}{2}\gamma\right)} \cdot \frac{W_\gamma[\phi_0,\phi_1,\cdots,\phi_{s-1},\phi_n](x)}{W_\gamma[\phi_0,\phi_1,\cdots,\phi_{s-1}]\left(x-\mathrm{i}\frac{\gamma}{2}\right)},$$
(5.2.128)

rdQM: $\quad W_C[f_1,\cdots,f_m](x) \stackrel{\text{def}}{=} \det\left[f_k(x+j-1)\right]_{1\leqslant j,k\leqslant m}$

(Casoratian), (5.2.129)

$$\phi_n^{[s]}(x) = (-1)^s \prod_{k=0}^{s-1}\sqrt{B^{[k]}(x)} \cdot \frac{W_C[\phi_0,\phi_1,\cdots,\phi_{s-1},\phi_n](x)}{W_C[\phi_0,\phi_1,\cdots,\phi_{s-1}](x+1)}$$
(5.2.130)

$$= (-1)^s \prod_{k=0}^{s-1}\sqrt{D^{[k]}(x+s-k)} \cdot \frac{W_C[\phi_0,\phi_1,\cdots,\phi_{s-1},\phi_n](x)}{W_C[\phi_0,\phi_1,\cdots,\phi_{s-1}](x)}.$$
(5.2.131)

The norm of the s-th step eigenfunctions have the same uniform expression as in the ordinary QM (5.2.34):

$$(\phi_n^{[s]},\phi_m^{[s]}) = \prod_{j=0}^{s-1}(\mathcal{E}_n - \mathcal{E}_j) \cdot (\phi_n,\phi_m).$$

2. Shape Invariance

The Hamiltonians in the ordinary QM (5.1.11) have a fixed overall scale as the coefficient of the p^2 terms, but those of dQM have no such scale as the momenta are exponentiated. Thus the shape invariance relation in dQM has one extra factor (κ) corresponding to the overall scaling:

$$\mathcal{A}(\boldsymbol{\lambda})\mathcal{A}(\boldsymbol{\lambda})^\dagger = \kappa\mathcal{A}(\boldsymbol{\lambda}+\boldsymbol{\delta})^\dagger\mathcal{A}(\boldsymbol{\lambda}+\boldsymbol{\delta}) + \mathcal{E}_1(\boldsymbol{\lambda}),\qquad(5.2.132)$$

which is equivalent to the following conditions:

idQM: $\quad V\left(x-\mathrm{i}\frac{\gamma}{2};\boldsymbol{\lambda}\right)V^*\left(x-\mathrm{i}\frac{\gamma}{2};\boldsymbol{\lambda}\right) = \kappa^2\, V(x;\boldsymbol{\lambda}+\boldsymbol{\delta})V^*(x-\mathrm{i}\gamma;\boldsymbol{\lambda}+\boldsymbol{\delta}),$ (5.2.133)

$$V\left(x+\mathrm{i}\frac{\gamma}{2};\boldsymbol{\lambda}\right) + V^*\left(x-\mathrm{i}\frac{\gamma}{2};\boldsymbol{\lambda}\right) = \kappa\bigl[V(x;\boldsymbol{\lambda}+\boldsymbol{\delta}) + V^*(x;\boldsymbol{\lambda}+\boldsymbol{\delta})\bigr] - \mathcal{E}_1(\boldsymbol{\lambda}),$$
(5.2.134)

rdQM: $\quad B(x+1;\boldsymbol{\lambda})D(x+1;\boldsymbol{\lambda}) = \kappa^2\, B(x;\boldsymbol{\lambda}+\boldsymbol{\delta})D(x+1;\boldsymbol{\lambda}+\boldsymbol{\delta}),$ (5.2.135)

$$B(x;\boldsymbol{\lambda}) + D(x+1;\boldsymbol{\lambda}) = \kappa\bigl[B(x;\boldsymbol{\lambda}+\boldsymbol{\delta}) + D(x+1;\boldsymbol{\lambda}+\boldsymbol{\delta})\bigr] + \mathcal{E}_1(\boldsymbol{\lambda}).$$ (5.2.136)

Owing to the extra parameter κ in (5.2.132), the eigenvalue formula (5.2.63) is changed to

$$\mathcal{E}_n(\boldsymbol{\lambda}) = \sum_{s=0}^{n-1}\kappa^s \mathcal{E}_1(\boldsymbol{\lambda}^{[s]}),\qquad \boldsymbol{\lambda}^{[s]} \stackrel{\text{def}}{=} \boldsymbol{\lambda} + s\boldsymbol{\delta},$$
(5.2.137)

whereas the eigenfunction formula (5.2.64) stays the same. All the (q)-hypergeometric orthogonal polynomials belonging to the Askey scheme are shape invariant. It is straightforward to verify the shape invariant condition (5.2.132) and to derive the eigenvalues for the six explicit examples (5.2.103)~(5.2.108):

idQM : MP : $\quad \lambda = a, \quad \delta = \dfrac{1}{2}, \quad \kappa = 1, \quad \mathcal{E}_n(\lambda) = 2n,$ (5.2.138)

W : $\quad \lambda = (a_1, a_2, a_3, a_4), \quad \delta = \left(\dfrac{1}{2}, \dfrac{1}{2}, \dfrac{1}{2}, \dfrac{1}{2}\right), \quad \kappa = 1,$

$\mathcal{E}_n(\lambda) = 4n(n + b_1 - 1),$ (5.2.139)

AW : $\quad q^\lambda = (a_1, a_2, a_3, a_4), \quad \delta = \left(\dfrac{1}{2}, \dfrac{1}{2}, \dfrac{1}{2}, \dfrac{1}{2}\right), \quad \kappa = q^{-1},$

$\mathcal{E}_n(\lambda) = (q^{-n} - 1)(1 - b_4 q^{n-1}),$ (5.2.140)

rdQM : M : $\quad \lambda = (\beta, c), \quad \delta = (1, 0), \quad \kappa = 1, \quad \mathcal{E}(n; \lambda) = n,$ (5.2.141)

R : $\quad \lambda = (a, b, c, d), \quad \delta = (1, 1, 1, 1), \quad \kappa = 1, \quad \mathcal{E}_n(\lambda) = 4n(n + \tilde{d}),$ (5.2.142)

qR : $\quad q^\lambda = (a, b, c, d), \quad \delta = (1, 1, 1, 1), \quad \kappa = q^{-1},$

$\mathcal{E}_n(\lambda) = (q^{-n} - 1)(1 - \tilde{d}q^{n-1}),$ (5.2.143)

where $q^{(\lambda_1, \lambda_2, \cdots)} = (q^{\lambda_1}, q^{\lambda_2}, \cdots)$.

3. Solvability in the Heisenberg Picture

The discrete QM systems of all (q)-hypergeometric orthogonal polynomials belonging to the Askey scheme are exactly solvable in the Heisenberg picture. Based on the data of the closure relation (5.2.75), the eigenvalues (5.2.81)、(5.2.82) as well as the creation/annihilation operators (5.2.89)~(5.2.92) are constructed. The data for the three examples in idQM (5.2.103)~(5.2.105) are:

MP : $\quad R_1(y) = 0, \quad R_0(y) = 4, \quad R_{-1}(y) = 0,$ (5.2.144)

W : $\quad R_1(y) = 2,$

$R_0(y) = 4y + b_1(b_1 - 2), \quad R_{-1}(y) = -2y^2 + (b_1 - 2b_2)y + (2 - b_1)b_3,$

$b_2 \stackrel{\text{def}}{=} \displaystyle\sum_{1 \leqslant j < k \leqslant 4} a_j a_k, \quad b_3 \stackrel{\text{def}}{=} \displaystyle\sum_{1 \leqslant j < k < l \leqslant 4} a_j a_k a_l,$ (5.2.145)

AW : $\quad R_1(y) = (q^{-\frac{1}{2}} - q^{\frac{1}{2}})^2 y', \quad R_0(y) = (q^{-\frac{1}{2}} - q^{\frac{1}{2}})^2 \left[y'^2 - (1 + q^{-1})^2 b_4\right],$

$$R_{-1}(y) = \frac{1}{2}(q^{-\frac{1}{2}} - q^{\frac{1}{2}})^2 \big[(b_1 + q^{-1}b_3)y' - (1 + q^{-1})(b_3 + q^{-1}b_1b_4)\big] - R_0(y),$$

$$y' \stackrel{\text{def}}{=} y + 1 + q^{-1}b_4, \quad b_1 \stackrel{\text{def}}{=} \sum_{j=1}^{4} a_j, \quad b_3 \stackrel{\text{def}}{=} \sum_{1 \leqslant j < k < l \leqslant 4} a_j a_k a_l.$$

(5.2.146)

The data for the three examples in rdQM (5.2.106)~(5.2.108) are:

$$\text{M}: \quad R_1(y) = 0, \quad R_0(y) = 1, \quad R_{-1}(y) = -\frac{1+c}{1-c}y - \frac{\beta c}{1-c}, \tag{5.2.147}$$

$$\text{R}: \quad R_1(y) = 2, \quad R_0(y) = 4y + \tilde{d}^2 - 1,$$

$$R_{-1}(y) = 2y^2 + [2(ab + bc + ca) - (1+d)(1+\tilde{d})]y + abc(\tilde{d} - 1), \tag{5.2.148}$$

$$q\text{R}: \quad R_1(y) = (q^{-\frac{1}{2}} - q^{\frac{1}{2}})^2 y', \quad R_0(y) = (q^{-\frac{1}{2}} - q^{\frac{1}{2}})^2 \big[y'^2 - (q^{-\frac{1}{2}} + q^{\frac{1}{2}})^2 \tilde{d}\big],$$

$$y' \stackrel{\text{def}}{=} y + 1 + \tilde{d},$$

$$R_{-1}(y) = (q^{-\frac{1}{2}} - q^{\frac{1}{2}})^2 \Big\{(1+d)y'^2 - \big[a + b + c + d + \tilde{d} + (ab + bc + ca)q^{-1}\big]y'$$

$$+ (1-a)(1-b)(1-c)(1-\tilde{d}q^{-1})$$

$$+ \big[a + b + c - 1 - d\tilde{d} + (ab + bc + ca)q^{-1}\big](1 + \tilde{d})\Big\}. \tag{5.2.149}$$

Exercise Solve the Hamilton equation for the Meixner-Pollaczek (MP) system (5.2.103) and derive

$$x(t) = x(0)\cos 2t + \sqrt{a^2 + x^2(0)} \sinh p(0) \sin 2t.$$

It should be emphasised that the classical solutions of the sinusoidal coordinate of those exactly solvable QM systems have simple, i.e. sinusoidal, time-dependence and the frequencies are simple functions of the energy. See various examples in [14].

5.2.7 From One Particle to Many Particles (从单粒子到多粒子)

A naive idea to construct multi-particle exactly solvable QM is to use the mirror image method. In the r-dimensional euclidean space $\mathbf{R}^r$, we place r mirrors ($r-1$ dimensional hyperplane) intersecting at the origin. The interactions introduced by mirroring a one dimensional exactly solvable potential, if properly managed, could produce its multi-particle version. In fact, most of the known exactly solvable QM are obtained in this fashion and the resulting systems depend on two elements; the choice of one particle potentials and the choice of the mirror configurations.

As we have seen, there are three types of one dimensional QM; the ordinary QM with differential Schrödinger equations, the two kinds (the pure imaginary shifts and

5.2 1-Degree of Freedom System

the real shifts) of the discrete QM with difference Schrödinger equations. Multiparticle versions of exactly solvable ordinary QM obtained by the mirror image method are known as Calogero-Sutherland [27–53] systems. They will be discussed in section 5.3. Many examples of exactly solvable multi-particle difference Schrödinger equations with pure imaginary shifts [54–60] have been constructed by the mirror image method, too. Section 5.4 will be devoted to the subject. In contrast, the mirror method has not been successful in discrete QM with real shifts. There is a limited number of reports on solvable multi-particle multi-particle difference Schrödinger equations with real shifts [61, 62].

Let us now discuss the 'mirror' configurations. A 'mirror' (an $r-1$ dimensional hyperplane) located at the origin can be represented by its normal vector α at the origin (Fig. 5.2). The reflection of a point x by the 'mirror' α is

$$s_\alpha(x) \stackrel{\text{def}}{=} x - (\alpha^\vee \cdot x)\alpha, \quad \alpha^\vee \stackrel{\text{def}}{=} \frac{2\alpha}{\alpha \cdot \alpha}. \tag{5.2.150}$$

Obviously α and $-\alpha$ define the same 'mirror' (hyperplane) and $s_\alpha \equiv s_{-\alpha}$ and s_α^2 is the identity transformation in $\mathbf{R}^r$. The reflection does not change the length of the vectors $s_\alpha(x) \cdot s_\alpha(x) = x \cdot x$ and it is an orthogonal transformation in $\mathbf{R}^r$. A 'mirror' β reflected by a 'mirror' α, $s_\alpha(\beta)$ is also a 'mirror'. Let us use the term root instead of 'mirror'. A configuration of a finite number of root vectors is called a root system Δ if

$$s_\alpha(\Delta) = \Delta, \quad \forall \alpha \in \Delta. \tag{5.2.151}$$

Obviously $s_\alpha(\alpha) = -\alpha$. That is, if $\alpha \in \Delta$ then $-\alpha \in \Delta$. $\alpha^\vee$ is called the coroot of α and the coroot system $\Delta^\vee$ is $\{\alpha^\vee : \alpha \in \Delta\}$.

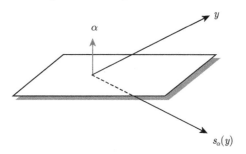

Fig. 5.2 Reflection by a vector α

5.2.8 Reflection Groups and Root Systems (反射群和根系)

The set of reflections $\{s_\alpha | \alpha \in \Delta\}$ generates a group G_Δ, known as a Coxeter group, or finite reflection group. The orbit of $\beta \in \Delta$ is the set of root vectors resulting from the action of the Coxeter group on it. They all have the same length. The set of positive roots Δ_+ may be defined in terms of a vector $U \in \mathbf{R}^r$, with $\alpha \cdot U \neq 0$, $\forall \alpha \in \Delta$, as those roots $\alpha \in \Delta$ such that $\alpha \cdot U > 0$. Given Δ_+, there is a unique set of r simple roots $\Pi = \{\alpha_j | j = 1, \cdots, r\}$ defined such that they span the root space and the coefficients $\{a_j\}$ in $\beta = \sum_{j=1}^{r} a_j \alpha_j$ for $\beta \in \Delta_+$ are all non-negative. In the 'mirror' language, these are the original r 'mirrors' and the rest are image 'mirrors'. The highest root α_h, for which $\sum_{j=1}^{r} a_j$ is maximal, is then also determined uniquely. The subset of reflections $\{s_\alpha | \alpha \in \Pi\}$ in fact generates the Coxeter group G_Δ. The products of s_α, with $\alpha \in \Pi$, are subject solely to the relations $(s_\alpha s_\beta)^{m(\alpha,\beta)} = 1$, $\alpha, \beta \in \Pi$. The interpretation is that $s_\alpha s_\beta$ is a rotation in some plane by $2\pi/m(\alpha,\beta)$. The set of positive integers $m(\alpha,\beta)$ (with $m(\alpha,\alpha) = 1$, $\forall \alpha \in \Pi$) uniquely specify the Coxeter group. The root lattice $Q(\Delta)$ and its positive octant Q^+ are defined by

$$Q \stackrel{\text{def}}{=} \sum_{j=1}^{r} \mathbf{Z}\alpha_j, \quad Q^+ \stackrel{\text{def}}{=} \sum_{j=1}^{r} \mathbf{N}\alpha_j. \tag{5.2.152}$$

The fundamental weights $\{\lambda_j\}$ are the dual basis of the co-simple roots

$$\text{fundamental weight} \quad \alpha_j^\vee \cdot \lambda_k = \delta_{jk}, \quad \forall \alpha_j \in \Pi. \tag{5.2.153}$$

Correspondingly the weight lattice $P(\Delta)$ is defined as the $\mathbf{Z}$-span of fundamental weights and $P^{++}(\Delta)$ is the cone of dominant weights:

$$P \stackrel{\text{def}}{=} \sum_{j=1}^{r} \mathbf{Z}\lambda_j, \quad P^{++} \stackrel{\text{def}}{=} \sum_{j=1}^{r} \mathbf{N}\lambda_j. \tag{5.2.154}$$

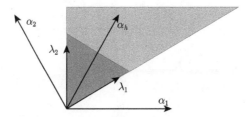

Fig. 5.3 Simple roots, the highest root, fundamental weights and the principal Weyl alcove (grey) and the principal Weyl chamber (light grey, extending to infinity) in a two dimensional root system

The root systems for finite reflection groups may be divided into two types: crystallographic and non-crystallographic. Crystallographic root systems satisfy the

5.2 1-Degree of Freedom System

additional condition
$$\alpha^\vee \cdot \beta = \frac{2\alpha \cdot \beta}{\alpha \cdot \alpha} \in \mathbf{Z}, \quad \forall \alpha, \beta \in \Delta. \tag{5.2.155}$$

They are related to the Lie algebras. The reflection groups generated by crystallographic root systems are also called the Weyl Groups. The remaining non-crystallographic root systems are H_3, H_4, whose Coxeter groups are the symmetry groups of the icosahedron and four-dimensional 600-cell, respectively, and the dihedral group of order $2m$, $\{I_2(m), m \geqslant 4\}$.

The explicit examples of the classical root systems, i.e. A, B, C and D are given below. For the exceptional and non-crystallographic root systems, see for example Humphrey's book [63]. In all cases $\{e_j\}$ denotes an orthonormal basis in $\mathbf{R}^r$, $e_j \cdot e_k = \delta_{jk}$.

(1) A_{r-1}: This root system is related with the Lie algebra $su(r)$.
$$\Delta = \cup_{1 \leqslant j < k \leqslant r} \{\pm(e_j - e_k)\}, \; \Pi = \cup_{j=1}^{r-1} \{\alpha_j\}, \; \alpha_j \stackrel{\text{def}}{=} e_j - e_{j+1}, \; j = 1, \cdots, r-1. \tag{5.2.156}$$

All roots have the same length and the number of roots is $r(r-1)$. For $\alpha = e_j - e_k$, $s_\alpha(x)$ interchanges $x_j \leftrightarrow x_k$. The reflection group by A_{r-1} is the symmetric group S_r.

(2) B_r: This root system is associated with Lie algebra $so(2r+1)$. The long roots have $(\text{length})^2 = 2$ and short roots have $(\text{length})^2 = 1$:
$$\Delta = \cup_{1 \leqslant j < k \leqslant r} \{\pm e_j \pm e_k\} \cup_{j=1}^r \{\pm e_j\},$$
$$\Pi = \cup_{j=1}^r \{\alpha_j\}, \quad \alpha_j \stackrel{\text{def}}{=} e_j - e_{j+1}, \quad j = 1, \cdots, r-1, \; \alpha_r \stackrel{\text{def}}{=} e_r. \tag{5.2.157}$$

The number of roots is $2r^2$. For $\alpha = e_j + e_k$, $s_\alpha(x)$ interchanges $x_j \leftrightarrow -x_k$ and for $\alpha = \pm e_j$, $s_\alpha(x)$ interchanges $x_j \leftrightarrow -x_j$.

(3) C_r: This root system is associated with Lie algebra $sp(2r)$. The long roots have $(\text{length})^2 = 4$ and short roots have $(\text{length})^2 = 2$:
$$\Delta = \cup_{1 \leqslant j < k \leqslant r} \{\pm e_j \pm e_k\} \cup_{j=1}^r \{\pm 2e_j\},$$
$$\Pi = \cup_{j=1}^r \{\alpha_j\}, \quad \alpha_j \stackrel{\text{def}}{=} e_j - e_{j+1}, \quad j = 1, \cdots, r-1, \; \alpha_r \stackrel{\text{def}}{=} 2e_r. \tag{5.2.158}$$

This is the coroot system of B_r, $B_r^\vee = C_r$ and they generate the same Weyl group.

(4) D_r: This root system is associated with Lie algebra $so(2r)$:
$$\Delta = \cup_{1 \leqslant j < k \leqslant r} \{\pm e_j \pm e_k\},$$
$$\Pi = \cup_{j=1}^r \{\alpha_j\}, \quad \alpha_j \stackrel{\text{def}}{=} e_j - e_{j+1}, \quad j = 1, \cdots, r-1, \; \alpha_r \stackrel{\text{def}}{=} e_{r-1} + e_r. \tag{5.2.159}$$

All roots have the same length and the number of roots is $2r(r-1)$.

In the root systems A_{r-1}, D_r, E_6, E_7 and E_8, all the roots have the same length. They are called simply laced root systems (the A, D, E series).

Exercise Derive the following expressions of the fundamental weights.

$$A_{r-1}: \quad \lambda_j = e_1 + \cdots + e_j - \frac{j}{r}\xi, \quad j = 1, \cdots, r, \quad \xi \stackrel{\text{def}}{=} \sum_{j=1}^{r} e_j, \quad (5.2.160)$$

$$B_r: \quad \lambda_j = e_1 + \cdots + e_j, \quad j = 1, \cdots, r-1, \quad \lambda_r = \frac{1}{2}\sum_{j=1}^{r} e_j, \quad (5.2.161)$$

$$C_r: \quad \lambda_j = e_1 + \cdots + e_j, \quad j = 1, \cdots, r, \quad (5.2.162)$$

$$D_r: \quad \lambda_j = e_1 + \cdots + e_j, \quad j = 1, \cdots, r-2,$$

$$\lambda_{r-1} = \frac{1}{2}(e_1 + \cdots + e_{r-1} - e_r),$$

$$\lambda_r = \frac{1}{2}(e_1 + \cdots + e_{r-1} + e_r). \quad (5.2.163)$$

The Weyl orbit of λ_1 in all four cases is called the vector weights. They are $\{e_j - \frac{1}{r}\xi\}$, $j = 1, \cdots, r$ for A_{r-1} and $\{\pm e_j\}$, $j = 1, \cdots, r$ for B_r, C_r and D_r. The Weyl orbits of λ_{r-1} and λ_r of D_r are called (anti-)spinor weights. They are $\frac{1}{2}(\pm e_1 \pm e_2 \pm \cdots \pm e_r)\}$, odd (even) number of $-$. Note that in A_{r-1} Lie algebra theory, there are only $r-1$ fundamental weights $\lambda_1, \cdots, \lambda_{r-1}$. The last one, λ_r is redundant, since $\lambda_r \stackrel{\text{def}}{=} e_1 + \cdots + e_r - \xi \equiv 0$. But it is useful for making connection with the partitions.

Exercise Show that the following fundamental weights $\{\lambda_j\}$, $j = 1, \cdots, r-1$ of A_{r-1}, λ_1 of B_r, C_r and $\lambda_1, \lambda_{r-1}, \lambda_r$ of D_r have the following property

for A_{r-1}, B_r, D_r $\mu \cdot \alpha = 0, \pm 1$, $\forall \alpha \in \Delta$, for C_r $\mu \cdot \alpha^\vee = 0, \pm 1$, $\forall \alpha \in \Delta$. (5.2.164)

These fundamental weights is called minuscule (minimal) weights.

Exercise Calculate the highest roots of the classical root systems.

$$A_{r-1}: \quad \alpha_h = e_1 - e_r = \alpha_1 + \cdots + \alpha_{r-1}, \quad (5.2.165)$$

$$B_r: \quad \alpha_h = e_1 + e_2 = \alpha_1 + 2(\alpha_2 + \cdots + \alpha_r), \quad (5.2.166)$$

$$C_r: \quad \alpha_h = 2e_1 = 2(\alpha_1 + \cdots + \alpha_{r-1}) + \alpha_r, \quad (5.2.167)$$

$$D_r: \quad \alpha_h = e_1 + e_2 = \alpha_1 + 2(\alpha_2 + \cdots + \alpha_{r-2}) + \alpha_{r-1} + \alpha_r. \quad (5.2.168)$$

Convince yourself that the minuscule weights in A, B and D are dual to the simple roots having unit coefficients in the highest roots α_h above.

5.3 Calogero-Sutherland Systems
(Calogero-Sutherland 系统)

The first example of multi-particle (classical) integrable system was discovered by Toda in 1967 [64]. The system has a repulsive exponential potential $U(x) = e^{-x}$ governed by the simple roots of A_{r-1} and its affine version. Since the one particle repulsive exponential potential is not exactly solvable quantum mechanically, its quantum integrability has not yet been established in spite of several interesting works[65]. In this connection, it is interesting to note that non-analytic attractive exponential potential $U(x) = -g^2 \exp(-|x|)$ is exactly solvable [66]. The first example of quantum multi-particle exactly solvable system was found by Calogero in 1971 [27]. It is a multi-particle version of the radial oscillator $(x^2 + 1/x^2)$ potential (5.2.36). One year later, another example with $1/\sin^2 x$ potential (5.2.37) was announced by Sutherland [28]. They demonstrated that all the discrete eigenvalues and the corresponding eigenfunctions can be obtained explicitly. Several years later Moser showed the classical integrability of the multi-particle $1/x^2$ potential system by using the Lax pair method [29]. About the same time the quantum solvability of the multi-particle hyperbolic $1/\sinh^2 x$ and elliptic $\wp(x)$ (Weierstrass $\wp$ function) potentials was shown[30, 31]. Within a few years, the possibility of extending the original systems to those based on other root systems was realised by Olshanetsky and Perelomov[34−36].

Now let us start with the simplest cases.

5.3.1 Simplest Cases (Based on A_{r-1} Root System) (几个简单的例子 (基于 A_{r-1} 根系))

The simplest example of a C-M system consists of r particles of equal mass (normalised to unity) on a line with pair-wise $1/\text{distance}^2$ interactions described by the following Hamiltonian:

$$\hat{\mathcal{H}} = \frac{1}{2} \sum_{j=1}^{r} p_j^2 + g(g - \hbar) \sum_{j<k}^{r} \frac{1}{(x_j - x_k)^2}, \qquad (5.3.1)$$

in which g is a real positive coupling constant. Here $x = (x_1, \cdots, x_r)$ are the coordinates and $p = (p_1, \cdots, p_r)$ are the conjugate canonical momenta obeying the canonical commutation relations: $[x_j, p_k] = i\hbar \delta_{jk}$, $[x_j, x_k] = [p_j, p_k] = 0$, $j, k = 1, \cdots, r$. Here we have retained $\hbar$ in order to show both classical ($\hbar \to 0$) and quantum

($\hbar=1$) simultaneously. The Heisenberg equations of motion are $\dot{x}_j = \frac{i}{\hbar}[\hat{\mathcal{H}}, x_j] = p_j$, $\ddot{x}_j = \dot{p}_j = \frac{i}{\hbar}[\hat{\mathcal{H}}, p_j] = 2g(g-\hbar)\sum_{k\neq j} 1/(x_j - x_k)^3$. The repulsive $1/(\text{distance})^2$ potential cannot be surmounted classically or quantum mechanically and the relative position of the particles on the line is not changed during the time evolution. Classically it means that if a motion starts at a configuration $x_1 > x_2 > \cdots > x_r$, then the inequalities remain true throughout the entire time evolution. At the quantum level, the wavefunctions vanish at the boundaries and the configuration space can be naturally limited to $x_1 > x_2 > \cdots > x_r$ (the principal Weyl chamber).

Similar solvable (integrable) quantum many-particle dynamics are obtained by replacing the inverse square potential in (5.3.1) by the trigonometric (hyperbolic) counterpart (see Fig. 5.4) $1/(x_j - x_k)^2 \to a^2/\sinh^2 a(x_j - x_k)$, in which $a > 0$ is a real parameter. The $1/\sin^2 x$ potential case (the Sutherland system) corresponds to the $1/\text{distance}^2$ interaction on a *circle* of radius $1/2a$, see Fig.5.5. A harmonic confining potential $\omega^2 \sum_{j=1}^{r} x_j^2/2$ can be added to the rational Hamiltonian (5.3.1) without breaking the solvability (integrability) (the Calogero system, see Fig. 5.4). At the classical level, the trigonometric (hyperbolic) and rational C-M systems are obtained from the elliptic potential ones (with the Weierstrass $\wp$ function) as the *degenerate* limits: $\wp(x_1 - x_2) \to a^2/\sinh^2 a(x_1 - x_2) \to 1/(x_1 - x_2)^2$, namely as one (two) period(s) of the $\wp$ function tends to infinity.

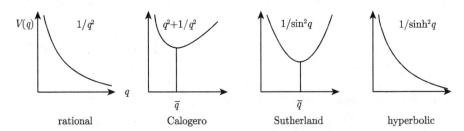

Fig. 5.4 Four different types of quantum C-S potentials

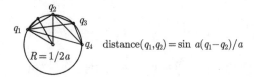

Fig. 5.5 Sutherland potential is $1/\text{distance}^2$ interaction on a circle. The large radius limit $a \to 0$ gives the rational potential

5.3 Calogero-Sutherland Systems

It is remarkable that these equations of motion can be expressed in a matrix form (Lax pair): $i/\hbar[\hat{\mathcal{H}}, L] = dL/dt = LM - ML = [L, M] \Leftrightarrow$ Heisenberg eq. of motion, in which L and M are given by

$$L = \begin{pmatrix} p_1 & \dfrac{ig}{x_1-x_2} & \cdots & \dfrac{ig}{x_1-x_r} \\ \dfrac{ig}{x_2-x_1} & p_2 & \cdots & \dfrac{ig}{x_2-x_r} \\ \vdots & \vdots & & \vdots \\ \dfrac{ig}{x_r-x_1} & \dfrac{ig}{x_r-x_2} & \cdots & p_r \end{pmatrix},$$

$$M = \begin{pmatrix} m_1 & -\dfrac{ig}{(x_1-x_2)^2} & \cdots & -\dfrac{ig}{(x_1-x_r)^2} \\ -\dfrac{ig}{(x_2-x_1)^2} & m_2 & \cdots & -\dfrac{ig}{(x_2-x_r)^2} \\ \vdots & \vdots & & \vdots \\ -\dfrac{ig}{(x_r-x_1)^2} & -\dfrac{ig}{(x_r-x_2)^2} & \cdots & m_r \end{pmatrix}. \qquad (5.3.2)$$

The diagonal element m_j of M is given by $m_j = ig \sum_{k \neq j} 1/(x_j - x_k)^2$. The matrix M has a special property $\sum_{j=1}^{r} M_{jk} = \sum_{k=1}^{r} M_{jk} = 0$, which ensures the quantum conserved quantities as the total sum of powers of Lax matrix L: $dK_n/dt = \dfrac{i}{\hbar}[\hat{\mathcal{H}}, K_n] = 0$, $K_n \equiv \mathrm{Ts}(L^n) = \sum_{j,k}(L^n)_{jk}$, $(n = 1, \cdots, r)$, $[K_n, K_m] = 0$. It is obvious that K_1 is the total momentum. K_n is a degree n polynomial in p and is independent from K_m. It should be stressed that the trace of L^n is conserved classically but not quantum mechanically because of the non-commutativity of x and p. The Hamiltonian is equivalent to K_2, $\hat{\mathcal{H}} \propto K_2 + \mathrm{const}$. In other words, the Lax matrix L is a kind of "square root" of the Hamiltonian. The quantum equations of motion for the Sutherland and hyperbolic potentials are again expressed by Lax pairs if the following replacements are made: $1/(x_j - x_k) \rightarrow a\cot(h)a(x_j - x_k)$ in L and $1/(x_j - x_k)^2 \rightarrow a^2/\sin(h)^2 a(x_j - x_k)$ in M. We obtain quantum conserved quantities in the same manner as above for the systems with the trigonometric and hyperbolic interactions.

Exercise Verify the Lax equation $dL/dt = [L, M]$ for the above Lax pair (5.3.2). The diagonal part gives the Heisenberg eq. $\ddot{x}_j = \dot{p}_j = \dfrac{i}{\hbar}[\hat{\mathcal{H}}, p_j] = 2g(g - \hbar)$

$\sum_{k\neq j} 1/(x_j - x_k)^3$. For the off diagonal part, the functional relation (5.3.6) is useful. Next, verify the same equation with the trigonometric and hyperbolic potentials given in Table 5.1.

Table 5.1 Functions appearing in the prepotential and Lax pair

potential	$\Theta(u)$	$\Lambda(u)$	$\Omega(u)$
rational	u	$1/u$	$-1/u^2$
hyperbolic	$\sinh au$	$a \coth au$	$-a^2/\sinh^2 au$
trigonometric	$\sin au$	$a \cot au$	$-a^2/\sin^2 au$

Exercise Convince yourself that $dL/dt = [L, M]$ means $dL^n/dt = [L^n, M]$.

The main goal is to find all the eigenvalues $\{\mathcal{E}\}$ and eigenfunctions $\{\psi(x)\}$ of the Hamiltonians with the rational, Calogero, Sutherland, hyperbolic potentials: $\hat{\mathcal{H}}\psi(x) = \mathcal{E}\psi(x)$. The momentum operator p_j acts as a differential operator $p_j = -i\hbar\partial/\partial x_j$. For example, for the rational model Hamiltonian (5.3.1) the eigenvalue equation reads

$$\left[-\frac{\hbar^2}{2}\sum_{j=1}^{r}\frac{\partial^2}{\partial x_j^2} + g(g-\hbar)\sum_{j<k}^{r}\frac{1}{(x_j - x_k)^2}\right]\psi(x) = \mathcal{E}\psi(x), \quad (5.3.3)$$

which is a second order Fuchsian differential equation for each variable $\{x_j\}$ with a regular singularity at each hyperplane $x_j = x_k$ whose characteristic exponents are the same $g/\hbar$, $1 - g/\hbar$. Any solution ψ of (5.3.3) is regular at all points except for those on the union of hyperplanes $x_j = x_k$. Since the structure of the singularity is the same for the other three types of potentials, the same assertion for the regularity and singularity of the solution ψ holds for these cases, too. For the trigonometric (Sutherland) case there are other singularities at $x_j - x_k = \ell\pi/a$, $\ell \in \mathbf{Z}$, due to the periodicity of the potential. As is clear from the shape of the potentials, see Fig.5.4, the rational and hyperbolic Hamiltonians have continuous spectra only, whereas the Calogero and Sutherland Hamiltonians have only discrete spectra.

The solvability or more precisely the triangularity of the quantum C-S Hamiltonian was first discovered by Calogero [27] for particles on a line with inverse square potential plus a confining harmonic force and by Sutherland [28] for the particles on a circle with the trigonometric potential. Later, classical integrability of the models in terms of Lax pairs was proved by Moser [29]. Olshanetsky and Perelomov [36] showed that these systems were based on A_{r-1} root systems, i.e. $x_j - x_k = \alpha \cdot x$ and α is one of the root vectors of A_{r-1} root system (5.2.156). They also introduced generalisations

5.3 Calogero-Sutherland Systems

of the C-M systems based on any root system including the non-crystallographic ones.

As shown by Heckman-Opdam [40, 41] and Sasaki and collaborators [50, 53], quantum Calogero-Sutherland systems with degenerate potentials, that is the rational with/without harmonic force, the hyperbolic and the trigonometric potentials, based on any root system can be formulated and solved universally. To be more precise, the rational and Calogero systems are solvable for all root systems, the crystallographic and non-crystallographic. The hyperbolic and trigonometric (Sutherland) systems are solvable for any crystallographic root system. The universal formulae for the Hamiltonians, Lax pairs, ground state wave functions, conserved quantities, the triangularity, the discrete spectra for the Calogero and Sutherland systems, the creation and annihilation operators etc are equally valid for any root system. This will be shown in the next section.

5.3.2 Universal Formalism (普适形式)

A Calogero-Sutherland system is a Hamiltonian dynamics associated with a root system Δ of rank r, which is a set of vectors in $\mathbf{R}^r$ with its standard inner product.

Here we mainly follow [49]~[51],[53]. The Hamiltonian for the quantum Calogero-Sutherland system can be written in terms of a prepotential $w(x)$ (5.2.8) in a 'factorised form':

$$\mathcal{H} = \frac{1}{2}\sum_{j=1}^{r}\left[p_j - i\frac{\partial w(x)}{\partial x_j}\right]\left[p_j + i\frac{\partial w(x)}{\partial x_j}\right]. \tag{5.3.4}$$

The prepotential is a sum over positive roots:

$$w(x) = \sum_{\alpha \in \Delta_+} g_\alpha \ln|\Theta(\alpha \cdot x)| + \left(-\frac{\omega}{2}x^2\right). \tag{5.3.5}$$

The real *positive* coupling constants g_α are defined on orbits of the corresponding Coxeter group, i.e. they are identical for roots in the same orbit. That is, for the simple Lie algebra cases one coupling constant $g_\alpha = g$ for all roots in simply-laced models and two independent coupling constants, $g_\alpha = g_L$ for long roots and $g_\alpha = g_S$ for short roots in non-simply laced models. The function $w(u)$ and the other functions $\Lambda(u)$ and $\Omega(u)$ appearing in the Lax pair L and M (5.3.2), (5.3.22), (5.3.23) are listed in Table 5.1 for each type of degenerate potentials:

Exercise Show the following functional relation for all three types of potentials in Table 5.1.

$$\Omega(u)\Lambda(v) - \Omega(v)\Lambda(u) = \Lambda(u+v)[\Omega(u) - \Omega(v)]. \tag{5.3.6}$$

This is essential for demonstrating the universal Lax pair (5.3.22), (5.3.23) in the off diagonal part.

The dynamics of the prepotentials $w(x)$ (5.3.5) has been discussed by Dyson from a different point of view (random matrix model). The above factorised Hamiltonian (5.3.4) consists of an operator part $\hat{\mathcal{H}}$, which is the Hamiltonian in the usual definition (see the Hamiltonians in the previous section (5.3.1)), and a constant $\mathcal{E}_0$ which is the ground state energy, $\mathcal{H} = \hat{\mathcal{H}} - \mathcal{E}_0$. The factorised Hamiltonian $\mathcal{H}$ (5.3.4) is the cornerstone of the exactly solvable QM as explained in section 5.2.

The prepotential and the Hamiltonian are invariant under reflection of the phase space variables in the hyperplane perpendicular to any root $w(s_\alpha(x)) = w(x)$, $\mathcal{H}(s_\alpha(p), s_\alpha(x)) = \mathcal{H}(p, x)$, $\forall \alpha \in \Delta$, with s_α defined by (5.2.150). The above Coxeter (Weyl) invariance is the only (discrete) symmetry of the Calogero-Sutherland systems. It should be stressed that the Weyl invariance does not mean the corresponding Lie algebra symmetry, which is the invariance under continuous transformations. In the early years of Calogero-Sutherland systems there were some confused attempts to construct Lax pairs based on Lie algebra representations, which usually do not work when the zero weights are contained. The main problem is, as in the A_{r-1} case, to find all the eigenvalues $\{\mathcal{E}\}$ and eigenfunctions $\{\psi(x)\}$ of the above Hamiltonian $\mathcal{H}\psi(x) = \mathcal{E}\psi(x)$.

For any root system and for any choice of potentials, the C-S system has a hard repulsive potential $\sim 1/(\alpha \cdot x)^2$ near the reflection hyperplane $H_\alpha = \{x \in \mathbf{R}^r, \alpha \cdot x = 0\}$. The C-S eigenvalue equation is a second order Fuchsian differential equation with regular singularity at each reflection hyperplane H_α and those arising from the periodicity in the case of the Sutherland potential. Near the reflection hyperplane H_α, the solution behaves

$$\psi \sim (\alpha \cdot q)^{g_\alpha/\hbar}(1 + \text{reg. terms}), \quad \text{or} \quad \psi \sim (\alpha \cdot q)^{1-g_\alpha/\hbar}(1 + \text{reg. terms}).$$

The former solution is chosen for the square integrability. Because of the singularities, the configuration space is restricted to the principal Weyl chamber PW or the principal Weyl alcove PW_T for the trigonometric potential (see Fig.5.3): $PW = \{x \in \mathbf{R}^r | \alpha \cdot x > 0, \alpha \in \Pi\}$, $PW_T = \{x \in \mathbf{R}^r | \alpha \cdot x > 0, \alpha \in \Pi, \alpha_h \cdot x < \pi/a\}$, ($\Pi$: set of simple roots, see (5.2.156)~(5.2.159) section 5.2.8). Here α_h is the highest root.

1. Ground state wavefunction and energy

One straightforward outcome of the factorised Hamiltonian (5.3.4) is the universal ground state wavefunction which is given by

5.3 Calogero-Sutherland Systems

$$\Phi_0(x) = e^{w(x)/\hbar} = \prod_{\alpha \in \Delta_+} |\Theta(\alpha \cdot x)|^{g_\alpha/\hbar} \times \left(e^{-\frac{\omega}{2\hbar} x^2}\right), \quad \mathcal{H}\Phi_0(x) = 0. \quad (5.3.7)$$

The exponential factor $e^{-\frac{\omega}{2\hbar} x^2}$ exists only for the Calogero systems. The ground state energy, i.e. the constant part of $\mathcal{H} = \hat{\mathcal{H}} - \mathcal{E}_0$ has a universal expression for each potential:

$$\mathcal{E}_0 = \begin{cases} 0, & \text{rational}, \\ \omega\left(\hbar r/2 + \sum_{\alpha \in \Delta_+} g_\alpha\right), & \text{Calogero}, \end{cases} \quad \mathcal{E}_0 = 2a^2 \varrho^2 \times \begin{cases} -1, & \text{hyperbolic}, \\ 1, & \text{Sutherland}, \end{cases}$$
(5.3.8)

in which $\varrho = 1/2 \sum_{\alpha \in \Delta_+} g_\alpha \alpha$ is called a 'deformed Weyl vector'. Obviously $\Phi_0(x)$ is square integrable in the configuration spaces for the Calogero and Sutherland systems and not square integrable for the rational and hyperbolic potentials.

Exercise Calculate the deformed Weyl vectors for the classical root systems:

$$A_{r-1}: \varrho = \frac{g}{2} \sum_{j=1}^{r} (r+1-2j)e_j, \quad B_r: \varrho = g_L \sum_{j=1}^{r} (r+1-j)e_j + \frac{g_S}{2} \sum_{j=1}^{r} e_j, \quad (5.3.9)$$

$$C_r: \varrho = g_S \sum_{j=1}^{r} (r+1-j)e_j + g_L \sum_{j=1}^{r} e_j, \quad D_r: \varrho = g \sum_{j=1}^{r} (r+1-j)e_j. \quad (5.3.10)$$

2. Excited states, triangularity and spectrum

The excited states of the Calogero-Sutherland systems can be easily obtained as eigenfunctions of a differential operator $\tilde{\mathcal{H}}$ obtained from $\mathcal{H}$ by a similarity transformation (5.2.47):

$$\tilde{\mathcal{H}} = e^{-w/\hbar} \mathcal{H} e^{w/\hbar} = -1/2 \sum_{j=1}^{r} \left[\hbar^2 \frac{\partial^2}{\partial x_j^2} + 2\hbar \left(\frac{\partial w}{\partial x_j}\right) \frac{\partial}{\partial x_j}\right], \quad (5.3.11)$$

then the eigenvalue equation for $\tilde{\mathcal{H}}$, $\tilde{\mathcal{H}}\Psi_\mathcal{E} = \mathcal{E}\Psi_\mathcal{E}$ is equivalent to that of the original Hamiltonian $\mathcal{H}\Psi_\mathcal{E} e^{w/\hbar} = \mathcal{E}\Psi_\mathcal{E} e^{w/\hbar}$. Since all the singularities of the Fuchsian differential equation $\mathcal{H}\psi(x) = \mathcal{E}\psi(x)$ are contained in the ground state wavefunction $e^{w/\hbar}$ the function $\Psi_\mathcal{E}$ above must be regular at finite x including all the reflection boundaries. As for the rational and hyperbolic potential cases, the energy eigenvalues are continuous spectrum only. For the rational case, the eigenfunctions are multivariable generalisation of Bessel functions.

Calogero systems The similarity transformed Hamiltonian $\tilde{\mathcal{H}}$ reads

$$\tilde{\mathcal{H}} = \hbar\omega\, x \cdot \frac{\partial}{\partial x} - \frac{\hbar^2}{2} \sum_{j=1}^{r} \frac{\partial^2}{\partial x_j^2} - \hbar \sum_{\alpha \in \Delta_+} \frac{g_\alpha}{\alpha \cdot x} \alpha \cdot \frac{\partial}{\partial x}, \qquad (5.3.12)$$

which maps a Coxeter invariant polynomial in x of degree d to another of degree d. Thus the Hamiltonian $\tilde{\mathcal{H}}$ (5.3.12) is triangular in the basis of Coxeter invariant polynomials and the diagonal elements are $\hbar\omega\times$degree as given by the first term. Independent Coxeter invariant polynomials exist at the degrees f_j listed in Table 5.2: $f_j = 1 + e_j$, $j = 1, \cdots, r$, in which $\{e_j\}$, $j = 1, \cdots, r$, are the exponents of Δ, see Table 3.1 in [63].

Table 5.2 The degrees f_j in which independent Coxeter invariant polynomials exist

Δ	$f_j = 1+e_j$	Δ	$f_j = 1+e_j$
A_r	$2, 3, 4, \cdots, r+1$	E_8	$2, 8, 12, 14, 18, 20, 24, 30$
B_r	$2, 4, 6, \cdots, 2r$	F_4	$2, 6, 8, 12$
C_r	$2, 4, 6, \cdots, 2r$	G_2	$2, 6$
D_r	$2, 4, \cdots, 2r-2; r$	$I_2(m)$	$2, m$
E_6	$2, 5, 6, 8, 9, 12$	H_3	$2, 6, 10$
E_7	$2, 6, 8, 10, 12, 14, 18$	H_4	$2, 12, 20, 30$

The spectrum of the Hamiltonian $\mathcal{H}$ is $\hbar\omega N$ with a non-negative integer N which can be expressed as $N = \sum_{j=1}^{r} n_j f_j$, $n_j \in \mathbf{N}$, and the degeneracy of the above eigenvalue is the number of partitions of N. It is remarkable that the coupling constant dependence appears only in the ground state energy $\mathcal{E}_0$. This is a deformation of the isotropic harmonic oscillator confined in the principal Weyl chamber. The eigenpolynomials are generalisation of multivariable Laguerre (Hermite) polynomials. One immediate consequence of the above spectrum is the periodicity of the quantum motion. If a system has a wavefunction $\psi(x; 0)$ at $t = 0$, then at $t = T = 2\pi/\omega$ the system has physically the same wavefunction as $\psi(x; 0)$, i.e. $\psi(x; T) = \mathrm{e}^{-i\mathcal{E}_0 T/\hbar}\psi(x; 0)$. The same assertion holds at the classical level, too.

Sutherland Systems The periodicity of the trigonometric potential dictates that the wavefunction should be a Bloch factor $\mathrm{e}^{2ia\mu \cdot x}$, in which μ is a weight, times a Fourier series in terms of simple roots. The basis of the Weyl invariant wave functions is specified by a dominant weight $\lambda = \sum_{j=1}^{r} m_j \lambda_j$, $m_j \in \mathbf{N}$, $\phi_\lambda(x) \equiv \sum_{\mu \in O_\lambda} \mathrm{e}^{2ia\mu \cdot x}$, in which O_λ is the orbit of λ by the action of the Weyl group: $O_\lambda = \{g(\lambda)|g \in G_\Delta\}$.

5.3 Calogero-Sutherland Systems

The set of functions $\{\phi_\lambda\}$ has an order $\succ$, $|\lambda|^2 > |\lambda'|^2 \Rightarrow \phi_\lambda \succ \phi_{\lambda'}$. The similarity transformed Hamiltonian $\tilde{\mathcal{H}}$

$$\tilde{\mathcal{H}} = -\frac{\hbar^2}{2}\sum_{j=1}^{r}\frac{\partial^2}{\partial x_j^2} - a\hbar \sum_{\alpha \in \Delta_+} g_\alpha \cot(a\alpha \cdot x)\alpha \cdot \frac{\partial}{\partial x} \qquad (5.3.13)$$

is triangular in this basis: $\tilde{\mathcal{H}}\phi_\lambda = 2a^2(\hbar^2\lambda^2 + 2\hbar\varrho \cdot \lambda)\phi_\lambda + \sum_{|\lambda'|<|\lambda|} c_{\lambda'}\phi_{\lambda'}$. That is the eigenvalue is

$$\mathcal{E} = 2a^2(\hbar^2\lambda^2 + 2\hbar\varrho \cdot \lambda), \quad \text{or} \quad \mathcal{E} + \mathcal{E}_0 = 2a^2(\hbar\lambda + \varrho)^2. \qquad (5.3.14)$$

The universal proof for all root systems goes as follows [53]. By using (5.3.13) we obtain

$$\tilde{\mathcal{H}}\phi_\lambda = 2a^2\hbar^2\lambda^2\phi_\lambda - 2i\hbar a^2 \sum_{\rho \in \Delta_+}\sum_{\mu \in O_\lambda} g_\rho \cot(a\rho \cdot x)(\rho \cdot \mu)e^{2ia\mu \cdot x}. \qquad (5.3.15)$$

First let us fix one positive root ρ and a weight μ in O_λ such that $\rho \cdot \mu \neq 0$. Then

$$\mu' \equiv s_\rho(\mu) = \mu - (\rho^\vee \cdot \mu)\rho \in O_\lambda, \quad \rho \cdot \mu' = -\rho \cdot \mu. \qquad (5.3.16)$$

Without loss of generality we may assume

$$\rho^\vee \cdot \mu = k > 0, \quad k \in \mathbf{Z}_{>0}. \qquad (5.3.17)$$

The contribution of the pair (μ, μ') in the summation of (5.3.15) reads

$$|\rho \cdot \mu|e^{2ai\mu \cdot x}(1 - e^{-2aik\rho \cdot x})\cot(a\rho \cdot x)$$

$$= i|\rho \cdot \mu|\left(e^{2ai\mu \cdot x} + e^{2ai\mu' \cdot x} + 2\sum_{j=1}^{k-1}e^{2ai(\mu - j\rho)\cdot x}\right), \qquad (5.3.18)$$

which is the generalisation of Sutherland's fundamental identity eq(15) in [28] to arbitrary root systems. The summation in the expression correspond to $\phi_{\lambda'}$ with λ' being lower than λ. Thus (5.3.15) reads

$$\tilde{\mathcal{H}}\phi_\lambda = 2a^2\hbar^2\lambda^2\phi_\lambda + 2\hbar a^2\sum_{\rho \in \Delta_+}\sum_{\mu \in O_\lambda}g_\rho|\rho \cdot \mu|e^{2ia\mu \cdot x} + \sum_{|\lambda'|<|\lambda|}c_{\lambda'}\phi_{\lambda'}, \qquad (5.3.19)$$

in which $\{c_{\lambda'}\}$'s are constants. It is easy to see that $(\mu = g(\lambda), \exists g \in G_\Delta)$

$$\sum_{\rho \in \Delta_+}g_\rho|\rho \cdot \mu| = \sum_{\rho \in \Delta_+}g_\rho|g(\rho) \cdot \lambda| = \left(\sum_{\rho \in \Delta_+}g_\rho\rho\right) \cdot \lambda = 2\varrho \cdot \lambda, \qquad (5.3.20)$$

which is independent of μ. Thus we have demonstrated the triangularity of $\tilde{\mathcal{H}}$:

$$\tilde{\mathcal{H}}\phi_\lambda = (2a^2\hbar^2\lambda^2 + 2\hbar\varrho\cdot\lambda)\phi_\lambda + \sum_{|\lambda'|<|\lambda|} c_{\lambda'}\phi_{\lambda'}. \tag{5.3.21}$$

Exercise Convince yourself that $(\mu - j\rho)^2 < \mu^2 = \lambda^2$, $j = 1, \cdots, k-1$, $\rho^\vee \cdot \mu = k$.

Again the coupling constant dependence comes solely from the deformed Weyl vector ϱ. This spectrum is a deformation of that of the free motion with momentum $2\hbar a\lambda$ in the principal Weyl alcove. The corresponding eigenfunction is called a generalised Jack polynomial [37] or Heckman-Opdam's Jacobi polynomial. For the rank two ($r = 2$) root systems, A_2, $B_2 \cong C_2$ and $I_2(m)$ (the dihedral group), the complete set of eigenfunctions are known explicitly [32, 33, 53].

3. Quantum Lax pair and quantum conserved quantities

The universal Lax pair for Calogero-Sutherland systems is given in terms of the representations of the Coxeter (Weyl) group in stead of the Lie algebra. The Lax operators without spectral parameter for the rational, trigonometric and hyperbolic potentials are

$$L(p,x) = p\cdot\hat{H} + X(x), \quad X(x) = i\sum_{\alpha\in\Delta_+} g_\alpha(\alpha\cdot\hat{H})\Lambda(\alpha\cdot x)\hat{s}_\alpha, \tag{5.3.22}$$

$$M(x) = \frac{i}{2}\sum_{\alpha\in\Delta_+} g_\alpha\alpha^2\,\Omega(\alpha\cdot x)(\hat{s}_\alpha - I), \tag{5.3.23}$$

in which I is the identity operator $(I)_{\mu\nu} = \delta_{\mu\nu}$ and $\{\hat{s}_\alpha|\alpha\in\Delta\}$ are the reflection operators of the root system. It should be stressed that only the positive roots (Δ_+) are summed in L and M, (5.3.22), (5.3.23) for avoiding the double counting, since if $(\hat{s}_\alpha)_{\mu\nu} = 1$ then $(\hat{s}_{-\alpha})_{\mu\nu} = 1$, too. They act on a set of $\mathbf{R}^r$ vectors $\mathcal{R} = \{\mu^{(k)} \in \mathbf{R}^r | k = 1, \cdots, d\}$, permuting them under the action of the reflection group. The vectors in $\mathcal{R}$ form a basis for the representation space $\mathbf{V}$ of dimension d. The matrix elements of the operators $\{\hat{s}_\alpha|\alpha\in\Delta\}$ and $\{\hat{H}_j|j=1,\cdots,r\}$ are defined as follows:

$$(\hat{s}_\alpha)_{\mu\nu} = \delta_{\mu,s_\alpha(\nu)} = \delta_{\nu,s_\alpha(\mu)}, (\hat{H}_j)_{\mu\nu} = \mu_j\delta_{\mu\nu}, \alpha\in\Delta, \mu,\nu\in\mathcal{R}. \tag{5.3.24}$$

The simplest and the most natural representation spaces of the Lax pair operators are provided by the set of all roots Δ for the simply-laced root systems, and the set of short roots Δ_S or the set of long roots Δ_L for non-simply laced root systems. These give root type Lax pairs, [49]. For the Calogero-Sutherland systems based on the E_8

5.3 Calogero-Sutherland Systems

root system, the root type Lax pairs was the first realisation [49]. Another class of simple representations are the so-called minuscule (minimal) type representations, for which $\mathcal{R}$ consists of the weights belonging to a minuscule (minimal) representation, and which give minuscule (minimal) type Lax pairs [49]. The forms of the functions Λ, Ω depend on the chosen potential as given in Table 5.1. Then the equations of motion can be expressed in a matrix form

$$\mathrm{d}L/\mathrm{d}t = \frac{i}{\hbar}[\mathcal{H}, L] = [L, M].$$

The operator M satisfies the relations

$$\sum_{\mu \in \mathcal{R}} M_{\mu\nu} = \sum_{\nu \in \mathcal{R}} M_{\mu\nu} = 0, \qquad (5.3.25)$$

which are essential for deriving quantum conserved quantities as the total sum (Ts) of all the matrix elements of L^n: $K_n = \mathrm{Ts}(L^n) \equiv \sum_{\mu,\nu \in \mathcal{R}} (L^n)_{\mu\nu}$, $[\mathcal{H}, K_n] = 0$, $[K_m, K_n] = 0$, $n, m = 1, \cdots$. In particular, the power 2 is universal to all the root systems and the quantum Hamiltonian is given by $\mathcal{H} \propto K_2 + \mathrm{const.}$ As in the affine Toda molecule systems, a Lax pair with a spectral parameter can also be introduced universally for all the above potentials. The Dunkl operators [39], or the commuting differential-difference operators are also used to construct quantum conserved quantities for some root systems. This method is essentially equivalent to the universal Lax operator formalism. As the Lax operators do not contain the Planck's constant, the quantum Lax pair is essentially of the same form as the classical Lax pair. The difference between the trace (Tr) and the total sum (Ts) vanishes as $\hbar \to 0$.

Exercise Calculate the following three examples of minimal type Lax pairs.

(1) For A_{r-1} Lax pair, choose $\mathcal{R}$ the vector representation weights (the Weyl orbit of the fundamental weight λ_1), $\mathcal{R} = \left\{ e_j - \frac{1}{r}\xi, j = 1, \cdots, r \right\}$ and the rational potential. Show the universal Lax pair reduces to the one given in (5.3.2).

(2) D_r Lax pair. Calculate the Lax pair by choosing for $\mathcal{R}$ the vector representation weights (the Weyl orbit of the fundamental weight λ_1), $\mathcal{R} = \{\pm e_j, j = 1, \cdots, r\}$ and the rational potential.

(3) B_r Lax pair. Calculate the Lax pair by choosing for $\mathcal{R}$ the set of short roots $\mathcal{R} = \{\pm e_j, j = 1, \cdots, r\}$ and the rational potential.

Lax pair for Calogero systems The quantum Lax pair for the Calogero systems is obtained from the universal Lax pair (5.3.22) by replacement $L \to L^{\pm} =$

$L \pm i\omega Q$, $Q \equiv x \cdot \hat{H}$, which correspond to the creation and annihilation operators of a harmonic oscillator. The equations of motion are rewritten as $dL^{\pm}/dt = \frac{i}{\hbar}[\mathcal{H}, L^{\pm}] = [L^{\pm}, M] \pm i\omega L^{\pm}$. Then $\mathcal{L}^{\pm} = L^{\pm}L^{\mp}$ satisfy the Lax type equation $d\mathcal{L}^{\pm}/dt = \frac{i}{\hbar}[\mathcal{H}, \mathcal{L}^{\pm}]$, giving rise to conserved quantities $\text{Ts}(\mathcal{L}^{\pm})^n$, $n = 1, 2, \cdots$. The Calogero Hamiltonian is given by $\mathcal{H} \propto \text{Ts}(\mathcal{L}^{\pm})$.

All the eigenstates of the Calogero Hamiltonian $\mathcal{H}$ with eigenvalues $\hbar\omega N$, $N = \sum_{j=1}^{r} n_j f_j$, $n_j \in \mathbf{N}$ are simply constructed in terms of $L^{\pm}$: $\prod_{j=1}^{r} (B_{f_j}^+)^{n_j} e^w$. Here the integers $\{f_j\}$, $j = 1, \cdots, r$ are listed in Table 5.2. The creation operators $B_{f_j}^+$ and the corresponding annihilation operators $B_{f_j}^-$ are defined by $B_{f_j}^{\pm} = \text{Ts}(L^{\pm})^{f_j}$, $j = 1, \cdots, r$. They are hermitian conjugate to each other $(B_{f_j}^{\pm})^{\dagger} = B_{f_j}^{\mp}$ with respect to the standard hermitian inner product of the states defined in PW. They satisfy commutation relations $[\mathcal{H}, B_k^{\pm}] = \pm \hbar k \omega B_k^{\pm}$, $[B_k^+, B_l^+] = [B_k^-, B_l^-] = 0$, $k, l \in \{f_j|\ j = 1, \cdots, r\}$. The ground state is annihilated by all the annihilation operators $B_{f_j}^- e^w = 0$, $j = 1, \cdots, r$.

A few remarks on quantum elliptic C-S systems. As is well known one particle Schrödinger equation with the Weierstrass $\wp$ function potential, or the Lamé equation[76]

$$\left[-\frac{d^2}{dx^2} + g(g-1)\wp(x)\right]\psi(x) = \mathcal{E}\psi(x),$$

is not exactly solvable for generic g. The groundstate eigenfunction is not known, either. For multi-particle systems, the quantum Lax pair is not known but Weyl invariant commuting differential operators as many as the degrees of freedom have been constructed [43]. For the classical elliptic C-S systems, Lax pairs of various types are known for all crystallographic root systems [49, 50], which reduce to those of the degenerate potentials, rational, trigonometric and hyperbolic. The quantum C-S system with the elliptic potential are integrable, but not exactly solvable.

Rational potentials: super-integrability The systems with the rational potential (i.e. without the quadratic confining terms) have a remarkable property: super-integrability [67]. Rational C-S systems based on a rank r root system have $2r - 1$ independent conserved quantities. Roughly speaking they are of the forms $K_n = \text{Ts}(L^n)$, $J_m = \text{Ts}(QL^m)$, $Q \equiv x \cdot \hat{H}$, among which only r are involutive. At the classical level, super-integrability can be characterised as algebraic linearisability [68]. Since a commutator of any conserved quantities is again a conserved quantity, these conserved

5.3 Calogero-Sutherland Systems

quantities form a non-linear algebra called a quadratic algebra [69]. It can be considered as a finite-dimensional analogue of the W-algebra appearing in certain conformal field theory.

Quantum vs classical integrability Usually quantum-classical correspondence/contrast is discussed in the "quasi-classical" e.g. the WKB or "quasi-macroscopic" regime of quantum mechanics in which the expectation values are good representations of the classical variables. These are exemplified in Ehrenfest's well-known theorem or in the WKB method.

We will ask and answer in simple terms the following universal question in multi-particle quantum mechanics:

How can we relate the knowledge of the eigenfunctions and eigenvalues of a multi-particle quantum mechanical system to the properties of the corresponding classical system, in particular, at equilibrium?

Let us start with the famous pioneering problem. In 1886 Stieltjes [71] showed that the maximal point of a certain multi-variable function:

$$T(x) = \prod_{j=1}^{n}(1-x_j)^p(1+x_j)^q \cdot \prod_{1 \leqslant j < k \leqslant n} |x_j - x_k|,$$

was given by the zeros of the Jacobi polynomial $P_n^{(\alpha,\beta)}(x)$, $\alpha = 2p-1$, $\beta = 2q-1$. It was explained as equilibrium of logarithmic electrostatic potential in Section 6.6 of Szegö[72]. In the present language it is the maximum of the ground state eigenfunction $\phi_0(x) = e^{w(x)}$ of the Sutherland system, which can be rephrased as the minimum (equilibrium) of the corresponding classical potential $U(x) = 1/2 \sum_{j=1}^{r}[\partial w(x)/\partial x_j]^2$ for the B_r, BC_r or D_r.

Let us amplify this point further. As integrable systems the quantum spectra of the Calogero and Sutherland systems are nicely quantised. We will show that the classical eigenvalues of the Calogero and Sutherland systems are also nicely quantised.

What are the the classical eigenvalues? We know that any classical multi-particle system near an equilibrium point can be reduced to a set of coupled harmonic oscillators, which can be diagonalised giving rise to classical eigenvalues as many as the degrees of freedom.

Here we discuss the properties of the classical potential

$$U_C = \frac{1}{2}\sum_{j=1}^{N}\left(\frac{\partial w}{\partial x_j}\right)^2, \qquad (5.3.26)$$

the pre-potential w, and Lax matrices L, M, near the classical equilibrium point:

$$p = 0, \quad x = \bar{x}. \tag{5.3.27}$$

For the classical potential the point $\bar{x}$ is characterised as its minimum point:

$$\left.\frac{\partial U_C}{\partial x_j}\right|_{\bar{x}} = 0, \quad j = 1, \cdots, N, \tag{5.3.28}$$

whereas it is a maximal point of the pre-potential w and of the ground state wavefunction $\phi_0 = e^w$:

$$\left.\frac{\partial w}{\partial x_j}\right|_{\bar{x}} = 0, \quad j = 1, \cdots, N. \tag{5.3.29}$$

In this connection, it should be noted that the condition $(p + i\partial w/\partial x_j)e^w = 0$ is also satisfied classically at this point. In the Lax representation it is a point at which two Lax matrices commute:

$$0 = [\bar{L}, \bar{M}], \tag{5.3.30}$$

in which $\bar{L} = L(0, \bar{x})$, $\bar{M} = M(\bar{x})$ etc and $d\bar{L}/dt = 0$, etc at the equilibrium point. The value of a quantity A at the equilibrium is expressed by $\bar{A}$.

By differentiating (5.3.26), we obtain

$$\frac{\partial U_C}{\partial x_j} = \sum_{l=1}^{N} \frac{\partial^2 w}{\partial x_j \partial x_l} \frac{\partial w}{\partial x_l}. \tag{5.3.31}$$

Since $\partial^2 w/\partial x_j \partial x_k$ is negative definite everywhere, we find the equilibrium point of w is a maximum and that the two conditions (5.3.28) and (5.3.29) are equivalent:

$$\left.\frac{\partial U_C}{\partial x_j}\right|_{\bar{x}} = 0, \quad j = 1, \cdots, N, \iff \left.\frac{\partial w}{\partial x_j}\right|_{\bar{x}} = 0, \quad j = 1, \cdots, N. \tag{5.3.32}$$

By differentiating (5.3.31) again, we obtain

$$\frac{\partial^2 U_C}{\partial x_j \partial x_k} = \sum_{l=1}^{N} \frac{\partial^2 w}{\partial x_j \partial x_l} \frac{\partial^2 w}{\partial x_l \partial x_k} + \sum_{l=1}^{N} \frac{\partial^3 w}{\partial x_j \partial x_k \partial x_l} \frac{\partial w}{\partial x_l}.$$

Thus at the equilibrium point of the classical potential U_C, the following relation holds:

$$\left.\frac{\partial^2 U_C}{\partial x_j \partial x_k}\right|_{\bar{x}} = \sum_{l=1}^{N} \left.\frac{\partial^2 w}{\partial x_j \partial x_l}\right|_{\bar{x}} \left.\frac{\partial^2 w}{\partial x_l \partial x_k}\right|_{\bar{x}}. \tag{5.3.33}$$

If we define the following two symmetric $N \times N$ matrices $\tilde{U}$ and $\tilde{w}$,

$$\tilde{U} = \text{Matrix}\left[\left.\frac{\partial^2 U_C}{\partial x_j \partial x_k}\right|_{\bar{x}}\right], \quad \tilde{w} = \text{Matrix}\left[\left.\frac{\partial^2 w}{\partial x_j \partial x_k}\right|_{\bar{x}}\right], \tag{5.3.34}$$

5.3 Calogero-Sutherland Systems

we have
$$\tilde{U} = \tilde{w}^2, \tag{5.3.35}$$

and
$$\begin{aligned} \text{Eigenvalues}(\tilde{U}) &= \{w_1^2, \cdots, w_N^2\}, \\ \text{Eigenvalues}(-\tilde{w}) &= \{w_1, \cdots, w_N\}, \quad w_j > 0, \quad j = 1, \cdots, N. \end{aligned} \tag{5.3.36}$$

That is $\tilde{U}$ is positive definite and the point $\bar{x}$ is actually a minimal point of U_C.

Calogero case For the Calogero system, the equations determining the maximum of the ground state wavefunction read $\left(\left\{y_j \stackrel{\text{def}}{=} \sqrt{\frac{\omega}{g}}\bar{x}_j\right\}\right)$

$$\sum_{k \neq j}^{N} \frac{1}{\bar{x}_j - \bar{x}_k} = \frac{\omega}{g}\bar{x}_j \Rightarrow \sum_{k \neq j}^{N} \frac{1}{y_j - y_k} = y_j, \quad j = 1, \cdots, N. \tag{5.3.37}$$

It is very interesting to note that $\{y_j\}$ are the zeros of the Hermite polynomial $H_N(x)$ (Stieltjes, Calogero). The matrix $\tilde{w}$ is given by

$$\tilde{w}_{jk} = -\left(\omega + g\sum_{l \neq j}\frac{1}{(\bar{x}_j - \bar{x}_l)^2}\right)\delta_{jk} + g\frac{1}{(\bar{x}_j - \bar{x}_k)^2}, \tag{5.3.38}$$

and is equal to $(-\omega I + i\bar{M})_{jk}$ for the representation of the Lax matrix $\bar{M}$ (5.3.2). We obtain
$$\text{Spec}(-\tilde{w}) = \omega\{1, 2, \cdots, N\}. \tag{5.3.39}$$

Sutherland case The equations determining the equilibrium position (5.3.28) and (5.3.29) read:

$$\sum_{k \neq j}^{N} \frac{\cos(\bar{x}_j - \bar{x}_k)}{\sin^3(\bar{x}_j - \bar{x}_k)} = 0, \quad \sum_{k \neq j}^{N} \cot(\bar{x}_j - \bar{x}_k) = 0, \quad j = 1, \cdots, N,$$

and the equilibrium position is "equally-spaced"

$$\bar{q} = \pi(0, 1, \cdots, N-2, N-1)/N + \xi(1, 1, \cdots, 1), \quad \xi \in \mathbf{R} : \text{arbitrary}, \tag{5.3.40}$$

due to the well-known trigonometric identities:

$$\sum_{k \neq j}^{N} \frac{\cos[\pi(j-k)/N]}{\sin^3[\pi(j-k)/N]} = 0, \quad \sum_{k \neq j}^{N} \cot[\pi(j-k)/N] = 0, \quad j = 1, \cdots, N.$$

We choose this constant shift ξ such that the "center of mass" coordinate vanishes, $\sum_{j=1}^{N} \bar{x}_j = 0$:

$$\bar{x}_j = \frac{\pi(N+1-j)}{N} - \frac{\pi(N+1)}{2N} = \frac{\pi}{2} - \frac{\pi(2j-1)}{2N} = -\bar{x}_{N+1-j}, \quad j = 1, \cdots, N. \tag{5.3.41}$$

Then the degree N polynomial in x, having zeros at $\{\sin \bar{x}_j\}$,

$$2^{N-1} \prod_{j=1}^{N} (x - \sin \bar{x}_j) = 2^{N-1} \prod_{j=1}^{N} \left[x - \cos \frac{\pi(2j-1)}{2N} \right] \stackrel{\text{def}}{=} T_N(x), \tag{5.3.42}$$

is the Chebyshev polynomial of the first kind, $T_n(\cos \varphi) = \cos(n\varphi)$.

This enables us to calculate most quantities exactly. For example, we have

$$\widetilde{w}_{jk} = g \frac{(1 - \delta_{jk})}{\sin^2[(j-k)\pi/N]} - g \delta_{jk} \sum_{l \neq j} \frac{1}{\sin^2[(j-l)\pi/N]}, \quad j, k = 1, \cdots, N \tag{5.3.43}$$

and

$$\text{Spec}(-\widetilde{w}) = 2g \{N-1, (N-2)2, \cdots, (N-j)j, \cdots, 2(N-2), N-1\} \tag{5.3.44}$$
$$= 4(\lambda_1 \cdot \varrho, \lambda_2 \cdot \varrho, \cdots, \lambda_{N-1} \cdot \varrho), \tag{5.3.45}$$

in which the trivial eigenvalue 0, coming from the translational invariance, is removed. This agrees with the $o(\hbar)$ part of the exact energy eigenvalue (5.3.14) the j-th entry is $4\lambda_j \cdot \varrho$. The spectrum (5.3.44) is symmetric with respect to the middle point, $\lambda_j \leftrightarrow \lambda_{N-j}$. It is easy to see that $\widetilde{W}$ is essentially the same as the Lax matrix $\bar{M}$, $\bar{M} = -i\widetilde{w}$.

As a summary, we have: For the A-type Calogero, the equilibrium positions are described by the zeros of the Hermite polynomial, for the B, C, D-type Calogero they are the zeros the Laguerre polynomial and the Chebyshev polynomial for the A-type Sutherland [70, 73, 74]. For the exceptional root systems the corresponding polynomials had not been known.

The minimum energy of the classical potential $U_C(x)$ at the equilibrium is the quantum ground state energy $\lim_{\hbar \to 0} \mathcal{E}_0$ itself. It is also coupling constants times integer for both Calogero and Sutherland cases. Near a classical equilibrium, a multiparticle dynamical system is always reduced to a system of coupled harmonic oscillators. For Calogero systems the eigenfrequencies of these small oscillations are, in fact, exactly the same as the quantum eigenfrequencies, $\omega f_j = \omega(1 + e_j)$. For Sutherland systems, the classical eigenfrequencies are the same as the $o(\hbar)$ part of the quantum spectra corresponding to all the fundamental weights λ_j: $2a^2\lambda_j \cdot \varrho$. Moreover, the eigenvalues of various Lax matrices L and M at the equilibrium take many "interesting values"[70]. These results provide ample explicit examples of the general.

Theorem 2 (quantum-classical correspondence, Loris-Sasaki [75]). The lowest order term i.e. the $\mathcal{O}(\hbar)$ part, of the quantum energy eigenvalue is determined solely by the corresponding classical data, i.e. the eigenfrequencies of the normal mode oscillations at the classical equilibrium.

Outline of the proof:
$$\phi_n(x) = \psi_n(x)\phi_0(x), \quad n = 0, 1, \cdots,$$
$$\tilde{H}\psi_n = E_n\psi_n, \quad E_n = \hbar\mathcal{E}_n + \mathcal{O}(\hbar^2),$$
$$\tilde{H} \stackrel{\text{def}}{=} e^{-\frac{1}{\hbar}w} H e^{\frac{1}{\hbar}w} = -\frac{\hbar^2}{2}\sum_{j=1}^{N}\frac{\partial^2}{\partial x_j^2} - \hbar\sum_{j=1}^{N}\frac{\partial w}{\partial x_j}\frac{\partial}{\partial x_j}.$$

Let us introduce "classical" eigenfunctions $\{\varphi_n\}$, $\lim_{\hbar \to 0}\psi_n(x) = \varphi_n(x)$. They satisfy
$$-\sum_{j=1}^{N}\frac{\partial w}{\partial x_j}\frac{\partial \varphi_n}{\partial x_j} = \mathcal{E}_n\varphi_n. \tag{5.3.46}$$

At the equilibrium they pick up the eigenvalues of the small oscillations:
$$-\sum_{j=1}^{N}\frac{\partial^2 w(\bar{x})}{\partial x_k \partial x_j}\frac{\partial \varphi_n(\bar{x})}{\partial x_j} = \mathcal{E}_n\frac{\partial \varphi_n(\bar{x})}{\partial x_k}. \tag{5.3.47}$$

The theorem is true for *non-integrable systems*, too.

5.3.3 Jack Polynomials (Jack 多项式)

The (generalised) Jack polynomials are the eigenpolynomials for the Sutherland systems based on various root systems. Here we present simple introduction to original Jack polynomials based on the A_{r-1} root system, for which various concepts and methods related to the Young diagrams and partitions can be employed.

1. Partitions or Young diagrams

Let $\lambda = (\lambda_1, \lambda_2, \cdots)$ be a partition, that is, a sequence of non-negative integers in decreasing order $\lambda_1 \geqslant \lambda_2 \geqslant \cdots \lambda_j \geqslant \cdots$ and $\lambda' = (\lambda'_1, \lambda'_2, \cdots)$ the conjugate partition to λ. The length of λ, denoted by $\ell(\lambda)$, is the total number of non-vanishing λ_j's and it is equal to $\lambda'_1 = \ell(\lambda)$. The sum of the parts is the *weight* of λ denoted by $|\lambda|$:
$$|\lambda| = \lambda_1 + \lambda_2 + \cdots.$$

The partition can be identified with its diagram

$$\lambda = \{(j,k) : 1 \leqslant j \leqslant \ell(\lambda),\ 1 \leqslant k \leqslant \lambda_j\}. \tag{5.3.48}$$

The partition λ can also be specified by the number of times each integer occurs as a part

$$\lambda = (1^{m_1} 2^{m_2} \cdots j^{m_j} \cdots), \tag{5.3.49}$$

and we have

$$|\lambda| = \sum_j j\, m_j(\lambda), \quad m_j(\lambda) = \lambda'_j - \lambda'_{j+1}, \Rightarrow \sum_j m_j(\lambda) = \ell(\lambda). \tag{5.3.50}$$

For example, has $|\lambda| = 18$, $\ell(\lambda) = 7$ and $\lambda_1 = \lambda_2 = 4$, $\lambda_3 = 3$, $\lambda_4 = \lambda_5 = \lambda_6 = 2$, $\lambda_7 = 1$. It is also expressed as $\lambda = (1, 2^3, 3, 4^2)$. Its conjugate is $\lambda'_1 = 7$, $\lambda'_2 = 6$, $\lambda'_3 = 3$, $\lambda'_4 = 2$. The sum $\lambda + \mu$ and product $\lambda\mu$ are defined as follows:

$$(\lambda + \mu)_j = \lambda_j + \mu_j, \quad (\lambda\mu)_j = \lambda_j \mu_j. \tag{5.3.51}$$

Among the partitions of n $n = |\lambda| = |\mu|$, the following dominance (total) ordering $\lambda \geqslant \mu$ can be introduced

$$\lambda \geqslant \mu \Leftrightarrow |\lambda| = |\mu|, \quad \text{first non-vanishing } \lambda_j - \mu_j > 0. \tag{5.3.52}$$

For example

$$(5) \geqslant (41) \geqslant (32) \geqslant (31^2) \geqslant (2^2 1) \geqslant (21^3) \geqslant (1^5),$$

There is another natural (partial) ordering $\lambda \succeq \mu$:

$$\lambda \succeq \mu \Leftrightarrow |\lambda| = |\mu|, \quad \lambda_1 + \cdots + \lambda_i = \mu_1 + \cdots + \mu_i \quad (\forall i). \tag{5.3.53}$$

5.3 Calogero-Sutherland Systems

Remark: These are orderings of partitions in general. However, in the discussion of Jaclk polynomials for particle number r, we only consider partitions of length not larger than r, $\ell(\lambda) \leqslant r$. Then the conditions for $i > r$ in (5.3.53) are redundant. In fact, the actual ordering appearing in Sutherland systems are different from these, $|\mu|^2 < |\lambda|^2$ (5.3.21).

Exercise Show that there is a one to one correspondence between an A_{r-1} dominant weight λ^D and a Young diagram. The A_{r-1} fundamental weights λ_j^A are given in (5.2.160), and the $\boldsymbol{\xi}$ part should be dropped:

$$\lambda^D = \sum_{j=1}^r u_j \lambda_j^A, \quad u_j \in \mathbf{N}, \quad \Longleftrightarrow \quad \lambda = (\lambda_1, \lambda_2, \cdots, \lambda_r), \quad \lambda_j = \sum_{k=j}^r u_k. \quad (5.3.54)$$

In other words, λ_j is the coefficient of e_j in λ^D.

Symmetric Functions For a partition of length $\leqslant n$, a monomial symmetric polynomial is defined by

$$m_\lambda(x_1,\cdots,x_n) \stackrel{\text{def}}{=} \sum_\alpha x^\alpha, \quad x^\alpha \stackrel{\text{def}}{=} x_1^{\alpha_1}\cdots x_n^{\alpha_n}, \quad (5.3.55)$$

in which the summation is over all distinct permutations α of $\lambda = (\lambda_1,\cdots,\lambda_n)$. For a partition λ a power sum symmetric polynomial p_λ is defined by

$$p_\lambda(x_1,\cdots,x_n) \stackrel{\text{def}}{=} p_{\lambda_1}(x)p_{\lambda_2}(x)\cdots p_{\lambda_{\ell(\lambda)}}(x), \quad p_k(x) \stackrel{\text{def}}{=} \sum_{j=1}^n x_j^k. \quad (5.3.56)$$

Both $m_\lambda(x)$ and $p_\lambda(x)$ are polynomials of homogeneous degree $|\lambda|$ in $x=(x_1,\cdots,x_n)$.

Exercise Show for $\lambda=(1^k)$, $k \leqslant n$ that $m_{1^k}(x)$ is an elementary symmetric polynomial of degree k:

$$m_{1^k}(x) = e_k(x), \quad e_1(x) = x_1 + x_2 + \cdots + x_n,$$
$$e_2(x) = x_1 x_2 + x_2 x_3 + \cdots, \quad e_3(x) = x_1 x_2 x_3 + \cdots.$$

Show for $\lambda = (k)$ that $m_k(x)$ is a power sum symmetric polynomial

$$m_{(k)}(x) = \sum_{j=1}^n x_j^k = p_{(k)}(x) = p_k(x).$$

These examples show that if $\lambda \geqslant \mu$ ($|\lambda|=|\mu|$) then the highest power of each variable x_j in $m_\lambda(x)$ is higher than that in $m_\mu(x)$.

Among power sum symmetric polynomials, we adopt the following inner product[37]

$$\langle p_\lambda, p_\mu \rangle = \delta_{\lambda\mu} \prod_j j^{m_j} m_j! \cdot \beta^{-\sum_j m_j}. \tag{5.3.57}$$

2. A_{r-1} Sutherland System

Here we rewrite the A_{r-1} Sutherland system in notation suitable for Jack polynomials,

$$\hat{\mathcal{H}}_S = \sum_{j=1}^r \frac{1}{2} p_j^2 + \beta(\beta - 1) \sum_{1 \leq j < k \leq r} \frac{1}{\sin^2(q_j - q_k)}, \tag{5.3.58}$$

$$= \sum_{j=1}^r A_j^\dagger A_j + \beta^2 \frac{r(r^2 - 1)}{6}, \quad A_j \stackrel{\text{def}}{=} p_j + i\beta \sum_{k \neq j} \cot(q_j - q_k), \tag{5.3.59}$$

in which the notation in section 5.3 are changed $x_j \to q_j$, $\hbar \to 1$, $a \to 1$, $g \to \beta$. For each q_j the Hamiltonian is a periodic function of period π. The constant term is the ground state energy $\mathcal{E}_0$ calculated by the expression (5.3.9) of the deformed Weyl vector. The ground state eigenfunction ϕ_0 is

$$\phi_0 \propto \left[\prod_{j<k} \sin(q_j - q_k)\right]^\beta \propto \left[\prod_{j<k} \sqrt{\frac{x_j}{x_k}} \left(1 - \frac{x_k}{x_j}\right)\right]^\beta, \quad x_j \stackrel{\text{def}}{=} e^{2iq_j}. \tag{5.3.60}$$

The similarity transformed Hamiltonian (5.3.11) reads:

$$\phi_0^{-1} \circ \hat{\mathcal{H}}_S \circ \phi_0 = 2\left[H_\beta + \beta^2 \frac{r(r^2 - 1)}{12}\right], \tag{5.3.61}$$

$$H_\beta \stackrel{\text{def}}{=} \sum_{i=1}^r D_i^2 + \beta \sum_{1 \leq j < k \leq r} \frac{x_j + x_k}{x_j - x_k}(D_j - D_k), \quad D_j \stackrel{\text{def}}{=} x_j \frac{\partial}{\partial x_j}. \tag{5.3.62}$$

From this expression, the periodicity of H_β is obvious. By construction H_β is invariant under permutations $x_j \leftrightarrow x_k$, implying that the eigenfunctions are also symmetric functions of $\{x_j\}$. Since the total degree $|\alpha| \stackrel{\text{def}}{=} \sum_{j=1}^r \alpha_j$ of a monomial $x^\alpha \stackrel{\text{def}}{=} \prod_{j=1}^r x_j^{\alpha_j}$ ($\alpha_j \in \mathbb{N}$) of a symmetric polynomial is preserved, the eigenfunctions of H_β can be constructed among those symmetric polynomials of a fixed total degree.

Exercise Calculate for lower r and $k \leq r$ that the elementary symmetric polynomials $m_{(1^k)}(x)$ are the eigenpolynomials of H_β:

$$H_\beta m_{(1^k)}(x) = k[1 + \beta(r - k)] m_{(1^k)}(x). \tag{5.3.63}$$

5.3 Calogero-Sutherland Systems

Exercise Calculate the action of H_β on $m_{(k)}(x)$, i.e. the power sum symmetric polynomials $p_k(x) = \sum_{j=1}^{r} x_j^k$, for lower r and k:

$$H_\beta p_k(x) = k\big[k + \beta(r-1)\big]p_k(x) + \text{lower order terms.} \tag{5.3.64}$$

Exercise Calculate the action of H_β on $m_{(k^r)}(x) = \prod_{j=1}^{r} x_j^k$:

$$H_\beta m_{(k^r)}(x) = k^2 r\, m_{(k^r)}(x) \quad \Rightarrow \quad J_{(k^r)}(x) = m_{(k^r)}(x). \tag{5.3.65}$$

In this case the second term in H_β (5.3.62) simply vanishes.

Exercise Schur polynomial for $\beta = 1$ Show for $\beta = 1$ the Schur polynomial $s_\lambda(x) \stackrel{\text{def}}{=} \det(x_j^{\lambda_k + r - k})_{1 \leqslant j < k \leqslant r} / \det(x_j^{r-k})_{1 \leqslant j < k \leqslant r}$ is the eigenpolynomial of H_1 (i.e. H_β with $\beta = 1$) with:

$$H_1 s_\lambda(x) = \sum_{j=1}^{r} \Big[\lambda_j^2 + (r + 1 - 2j)\lambda_j\Big] \cdot s_\lambda(x). \tag{5.3.66}$$

In this case the Sutherland Hamiltonian (5.3.58) is a free one $\hat{\mathcal{H}}_S = \frac{1}{2}\sum_{j=1}^{r} p_j^2$. We have to show $\psi_S(q) = s_\lambda(x)\phi_0(q)$ is the eigenfunction of the free Hamiltonian. The denominator of the Schur polynomial is the vanderMond determinant

$$\det(x_j^{r-k})_{1 \leqslant j < k \leqslant r} \propto \prod_{j<k}(x_j - x_k) \propto e^{i(r-1)\sum_{j=1}^{r} q_j} \times \phi_0(q), \qquad \phi_0(q) \propto \prod_{j<k}\sin(q_j - q_k).$$

Thus $\psi_S(q) \propto \exp\!\left[-i(r-1)\sum_{j=1}^{r} q_j\right] \det(x_j^{\lambda_k + r - k})_{1 \leqslant j < k \leqslant r}$, which is a sum of the terms of the form $\exp\!\left(i \sum_{j=1}^{r} u_j q_j\right)$. The set of the exponents $\{u_j\}$ is $\{2\lambda_j + r - 2j + 1\}$.

This is an eigenfunction of the free Hamiltonian $\hat{\mathcal{H}}_S = \frac{1}{2}\sum_{j=1}^{r} p_j^2$. For determining the eigenvalue, the A_{r-1} deformed Weyl vector (5.3.9) and the ground state energy $r(r^2 - 1)/12$ (5.3.61) can be used.

Definition of the Jack polynomials The Jack symmetric polynomial $J_\lambda = J_\lambda(x; \beta)$ is uniquely determined by the following two conditions [37, 55],

(i) (triangularity) $\quad J_\lambda(x) = \sum_{\mu \leqslant \lambda} u_{\lambda,\mu} m_\mu(x), \quad u_{\lambda,\lambda} = 1,$ \hfill (5.3.67)

(ii) (orthogonality) $\quad \langle J_\lambda, J_\mu \rangle_\beta = 0 \quad \text{if } \lambda \neq \mu,$ \hfill (5.3.68) in which the inner product (5.3.57) is used. The orthogonality condition (ii) can be replaced by the eigenvalue equation (ii)',

$$\text{(ii)' (eigenvalue eq.)} \quad H_\beta J_\lambda = \varepsilon_{\beta,\lambda} J_\lambda, \tag{5.3.69}$$

$$\varepsilon_{\beta,\lambda} = \sum_{j=1}^{r} \left[\lambda_j^2 + \beta(r+1-2j)\lambda_j\right] = \lambda^2 + 2\varrho \cdot \lambda, \tag{5.3.70}$$

in which ϱ is the deformed Weyl vector for A_{r-1} (5.3.9). In section 5.3.2 part 2 the excitation energy of the Sutherland system is expressed in terms of the dominant weights. The above formula uses the partition (Young diagram).

Exercise Verify, by using the correspondence between the dominant weight and the Young diagram (5.3.54) together with the expression of the deformed Weyl vector ϱ (5.3.9), that the excited state energy (5.3.14) gives the above $\varepsilon_{\beta,\lambda}$ for the A_{r-1} Sutherland system.

There exists another inner product among symmetric functions of l variables $x = (x_1, \cdots, x_l)$:

$$\langle f, g \rangle'_{l,\beta} = \frac{1}{l!} \oint \prod_{j=1}^{l} \mathrm{d}x_j \cdot \bar{\Delta}(x) f(\bar{x}) g(x), \tag{5.3.71}$$

$$\bar{\Delta}(x) = \prod_{j \neq k} \left(1 - \frac{x_k}{x_j}\right)^\beta, \quad f(\bar{x}) = f\left(\frac{1}{x_1}, \frac{1}{x_2}, \cdots\right), \quad \underline{\mathrm{d}x_j} = \frac{\mathrm{d}x_j}{2\pi\mathrm{i}x_j}. \tag{5.3.72}$$

which is the inner product in QM among functions of x with real coefficients. To be more precise, $\bar{\Delta}(x) \propto \phi_0(q)^2$ and the integration is taken in q space for one period $\left(\prod_j \int_0^\pi \mathrm{d}q_j\right)$ restricted to the principal Weyl chamber due to the factor $1/l!$. These two inner products are proportional,

$$\langle\,,\,\rangle'_{l,\beta} \propto \langle\,,\,\rangle_\beta. \tag{5.3.73}$$

It should be stressed that the A_{r-1} Jack polynomials are not exactly the same as the eigenpolynomials of the A_{r-1} Sutherland system, as is clear that the Jack polynomial $J_\lambda(x_1, \cdots, x_r)$ of a partition λ is a homogeneous polynomial of degree $|\lambda|$ in $x_1, \cdots, x_r$. That is, each term is of the form $\exp\left[2\mathrm{i} \sum_{j=1}^{r} m_j q_j\right]$, in which $m = (m_1, \cdots, m_r)$, $(m_i \geq 0)$, is a permutation of a partition $\mu = (\mu_1, \cdots, \mu_r)$, $|\mu| = |\lambda|$. In the original $\{q_j\}$ space, each term has a positive momentum and the total momentum $P \stackrel{\text{def}}{=} \sum_{j=1}^{r} p_j = -\mathrm{i} \sum_{j=1}^{r} \frac{\partial}{\partial q_j}$ is $2 \sum_{j=1}^{r} m_j = 2 \sum_{j=1}^{r} \mu_j = 2|\mu| = 2|\lambda|$. In contrast, in the Sutherland system in general, we consider the eigenfunctions in the rest frame, that is the zero total momentum $P = 0$. Bringing back to the rest frame is achieved by multiplying $\left(\prod_{j=1}^{r} x_j \right)^{-|\lambda|/r} = \exp\left[-\mathrm{i} \frac{2|\lambda|}{r} \sum_{j=1}^{r} q_j \right]$ to the Jack polynomial $J_\lambda(x_1, \cdots, x_r)$, which makes it a non polynomial. We have the following one to one correspondence

$$J_\lambda(x_1, \cdots, x_r) \times \left(\prod_{j=1}^{r} x_j \right)^{-|\lambda|/r} \Leftrightarrow \text{eigen polynomial} \quad \psi_{\lambda^D}(q_1, \cdots, q_r), \quad (5.3.74)$$

in which the dominant weight λ^D is related to the partition λ by the formula (5.3.54). Since $\left(\left(\prod_{j=1}^{r} x_j \right)^{-|\lambda|/r} \right)^* \left(\left(\prod_{j=1}^{r} x_j \right)^{-|\lambda|/r} \right) = 1$, we have

$$\left[J_\lambda(x_1, \cdots, x_r) \times \left(\prod_{j=1}^{r} x_j \right)^{-|\lambda|/r} \right]^* \left[J_\mu(x_1, \cdots, x_r) \times \left(\prod_{j=1}^{r} x_j \right)^{-|\lambda|/r} \right]$$
$$= J_\lambda(\bar{x}_1, \cdots, \bar{x}_r) J_\mu(x_1, \cdots, x_r),$$

and the inner product in (5.3.71) is in fact the same as in QM:

$$J_\lambda(\bar{x}_1, \cdots, \bar{x}_r) J_\mu(x_1, \cdots, x_r) \bar{\Delta}(x) \propto \psi_{\lambda^D}(q_1, \cdots, q_r)^* \psi_{\mu^D}(q_1, \cdots, q_r) \phi_0(\{q\})^2. \quad (5.3.75)$$

The orthogonality of Jack polynomials corresponding to different weights $|\lambda| \neq |\mu|$ can be understood as the orthogonality between two eigenstates having different total momenta.

Exercise Convince yourself that the multiplication of $(\prod_{j=1}^{r} x_j)^{-|\lambda|/r}$ corresponds to using the exact form of the fundamental weight $\lambda_j^A = \sum_{k=1}^{j} e_k - \frac{j}{r} \xi$ (5.2.160) instead of the simplified integer form $\sum_{k=1}^{j} e_k$. See the Exercise before (5.3.54).

5.4 Multi-Particle QM with Difference Schrödinger Equations (差分薛定谔方程中的多粒子量子力学)

Now we discuss the difference equation version of the quantum C-S systems. For this deformation, there are several paths. One well known deformation to physicists is called 'relativistic' Calogero-Sutherland systems due to Ruijsenaars and Schneider [54], which is the difference equation version of the classical Calogero-Sutherland systems. Construction of complete sets of classical conserved quantities for the four types of interaction potentials, rational, trigonometric, hyperbolic and elliptic potentials, via Lax pair like arguments was the main result. For the classical root systems, van Diejen[58–60] formulated the quantum version of Ruijsenaars-Schneider systems for the degenerate potentials. It is usually asserted that in Ruijsenaars-Schneider theory the shifts are inversely proportional to the 'speed of light' and that the ordinary C-S systems are obtained in the limit of infinite light speed. This is rather misleading and conceptually wrong as explained in [77]. The multi-particle interaction mechanism is, being based on instantaneous interactions (action at a distance), essentially non-relativistic. As is well known relativistic formulation requires field theory. Due to lack of time we have not discussed various interesting results of classical Calogero-Sutherland systems, we also have to skip many aspects of the Ruijsenaars-Schneider systems.

Another path is to deform the the quantum Sutherland systems for various root systems. This is a way developed by Macdonald [55, 56]. To be more precise, he deformed the eigenpolynomials of the Sutherland system, the Jack polynomials [37] for the A type root systems, which are the multi-particle Gegenbauer (ultraspherical) polynomials. Thus he has derived multi-particle q-ultraspherical polynomials which are invariant under various Weyl groups (reflection groups of crystallographic root systems) including the exceptional ones [56]. The deformation parameter is q in the range $0 < q < 1$ and $q \to 1$ recovers the original theory (polynomials). For the A_{r-1} system which has one parameter t governing the multi-particle interactions (corresponding to g in $1/\sin^2 x$ potential) on top of the deformation parameter q, commuting difference operators corresponding to the minuscule weights (5.2.164) are explicitly presented.

A third way is to construct the multi-particle versions of the 1-d discrete QM. Koornwinder [57] introduced multi-particle Askey-Wilson polynomials (5.2.107) according to Macdonald recipe. For the simply laced root systems, i.e. A, D and E type root systems, the theory has four parameters of the Askey-Wilson polynomial on top of t which governs the multi-particle interaction and q. For the non-simply laced root systems, i.e. B, C, F and G type root systems, the interaction parameter t is replaced by two independent ones, t_L for the long roots and t_S for the short roots. Both in Macdonald and Koornwindr deformations, the difference operators acting on

5.4 Multi-Particle QM with Difference Schrödinger Equations

the polynomials for A, B, C, D, E_6 and E_7 theories are constructed in terms of the Weyl orbit of minuscule weights. In E_8 theory the set of roots, which is the smallest orbit of 240 roots in E_8 weight lattice, is used. For F_4 the set of short roots with 24 roots is employed. For G_2 the 6 members of the set of long roots are employed. These are called quasi-minuscule weights. So far as I am aware, there was no mention of Lax pair like objects. An important point of these deformations by Macdonald and Koornwinder is that the language of QM is not used. The reflection invariant orthogonality measures for the eigenpolynomials are constructed ad hoc, not as the square of the ground state eigenfunctions (5.2.46). A quantum mechanical version of Macdonald-Koornwinder method was developed by van Diejen[58–60] including the rational and elliptic interactions. The multi-variable Wilson polynomials are obtained as the eigenpolynomials for the multi-particle interactions with the rational potential[59].

5.4.1 Ruijsenaars-Schneider Systems (Ruijsenaars-Schneider 系统)

The Ruijsenaars-Schneider [54] system was introduced as a classical multi-particle dynamical system based on the A_{r-1} root system. In the classical system the dynamical variables are commuting and their order is immaterial. They are integrable, as in the Calogero-Sutherland systems, for the elliptic interactions and the degenerate versions, the trigonometric, hyperbolic and rational. Here we rewrite the variables in the Hamiltonian and the conserved quantities in proper order so that they are correct quantum objects.

The system has many involutive difference operators $H_{\pm k}$ ($k = 1, \cdots, r$) (conserved quantities) [54] corresponding to the fundamental (minuscule) weights $\{\lambda_k\}$ of A_{r-1}. In the original notation they are

$$H_{\pm k} = \sum_{\substack{I \subset \{1,\cdots,r\} \\ |I|=k}} \prod_{\substack{i \in I \\ j \notin I}} h[\pm(q_i - q_j)]^{\frac{1}{2}} \cdot e^{\mp \sum_{i \in I} \theta_i} \cdot \prod_{\substack{i \in I \\ j \notin I}} h[\mp(q_i - q_j)]^{\frac{1}{2}}, \quad (5.4.1)$$

$$[H_k, H_l] = 0, \quad k, l = -r, \cdots, r, \quad (5.4.2)$$

where θ_i ($i = 1, \cdots, r$) is the rapidity and $\bar{q}_i$ is its conjugate variable and $q_i = \dfrac{\bar{q}_i}{mc}$ is the coordinate. The operator θ_j is

$$\theta_j = \frac{\hbar}{i} \frac{\partial}{\partial \bar{q}_j} = \frac{1}{i} \frac{\hbar}{mc} \frac{\partial}{\partial q_j}, \quad (5.4.3)$$

and the function $h(q)$ for the most general elliptic case is

$$h(q) = \frac{\sigma(\bar{q} + i\bar{\beta})}{\sigma(\bar{q})}, \tag{5.4.4}$$

where $\bar{\beta}$ is a coupling constant and $\sigma(z)$ is the Weierstrass σ function. The combination of the difference operators $H_{\pm 1}$ and others satisfy Poincaré algebra:

$$H = mc^2 \tfrac{1}{2}(H_{-1} + H_1), \quad [H, P] = 0, \tag{5.4.5}$$

$$P = mc(H_{-1} - H_1), \quad [H, B] = i\hbar P, \tag{5.4.6}$$

$$B = -\frac{1}{c}\sum_{i=1}^{r} \bar{q}_i, \quad [P, B] = i\hbar \frac{1}{c^2} H. \tag{5.4.7}$$

Based on this it is claimed that the 'non-relativistic' limit (the speed of light $c \to \infty$), the Hamiltonian H reduces to the elliptic Sutherland system

$$\lim_{c\to\infty}(H - rmc^2) = \sum_{j=1}^{r}\frac{1}{2m}\left(\frac{\hbar}{i}\frac{\partial}{\partial q_j}\right)^2 + \frac{1}{m}\bar{\beta}(\bar{\beta} - \hbar)\sum_{1\leq j<k\leq r}\wp(q_j - q_k), \tag{5.4.8}$$

where $\wp(z)$ is the Weierstrass $\wp$ function and the periods of $\sigma(z)$ have been rescaled. In the trigonometric case the function h is

$$h(q) = \frac{\sin\frac{\pi}{mcL}(\bar{q} + i\bar{\beta})}{\sin\frac{\pi}{mcL}\bar{q}}, \tag{5.4.9}$$

and the $c \to \infty$ limit of the corresponding Hamiltonian is the Sutherland Hamiltonian (5.3.58)

$$\lim_{c\to\infty}(H - rmc^2) = \hat{\mathcal{H}}_S. \tag{5.4.10}$$

The above arguments show clearly that the Ruijsenaars-Schneider systems are one parameter deformations of the corresponding Calogero-Sutherland systems. However, as we have stressed earlier, the interpretation of the deformed system as a relativistic one is untenable. The multi-particle interaction Hamiltonian like (5.4.1) is an action at a distance and it is inherently non-relativistic.

5.4.2 Macdonald Polynomials (Macdonald 多项式)

In order to discuss the quantum theory, let us rewrite the Hamiltonian (5.3.58) with dimensionless quantities. The deformation parameter is q, $0 < q < 1$, and $q \to 1$ reduces to the ordinary quantum mechanical systems with differential Schrödinger equations. The corresponding eigenpolynomials are the Macdonald polynomials [55].

5.4 Multi-Particle QM with Difference Schrödinger Equations

The starting point is rewriting $\mathcal{H}_{\pm 1}$ (5.3.58) corresponding to the vector representation, the fundamental weight λ_1:

$$\mathcal{H}_1 = \sum_{j=1}^{r} \sqrt{V_j^*(\{q\})}\, e^{-\gamma \theta_j} \sqrt{V_j(\{q\})}, \quad \mathcal{H}_{-1} = \sum_{j=1}^{r} \sqrt{V_j(\{q\})}\, e^{\gamma \theta_j} \sqrt{V_j^*(\{q\})}, \quad (5.4.11)$$

$$\theta_j = -i\frac{\partial}{\partial q_j},\ q \stackrel{\text{def}}{=} e^{-2\gamma},\ 0 < \gamma, \quad (5.4.12)$$

$$V(u) \stackrel{\text{def}}{=} \frac{\sin(u - i\gamma\beta)}{\sin u},\ V^*(u) \stackrel{\text{def}}{=} \frac{\sin(u + i\gamma\beta)}{\sin u},\ 0 < \beta, \quad (5.4.13)$$

$$V_j(\{q\}) \stackrel{\text{def}}{=} \prod_{k \neq j} V(q_j - q_k),\quad V_j^*(\{q\}) \stackrel{\text{def}}{=} \prod_{k \neq j} V^*(q_j - q_k), \quad (5.4.14)$$

which are not exactly the multi-particle version of the Hamiltonian of discrete QM with pure imaginary shifts. As in exactly solvable 1-d QM we adopt a factorised and positive semi-definite Hamiltonian instead of (5.4.5):

$$\mathcal{H}_v \stackrel{\text{def}}{=} \mathcal{H}_1 + \mathcal{H}_{-1} - \sum_{j=1}^{r} [V_j(\{q\}) + V_j^*(\{q\})] = \sum_{j=1}^{r} \mathcal{A}_j^\dagger \mathcal{A}_j, \quad (5.4.15)$$

$$\mathcal{A}_j \stackrel{\text{def}}{=} i\left[e^{\frac{\gamma}{2}\theta_j}\sqrt{V_j^*(\{q\})} - e^{-\frac{\gamma}{2}\theta_j}\sqrt{V_j(\{q\})}\right], \quad (5.4.16)$$

$$\mathcal{A}_j^\dagger = -i\left[\sqrt{V_j(\{q\})}\,e^{\frac{\gamma}{2}\theta_j} - \sqrt{V_j^*(\{q\})}\,e^{-\frac{\gamma}{2}\theta_j}\right]. \quad (5.4.17)$$

Exercise Show that the additional term is a constant

$$\sum_{j=1}^{r} V_j(\{q\}) = \sum_{j=1}^{r} V_j^*(\{q\}) = \begin{cases} 1 + 2\{\cosh 2\gamma\beta + \cdots + \cosh[(r-1)\gamma\beta]\} & r: \text{odd} \\ 2\{\cosh \gamma\beta + \cdots + \cosh[(r-1)\gamma\beta]\} & r: \text{even} \end{cases}, \quad (5.4.18)$$

and the dynamics is unchanged.

Exercise Derive the corresponding result for the D_r root system, i.e. the first fundamental weight λ_1 (5.2.163) corresponding to the vector representation:

$$V_j(\{q\}) \stackrel{\text{def}}{=} \prod_{k \neq j} V(q_j - q_k)V(q_j + q_k),$$

$$V_j^*(\{q\}) \stackrel{\text{def}}{=} \prod_{k \neq j} V^*(q_j - q_k)V^*(q_j + q_k), \quad (5.4.19)$$

$$\sum_{j=1}^{r} [V_j(\{q\}) + V_j^*(\{q\})]$$

$$= 2\{1 + \cosh 2\gamma\beta + \cosh 4\gamma\beta + \cdots + \cosh[2(r-1)\gamma\beta]\}. \quad (5.4.20)$$

In this case the individual sum $\sum_{j=1}^{r} V_j(\{q\})$ is not real nor a constant. Note that $\cosh 2\gamma\beta = (t+t^{-1})/2$ with the parameter t defined in (5.4.24).

The ground state eigenfunction $\phi_0(\{q\})$ is obtained from the zero mode equation

$$\mathcal{A}_j \phi_0(\{q\}) = 0 \quad j=1,\cdots,r \quad \Rightarrow \mathcal{H}\phi_0(\{q\}) = 0, \tag{5.4.21}$$

$$\phi_0^2(\{q\}) = \prod_{j=1}^{r}\prod_{k\neq j} \frac{(e^{2i(q_j-q_k)};q)_\infty (e^{-2i(q_j-q_k)};q)_\infty}{(t\,e^{2i(q_j-q_k)};q)_\infty (t\,e^{-2i(q_j-q_k)};q)_\infty} \tag{5.4.22}$$

$$= \prod_{j=1}^{r}\prod_{k\neq j} \frac{(x_j/x_k;q)_\infty (x_k/x_j;q)_\infty}{(t\,x_j/x_k;q)_\infty (t\,x_k/x_j;q)_\infty} = \prod_{\substack{i,j=1 \\ i\neq j}}^{r} \frac{\left(\frac{x_i}{x_j};q\right)_\infty}{\left(t\frac{x_i}{x_j};q\right)_\infty} \stackrel{\text{def}}{=} \Delta(x;q,t), \tag{5.4.23}$$

in which the parameter t and x_j are defined by

$$t \stackrel{\text{def}}{=} q^\beta = e^{-2\gamma\beta}, \quad x_j \stackrel{\text{def}}{=} e^{2iq_j}, \quad j=1,\cdots,r. \tag{5.4.24}$$

This is essentially the same process of solving the zero mode equation (5.2.101) for the Askey-Wilson system, for each pair $q_j - q_k$. As in other QM systems, the square of the ground state eigenfunction provides the orthogonality measure $\Delta(x;q,t)$ of the eigenpolynomials, *i.e.* the Macdonald polynomials.

Let us write down the operators (5.4.1) belonging to the second fundamental weight λ_2 (5.2.160), which is called the tensor representation:

$$\mathcal{H}_2 = \sum_{j\neq k=1}^{r} \sqrt{V^*_{(j,k)}(\{q\})}\, e^{-\gamma(\theta_j+\theta_k)} \sqrt{V_{(j,k)}(\{q\})}, \tag{5.4.25}$$

$$\mathcal{H}_{-2} = \sum_{j\neq k=1}^{r} \sqrt{V_{(j,k)}(\{q\})}\, e^{\gamma(\theta_j+\theta_k)} \sqrt{V^*_{(j,k)}(\{q\})}, \tag{5.4.26}$$

$$V_{(j,k)}(\{q\}) \stackrel{\text{def}}{=} \prod_{l\neq j,k} V(q_j-q_l)V(q_k-q_l), \quad V^*_{(j,k)}(\{q\}) \stackrel{\text{def}}{=} \prod_{l\neq j,k} V^*(q_j-q_l)V^*(q_k-q_l), \tag{5.4.27}$$

$$\mathcal{H}_t = \mathcal{H}_2 + \mathcal{H}_{-2} - \sum_{j\neq k=1}^{r}\left[V_{(j,k)}(\{q\}) + V^*_{(j,k)}(\{q\})\right] = \sum_{j\neq k=1}^{r} \mathcal{A}^\dagger_{(j,k)}\mathcal{A}_{(j,k)}, \tag{5.4.28}$$

$$\mathcal{A}_{(j,k)} \stackrel{\text{def}}{=} i\left[e^{\frac{\gamma}{2}(\theta_j+\theta_k)}\sqrt{V^*_{(j,k)}(\{q\})} - e^{-\frac{\gamma}{2}(\theta_j+\theta_k)}\sqrt{V_{(j,k)}(\{q\})}\right], \tag{5.4.29}$$

$$\mathcal{A}^\dagger_{(j,k)} = -i\left[\sqrt{V_{(j,k)}(\{q\})}\,e^{\frac{\gamma}{2}(\theta_j+\theta_k)} - \sqrt{V^*_{(j,k)}(\{q\})}\,e^{-\frac{\gamma}{2}(\theta_j+\theta_k)}\right]. \tag{5.4.30}$$

5.4 Multi-Particle QM with Difference Schrödinger Equations

Exercise Show that the additional term in $\mathcal{H}_t$ (5.4.28) is a constant

$$\sum_{j\neq k=1}^{r} V_{(j,k)}(\{q\})$$

$$= \sum_{j\neq k=1}^{r} V^*_{(j,k)}(\{q\}) = \text{constant}$$

$$= \begin{cases} 2(1+\cosh 2\gamma\beta+\cosh 4\gamma\beta), & r=4 \\ 2(1+\cosh 2\gamma\beta+\cosh 4\gamma\beta+\cosh 6\gamma\beta), & r=5 \\ 3+4(\cosh 2\gamma\beta+\cosh 4\gamma\beta)+2(\cosh 6\gamma\beta+\cosh 8\gamma\beta), & r=6 \\ \cdots & \end{cases} \quad (5.4.31)$$

The ground state eigenfunction $\phi_0(\{q\})$ (5.4.22), (5.4.23) is also annihilated by the operators $\mathcal{A}_{(j,k)}$ (5.4.29):

$$\mathcal{A}_{(j,k)}\phi_0(\{q\}) = 0, \quad j,k=1,\cdots,r \quad \Rightarrow \quad \mathcal{H}_t\phi_0(\{q\}) = 0. \quad (5.4.32)$$

Let us introduce the similarity transformed operators in terms of the ground state eigenfunction $\phi_0 \equiv \Delta^{\frac{1}{2}}$ and express them in variables $\{x_j\}$ instead of $\{q_j\}$:

$$\Delta^{-\frac{1}{2}} \circ \mathcal{H}_1 \circ \Delta^{\frac{1}{2}} = \sum_{j=1}^{r} V_j^*(\{q\}) e^{-\gamma\theta_j} = t^{-\frac{1}{2}(r-1)} \cdot \sum_{j=1}^{r} \prod_{k\neq j=1}^{r} \frac{t\, x_j - x_k}{x_j - x_k} \cdot q^{D_j}, \quad (5.4.33)$$

$$\Delta^{-\frac{1}{2}} \circ \mathcal{H}_{-1} \circ \Delta^{\frac{1}{2}} = \sum_{j=1}^{r} V_j(\{q\}) e^{\gamma\theta_j} = t^{\frac{1}{2}(r-1)} \sum_{j=1}^{r} \prod_{k\neq j=1}^{r} \frac{t^{-1}x_j - x_k}{x_j - x_k} \cdot q^{-D_j}, \quad (5.4.34)$$

$$\Delta^{-\frac{1}{2}} \circ \mathcal{H}_2 \circ \Delta^{\frac{1}{2}} = \sum_{j\neq k=1}^{r} V^*_{(j,k)}(\{q\}) e^{-\gamma(\theta_j+\theta_k)}$$

$$= t^{-(r-2)} \sum_{j\neq k=1}^{r} \prod_{l\neq j,k=1}^{r} \frac{t\, x_j - x_l}{x_j - x_l} \cdot \frac{t\, x_k - x_l}{x_k - x_l} \cdot q^{D_j+D_k}, \quad (5.4.35)$$

$$\Delta^{-\frac{1}{2}} \circ \mathcal{H}_{-2} \circ \Delta^{\frac{1}{2}} = \sum_{j\neq k=1}^{r} V_{(j,k)}(\{q\}) e^{\gamma(\theta_j+\theta_k)}$$

$$= t^{(r-2)} \sum_{j\neq k=1}^{r} \prod_{l\neq j,k=1}^{r} \frac{t^{-1}x_j - x_l}{x_j - x_l} \cdot \frac{t^{-1}x_k - x_l}{x_k - x_l} \cdot q^{-(D_j+D_k)}, \quad (5.4.36)$$

in which, as before (5.3.62), $D_j \stackrel{\text{def}}{=} x_j \frac{\partial}{\partial x_j}$ and q^{D_j} is a q-shift operator

$$(q^{D_j}f)(x_1,\cdots,x_j,\cdots,x_r) = f(x_1,\cdots,qx_j,\cdots,x_r).$$

The above operators and higher operators are expressed succinctly:

$$\Delta^{-\frac{1}{2}} H_{\pm k} \Delta^{\frac{1}{2}} = t^{\mp \frac{1}{2}k(r-1)} \mathcal{D}_k(q^{\pm 1}, t^{\pm 1}), \tag{5.4.37}$$

in which $\mathcal{D}_k(q,t)$ $(k=1,\cdots,r)$ is the Macdonald operator

$$\mathcal{D}_k(q,t) \stackrel{\text{def}}{=} t^{\frac{1}{2}k(k-1)} \sum_{\substack{I \subset \{1,\cdots,r\} \\ |I|=k}} \prod_{\substack{i \in I \\ j \notin I}} \frac{tx_i - x_j}{x_i - x_j} \cdot \prod_{i \in I} q^{D_i}. \tag{5.4.38}$$

In particular,

$$\mathcal{D}_1(q,t) = \sum_{j=1}^{r} \prod_{k \neq j=1}^{r} \frac{tx_j - x_k}{x_j - x_k} \cdot q^{D_j},$$

$$\mathcal{D}_2(q,t) = t \sum_{j \neq k=1}^{r} \prod_{l \neq j,k=1}^{r} \frac{tx_j - x_l}{x_j - x_l} \cdot \frac{tx_k - x_l}{x_k - x_l} \cdot q^{D_j + D_k}.$$

The Macdonald symmetric polynomial $P_\lambda = P_\lambda(x; q, t)$ is a multivariable orthogonal polynomial with two parameters q and t, which is determined uniquely by the following conditions [55],

$$\text{(i): triangularity} \quad P_\lambda = \sum_{\mu \leqslant \lambda} u_{\lambda,\mu} m_\mu(x), \quad u_{\lambda,\lambda} = 1, \tag{5.4.39}$$

$$\text{(ii): orthogonality} \quad \langle P_\lambda, P_\mu \rangle_{q,t} = 0 \quad \text{if} \quad \lambda \neq \mu, \tag{5.4.40}$$

where the inner product is

$$\langle p_\lambda, p_\mu \rangle_{q,t} = \delta_{\lambda,\mu} \prod_j j^{m_j} m_j! \cdot \prod_{j=1}^{\ell(\lambda)} \frac{1 - q^{\lambda_j}}{1 - t^{\lambda_j}}. \tag{5.4.41}$$

The orthogonality condition (ii) can be replaced by the eigenvalue equation (ii)$'$,

$$\text{(ii)}': \text{eigenvalue eq.} \quad \mathcal{D}_1(q,t) P_\lambda = \sum_{i=1}^{r} t^{r-i} q^{\lambda_i} \cdot P_\lambda. \tag{5.4.42}$$

Moreover the Macdonald symmetric polynomial is the simultaneous eigenfunction of the Macdonald operators,

$$\sum_{k=0}^{r} (-u)^k \mathcal{D}_k(q,t) P_\lambda(x; q, t) = \prod_{j=1}^{r} (1 - ut^{r-j} q^{\lambda_j}) \cdot P_\lambda(x; q, t), \tag{5.4.43}$$

or

$$\mathcal{D}_k(q^{\pm 1}, t^{\pm 1}) P_\lambda(x; q, t) = \sum_{1 \leqslant i_1 < \cdots < i_k \leqslant r} \prod_{l=1}^{k} t^{r-i_l} q^{\lambda_{i_l}} \cdot P_\lambda(x; q, t). \tag{5.4.44}$$

Note that $P_\lambda(x;q,t) = P_\lambda(x;q^{-1},t^{-1})$ and $\langle f,g\rangle_{q^{-1},t^{-1}} = (q^{-1}t)^r \langle f,g\rangle_{q,t}$. In the 'conformal limit',
$$t = q^\beta, \quad q \to 1, \quad \beta : \text{fixed}, \tag{5.4.45}$$
which corresponds to the 'non-relativistic' limit ($c \to \infty$), the Macdonald polynomial reduces to the Jack polynomial
$$\lim_{\substack{q\to 1\\ t=q^\beta}} P_\lambda(x;q,t) = J_\lambda(x;\beta). \tag{5.4.46}$$

The $\beta \to 1$ limit of the Jack polynomial is the Schur polynomial $s_\lambda(x) = \lim_{\beta\to 1} J_\lambda(x;\beta)$. Likewise $t \to q$ limit of the Macdonald polynomial is also $s_\lambda(x) = \lim_{t\to q} P_\lambda(x;q,t)$, because the norm (5.4.41) reduces to that of the Jack polynomials with $\beta = 1$ (5.3.57).

There exists another inner product,
$$\langle f,g\rangle'_{l;q,t} = \frac{1}{l!}\oint \prod_{j=1}^{l} \underline{dx}_j \cdot \Delta(x;q,t)f(\bar{x})g(x), \quad \underline{dx}_j = \frac{dx_j}{2\pi i x_j}, \tag{5.4.47}$$
which satisfies
$$\langle\,,\,\rangle'_{l;q,t} \propto \langle\,,\,\rangle_{q,t}. \tag{5.4.48}$$

5.5 Comments and Discussion (总结和讨论)

Due to the lack of time we cannot cover many interesting facets of multi-particle classical Calogero-Sutherland systems including elliptic interactions; various Lax pairs, R-matrices, the involution of the conserved quantities. The same applies to the classical Ruijsenaars-Schneider systems including their quantum-classical correspondences[78, 79].

In recent years there were remarkable discoveries in 1-d QM including the two types of discrete QM[80–85]. Infinitely many exactly solvable QM systems of new types were constructed and their eigenfunctions consisted of the exceptional and multi-indexed orthogonal polynomials starting at degrees at $\ell \geqslant 1$. They do not satisfy the three term recurrence relations but they form complete sets. It is a very interesting challenge to construct their multi-particle versions.

5.6 Appendix: Symbols, Definitions & Formulas
(附录: 符号、定义和公式)

○ shifted factorial (Pochhammer symbol) $(a)_n$:

$$(a)_n \stackrel{\text{def}}{=} \prod_{k=1}^{n}(a+k-1) = a(a+1)\cdots(a+n-1) = \frac{\Gamma(a+n)}{\Gamma(a)}. \tag{5.6.1}$$

○ q-shifted factorial (q-Pochhammer symbol) $(a\,;q)_n$:

$$(a\,;q)_n \stackrel{\text{def}}{=} \prod_{k=1}^{n}(1-aq^{k-1}) = (1-a)(1-aq)\cdots(1-aq^{n-1}). \tag{5.6.2}$$

○ hypergeometric function ${}_rF_s$:

$${}_rF_s\!\left(\begin{matrix}a_1,\cdots,a_r\\ b_1,\cdots,b_s\end{matrix}\,\bigg|\,z\right) \stackrel{\text{def}}{=} \sum_{n=0}^{\infty}\frac{(a_1,\cdots,a_r)_n}{(b_1,\cdots,b_s)_n}\frac{z^n}{n!}, \tag{5.6.3}$$

where $(a_1,\cdots,a_r)_n \stackrel{\text{def}}{=} \prod_{j=1}^{r}(a_j)_n = (a_1)_n\cdots(a_r)_n$.

○ q-Hypergeometric series (the basic hypergeometric series) ${}_r\phi_s$:

$${}_r\phi_s\!\left(\begin{matrix}a_1,\cdots,a_r\\ b_1,\cdots,b_s\end{matrix}\,\bigg|\,q\,;z\right) \stackrel{\text{def}}{=} \sum_{n=0}^{\infty}\frac{(a_1,\cdots,a_r\,;q)_n}{(b_1,\cdots,b_s\,;q)_n}(-1)^{(1+s-r)n}q^{(1+s-r)n(n-1)/2}\frac{z^n}{(q\,;q)_n}, \tag{5.6.4}$$

where $(a_1,\cdots,a_r\,;q)_n \stackrel{\text{def}}{=} \prod_{j=1}^{r}(a_j\,;q)_n = (a_1\,;q)_n\cdots(a_r\,;q)_n$.

○ differential equations

$$\text{H}: \quad \partial_x^2 H_n(x) - 2x\partial_x H_n(x) + 2nH_n(x) = 0, \tag{5.6.5}$$

$$\text{L}: \quad x\partial_x^2 L_n^{(\alpha)}(x) + (\alpha+1-x)\partial_x L_n^{(\alpha)}(x) + nL_n^{(\alpha)}(x) = 0, \tag{5.6.6}$$

$$\text{J}: \quad (1-x^2)\partial_x^2 P_n^{(\alpha,\beta)}(x) + \bigl[\beta-\alpha-(\alpha+\beta+2)x\bigr]\partial_x P_n^{(\alpha,\beta)}(x)$$
$$+n(n+\alpha+\beta+1)P_n^{(\alpha,\beta)}(x) = 0. \tag{5.6.7}$$

○ Rodrigues formulas

$$\text{H}: \quad H_n(x) = (-1)^n e^{x^2}\!\left(\frac{\mathrm{d}}{\mathrm{d}x}\right)^{\!n}\! e^{-x^2}, \tag{5.6.8}$$

$$\text{L}: \quad L_n^{(\alpha)}(x) = \frac{1}{n!}\frac{1}{e^{-x}x^\alpha}\!\left(\frac{\mathrm{d}}{\mathrm{d}x}\right)^{\!n}\!\bigl(e^{-x}x^{n+\alpha}\bigr). \tag{5.6.9}$$

$$\text{J}: \quad P_n^{(\alpha,\beta)}(x) = \frac{(-1)^n}{2^n n!}\frac{1}{(1-x)^\alpha(1+x)^\beta}\!\left(\frac{\mathrm{d}}{\mathrm{d}x}\right)^{\!n}\!\bigl[(1-x)^{n+\alpha}(1+x)^{n+\beta}\bigr]. \tag{5.6.10}$$

参考文献

[1] Landau L D, Lifshitz L M. Mechanics. 2nd ed. Pergamon Press, 1960.

[2] Dirac P A M. The Principles of Quantum Mechanics. 4th edition. Oxford: Oxford Univ. Press, 1963.

[3] Landau L D, Lifshitz L M. Quantum Mechanics: Non-Relativistic Theory. 3rd edition. Oxford: Pergamon Press, 1965.

[4] Arnold V. Mathematical Methods of Classical Mechanics. New York, Berlin: Springer Verlag, 1978: 49.

[5] Hille E. Ordinary Differential Equations in the Complex Domain. New York: Wiley, 1976.

[6] Infeld L, Hull T E. The factorization method. Rev. Mod. Phys., 1951, 23: 21-68.

[7] Darboux G. Sur une proposition relative aux équations linéaires. C. R. Acad. Paris, 1882, 94: 1456-1459.

[8] Crum M M. Associated Sturm-Liouville systems. Quart. J. Math. Oxford Ser., 1955, 6: 121-127, arXiv:physics/9908019.

[9] Flügge S. Practical Quantum Mechanics. Springer-Verlag, New York-Heidelberg, 1974.

[10] Cooper F, Khare A, Sukhatme U P. Supersymmetry and quantum mechanics. Phys. Rep., 1995, C251: 267-385.

[11] Coddington E A, Levinson N. Theory of Ordinary Differential Equations, MaGrawhill, 1955.

[12] Gendenshtein L E. Derivation of exact spectra of the Schrodinger equation by means of supersymmetry. JETP Lett., 1983, 38: 356-359. [Pisma Zh. Eksp. Teor. Fiz., 1983, 38: 299].

[13] Dabrowska J W, Khare A, Sukhatme U P. Explicit wavefunctions for shape-invariant potentials by operator technique. J. Phys. A, 1988, 21: L195-L200.

[14] Odake S, Sasaki R. Unified theory of annihilation-creation operators for solvable ('discrete') quantum mechanics. J. Math. Phys., 2006, 47: 102102 (33pp), arXiv:quant-ph/0605215.

[15] Odake S, Sasaki R. Exact solution in the Heisenberg picture and annihilation-creation operators. Phys. Lett. B, 2006, 641: 112-117, arXiv:quant-ph/0605221.

[16] Andrews G E, Askey R, Roy R. Special Functions, Encyclopedia of Mathematics and its Applications. Cambridge: Cambridge Univ. Press, 1999.

[17] Ismail M E H. Classical and Quantum Orthogonal Polynomials in One Variable, Encyclopedia of Mathematics and Its Applications. Cambridge: Cambridge Univ. Press, 2005.

[18] Chihara T S. An Introduction to Orthogonal Polynomials. New York: Gordon and Breach, 1978.

[19] Koekoek R, Swarttouw R F. The Askey-scheme of hypergeometric orthogonal polynomials and its q-analogue. arXiv:math.CA/9602214; Koekoek R, Lesky P A, Swarttouw R F. Hypergeometric orthogonal polynomials and their q-analogues. Springer-Verlag, 2010.

[20] Odake S, Sasaki R. Exact Heisenberg operator solutions for multi-particle quantum mechanics. J. Math. Phys., 2007, 48: 082106 (12 pp), arXiv:0706.0768[quant-ph].

[21] Odake S, Sasaki R. Orthogonal Polynomials from Hermitian Matrices. J. Math. Phys., 2008, 49: 053503 (43pp), arXiv:0712.4106[math.CA].

[22] Odake S, Sasaki R. Exactly solvable 'discrete' quantum mechanics; shape invariance, Heisenberg solutions, annihilation-creation operators and coherent states. Prog. Theor. Phys. 2008, 119: 663-700, arXiv:0802.1075[quant-ph].

[23] Gaillard P, Matveev V B. Wronskian and Casorati determinant representation for Darboux-Pöschel-Teller potentials and their difference extensions. J. Phys. A, 2009, 42: 404009 (16pp).

[24] Odake S, Sasaki R. Crum's theorem for 'discrete' quantum mechanics. Prog. Theor. Phys. 2009, 122: 1067-1079, arXiv:0902.2593[math-ph].

[25] García-Gutiérrez L, Odake S, Sasaki R. Modification of Crum's Theorem for 'Discrete' Quantum Mechanics. Prog. Theor. Phys., 2010, 124: 1-26, arXiv:1004.0289 [math-ph].

[26] Odake S, Sasaki R. Dual Christoffel transformations. Prog. Theor. Phys., 2011, 126: 1-34, arXiv:1101.5468[math-ph].

[27] Calogero F. Solution of the one-dimensional N-body problem with quadratic and/or inversely quadratic pair potentials. J. Math. Phys., 1971, 12: 419-436.

[28] Sutherland B. Exact results for a quantum many-body problem in one-dimension. II, Phys. Rev., 1972, A5: 1372-1376.

[29] Moser J. Three integrable Hamiltonian systems connected with isospectral deformations. Adv. Math. 1975, 16: 197-220; Integrable systems of non-linear evolution equations//in Dynamical Systems, Theory and Applications. J. Moser, ed., Lecture Notes in Physics, 1975, 38, Springer-Verlag.

[30] Calogero F, Marchioro C, Ragnisco O. Exact solution of the classical and quantal one-dimensional many body problems with the two body potential $V_a(x) = g^2 a^2 / \sinh^2 ax$. Lett. Nuovo Cim., 1975, 13: 383-387.

[31] Calogero F. Exactly solvable one-dimensional many body problems. Lett. Nuovo. Cim., 1975, 13: 411-416.

[32] Calogero F. Solution of a three body problem in one dimension. J. Math. Phys., 1969, 10: 2191-2196; Ground state of a one-dimensional N-body problem. J. Math. Phys.,

1969, 10: 2197-2200.

[33] Wolfes J. On the three-body linear problem with three-body interaction. J. Math. Phys., 1974, 15: 1420-1424; Calogero F, Marchioro C. Exact solution of a one-dimensional three-body scattering problem with two-body and/or three-body inverse-square potential. J. Math. Phys., 1974, 15: 1425-1430.

[34] Olshanetsky M A, Perelomov A M. Completely integrable Hamiltonian systems connected with semisimple Lie algebras. Inventions Math., 1976, 37: 93-108.

[35] Olshanetsky M A, Perelomov A M. Classical integrable finite-dimensional systems related to Lie algebras. Phys. Rep., 1981, 71: 543-400.

[36] Olshanetsky M A, Perelomov A M. Quantum integrable systems related to Lie algebras. Phys. Rep., 1983, 94: 313-404.

[37] Stanley R. Some combinatorial properties of Jack symmetric function. Adv. Math., 1989, 77: 76-115.

[38] Freedman D Z, Mende P F. An exactly solvable N-particle system in supersymmetric quantum mechanics. Nucl. Phys. B, 1990, 344: 317-343.

[39] Dunkl C F. Differential-difference operators associated to reflection groups. Trans. Am. math. Soc., 1989, 311: 167-183; Dunkl C F, Xu Y. Orthogonal polynomials of several variables. Cambridge Univ. Press, 2001.

[40] Heckman G J. A remark on the Dunkl differential-difference operators//in Harmonic analysis on reductive groups. Birkhäuser, Basel, 1991.

[41] Heckman G J, Opdam E M. Root systems and hypergeometric functions I. Comp. Math., 1987, 64: 329-352; Heckman G J. Root systems and hypergeometric functions II, Comp. Math., 1987, 64: 353-373; Opdam E M. Root systems and hypergeometric functions III. Comp. Math., 1988, 67: 21-49; Root systems and hypergeometric functions IV. Comp. Math., 1988, 67: 191-209.

[42] Shastry B S, Sutherland B. Superlax pairs and infinite symmetries in the $1/r^2$ system. Phys. Rev. Lett., 1993, 70: 4029-4033.

[43] Ochiai H, Oshima T, Sekiguchi H. Commuting families of symmetric differential operators. Proc. Japan Acad. Ser. A Math. Sci. 1994, 70: 62-66; Oshima T, Sekiguchi H. Commuting families of differential operators invariant under the action of a Weyl group. J. Math. Sci. Univ. Tokyo, 1995, 2: 1-75.

[44] Ujino H, Wadati M, Hikami K. The quantum Calogero-Moser model: algebraic structures. J. Phys. Soc. Jpn., 1993, 62: 3035-3043; Ujino H, Wadati M. Rodrigues formula for Hi-Jack symmetric polynomials associated with the quantum Calogero model. J. Phys. Soc. Jpn., 1996, 65: 2423-2439, arXiv:cond-mat/9609041; Ujino H. Orthogonal symmetric polynomials associated with the Calogero model. J. Phys. Soc. Jpn, 1995, 64: 2703-2706, cond-mat/9706133; Nishino A, Ujino H, Wadati M. Symmetric Fock space and orthogonal symmetric polynomials associated with the Calogero model.

Chaos Solitons Fractals, 2000, 11: 657-674, arXiv: cond-mat/9803284.

[45] Mimachi K, Yamada Y. Singular vectors of the virasoro algebra in terms of jack symmetric polynomials. Comm. Math. Phys., 1995, 174: 447-455; RIMS Kokyuroku, 1995, 919: 68-78.

[46] Awata H, Matsuo Y, Odake S, et al. Excited States of Calogero-Sutherland Model and Singular Vectors of the W_N Algebra. Nucl. Phys. B, 1995, 449: 347-374, arXiv:hep-th/9503043.

[47] Lapointe L, Vinet L. Exact operator solution of the Calogero-Sutherland model. Commun. Math. Phys., 1996, 178: 425-452, arXiv:q-alg/9509003; Rodrigues formulas for the Macdonald polynomials. Adv. Math., 1997, 130: 261-279, arXiv: q-alg/9607025.

[48] Kakei S. Common algebraic structure for the Calogero-Sutherland models. J. Phys. A, 1996, 29: L619-L624; An orthogonal basis for the B_N-type Calogero model. J. Phys. A, 1997, 30: L535-L541.

[49] Bordner A J, Corrigan E, Sasaki R. Calogero-Moser models I: a new formulation. Prog. Theor. Phys., 1998, 100: 1107-1129, arXiv:hep-th/9805106; Bordner A J, Sasaki R, Takasaki K. Calogero-Moser models II: symmetries and foldings. Prog. Theor. Phys., 1999, 101: 487-518, arXiv:hep-th/9809068; Bordner A J, Sasaki R. Calogero-Moser models III: elliptic potentials and twisting. Prog. Theor. Phys., 1999, 101: 799-829, arXiv:hep-th/9812232; Khastgir S P, Sasaki R, Takasaki K. Calogero-Moser Models IV: Limits to Toda theory. Prog. Theor. Phys., 1999, 102: 749-776, arXiv:hep-th/9907102.

[50] Bordner A J, Corrigan E, Sasaki R. Generalised Calogero-Moser models and universal Lax pair operators. Prog. Theor. Phys., 1999, 102: 499-529, arXiv: hep-th/9905011.

[51] Bordner A J, Manton N S, Sasaki R. Calogero-Moser models V: Supersymmetry and Quantum Lax Pair. Prog. Theor. Phys., 2000, 103: 463-487, arXiv: hep-th/9910033.

[52] Hoker E D, Phong D H. Calogero-Moser Lax pairs with spectral parameter for general Lie algebras. Nucl. Phys. B, 1998, 530: 537-610, arXiv:hep-th/9804124.

[53] Khastgir S P, Pocklington A J, Sasaki R. Quantum Calogero-Moser Models: Integrability for all Root Systems. J. Phys. A, 2000, 33: 9033-9064, arXiv: hep-th/0005277.

[54] Ruijsenaars S N M, Schneider H. A new class of integrable systems and its relation to solitons. Annals Phys., 1986, 170: 370-405; Ruijsenaars S N M. Complete integrability of relativistic Calogero-Moser systems and elliptic function identities. Comm. Math. Phys., 1987, 110: 191-213.

[55] Macdonald I G. Symmetric functions and Hall polynomials. second edition. Oxford Univ. Press, 1995.

[56] Macdonald I G. Orthogonal polynomials associated with root systems. Séminaire Lotharingen de Combinatoire 2000, 45 Article B45a (40pp).

[57] Koornwinder T H. Askey-Wilson Polynomials for Root Systems of Type BC. Contemporary Math. 1992, 138: 189-204.

[58] van Diejen J F. Integrability of difference Calogero-Moser systems. J. Math. Phys., 1994, 35: 2983-3004.

[59] van Diejen J F. The relativistic Calogero model in an external field. arXiv: solv-int/9509002; Multivariable continuous Hahn and Wilson polynomials related to integrable difference systems. J. Phys. A,1995, 28: L369-L374.

[60] van Diejen J F. Difference Calogero-Moser systems and finite Toda chains. J. Math. Phys., 1995, 36: 1299-1323.

[61] Tratnik M V. Multivariable meixner, krawtchouk, and meixner-pollaczek polynomials. J. Math. Phys., 1989, 30: 2740-2749; Some multivariable orthogonal polynomials of the Askey tableau-discrete families. J. Math. Phys., 1991, 32: 2337-2342.

[62] Geronimo J S, Illiev P. Bispectrality of multivariable Racah-Wilson polynomials. Constr. Approx., 2010, 31: 417-457, arXiv:0705.1469[math.CA].

[63] Humphreys J E. Reflection groups and Coxeter groups. Cambridge Univ. Press, 1990.

[64] Toda M. Vibration of a Chain with Nonlinear Interaction. J. Phys. Soc. Japan, 1967, 22: 431-436.

[65] Sklyanin E K. The quantum toda chain. Lect. Note. Phys., 2005, 266: 196-233.

[66] Sasaki R, Znojil M. One-dimensional Schrödinger equation with non-analytic potential $V(x) = -g^2 \exp(-|x|)$ and its exact Bessel-function solvability. arXiv:1605.07310 [math-ph].

[67] Wojciechowski S. Involutive set of integrals for completely integrable many-body problems with pair interaction. Lett. Nuouv. Cim., 1976, 18: 103-107.

[68] Caseiro R, Françoise J P, Sasaki R. Algebraic Linearization of Dynamics of Calogero Type for any Coxeter Group. J. Math. Phys., 2000, 41: 4679-4689, arXiv: hep-th/0001074.

[69] Caseiro R, Françoise J P, Sasaki R. Quadratic Algebra associated with Rational Calogero-Moser Models. J. Math. Phys., 2001, 42: 5329-5340, arXiv:hep-th/0102153.

[70] Corrigan E, Sasaki R. Quantum vs Classical Integrability in Calogero-Moser Systems. J. Phys. A, 2002, 35: 7017-7061, arXiv:hep-th/0204039.

[71] Stieltjes T J. Ouvres Complétes, vol.2. Groningen: Noordhoff, 1918.

[72] Szegö G. Orthogonal polynomials, Fourth edition. New York: Amer. Math. Soc., 1975.

[73] Calogero F. On the zeros of the classical polynomials. Lett. Nuovo. Cim., 1977, 19: 505-507; Equilibrium configuration of one-dimensional many-body problems with quadratic and inverse quadratic pair potentials. Lett. Nuovo Cim. 1977, 22: 251-253; Eigenvectors of a matrix related to the zeros of Hermite polynomials. Lett. Nuovo

Cim., 1979, 24: 601-604; Matrices, differential operators and polynomials. J. Math. Phys., 1981, 22: 919-934.

[74] Calogero F, Perelomov A M. Properties of certain matrices related to the equilibrium configuration of one-dimensional many-body problems with pair potentials $V_1 = -\log|\sin x|$ and $V_2 = 1/\sin^2 x$. Commun. Math. Phys., 1978, 59: 109-116.

[75] Loris I, Sasaki R. Quantum vs Classical Mechanics, role of elementary excitations. Phys. Lett. A, 2004, 327: 152-157, arXiv:quant-ph/0308040; Quantum & classical eigenfunctions in calogero & sutherland systems. J. Phys. A, 2004, 37: 211-237, arXiv:hep-th/0308052.

[76] Whittaker E T, Watson G N. A course of Modern Analysis. 4th ed. Cambridge Univ. Press, 1927.

[77] Braden H W, Sasaki R. The Ruijsenaars-Schneider model. Prog. Theor. Phys., 1998, 97: 1003-1017, arXiv:hep-th/9702182.

[78] Ragnisco O, Sasaki R. Quantum vs classical integrability in Ruijsenaars-Schneider systems. J. Phys. A, 2004, 37: 469-479, arXiv:hep-th/0305120.

[79] Odake S, Sasaki R. Equilibria of 'discrete' integrable systems and deformations of classical orthogonal polynomials. J. Phys. A, 2004, 37: 11841, arXiv:hep-th/0407155; Shape invariant potentials in discrete quantum mechanics. JNMP 12 Suppl., 2005, 12: 507-521, arXiv:hep-th/0410102; Equilibrium positions, shape invariance and Askey-Wilson polynomials. J. Math. Phys., 2005, 46: 063513 (10pp) arXiv:hep-th/0410109.

[80] Gómez-Ullate D, Kamran N, Milson R. An extension of Bochner's problem: exceptional invariant subspaces. J. Approx Theory, 2010, 162: 987-1006, arXiv:0805.3376[math-ph]; An extended class of orthogonal polynomials defined by a Sturm-Liouville problem. J. Math. Anal. Appl., 2009, 359: 352-367, arXiv:0807.3939[math-ph].

[81] Quesne C. Exceptional orthogonal polynomials, exactly solvable potentials and supersymmetry. J. Phys. A, 2008, 41: 392001, arXiv:0807.4087[quant-ph].

[82] Odake S, Sasaki R. Infinitely many shape invariant potentials and new orthogonal polynomials. Phys. Lett. B, 2009, 679: 414-417, arXiv:0906.0142[math-ph].

[83] Odake S, Sasaki R. Exactly Solvable Quantum Mechanics and Infinite Families of Multi-indexed Orthogonal Polynomials. Phys. Lett. B, 2011, 702: 164-170, arXiv:1105.0508[math-ph].

[84] Odake S, Sasaki R. Infinitely many shape invariant discrete quantum mechanical systems and new exceptional orthogonal polynomials related to the Wilson and the Askey-Wilson polynomials. Phys. Lett. B, 2009, 682: 130-136, arXiv:0909.3668[math-ph]; Exceptional (X_ℓ) (q)-Racah polynomials, Prog. Theor. Phys., 2011, 125: 851-870; arXiv:1102.0812[math-ph].

[85] Odake S, Sasaki R. Multi-indexed (q-)Racah polynomials. J. Phys. A, 2012, 45: 385201 (21 pp). arXiv:1203.5868[math-ph]; Multi-indexed Wilson and Askey-Wilson polynomials. J. Phys. A, 2013, 46: 045204 (22 pp), arXiv:1207.5584[math-ph].

第6章 Quasi-Exactly Solvable Systems (准精确可解系统)

Yao-Zhong Zhang(张耀中)

6.1 Exact Solvability Versus Quasi-Exact Solvability (精确可解性与准精确可解性)

Exact solutions of a non-trivial model can provide valuable information about nonperturbative properties of complex quantum problems. Very often the knowledge of such exact solutions can allow one to develop constructive perturbation analysis and discover unexpected features of real-world systems. This has motivated the extensive study of exactly solvable (ES) models in the past few decades. It is widely accepted that a quantum mechanical system is ES if all the eigenvalues and the corresponding eigenfunctions of the system can be determined exactly.

However, the number of ES models is quite limited, and most interesting problems in nature are not ES. In 1980's, one class of systems occupying an intermediate place between exactly solvable and non-solvable ones, coined as "quasi-exactly solvable systems", were discovered in [1, 2]. Naturally, quasi-exact solvability is closely related to exact solvability. According to [1-4] a quantum mechanical system is quasi-exactly solvable (QES) if only a finite number of eigenvalues and the corresponding eigenfunctions can be obtained analytically and in a closed form while the remaining spectrum of the system can not be found exactly.

QES systems are not mathematical constructs. They are highly non-trivial models which can shed new light to delicate analytic properties of complex quantum systems. Many of the QES problems are of the anharmonic oscillator type and they have found applications in a wide range of fields.

The existence of some exact eigenvalues and eigenfunctions can be seen from the following simple fact [5]. Let $W = W(x)$ be real smooth function of r coordinates $\{x_j\}$. Then for $p_j = -\mathrm{i}\dfrac{\partial}{\partial x_j}$, one has

$$\sum_{j=1}^{r} p_j^2 \, \mathrm{e}^W = -\sum_{j=1}^{r}\left[\left(\frac{\partial W}{\partial x_j}\right)^2 + \frac{\partial^2 W}{\partial x_j^2}\right]\mathrm{e}^W. \tag{6.1.1}$$

This implies that e^W is an eigenfunction of the Hamiltonian H_0 with eigenvalue 0:

$$H_0\,\mathrm{e}^W = 0, \quad H_0 = \sum_{j=1}^{r}\left[p_j^2 + \left(\frac{\partial W}{\partial x_j}\right)^2 + \frac{\partial^2 W}{\partial x_j^2}\right] \tag{6.1.2}$$

as long as e^W is square integrable, i.e. $\int \mathrm{e}^{2W}\,\mathrm{d}^r x < \infty$. This is the simplest quasi-exact solvability.

6.2 Generalities: Characterization of QES Operators (概论: 准精确可解算符的特性)

Among various characterizations of solvability of a linear operator (e.g. Hamiltonian of a quantum system), the one about the existence of invariant polynomial subspaces is conceptually the simplest.

Let us point out that in [5], [6] a so-called prepotential approach was used to study QES models. We did not cover this topic in our lectures in this Summer School. Students who are interested in this method may consult the seminar talk material given by the author in[6] in the Summer School.

Definition 1. [4] A linear operator $\mathcal{H}$ is called QES if it has a finite-dimensional (f.d.) invariant subspace $\mathcal{V}_\mathcal{N}$ with explicitly-described basis, that is,

$$\mathcal{H}\mathcal{V}_\mathcal{N} \subset \mathcal{V}_\mathcal{N}, \quad \dim\mathcal{V}_\mathcal{N} < \infty, \quad \mathcal{V}_\mathcal{N} = \mathrm{span}\{\xi_1,\cdots,\xi_{\dim\mathcal{V}_\mathcal{N}}\}.$$

An immediate consequence of this characterization of quasi-exact solvablility for the operator $\mathcal{H}$ is that it can be diagonalized algebraically and exact, closed-form expressions of the corresponding spectra can be obtained in the (solvable) subspace $\mathcal{V}_\mathcal{N}$. However, the remaining part of the spectrum is not in general analytically accessible and can only be computed through approximations (though sometimes rather accurately). If the space $\mathcal{V}_\mathcal{N}$ is a subspace of a Bargmann-Hilbert space of entire functions in which $\mathcal{H}$ is a 2nd-order linear differential operator which can be naturally defined as the Hamiltonian of some quantum system, then the solvable spectra and the corresponding vectors in $\mathcal{V}_\mathcal{N}$ give the exactly-obtainable part of the energy eigenvalues and eigenfunctions of the system, respectively.

Consider an infinite set of f.d. linear spaces $\mathcal{V}_\mathcal{N}$ with property
$$\mathcal{V}_1 \subset \mathcal{V}_2 \subset \cdots \subset \mathcal{V}_\mathcal{N} \subset \cdots \subset \mathcal{V}, \tag{6.2.1}$$
where $\mathcal{V}$ is the completion space. Such a structure is called a flag $\mathcal{V}$. If a linear operator H satisfies
$$\mathcal{H}: \mathcal{V}_\mathcal{N} \longrightarrow \mathcal{V}_\mathcal{N}, \quad \text{for any } \mathcal{N} = 1, 2, \cdots, \tag{6.2.2}$$
then one says that $\mathcal{H}$ preserves the infinite flag $\mathcal{V}$.

Definition 2. A linear operator $\mathcal{H}$ is ES if it preserves an infinite flag of f.d. linear spaces $\mathcal{V}_\mathcal{N}$, $\mathcal{N} = 1, 2, \cdots$, whose bases admit explicit description, that is
$$\mathcal{H}\mathcal{V}_\mathcal{N} \subset \mathcal{V}_\mathcal{N} \subset \mathcal{V}_{\mathcal{N}+1}, \quad \text{for any } \mathcal{N} = 1, 2, \cdots,$$
$$\dim \mathcal{V}_\mathcal{N} < \infty, \quad \mathcal{V}_\mathcal{N} = \text{span}\{\xi_1, \cdots, \xi_{\dim \mathcal{V}_\mathcal{N}}\}.$$

Remark: There are alternative definitions for exact solvability. For example in [7], a linear operator is called ES if solutions of its eigenvalue equation can be expressed in terms of hypergeometric functions.

Take a Lie algebra $\mathcal{G}$ with generators $\{J_\alpha, \alpha = 1, 2, \cdots, \dim \mathcal{G}\}$. Assume that $\mathcal{G}$ has a f.d. representation and $\mathcal{V}_\mathcal{N}$ is its representation space, i.e. $J_\alpha \mathcal{V}_\mathcal{N} \subset \mathcal{V}_\mathcal{N}$, $\alpha = 1, 2, \cdots, \dim \mathcal{G}$. Then obviously any operator $P(J_\alpha)$ made of J_α which is an element of the universal enveloping algebra $U(\mathcal{G})$ of $\mathcal{G}$ has $\mathcal{V}_\mathcal{N}$ as its invariant subspace $P(J_\alpha): \mathcal{V}_\mathcal{N} \to \mathcal{V}_\mathcal{N}$. This leads to [4]:

Definition 3. If the invariant subspace of the QES operator coincides with a f.d. representation space of Lie algebra $\mathcal{G}$, then the QES operator is said to be $\mathcal{G}$ Lie-algebraic QES operator.

Definition 4. If the flag $\mathcal{V}$ consists of f.d. representation spaces of Lie algebra $\mathcal{G}$, then the ES operator is called $\mathcal{G}$ Lie-algebraic ES operator.

Proposition 1. If spaces $\mathcal{V}_\mathcal{N}$ are irreducible f.d. representation spaces of $\mathcal{G}$, then Lie-algebraic QES or ES operators can be expressed in terms of the generators of $\mathcal{G}$. In this case, one says that the QES or ES operators admit an algebraization or have a hidden Lie $\mathcal{G}$ algebra structure or symmetry.

6.3 Lie Algebra Approach (李代数方法)

A typical feature of many QES systems is the existence of a hidden algebraic structure. In this section we recall a general algebraic construction of QES differential equations[26].

6.3 Lie Algebra Approach

Consider a set of first-order differential operators in r variables $z_1, z_2, \cdots, z_r$,

$$J^a = \sum_j^r \alpha_j^a(z) \frac{\partial}{\partial z_j} + \beta^a(z), \quad a = 1, 2, \cdots, d, \tag{6.3.1}$$

where $\alpha_j^a(z)$, $\beta^a(z)$ are certain functions of z_j. Assume that J^a form a finite-dimensional Lie algebra $\mathcal{G} = \text{span}\{J^a, \ a = 1, 2, \cdots, d\}$. Then one says that linear differential operator $\mathcal{H}$ is in $\mathcal{G}$ Lie-algebraic form if it is an element of the universal enveloping algebra $U(\mathcal{G})$ of $\mathcal{G}$.

Thus a 2nd-order linear differential operator (e.g. Hamiltonian) $\mathcal{H}$ is $\mathcal{G}$ Lie-algebraic, or is said to have Lie $\mathcal{G}$ algebraization, if it can be written as quadratic combination of J^a,

$$\mathcal{H} = \sum C_{ab} J^a J^b + \sum C_a J^a + C_* \tag{6.3.2}$$

where C_{ab}, C_a, C_* are constant coefficients.

If the differential operators J^a realize a f.d. representation of $\mathcal{G}$, then the operator $\mathcal{H}$ possess a finite-dimensional invariant subspace coinciding with the f.d. representation space of $\mathcal{G}$. If a basis of this f.d. representation space can be constructed explicitly such that $\mathcal{H}$ can be represented in explicit block-diagonal form, then the associated eigenvalues and corresponding eigenfunctions can be computed by purely algebraic means. This is the main idea in [4] behind quasi-exact solvability of a linear differential operator $\mathcal{H}$.

Remark: There is an analogy between the above Lie algebraic form of QES hamiltonian and the generalized Sugawara construction of the energy-momentum tensor in 2D conformal field theories. This was noted in [8]. The author in [9] discovered a direct relation between 1D QES problems and 3- and 4-point conformal blocks in a conformal field theory with zero vector at the 2nd level. For more details about the connection, see review article [10].

Let us consider the 1st-order differential operators in single variable z:

$$J^+ = -z^2 \frac{d}{dz} + nz, \quad J^0 = z\frac{d}{dz} - \frac{n}{2}, \quad J^- = \frac{d}{dz}. \tag{6.3.3}$$

These differential operators satisfy the $sl(2)$ commutation relations for any value of the parameter n,

$$[J^0, J^\pm] = \pm J^\pm, \quad [J^+, J^-] = 2J^0. \tag{6.3.4}$$

If n is a non-negative integer, $n = 0, 1, 2, \cdots$, then (6.3.3) provide a $(n+1)$-dimensional irreducible representation $\mathcal{P}_{n+1}(z) = \text{span}\{1, z, z^2, \cdots, z^n\}$ of the $sl(2)$ algebra. It is evident that any differential operator which is a polynomial of the $sl(2)$ generators (6.3.3) with n being non-negative integer will have the space $\mathcal{P}_{n+1}(z)$ as its invariant subspace, i.e. possesses $(n+1)$ eigen-functions in the form of polynomial in z of degree n.

Now consider the 2nd order differential operator of the form

$$\mathcal{H} = X(z)\frac{d^2}{dz^2} + Y(z)\frac{d}{dz} + Z(z), \qquad (6.3.5)$$

where $X(z), Y(z), Z(z)$ are polynomials of degree at most 4, 3, 2 respectively,

$$X(z) = \sum_{k=0}^{4} a_k z^k, \quad Y(z) = \sum_{k=0}^{3} b_k z^k, \quad Z(z) = \sum_{k=0}^{2} c_k z^k.$$

The differential operator (6.3.5) is usually called the Heun operator. Then we have [11]

Proposition 2. The differential operator $\mathcal{H}$ allows for an $sl(2)$ algebraization, i.e. has a hidden $sl(2)$ algebraic structure, if and only if

$$b_3 = -2(n-1)a_4, \quad c_2 = n(n-1)a_4, \quad c_1 = -n[(n-1)a_3 + b_2]. \qquad (6.3.6)$$

Proof [11]. It suffices to prove that $\mathcal{H}$ is a quadratic combination of the $sl(2)$ generators (6.3.3) if and only if the relations (6.3.6) are satisfied.

Sufficiency. We have

$$\mathcal{H} = X(z)\frac{d^2}{dz^2} + \left[-2(n-1)a_4 z^3 + b_2 z^2 + b_1 z + b_0\right]\frac{d}{dz}$$
$$+ n(n-1)a_4 z^2 - n[(n-1)a_3 + b_2]z + c_0. \qquad (6.3.7)$$

It is easy to check that

$$a_4 J^+ J^+ - a_3 J^+ J^0 + a_2 J^0 J^0 + a_1 J^0 J^- + a_0 J^- J^-$$
$$= X(z)\frac{d^2}{dz^2} + \left[-2(n-1)a_4 z^3 - \frac{3n-2}{2}a_3 z^2 - (n-1)a_2 z - \frac{n}{2}a_1\right]\frac{d}{dz}$$
$$+ n(n-1)a_4 z^2 + \frac{n^2}{2}a_3 z + \frac{n^2}{2}a_2,$$
$$-b_2 J^+ + b_1 J^0 + b_0 J^-$$
$$= (b_2 z^2 + b_1 z + b_0)\frac{d}{dz} - n b_2 z - \frac{n}{2}b_1. \qquad (6.3.8)$$

6.3 Lie Algebra Approach

Substituting into (6.3.7) gives rise to

$$\mathcal{H} = a_4 J^+ J^+ - a_3 J^+ J^0 + a_2 J^0 J^0 + a_1 J^0 J^- + a_0 J^- J^- - \left(\frac{3n-2}{2} a_3 + b_2\right) J^+$$

$$+ [(n-1)a_2 + b_1] J^0 + \left(\frac{n}{2} a_1 + b_0\right) J^- + \frac{n}{2} \left[\left(\frac{n}{2} - 1\right) a_2 + b_1\right] + c_0 \quad (6.3.9)$$

Necessity. We take

$$\mathcal{H} = A_{++} J^+ J^+ - A_{+0} J^+ J^0 + A_{00} J^0 J^0 + A_{0-} J^0 J^-$$
$$+ A_{--} J^- J^- - A_+ J^+ + A_0 J^0 + A_- J^- + A_*, \quad (6.3.10)$$

where A_{++} etc are constant coefficients to be determined. Then by means of the expressions (6.3.3),

$$\mathcal{H} = \left(A_{++} z^4 - A_{+0} z^3 + A_{00} z^2 + A_{0-} z + A_{--}\right) \frac{d^2}{dz^2} + \Big\{ -2(n-1) A_{++} z^3$$
$$+ \left(-A_+ + \frac{3n-2}{2} A_{+0}\right) z^2 + [A_0 - (n-1) A_{00}] z + A_- - \frac{n}{2}\Big\} \frac{d}{dz}$$
$$+ n(n-1) A_{++} z^2 + n \left(-\frac{n}{2} A_{+0} + A_+\right) z + \frac{n}{2} \left(\frac{n}{2} - A_0\right) + A_*. \quad (6.3.11)$$

The r.h.s. of (6.3.11) can be written as $X(z)\frac{d^2}{dz^2} + Y(z)\frac{d}{dz} + Z(z)$ provided that we make the identification

$$\begin{cases} a_4 = A_{++}, \quad a_3 = -A_{+0}, \quad a_2 = A_{00}, \quad a_1 = A_{0-}, \quad a_0 = A_{--}, \\[4pt] b_3 = -2(n-1) A_{++}, \quad b_2 = -A_+ + \frac{3n-2}{2} A_{+0}, \quad b_1 = A_0 - (n-1) A_{00}, \\[4pt] b_0 = A_- - \frac{n}{2}, \quad c_2 = n(n-1) A_{++}, \quad c_1 = n\left(-\frac{n}{2} A_{+0} + A_+\right), \\[4pt] c_0 = \frac{n}{2}\left(\frac{n}{2} - A_0\right) + A_*. \end{cases} \quad (6.3.12)$$

It follows that

$$b_3 = -2(n-1)a_4, \quad c_2 = n(n-1)a_4, \quad c_1 = -n[(n-1)a_3 + b_2]. \quad (6.3.13)$$

This completes our proof. □

Note that the number of parameters in the most general $sl(2)$ Lie algebraic form is equal to 9. Under the conditions $A_{++} = A_{+0} = A_+ = 0$, the operator $\mathcal{H}$ becomes ES:

$$\mathcal{H} = A_{00} J^0 J^0 + A_{0-} J^0 J^- + A_{--} J^- J^- + A_0 J^0 + A_- J^- + A_*. \quad (6.3.14)$$

So the number of free parameters of a 2nd-order ES differential operator is 6.

6.4 Relationship Between 2nd-order Differential Operator and Schrödinger Operator (二阶差分算符和薛定谔算符的关系)

Every 2nd-order differential operator in single variable z is equivalent to a Schrödinger operator $H_0 = p^2 + V(x)$ with potential $V(x)$. This is achieved by a combination of change of independent variable $z = z(x)$ and gauge transformation $\mu^{-1}(x)\mathcal{H}\mu(x)$.

A known elementary result (which may not necessarily be valid in higher dimensions) is (e.g. [3])

Proposition 3. Let

$$\mathcal{H} = -P(z)\frac{d^2}{dz^2} - \left[Q(z) + \frac{1}{2}P'(z)\right]\frac{d}{dz} - R(z)$$

be a 2nd-order differential operator such that $P(z) > 0$. Then there exists a (local) change of variable $z = z(x)$ and gauge factor $\mu(x) = e^{-W(x)}$ which transform $\mathcal{H}$ to a Schrödinger operator $H_0 = p^2 + V(x)$ with potential $V(x)$ given by

$$V(x) = \frac{1}{2}Q'(z) + \frac{Q(z)[Q(z) - P'(z)]}{4P(z)} - R(z)$$

Proof. The proof is an elementary computation. Let $z = z(x)$ Then $\frac{d}{dx} = z'\frac{d}{dz}$ and $e^{W(x)}\frac{d}{dz}e^{-W(x)} = \frac{1}{z'}e^{W(x)}\frac{d}{dx}e^{-W(x)} = \frac{1}{z'}\left(\frac{d}{dx} - \frac{dW}{dx}\right)$. So

$$e^{W(x)}\frac{d^2}{dz^2}e^{-W(x)} = e^{W(x)}\frac{d}{dz}\cdot\frac{d}{dz}e^{-W(x)} = \frac{1}{z'}\left(\frac{d}{dx} - \frac{dW}{dx}\right)\frac{1}{z'}\left(\frac{d}{dx} - \frac{dW}{dx}\right)$$

$$= \frac{1}{z'^2}\left[\frac{d^2}{dx^2} - \left(2\frac{dW}{dx} + \frac{z''}{z'}\right)\frac{d}{dx} + \left(\frac{dW}{dx}\right)^2 - \frac{d^2W}{dx^2} + \frac{z''}{z'}\frac{dW}{dx}\right]$$

Equating $e^{W(x)}\mathcal{H}e^{-W(x)}$ to $-\frac{d^2}{dx^2} + V(x)$ gives rise to

$$P(z) = z'^2,$$

$$2\frac{dW}{dx} + \frac{z''}{z'} - \frac{Q(z) + \frac{1}{2}P'(z)}{z'} = 0,$$

$$V(x) = \frac{d^2W}{dx^2} - \left(\frac{dW}{dx}\right)^2 - \frac{z''}{z'}\frac{dW}{dx} + \frac{Q(z) + \frac{1}{2}P'(z)}{z'}\frac{dW}{dx} - R(z). \quad (6.4.1)$$

Solving the first equation, one gets

$$x = x(z) = \pm \int^z \frac{1}{\sqrt{P(y)}} dy. \tag{6.4.2}$$

Differentiating the first equation to get $z'' = \frac{1}{2}P'(z)$ and substituting the result into the second equation in (6.4.1), one has

$$2z'\frac{dW}{dx} = Q(z) \longrightarrow \frac{dW}{dz} = \frac{Q(z)}{2z'^2}, \tag{6.4.3}$$

which gives, by using the first equation in (6.4.1),

$$W(x) = \int^z \frac{Q(y)}{2P(y)} dy. \tag{6.4.4}$$

Now from (6.4.3), one obtain

$$\frac{d^2W}{dx^2} = \frac{d}{dx}\left[\frac{Q(z)}{2z'}\right] = z'\frac{d}{dz}\left[\frac{Q(z)}{2z'}\right] = \frac{z'^2 Q'(z) - Q(z)z''}{2z'^2}. \tag{6.4.5}$$

Substituting the above results into the 3rd equation of (6.4.1), one completes the proof. □

Note that the potential $V(x)$ is uniquely determined up to translation $V(x) \to V(x+\delta)$. The gauge factor $\mu(x) = e^{-W(x)}$ is not necessarily unitary, and hence does not preserve the normalizability properties of the associated eigenfunctions. However, the change of variables and gauge transformations do both preserve the Lie algebra structure. When $\mathcal{H}$ has Lie algebraization, all algebraic eigenvalues can be computed in the simpler gauged coordinates z. The only questions is whether or not these represent genuine physical spectrum of H_0, i.e whether or not the eigenfunctions are, in the physical coordinates x, normalizable (e.g. square integrable).

6.5 Examples of QES Systems with Lie Algebraization (准精确可解系统李代数化的例子)

6.5.1 Sextic Potential (六次势)

Consider the following class of Hamiltonians[1, 2]

$$H_0 = -\frac{d^2}{dx^2} + \frac{(2s-1/2)(2s-3/2)}{x^2} - (4s+4J-2)x^2 + x^6. \tag{6.5.1}$$

Here s is an arbitrary parameter: when s lies between $1/4$ and $3/4$, there is an attractive centrifugal term; for s outside this range the centrifugal term is repulsive. When $s = 1/4$ or $3/4$, the centrifugal core term disappears, leaving a non-singular sextic oscillator hamiltonian $H_0 = -\dfrac{d^2}{dx^2} - (4s + 4J - 2)x^2 + x^6$.

Let
$$\mu(x) = x^a\, e^{-\frac{b}{4}x^4}. \tag{6.5.2}$$

Then we have
$$\mu^{-1}(x) \cdot H_0 \cdot \mu(x) = -\frac{d^2}{dx^2} + \left(2bx^3 - \frac{2a}{x}\right)\frac{d}{dx} + \frac{(2s-1/4)(2s-3/4) - a(a-1)}{x^2}$$
$$- (4s + 4J - 2 - 2ab - 3b)x^2 + (1 - b^2)x^6. \tag{6.5.3}$$

Choose $a = 2s - 1/2$ and $b^2 = 1$ to eliminate the inverse square and x^6 terms. For square integrability we choose the $b = 1$ solution of $b^2 = 1$ and obtain
$$\mu^{-1}(x) \cdot H_0 \cdot \mu(x) = -\frac{d^2}{dx^2} + \left(2x^3 - \frac{4s-1}{x}\right)\frac{d}{dx} - 4(J-1)x^2. \tag{6.5.4}$$

Making a change of variable, $z = x^2$, we get
$$\mathcal{H} = \mu^{-1}(x) \cdot H_0 \cdot \mu(x) = -4z\frac{d^2}{dz^2} + 4\left(z^2 - 2s\right)\frac{d}{dx} - 4(J-1)z. \tag{6.5.5}$$

When J is a non-negative integer, $\mathcal{H}$ can be written in terms of the $sl(2)$ generators $J_{\pm,0}(z)$ as
$$\mathcal{H} = -4J^0 J^- - 4J^+ - (8s + 2J - 2)J^- \tag{6.5.6}$$

where
$$J^+ = -z^2\frac{d}{dz} + (J-1)z, \quad J^0 = z\frac{d}{dz} - \frac{J-1}{2}, \quad J^- = \frac{d}{dz}. \tag{6.5.7}$$

It follows that $\mathcal{H}$ preserves the $(J-1)+1 = J$ dimensional representation space $\mathcal{P}_J$ of the $sl(2)$ algebra. So when J is a non-negative integer, $\mathcal{H}$ or the Schrödinger operator H_0 is QES and the corresponding time-independent Schrödinger equation $\mathcal{H}P(z) = EP(z)$ has J exact, closed-form solutions for any values of s. The J eigenfunctions $P(z)$ are polynomials of degree $J-1$ in z,
$$\psi_i(x) = x^{2s-1/2}e^{-\frac{1}{4}x^4} \times P^{(i)}_{J-1}(z(x)), \quad z(x) = x^2, \quad i = 0, 1, 2, \cdots, J-1. \tag{6.5.8}$$

Note that the degree of the polynomials in $\psi_i(x)$ above is $2(J-1)$ in terms of variable x for fixed J.

6.5.2 Harmonic Oscillator (谐振子)

The hamiltonian of the harmonic oscillator reads

$$H_0 = -\frac{d^2}{dx^2} + x^2. \tag{6.5.9}$$

Let $\mu(x) = e^{-\frac{1}{2}x^2}$. We obtain the gauge-transformed hamiltonian

$$\mathcal{H} = \mu^{-1}(x) H_0 \mu(x) = -\frac{d^2}{dx^2} + 2x\frac{d}{dx} + 1 \tag{6.5.10}$$

which can be written as

$$\mathcal{H} = -J^- J^- + 2J^0 + J + 1,$$
$$J^+ = -x^2 \frac{d}{dx} + Jx, \quad J^0 = x\frac{d}{dx} - \frac{J}{2}, \quad J^- = \frac{d}{dx}. \tag{6.5.11}$$

Here J is an arbitrary parameter, i.e not necessarily an integer! So $\mathcal{H}$ preserves an infinite flag of finite-dimensional polynomial spaces $\mathcal{V}_n$, $n = 0, 1, 2, \cdots$, independent of J. It follows that $\mathcal{H}$ (or H_0) is ES, as expected.

The time-independent Schrödinger equation $\mathcal{H}\psi = E\psi$ is of the form

$$\psi'' - 2x\psi' + (E-1)\psi = 0. \tag{6.5.12}$$

Comparing with the Hermite differential equation

$$H_n'' - 2x H_n' + 2n H_n = 0, \quad n = 0, 1, 2, \cdots \tag{6.5.13}$$

we see that solutions to the Schrödinger equation above are given by Hermite polynomials of degree n, $n = 0, 1, 2, \cdots$, with eigenvalues E given by $E = 2n + 1$.

6.5.3 Lamé Equation (Lamé 方程)

The Lamé equation reads

$$\left[-\frac{d^2}{dx^2} + m(m+1)\wp(x) \right] \psi(x) = E\psi(x), \tag{6.5.14}$$

where $m = 1, 2, \cdots$, and $\wp(x)$ is the Weierstrass function. $\wp(x)$ is a double-periodic meromorphic function satisfying the equation

$$\wp'^2 = 4(\wp - e_1)(\wp - e_2)(\wp - e_3), \quad \sum_{i=1}^{3} e_i = 0. \tag{6.5.15}$$

Differentiating we obtain

$$\wp'' = 2\left[(\wp - e_1)(\wp - e_2) + (\wp - e_1)(\wp - e_3) + (\wp - e_2)(\wp - e_3)\right]. \tag{6.5.16}$$

Introduce new variable

$$z = \wp(x) + \frac{1}{3}\sum_{i=1}^{3} a_i \tag{6.5.17}$$

with a_i obeying the relations $e_i = a_i - \frac{1}{3}\sum a_i$. We have

$$\frac{d^2}{dx^2} = \wp \frac{d}{dz} + \wp'^2 \frac{d^2}{dz^2}. \tag{6.5.18}$$

Substituting into (6.5.14) gives the other form of the Lamé equation

$$\left[\frac{d^2}{dz^2} + \frac{1}{2}\left(\frac{1}{z - a_1} + \frac{1}{z - a_2} + \frac{1}{z - a_3}\right)\frac{d}{dz} - \frac{1}{4}\frac{m(m+1)z + \epsilon}{(z - a_1)(z - a_2)(z - a_3)}\right]\psi = 0, \tag{6.5.19}$$

where $\epsilon = E - \frac{1}{3}m(m+1)\sum a_i$.

Eq.(6.5.19) is a Fuchsian differential equation with four regular singularities $z = a_i$ and $z = \infty$. The exponents at $z = a_i$ are equal to 0 and $1/2$. So solutions to the Lamé equation (6.5.19) have the following form

$$\psi = (z - a_1)^{a/2}(z - a_2)^{b/2}(z - a_3)^{c/2}\phi, \quad a, b, c = 0 \text{ or } 1. \tag{6.5.20}$$

Substituting (6.5.20) into the Lamé equation and noting that one always has $a^2 = a$, $b^2 = b$ and $c^2 = c$, we get

$$\mathcal{H}\phi = \epsilon\phi, \tag{6.5.21}$$

where

$$\mathcal{H} = \left[4z^3 - 4\sum_{i=1}^{3} a_i z^2 + 4\sum_{i \neq j} a_i a_j z - 4a_1 a_2 a_3\right]\frac{d^2}{dz^2}$$

$$+ \left\{[4(a + b + c) + 6]z^2 - 4\left[a_1(b + c) + a_2(a + c) + a_3(a + b) + \sum_{i=1}^{3} a_i\right]z\right.$$

$$+ 2\sum_{i \neq j} a_i a_j + 4(aa_2a_3 + ba_1a_3 + ca_1a_2)\right\}\frac{d}{dz}$$

$$+ \left[(a + b)^2 + (a + c)^2 + (b + c)^2 - m(m+1)\right]z$$

$$- a_1(b + c)^2 - a_2(a + c)^2 - a_3(a + b)^2. \tag{6.5.22}$$

6.5 Examples of QES Systems with Lie Algebraization

Split m into integers and half integers according to the values of a, b, and c as follows,

$$m = 2n + a + b + c, \quad n = \begin{cases} 1, 2, \cdots, & \text{if } a = b = c = 0, \\ 0, 1, 2, \cdots, & \text{otherwise.} \end{cases} \quad (6.5.23)$$

Then the requirements in Proposition (2) are fulfilled since

$$\begin{aligned} c_1 &\equiv (a+b)^2 + (a+c)^2 + (b+c)^2 - m(m+1) \\ &= -n[4(n+a+b+c) + 2] \equiv -n[(n-1)a_3 + b_2]. \end{aligned} \quad (6.5.24)$$

With m written in form (6.5.23), $\mathcal{H}$ is dependent on the integer parameter n and can be expressed as the quadratic combination of the $sl(2)$ generators as follows,

$$\begin{aligned} \mathcal{H} = &-4J^+J^0 - 4\sum a_i J^0 J^0 + 4\sum_{i \neq j} a_i a_j J^0 J^- - 4a_1 a_2 a_3 J^- J^- \\ &-2[3n + 1 + 2(a+b+c)]J^+ \\ &-4\left[n\sum a_i + a_1(b+c) + a_2(a+c) + a_3(a+b)\right]J^0 \\ &+4\left[\frac{n+1}{2}\sum_{i \neq j} a_i a_j + aa_2 a_3 + ba_1 a_3 + ca_1 a_2\right]J^- \\ &- a_1(b+c)(2n+b+c) - a_2(a+c)(2n+a+c) \\ &- a_3(a+b)(2n+a+b) - n^2 \sum a_i. \end{aligned} \quad (6.5.25)$$

So eigenfunction ϕ is polynomial of degree n. For a fixed n, there are $n+1$ such polynomials (of degree n)

$$\phi_i(z) = \text{Pol}_n^{(i)}(z), \quad i = 0, 1, 2, \cdots, n. \quad (6.5.26)$$

Eq.(6.5.25) provides an unified expression for $sl(2)$ algebraization of the Lamé operator for both even and odd m values.

6.5.4 QES Quartic Potential (准精确可解的四次势)

Until the work by Bender and Boettcher on QES quartic potential, it was believed that the lowest-degree 1D QES polynomial potential is sextic. Let us proceed with the usual algebraic approach for quartic potential to see how far one can go in parallel with the sextic potential case.

Consider the following family of Hamiltonians[12]

$$H_0 = -\frac{d^2}{dx^2} + a^2 x^4 + 2abx^3 + (b^2 + 2ac)x^2 + 2[bc - (n+1)a]x, \quad (6.5.27)$$

where a, b, c are parameters and n is a non-negative integer. Let $\mu(x) = e^{-\frac{a}{3}x^3 - \frac{b}{2}x^2 - cx}$. Then

$$\mathcal{H} = \mu^{-1}(x) H_0 \mu(x) = -\frac{d^2}{dx^2} + 2(ax^2 + bx + c)\frac{d}{dx} - 2anx + b - c^2. \quad (6.5.28)$$

This $\mathcal{H}$ can be expressed in terms of the $sl(2)$ generators (6.3.3) as

$$\mathcal{H} = -J^- J^- - 2aJ^+ + 2bJ^0 + 2cJ^- + (n+1)b - c^2. \quad (6.5.29)$$

So obviously $\mathcal{H}$ preserves a polynomial space $V_{n+1} = \text{span}\{1, x, x^2, \cdots, x^n\}$ and solutions to the eigenvalue problem $\mathcal{H}P(x) = EP(x)$ are polynomials of degree n for fixed n.

However, the eigenfunctions, $\psi \sim e^{-\frac{a}{3}x^3 - \frac{b}{2}x^2 - cx} P_n(x)$, of H_0 is not square integrable in the region $(-\infty, \infty)$ for whatever choice of the sign of $a \neq 0$. So this is an interesting example which fails to achieve square integrability of eigenfunctions for real parameters a, b and c.

The problem was solved by the authors in [12]. The key is to allow the model parameters to be complex. Let us take

$$a = i, \ b = \alpha, \ c = i\beta, \quad (6.5.30)$$

where α, β are real parameters and moreover $\alpha > 0$. Then obviously $\mu(x) = e^{-\frac{i}{3}x^3 - \frac{\alpha}{2}x^2 - i\beta x}$ is square integrable on the whole line (for $\alpha > 0$). However, now the hamiltonian H_0

$$H_0 = -\frac{d^2}{dx^2} - x^4 + 2i\alpha x^3 + (\alpha^2 - 2\beta)x^2 + 2i[\alpha\beta - (n+1)]x \quad (6.5.31)$$

is non-Hermitian! The gauge transformed Hamiltonian

$$\mathcal{H} = \mu^{-1}(x) H_0 \mu(x) = -\frac{d^2}{dx^2} + 2(ix^2 + \alpha x + i\beta)\frac{d}{dx} - 2inx + \alpha + \beta^2 \quad (6.5.32)$$

still can be expressed in terms of the $sl(2)$ generators as

$$\mathcal{H} = -J^- J^- - 2iJ^+ + 2\alpha J^0 + 2i\beta J^- + (n+1)\alpha + \beta^2. \quad (6.5.33)$$

So the system (6.5.31) is QES. It was argued by the authors in [12] that the spectrum of this family of non-Hermitian Hamiltonians with quartic potentials is real, discrete and bounded from below.

6.5.5 Quantum (Driven) Rabi Model (量子驱动的 Rabi 模型)

The quantum Rabi model describes the interaction of a two-level atom with a single harmonic mode of electromagnetic field. It is perhaps the simplest system for modeling the ubiquitous matter-light interactions in modern physics, and has applications in a variety of physical fields.

Recently Braak[13] presented a transcendental function defined as an infinite power series with coefficients satisfying a three-term recursive relation, and argued that the spectrum of the Rabi model is given by the zeros of the transcendental function. This theoretical progress has renewed the interest in the Rabi and related models (e.g. [14]~[16]) However, since Braak transcendental function is given as an infinite power series, unless the model parameters satisfy certain constraints for which the infinite series truncates, its exact zeros and therefore closed-form expressions for the energies of the Rabi model can not be obtained even for those corresponding to the low-lying spectrum. This indicates that the Rabi model is not exactly solvable.

Quasi-exact solvability of the Rabi model has recently been noted[17, 18]. Special exact spectrum of the model was obtained in [19]~[21]. Here we consider the driven Rabi model and show that it has an $sl(2)$ hidden structure. This manifests the first appearance of a hidden algebraix structure in quantum spin-boson systems without $U(1)$ symmetry.

The Hamiltonian of the driven Rabi model is

$$H = \omega a^\dagger a + \Delta \sigma_z + g\,\sigma_x \left[a^\dagger + a\right] + \delta\,\sigma_x, \qquad (6.5.34)$$

where g is the interaction strength, σ_z, σ_x are the Pauli matrices describing the two atomic levels separated by energy difference 2Δ, and $a^\dagger$ (a) are creation (annihilation) operators of a boson mode with frequency ω. Here $a^\dagger$ (a) satisfy the Heisenberg algebra relations $[a, a^\dagger] = 1$, $[a, a] = 0 = [a^\dagger, a^\dagger]$. The addition of the driving term $\delta \sigma_x$ breaks the Z_2 symmetry of the Rabi model. The driven Rabi model (6.5.34) is relevant to the description of some hybrid mechanical systems.

By means of the Fock-Bargmann correspondence $a^\dagger \to z$, $a \to \dfrac{\mathrm{d}}{\mathrm{d}z}$, the Hamiltonian becomes a matrix differential operator

$$H = \omega z \frac{\mathrm{d}}{\mathrm{d}z} + \Delta \sigma_z + g\,\sigma_x \left(z + \frac{\mathrm{d}}{\mathrm{d}z}\right) + \delta\,\sigma_x. \qquad (6.5.35)$$

Working in a representation defined by σ_x diagonal and in terms of the two-component wavefunction $\psi(z) = \begin{pmatrix} \psi_+(z) \\ \psi_-(z) \end{pmatrix}$, the time-independent Schrödinger equation $H_R \psi(z) = E\psi(z)$ gives rise to a coupled system of two 1st-order differential equations:

$$(\omega z + g)\frac{d}{dz}\psi_+(z) + [gz - (E - \delta)]\psi_+(z) + \Delta\psi_-(z) = 0,$$
$$(\omega z - g)\frac{d}{dz}\psi_-(z) - [gz + (E + \delta)]\psi_-(z) + \Delta\psi_+(z) = 0. \qquad (6.5.36)$$

If $\Delta = 0$ these two equations decouple and reduce to the differential equations of two uncoupled displaced harmonic oscillators[22]. For this reason we will concentrate on the non-trivial $\Delta \neq 0$ case.

With the substitution $\psi_\pm(z) = e^{-gz/\omega}\phi_\pm(z)$, it follows

$$\left[(\omega z + g)\frac{d}{dz} - \left(\frac{g^2}{\omega} - \delta + E\right)\right]\phi_+(z) = -\Delta\phi_-(z),$$
$$\left[(\omega z - g)\frac{d}{dz} - \left(2gz - \frac{g^2}{\omega} + \delta + E\right)\right]\phi_-(z) = -\Delta\phi_+(z). \qquad (6.5.37)$$

Eliminating $\phi_-(z)$ from the system we obtain the 2nd-order differential equation for $\phi_+(z)$,

$$\mathcal{H}\phi_+(z) = \Delta^2 \phi_+(z), \qquad (6.5.38)$$

where

$$\mathcal{H} = (\omega z - g)(\omega z + g)\frac{d^2}{dz^2} + \left[-2\omega g z^2 + (\omega^2 - 2g^2 - 2E\omega)z - g\omega \right.$$
$$\left. + 2g\left(\frac{g^2}{\omega} - \delta\right)\right]\frac{d}{dz} + 2g\left(\frac{g^2}{\omega} - \delta + E\right)z + E^2 - \left(\delta - \frac{g^2}{\omega}\right)^2. \qquad (6.5.39)$$

By Proposition 2, $\mathcal{H}_R$ allows for an $sl(2)$ algebraization if

$$2g\left(E + \frac{g^2}{\omega} - \delta\right) \equiv c_1 = -n[(n-1)a_3 + b_2] \equiv 2g\omega n, \qquad (6.5.40)$$

which gives one set of the exact (exceptional) energies of the driven Rabi model

$$E = \omega n + \delta - \frac{g^2}{\omega}, \quad n = 0, 1, 2, \cdots. \qquad (6.5.41)$$

Indeed, for such E values, $\mathcal{H}$ is dependent on the integer parameter n and can be expressed as the quadratic combination of the $sl(2)$ generators (6.3.3)

6.5 Examples of QES Systems with Lie Algebraization

$$\mathcal{H} = \omega^2 J^0 J^0 - g^2 J^- J^- + 2g\omega J^+ + (n\omega^2 - 2g^2 - 2\omega E)J^0$$
$$-g\left[\omega + 2\left(\delta - \frac{g^2}{\omega}\right)\right]J^- + n\left(\frac{n}{4}\omega^2 - g - \omega E\right) + E^2 - \left(\delta - \frac{g^2}{\omega}\right)^2, \quad (6.5.42)$$

where E is given by (6.5.41).

Similarly for the other set of solutions of the driven Rabi model, we set $\psi_\pm(z) = e^{gz/\omega}\varphi_\pm(z)$ and get from (6.5.36)

$$\begin{cases} \left[(\omega z + g)\dfrac{d}{dz} + \left(2gz + \dfrac{g^2}{\omega} + \delta - E\right)\right]\varphi_+(z) = -\Delta\varphi_-(z), \\ \left[(\omega z - g)\dfrac{d}{dz} - \left(\dfrac{g^2}{\omega} + \delta + E\right)\right]\varphi_-(z) = -\Delta\varphi_+(z). \end{cases} \quad (6.5.43)$$

Eliminating $\varphi_+(z)$ from the system we obtain the 2nd-order differential equation for $\varphi_-(z)$,

$$\tilde{\mathcal{H}}\varphi_-(z) = \Delta^2\varphi_-(z), \quad (6.5.44)$$

where

$$\tilde{\mathcal{H}} = (\omega z - g)(\omega z + g)\frac{d^2}{dz^2} + [2\omega g z^2 + (\omega^2 - 2g^2 - 2E\omega)z + g\omega]$$
$$-2g\left(\frac{g^2}{\omega} + \delta\right)\frac{d}{dz} + 2g\left(\frac{g^2}{\omega} + \delta + E\right)z + E^2 - \left(\delta + \frac{g^2}{\omega}\right)^2. \quad (6.5.45)$$

The operator $\tilde{\mathcal{H}}$ allows for an $sl(2)$ algebraization if

$$-2g\left(E + \frac{g^2}{\omega} + \delta\right) \equiv c_1 = -n[(n-1)a_3 + b_2] \equiv -2g\omega n, \quad (6.5.46)$$

which gives the other set of the exact (exceptional) energies of the driven Rabi model

$$E = \omega n - \delta - \frac{g^2}{\omega}, \quad n = 0, 1, 2, \cdots. \quad (6.5.47)$$

For such E values, $\tilde{\mathcal{H}}$ is dependent on the integer parameter n and can be expressed as the quadratic combination of the $sl(2)$ generators (6.3.3)

$$\tilde{\mathcal{H}} = \omega^2 J^0 J^0 - g^2 J^- J^- - 2g\omega J^+ + (n\omega^2 - 2g^2 - 2\omega E)J^0$$
$$+g\left[\omega - 2\left(\delta + \frac{g^2}{\omega}\right)\right]J^- + n\left(\frac{n}{4}\omega^2 - g - \omega E\right) + E^2 - \left(\delta + \frac{g^2}{\omega}\right)^2 \quad (6.5.48)$$

where E is given by (6.5.47).

Expressions (6.5.41) and (6.5.47) together form the full set of exceptional energies for the driven Rabi model (see e.g. the appendix of [23], and [24]). The $sl(2)$

algebraizations (6.5.42) and (6.5.48) mean that the corresponding spectral problems (6.5.38) and (6.5.44) possess $(n+1)$ eigenfunctionns, respectively, in the form of polynomials of degree n. This means that corresponding to every E value given above there exist $(n+1)$ allowed values for other model parameters which preserve the quasi-exact solvability of the driven Rabi model.

6.6 Stäckel Transform and Coupling Constant Metamorphosis (Stäckel 变换和耦合常数变形)

Let $\mathcal{H} = H(\boldsymbol{x},\boldsymbol{p}) - \alpha U(\boldsymbol{x})$ be the hamiltonian of a quantum mechanical system, where $H(\boldsymbol{x},\boldsymbol{p})$ is independent of the arbitrary parameter α. The time-independent Schrödinger equation takes the form

$$\mathcal{H}\psi(\boldsymbol{x}) = [H(\boldsymbol{x},\boldsymbol{p}) - \alpha U(\boldsymbol{x})]\psi(\boldsymbol{x}) = E\psi(\boldsymbol{x}). \tag{6.6.1}$$

Then we have

Proposition 4. Let

$$\mathcal{H}' = U^{-1}(\boldsymbol{x})\left[H(\boldsymbol{x},\boldsymbol{p}) - \alpha'\right] \tag{6.6.2}$$

be the Stäckel transformed hamiltonian describing certain quantum mechanical system with Schrödinger equation

$$\mathcal{H}'\psi(\boldsymbol{x}) = U^{-1}(\boldsymbol{x})\left[H(\boldsymbol{x},\boldsymbol{p}) - \alpha'\right]\psi(\boldsymbol{x}) = E'\psi(\boldsymbol{x}). \tag{6.6.3}$$

Then this system is equivalent to (or dual to) the system described by $\mathcal{H}$ under the coupling constant metamorphosis:

$$\alpha' \longleftrightarrow E, \quad E' \longleftrightarrow \alpha. \tag{6.6.4}$$

6.6.1 Two Electrons in External Oscillator Potential (谐振外势中的两电子体系)

Consider the Hamiltonian of two electrons in external oscillator potential,

$$H_0 = \sum_{i=1}^{2} \frac{1}{2}\left(\boldsymbol{p}_i^2 + \omega^2 \boldsymbol{r}_i^2\right) + \frac{Z}{||\boldsymbol{r}_1 - \boldsymbol{r}_2||}. \tag{6.6.5}$$

Introduce the relative coordinate $\boldsymbol{r} = \boldsymbol{r}_1 - \boldsymbol{r}_2$ and the center of mass coordinate

6.6 Stäckel Transform and Coupling Constant Metamorphosis

$R = \frac{1}{2}(r_1 + r_2)$, which give rise to new momentum operators

$$p = -i\nabla_r = \frac{1}{2}(p_2 - p_1), \quad P = -i\nabla_R = p_2 + p_1. \tag{6.6.6}$$

Then the hamiltonian can be written as

$$H_0 = 2\left(\frac{1}{2}p^2 + \frac{1}{2}\omega_r^2 r^2 + \frac{Z}{2r}\right) + \frac{1}{2}\left(\frac{1}{2}P^2 + \frac{1}{2}\omega_R^2 R^2\right) \equiv H_r + H_R, \tag{6.6.7}$$

where $\omega_R = 2\omega$ and $\omega_r = \frac{1}{2}\omega$.

The total wave function factorizes

$$\psi(1,2) = \phi(r)\xi(R) \tag{6.6.8}$$

and the Schrödinger equation $H_0\psi = E\psi$ separates into

$$H_r\phi(r) = \epsilon\phi(r), \quad H_R\xi(R) = \eta\xi(R) \tag{6.6.9}$$

with $E = \epsilon + \eta$ being the total energy of the system.

Introduce the spherical coordinates which separate the modulus r from the angular coordinates $\hat{r} = r/r$, giving rise to the Ansatz:

$$\phi(r) = \frac{u(r)}{r} Y_{lm}(\hat{r}), \tag{6.6.10}$$

where $Y_{lm}(\hat{r})$ are the spherical harmonics. Then the radial Schrödinger equation is

$$\left[-\frac{d^2}{dr^2} + \omega_r^2 r^2 + \frac{Z}{r} + \frac{l(l+1)}{r^2}\right] u(r) = \epsilon u(r). \tag{6.6.11}$$

Setting

$$u(r) = r^{l+1} e^{-\frac{\omega_r}{2} r^2} y(r), \tag{6.6.12}$$

we can write the radial Schrödinger equation in the form

$$\mathcal{H}y \equiv \left(H - \frac{\alpha}{r}\right) y = \mathcal{E} y \tag{6.6.13}$$

with $\alpha = Z$, $\mathcal{E} = -\epsilon$ and

$$H = \frac{d^2}{dr^2} + \left[-2\omega_r r + \frac{2(l+1)}{r}\right] \frac{d}{dr} - (2l+3)\omega_r. \tag{6.6.14}$$

Applying the Stäckel transform, we get

$$\mathcal{H}'y = Zy,$$

$$\mathcal{H}' = r(H - \mathcal{E}) = r\frac{d^2}{dr^2} + \left[-2\omega_r r^2 + 2(l+1)\right]\frac{d}{dr} + [\epsilon - (2l+3)\omega_r]r. \tag{6.6.15}$$

$\mathcal{H}'$ allows for an $sl(2)$ algebraization if

$$\epsilon - (2l+3)\omega_r \equiv c_1 = -n[(n-1)a_3 + b_2] \equiv 2\omega_r n \qquad (6.6.16)$$

which gives the energies obtained in [25]:

$$\epsilon = (2n + 2l + 3)\omega_r, \quad n = 0, 1, 2, \cdots. \qquad (6.6.17)$$

Indeed, for such ϵ values, $\mathcal{H}'$ is dependent on integer parameter n and can be expressed in terms of the $sl(2)$ generators as

$$\mathcal{H}' = J^0 J^- - 2\omega_r J^+ + \left[\frac{n}{2} + 2(l+1)\right] J^-. \qquad (6.6.18)$$

So for fixed n (i.e. fixed energy), there are $(n+1)$ solutions to model parameter Z corresponding to $(n+1)$ eigenfunctions.

We remark that the hidden $sl(2)$ symmetry of this model was first noted in [26] (see also [27]).

6.6.2 2D Hydrogen in Uniform Magnetic Field (均匀磁场中的二维氢原子)

The hamiltonian of 2D hydrogen in a uniform maganetic field is

$$H_0 = \frac{1}{2}\left(\boldsymbol{p} + \frac{1}{c}\boldsymbol{A}\right)^2 + \frac{Z}{r}, \qquad (6.6.19)$$

where c is the velocity of light and the vector potential in the symmetric gauge is given by $\boldsymbol{A} = \frac{1}{2}\boldsymbol{B} \times \boldsymbol{r}$. The magnetic field $\boldsymbol{B}$ is perpendicular to the plane in which the electron is located. In polar coordinates (r, θ) within the plane, the angular and radial part of the wavefunction $\phi(\boldsymbol{r})$ are decoupled through the Ansatz

$$\phi(\boldsymbol{r}) = \frac{e^{im\theta}}{\sqrt{2\pi}} \frac{u(r)}{\sqrt{r}}, \quad m = 0, \pm 1, \pm 2, \cdots \qquad (6.6.20)$$

The radial wavefunction $u(r)$ satisfies the radial Schrödinger equation

$$\left(-\frac{d^2}{dr^2} + \frac{m^2 - 1/4}{r^2} + \omega_L^2 r^2 + \frac{Z}{r}\right)u(r) = 2(E - m\omega_L)u(r), \qquad (6.6.21)$$

where $\omega_L = \frac{1}{2}\omega_c = B/2c$ is the Larmor frequency.

Setting

$$u(r) = r^{|m|+1/2} e^{-\frac{\omega_L}{2}r^2} y(r), \qquad (6.6.22)$$

we can write the radial Schrödinger equation in the form

$$\mathcal{H}y \equiv \left(H - \frac{\alpha}{r}\right)y = \mathcal{E}y \tag{6.6.23}$$

where $\alpha = Z$, $\mathcal{E} = -\epsilon\omega_L$ with $\epsilon = \dfrac{2E}{\omega_L} - 2m$ and

$$H = \frac{\mathrm{d}^2}{\mathrm{d}r^2} + \left(-2\omega_L r + \frac{2|m|+1}{r}\right)\frac{\mathrm{d}}{\mathrm{d}r} - 2(|m|+1)\omega_L. \tag{6.6.24}$$

Applying the Stäckel transform, we have

$$\mathcal{H}'y = Zy,$$
$$\mathcal{H}' = r(H - \mathcal{E}) = r\frac{\mathrm{d}^2}{\mathrm{d}r^2} + (-2\omega_L r^2 + 2|m|+1)\frac{\mathrm{d}}{\mathrm{d}r} + [\epsilon - 2(|m|+1)]\omega_L r. \tag{6.6.25}$$

$\mathcal{H}'$ allows for an $sl(2)$ algebraization if

$$[\epsilon - 2(|m|+1)]\omega_L \equiv c_1 = -n[(n-1)a_3 + b_2] \equiv 2\omega_L n \tag{6.6.26}$$

which gives the result in [25]:

$$\epsilon = 2(n + |m| + 1), \quad n = 0, 1, 2, \cdots. \tag{6.6.27}$$

Indeed, for such ϵ values, $\mathcal{H}'$ is dependent on integer parameter n and can be expressed in terms of the $sl(2)$ generators as

$$\mathcal{H}' = J^0 J^- - 2\omega_L J^+ + \left(\frac{n}{2} + 2|m| + 1\right)J^-. \tag{6.6.28}$$

6.6.3 Hooke Atom: Two Planar Charged Particles in Uniform Magnetic Field (Hooke 原子: 均匀磁场中的两个平面带电粒子)

Consider a system of two planar charged particles in a uniform magnetic field interacting through the combined Coulomb and harmonic potentials. The hamiltonian of the system is given by

$$H_0 = \sum_{i=1}^{2}\left[\frac{1}{2}\left(\boldsymbol{p}_i + \frac{1}{c}\boldsymbol{A}(\boldsymbol{r}_i)\right)^2 + \frac{1}{2}\omega_0^2 r_i^2\right] + \frac{Z}{\|\boldsymbol{r}_1 - \boldsymbol{r}_2\|} \tag{6.6.29}$$

where c is the speed of light and $\boldsymbol{A}(\boldsymbol{r}_i) = \dfrac{1}{2}\boldsymbol{B} \times \boldsymbol{r}_i$. Introduce relative and center of mass coordinates $r = r_1 - r_2$ and $R = \frac{1}{2}(r_1 + r_2)$, respectively, then the hamiltonian becomes

$$H_0 = 2\left[\frac{1}{2}\left(p + \frac{1}{c}A_r\right)^2 + \frac{1}{2}\omega_r^2 r^2 + \frac{Z}{r}\right] + \frac{1}{2}\left[\frac{1}{2}\left(P + \frac{1}{c}A_R\right)^2 + \frac{1}{2}\omega_R^2 R^2\right]$$
$$\equiv H_r + H_R, \tag{6.6.30}$$

where $\omega_r = \frac{1}{2}\omega_0$, $\omega_R = 2\omega_0$ and

$$\begin{cases} p = -i\nabla_r = \frac{1}{2}(p_2 - p_1), \quad P = -i\nabla_R = p_2 + p_1, \\ A_r = \frac{1}{2}A(r) = \frac{1}{2}[A(r_2) - A(r_1)], \quad A_R = 2A(R) = [A(r_2) + A(r_1)]. \end{cases} \tag{6.6.31}$$

The total wavefunction factorizes

$$\psi(1,2) = \xi(R)\phi(r) \tag{6.6.32}$$

and the Schrödinger equation $H_0\psi = E\psi$ separates into

$$H_r\phi(r) = \epsilon\phi(r), \quad H_R\xi(R) = \eta\xi(R), \tag{6.6.33}$$

with $E = \epsilon + \eta$ and the following Ansatz for the relative motion:

$$\phi(r) = \frac{e^{im\theta}}{\sqrt{2\pi}}\frac{u(r)}{\sqrt{r}}, \quad m = 0, \pm 1, \pm 2, \cdots \tag{6.6.34}$$

The radial wavefunction $u(r)$ satisfies the radial Schrödinger equation

$$\left(-\frac{d^2}{dr^2} + \frac{m^2 - 1/4}{r^2} + \tilde{\omega}_r^2 r^2 + \frac{Z}{r}\right)u(r) = (\epsilon - m\omega_L)u(r), \tag{6.6.35}$$

where $\omega_L = B/2c$ and $\tilde{\omega}_r = \frac{1}{2}\sqrt{\omega_L^2 + \omega_0^2}$ is the effective frequency.

The remaining analysis is quite similar to that in the last section for the 2D hydrogen in a maganetic field. Setting

$$u(r) = r^{|m|+1/2}e^{-\frac{\tilde{\omega}_r}{2}r^2}y(r), \tag{6.6.36}$$

we can write the radial Schrödinger equation in the form

$$\mathcal{H}y \equiv \left(H - \frac{\alpha}{r}\right)y = \mathcal{E}y \tag{6.6.37}$$

where $\alpha = Z$, $\mathcal{E} = m\omega_L - \epsilon$ and

$$\mathcal{H} = \frac{d^2}{dr^2} + \left(-2\tilde{\omega}_r r + \frac{2|m|+1}{r}\right)\frac{d}{dr} - 2(|m|+1)\tilde{\omega}_r. \tag{6.6.38}$$

6.6 Stäckel Transform and Coupling Constant Metamorphosis

Applying the Stäckel transform, we obtain

$$\mathcal{H}'y = Zy,$$
$$\mathcal{H}' = r(H - \mathcal{E})$$
$$= r\frac{d^2}{dr^2} + (-2\omega_L r^2 + 2|m| + 1)\frac{d}{dr} + [\epsilon - m\omega_L - 2(|m|+1)\tilde{\omega}_r]r. \quad (6.6.39)$$

$\mathcal{H}'$ allows for an $sl(2)$ algebraization if

$$\epsilon - m\omega_L - 2(|m|+1)\tilde{\omega}_r \equiv c_1 = -n[(n-1)a_3 + b_2] \equiv 2\tilde{\omega}_r n \quad (6.6.40)$$

which gives

$$\epsilon = m\omega_L + 2(n+|m|+1)\tilde{\omega}_r, \quad n = 0, 1, 2, \cdots \quad (6.6.41)$$

Indeed, for such ϵ values, $\mathcal{H}'$ is dependent on integer parameter n and can be expressed in terms of the $sl(2)$ generators as

$$\mathcal{H}' = J^0 J^- - 2\tilde{\omega}_r J^+ + \left(\frac{n}{2} + 2|m| + 1\right) J^-. \quad (6.6.42)$$

6.6.4 Two Coulombically Repelling Electrons on a Sphere (球面上的两个具有库仑排斥势的电子)

Consider a system of two electrons, interacting via a Coulomb potential, but constrained to remain on the surface of a D-dimensional sphere of radius R. The Hamiltonian of the system (in atomic units) is[28]

$$H_0 = -\frac{1}{2}\left(\nabla_1^2 + \nabla_2^2\right) - \frac{1}{u}, \quad (6.6.43)$$

where $u = |\mathbf{r}_1 - \mathbf{r}_2|$ is the inter-electronic distance. The Schrödinger wave function of the system can be separated as a product of spin, angular and inter-electron wave functions. The inter-electron wave function $\Psi(u)$ satisfies the ODE[28]

$$\left(\frac{u^2}{4R^2} - 1\right)\frac{d^2\Psi}{du^2} + \left(\frac{\delta u}{4R^2} - \frac{1}{\gamma u}\right)\frac{d\Psi}{du} + \frac{\Psi}{u} = E\Psi, \quad (6.6.44)$$

where δ and γ are parameters related to the dimension D of the sphere. Introduce dimensionless variable $z = \dfrac{u}{2R}$. Then the above ODE can be written as

$$\mathcal{H}\Psi = \left(H - \frac{\alpha}{z}\right)\Psi = \mathcal{E}\Psi \quad (6.6.45)$$

where $\alpha = -2R$, $\mathcal{E} = 4R^2 E$ and

$$H = (z^2 - 1)\frac{d^2}{dz^2} + \left(\delta z - \frac{1/\gamma}{z}\right)\frac{d}{dz} \quad (6.6.46)$$

Applying the Stäckel transform, we get

$$\mathcal{H}'\Psi = -2R\Psi,$$
$$\mathcal{H}' = z(H - \mathcal{E}) = z(z^2 - 1)\frac{d^2}{dz^2} + \left(\delta z^2 - \frac{1}{\gamma}\right)\frac{d}{dz} - 4R^2 E z. \quad (6.6.47)$$

$\mathcal{H}'$ allows for an $sl(2)$ algebraization if

$$-4R^2 E \equiv c_1 = -n[(n-1)a_3 + b_2] \equiv -n[n - 1 + \delta] \quad (6.6.48)$$

which gives the exact energies obtained in [28]:

$$E = \frac{1}{4R^2} n(n - 1 + \delta), \quad n = 0, 1, 2, \cdots. \quad (6.6.49)$$

Indeed, for such E values, $\mathcal{H}'$ is dependent on integer parameter n and can be expressed in terms of the $sl(2)$ generators (6.3.3) as

$$\mathcal{H}' = -J^+ J^0 - J^0 J^- - \left(\frac{3n-2}{2} + \delta\right) J^+ - \left(\frac{1}{\gamma} + \frac{n}{2}\right) J^-. \quad (6.6.50)$$

This provides an $sl(2)$ algebraization of the two electron system.

6.6.5 Inverse Sextic Power Potential (逆六次势)

Pais and Wu[29] studied the problem of scattering by the singular potential $d/r^{2+2n} + e/r^{2+n}$ ($n > 1$) in non-relativistic quantum mechanics. In this section we will consider the $n = 2$ case, i.e. the inverse sextic power potential

$$V(r) = \frac{e}{r^4} + \frac{d}{r^6}, \quad d > 0. \quad (6.6.51)$$

This potential has been used in atomic, molecular and nuclear physics. The corresponding radial Schrödinger equation is

$$\left[-\frac{d^2}{dr^2} + \frac{\ell(\ell+1)}{r^2} + \omega^2 r^2 + \frac{2e}{r^4} + \frac{2d}{r^6}\right]\Psi(r) = 2E\Psi(r). \quad (6.6.52)$$

We extract the appropriate asymptotic behaviour of the wave function $\Psi(r)$ by making the substitution[30]

6.6 Stäckel Transform and Coupling Constant Metamorphosis

$$\Psi(r) = r^{3/2+e/\sqrt{2d}} \exp\left[-\frac{\omega}{2}r^2 - \frac{\sqrt{2d}}{2}\frac{1}{r^2}\right] v(r), \quad 3/2 + e/\sqrt{2d} > 0. \tag{6.6.53}$$

We then obtain the differential equation for $v(r)$,

$$v''(r) + \frac{2}{r}\left(-\omega r^2 + \frac{3}{2} + \frac{e}{\sqrt{2d}} + \frac{\sqrt{2d}}{r^2}\right) v'(r) + 2\left[E - \omega\left(2 + \frac{e}{\sqrt{2d}}\right)\right] v(r)$$

$$= \frac{1}{r^2}\left[2\omega\sqrt{2d} + \left(\ell + \frac{1}{2}\right)^2 - \left(\frac{e}{\sqrt{2d}} + 1\right)^2\right] v(r). \tag{6.6.54}$$

A change of variable $z = r^2$ transforms (6.6.54) into the form,

$$\mathcal{H}v \equiv \left(H - \frac{\alpha}{z}\right) v = \mathcal{E}v, \tag{6.6.55}$$

where $\mathcal{E} = \frac{1}{2}\left[\omega\left(2 + \frac{e}{\sqrt{2d}}\right) - E\right]$ and

$$\alpha = \frac{1}{4}\left[2\omega\sqrt{2d} + \left(\ell + \frac{1}{2}\right)^2 - \left(\frac{e}{\sqrt{2d}} + 1\right)^2\right],$$

$$H = z\frac{d^2}{dz^2} + \left(-\omega z + 2 + \frac{e}{\sqrt{2d}} + \frac{\sqrt{2d}}{z}\right). \tag{6.6.56}$$

Applying the Stäckel transform, we have

$$\mathcal{H}'v = \alpha v,$$
$$\mathcal{H}' = z(H - \mathcal{E})$$
$$= z^2\frac{d^2}{dz^2} + \left[-\omega z^2 + \left(2 + \frac{e}{\sqrt{2d}}\right)z + \sqrt{2d}\right]\frac{d}{dz}$$
$$+ \frac{z}{2}\left[E - \omega\left(2 + \frac{e}{\sqrt{2d}}\right)\right]. \tag{6.6.57}$$

The operator $\mathcal{H}'$ allows for an $sl(2)$ algebraization if

$$\frac{1}{2}\left[E - \omega\left(2 + \frac{e}{\sqrt{2d}}\right)\right] \equiv c_1 = -n[(n-1)a_3 + b_2] \equiv \omega n \tag{6.6.58}$$

which gives the exact energies obtained in [30]:

$$E = \omega\left(2n + 2 + \frac{e}{\sqrt{2d}}\right), \quad n = 0, 1, 2, \cdots. \tag{6.6.59}$$

Indeed for such E values, $\mathcal{H}'$ depends on integer n and can be expressed in terms of the $sl(2)$ generators (6.3.3) as

$$\mathcal{H}' = J^0 J^0 + \omega J^+ + \left(n + 1 + \frac{e}{\sqrt{2d}}\right) J^0 + \sqrt{2d} J^- + \frac{n}{2}\left(\frac{n}{2} + 1 + \frac{e}{\sqrt{2d}}\right). \quad (6.6.60)$$

This showa that the inverse sextic potential is $sl(2)$ Lie-algebraic.

6.7 Solutions to QES: Bender-Dunne Polynomials (精确解: Bender-Dunne 多项式)

Consider a quantum mechanical system with an infinite but discrete spectrum E_n, $n = 0, 1, 2, \cdots$. Let $\mathcal{H}$ denote its hamiltonian. By means of the Lanczos algorithm [31] a self-adjoint hamiltonian can always be brought into a tridiagonal (or Jacobi) matrix form in suitable basis in the Hilbert space $\mathcal{V}$. Denote by $\{\xi_n\}$, $n = 0, 1, 2, \cdots$. Then tridiagonality of $\mathcal{H}$ implies that

$$\mathcal{H}\xi_n = A_n \xi_{n-1} + B_n \xi_n + C_n \xi_{n+1}, \quad n = 0, 1, 2, \cdots, \quad (6.7.1)$$

where A_n, B_n, C_n are algebraically computable coefficients.

If for certain non-negative integer $\mathcal{N}$, C_n satisfy

$$C_\mathcal{N} = 0, \quad C_n \neq 0 \text{ for } n \neq \mathcal{N}, \quad (6.7.2)$$

then the quantum system is QES. Indeed in this case, the basis vectors $\{\xi_0, \xi_1, \cdots, \xi_\mathcal{N}\}$ form a $(\mathcal{N}+1)$-dimensional invariant subspace $\mathcal{V}_\mathcal{N}$: $H\mathcal{V}_\mathcal{N} \subset \mathcal{V}$. This allows one to solve the Schrödinger equation $H\psi = E\psi$ exactly in $\mathcal{V}_\mathcal{N}$ and obtain $\mathcal{N}+1$ eigenvalues $E_0, E_1, \cdots, E_\mathcal{N}$ in closed form.

Expand the wavefunction ψ in terms of the basis $\{\xi_n\}$,

$$\psi(E) = \sum_{n=0}^{\infty} \xi_n P_n(E). \quad (6.7.3)$$

Substituting into the Schrödinger equation and using the tridiagonality condition (6.7.1), one derives the 3-term recurrence relation

$$EP_n(E) = A_{n+1} P_{n+1}(E) + B_n P_n(E) + C_{n-1} P_{n-1}(E), \quad (6.7.4)$$

subject to the initial conditions $P_{-1}(E) = 0$ and $P_0(E) = 1$. From the general theorem for orthogonal polynomials[32], it follows that $P_n(E)$ are orthogonal polynomials of degree n in energy parameter E and $\psi(E)$ is the generating function for the

6.7 Solutions to QES: Bender-Dunne Polynomials

polynomials. The polynomials $\{P_n(E)\}$ are orthogonal in the sense that there is a non-negative weight function $w(E)$ such that

$$\int_{-\infty}^{\infty} P_n(E) P_m(E) w(E) \mathrm{d}E = \delta_{nm}. \qquad (6.7.5)$$

The expression for the commpleteness is

$$w(E) \sum_{n=0}^{\infty} P_n(E) P_n(E') = \delta(E - E'). \qquad (6.7.6)$$

If the system is QES, then $C_{\mathcal{N}} = 0$. This means that if

$$P_n(E) = 0 \quad \text{for all } n \geqslant \mathcal{N} + 1, \qquad (6.7.7)$$

then the infinite power series truncates and becomes a polynomial of degree $\mathcal{N}$. Thus H has $\mathcal{N} + 1$ exactly calculable eigenvalues $E_0, E_1, \cdots, E_{\mathcal{N}}$ given by the roots of

$$P_{\mathcal{N}+1}(E_l) = 0, \quad l = 0, 1, 2, \cdots, \mathcal{N}. \qquad (6.7.8)$$

In other words,

$$P_{\mathcal{N}+1}(E) \sim \prod_{l=0}^{\mathcal{N}}(E - E_l). \qquad (6.7.9)$$

In this case one has the factorization property[33, 34]

$$P_{\mathcal{N}+1+n}(E) = P_{\mathcal{N}+1}(E) Q_n(E) \qquad (6.7.10)$$

where $Q_n(E)$ are polynomials of degree $n = 0, 1, 2, \cdots$.

As an example, let us recall the Bender-Dunne polynomial arising from the QES model with sextic potential. The hamiltonian of the model is given by (6.5.1). One seeks solution of the Schrödinger equation $H_0 \psi = E \psi$ of the form[33],

$$\psi(x) = x^{2s-1/2} e^{-\frac{1}{4}x^4} \sum_{n=0}^{\infty} \left(-\frac{1}{4}\right)^n \frac{P_n(E)}{n!\Gamma(n+2s)} x^{2n}. \qquad (6.7.11)$$

Substituting into the Schrödinger equation gives the 3-term recurrence relation

$$E P_n(E) = P_{n+1}(E) + C_{n-1} P_{n-1}(E), \quad C_{n-1} = -16n(n-J)(n+2s-1), \qquad (6.7.12)$$

supplemented by the initial conditions $P_{-1} = 0$ and $P_0 = 1$. The 3-term recurrence relation generates a set of orthogonal polynomials, the so-called the Bender-Dunnee polynomials:

$$P_1(E) = E,$$
$$P_2(E) = E^2 + (32 - 32J)s,$$
$$P_3(E) = E^3 + [(160 - 96J)s - 32J + 64]E,$$
$$\vdots$$

for positive integer J.

Obviously we have $C_\mathcal{N} \equiv C_{J-1} = 0$. So from the general theory above $P_{\mathcal{N}+1} \equiv P_J$ is a common factor of $P_n(E)$ for $n \geqslant \mathcal{N}+1 \equiv J$ and $P_{\mathcal{N}+1}(E) \equiv P_J(E) = 0$ gives the $\mathcal{N}+1 \equiv J$ exact energy eigenvalues of the QES hamiltonian. For instance, for $J = 3$ we have

$$P_0(E) = 1, \quad P_1(E) = E,$$
$$P_2(E) = E^2 - 64s,$$
$$P_3(E) = E^3 - (128s + 32)E,$$
$$P_4(E) = E[E^3 - (128s + 32)E] = EP_3(E),$$
$$\vdots$$

So $P_3(E)$ is a common factor of $P_n(E)$ for $n \geqslant 3$. The zeros of $P_3(E)$ are

$$E = 0, \quad \pm\sqrt{128s + 32} \tag{6.7.13}$$

which are the 3 exact energy eigenvalues. The corresponding exact eigenfunctions, obtained by evaluating $\psi(x)$ (6.7.11) at these values of E, are given by [33]:

$$\psi_0(x) = e^{-\frac{1}{4}x^4} \frac{x^{2s-1/2}}{\Gamma(2s)} \left(1 - \frac{x^4}{2s+1}\right),$$
$$\psi_+(x) = e^{-\frac{1}{4}x^4} \frac{x^{2s-1/2}}{\Gamma(2s)} \left(1 - \frac{\sqrt{128s+32}}{8s}x^2 + \frac{x^4}{2s}\right),$$
$$\psi_-(x) = e^{-\frac{1}{4}x^4} \frac{x^{2s-1/2}}{\Gamma(2s)} \left(1 + \frac{\sqrt{128s+32}}{8s}x^2 + \frac{x^4}{2s}\right). \tag{6.7.14}$$

6.8 3-term Recurrence Relation and Continued Fractions (3 项递归关系和连分式)

In section 6.7 we have seen that when the coefficients $C_n = 0$ for some $n = \mathcal{N} > 0$ then $\mathcal{H}$ has necessarily a finite-dimensional invariant subspace. So this corresponds to

6.8 3-term Recurrence Relation and Continued Fractions

QES case. However, there is a different scenario in which $C_n \neq 0$ for all $n \geq 0$. Then $\mathcal{H}$ has no finite dimensional invariant subspace and the infinite power series (6.7.3) would not truncate. In this case solutions to the 3-term recurrence relation is related to infinite continued fractions.

Given a 3-term recurrence relation (2nd-order difference equation),

$$y_{n+1} + a_n y_n + b_n y_{n-1} = 0, \quad n = 1, 2, 3, \cdots, \quad (6.8.1)$$

one may formally arrive at a continued fraction as follows. Introduces the ratios $r_n = y_{n+1}/y_n$, $n = 0, 1, 2, \cdots$. In terms of r_n, (6.8.1) becomes

$$r_{n-1} = -\frac{b_n}{a_n + r_n} \quad (6.8.2)$$

Apply this formula repeatedly, one obtains that the ratios of consecutive values of some solution y_n is related to the infinite continued fractions,

$$r_{n-1} = \frac{y_n}{y_{n-1}} = -\frac{b_n}{a_n-} \frac{b_{n+1}}{a_{n+1}-} \frac{b_{n+2}}{a_{n+2}-} \cdots \quad (6.8.3)$$

However, this derivation neither guarantees the convergence of the continued fractions, nor does it tells anything about what solution the ratios are to be formed. These issues are clarified by the Pincherle Theorem below.

Like 2nd-order differential equations, the 3-term recurrence relation (6.8.1) possess two independent solution sequences, denoted as $\{R_n, S_n \,|\, n = 1, 2, \cdots\}$. Then we have [35]:

Definition 5. Solution sequence $\{R_n \,|\, n = 1, 2, \cdots\}$ is referred to as minimal if there exists linearly independent sequence $\{S_n \,|\, n = 1, 2, \cdots\}$ of the same recurrence relation such that

$$\lim_{n \to \infty} R_n/S_n = 0.$$

Any non-minimal solution sequence $\{S_n \,|\, n = 1, 2, \cdots\}$ is referred to as dominant.

Note that dominant sequences are not unique, as any multiple of the minimal solution may be added to them without destroying their dominant property. We now state the Pincherle Theorem without proof.

Theorem 1. The continued fraction converges iff the recurrence relation possesses a minimal solution f_n, with $f_0 \neq 0$. In case of convergence, moreover, one has

$$\frac{f_n}{f_{n-1}} = -\frac{b_n}{a_n-} \frac{b_{n+1}}{a_{n+1}-} \frac{b_{n+2}}{a_{n+2}-} \cdots, \quad n = 1, 2, 3, \cdots,$$

provided $f_n \neq 0$ for $n = 0, 1, 2, \cdots$.

In application of the Pincherle Theorem, it is in general easier to recognize a given solution of a 3-term recurrence relation to be minimal than to establish convergence of the corresponding continued fraction. The former is helped by results from the asymptotic theory of linear 2nd-order difference equation (i.e. 3-term recurrence relation) as follows.

Assume that the coefficients a_n and b_n in the 3-term recurrence relation have the finite limits as $n \to \infty$: $a_n \to a$, $b_n \to b$, not excluding that $b = 0$. Then call

$$\Phi(t) = t^2 + at + b \tag{6.8.4}$$

the characteristic polynomial of the 3-term recurrence relation. We have the Poincare Theorem[35]:

Theorem 2. If the characteristic polynomial has roots t_1, t_2 of distinct moduli, $|t_1| > |t_2|$, then for every non-trivial solution y_n of the 3-term recurrence relation, one has

$$\lim_{n \to \infty} \frac{y_{n+1}}{y_n} = t_r, \quad r = 1 \text{ or } 2.$$

and the Perron Theorem:

Theorem 3. Under the assumption of the Poincare theorem, there exist two linearly independent solutions $y_{n,1}$ and $y_{n,2}$ of the 3-term recurrence relation such that

$$\lim_{n \to \infty} \frac{y_{n+1,r}}{y_{n,r}} = t_r, \quad r = 1 \text{ or } 2.$$

That is $f_n = y_{n,2}$ is a minimal solution.

One need to generalize the above results if the coefficients a_n, b_n satisfy

$$a_n \sim a\, n^\alpha, \quad b_n \sim b\, n^\beta, \quad ab \neq 0, \quad \alpha, \beta \text{ real}; \quad n \to \infty.$$

In this case, the asymptotic structure of the solutions depends on the so-called Newton-Puiseux diagram formed with the points $P_0(0,0), P_1(1,\alpha), P_2(2,\beta)$. Denote by σ the slope of $\overline{P_0 P_1}$, and by τ the slope of $\overline{P_1 P_2}$, so that $\sigma = \alpha, \tau = \beta - \alpha$. Then we have the following Perron-Kreuser Theorem[35]:

Theorem 4. (a) If $\sigma > \tau$, the difference equation has two linearly independent solutions $y_{n,1}$ and $y_{n,2}$, for which

$$\frac{y_{n+1,1}}{y_{n,1}} \sim -a\, n^\sigma, \quad \frac{y_{n+1,2}}{y_{n,2}} \sim -\frac{b}{a} n^\tau, \quad n \to \infty.$$

(b) If $\sigma = \tau = \alpha$, let t_1, t_2 be the roots of the characteristic equation $t^2 + at + b = 0$, and $|t_1| \geq |t_2|$. The difference equation has two solutions $y_{n,1}, y_{n,2}$ such that

$$\frac{y_{n+1,1}}{y_{n,1}} \sim t_1 n^\sigma, \qquad \frac{y_{n+1,2}}{y_{n,2}} \sim t_2 n^\tau, \qquad n \to \infty,$$

provided $|t_1| > |t_2|$. If $|t_1| = |t_2|$ then

$$\limsup_{n \to \infty} \left[\frac{|y_n|}{(n!)^\alpha} \right]^{1/n} = |t_1|$$

for all nontrivial solutions of the difference equation.

(c) If the point P_1 lies below the line segment $\overline{P_0 P_1}$ then

$$\limsup_{n \to \infty} \left[\frac{|y_n|}{(n!)^{\beta/2}} \right]^{1/n} = \sqrt{|b|}$$

for all nontrivial solutions of the difference equation.

Remark 1. In both case (a) and the first part of case (b) the solution $f_n = y_{n,2}$ is a minimal solution of the recurrence relation. The second part of (b) and part (c) of the theorem are somewhat inconclusive as they do not permit distinguishing two solutions with distinct asymptotic properties - need more analysis for our purposes.

Remark 2. Here we recall the definition of lim sup. Given a real sequence $\{u_\ell\}_{\ell=0}^\infty$, define $v_n = \sup\{u_\ell : \ell \geq n\}$. Then $\lim_{n \to \infty} \sup u_n = \lim_{n \to \infty} v_n$. The meaning of lim sup is as follows. Let $\{u_n\}_{n \geq 0}$ be a real sequence, Then $\lim_{n \to \infty} \sup u_n = u$ iff given $\epsilon > 0$, there exists $N(\epsilon) \in \mathbf{N}$, such that (i) $\forall n \geq N(\epsilon)$, $u_n < u + \epsilon$, and (ii) there exists an infinite set I of $\ell > N(\epsilon)$ such that $u_\ell > u - \epsilon$ for every $\ell \in I$.

6.8.1 Bargmann-Hilbert Spaces (Bargmann-Hilbert 空间)

Consider the algebra generated by $2n$ creation and annihilation operators $a_i, a_i^\dagger$ satisfying

$$[a_k, a_l^\dagger] = \delta_{kl}, \quad [a_k, a_l] = 0 = [a_k^\dagger, a_l^\dagger]. \tag{6.8.5}$$

This algebra has a representation on the symmetry algebra, i.e. the polynomial algebra $C(z_1, \cdots, z_n)$: $a_i^\dagger \to z_i$, $a_i \to \dfrac{\partial}{\partial z_i}$ satisfying the same commutation relations as above. This is called the Bargmann representation or correspondence.

We would like a representation on a Hilbert space and want $a_i, a_i^\dagger$ to be adjoint with respect to the inner product on the Hilber space. As shown by Bargmann[36], the Hilbert space can be chosen to be the following function space:

Definition 6. Given an identification $\mathbf{R}^{2n} = \mathbf{C}^n$, Fock space is the space of entire analytic functions on $\mathcal{C}^n$, with finite norm using the inner product

$$(f(z), g(z)) = \frac{1}{\pi^n} \int_{\mathcal{C}^n} \overline{f(z)} g(z) \mathrm{e}^{-|z|^2} \mathrm{d}^n z. \tag{6.8.6}$$

Such Hilbert spaces of entire functions are called Bargmann-Hilbert (BH) spaces. An orthonormal basis for the BH space is given by appropriately normalized monomials.

Let f be a function with power series

$$f(z) = \sum_{m_i} \alpha_{m_1 m_2 \cdots m_n} z_1^{m_1} z_2^{m_2} \cdots z_n^{m_n} \equiv \sum_m \alpha_{[m]} z^{[m]}. \tag{6.8.7}$$

By using polar coordinates $z_k = r_k \mathrm{e}^{\mathrm{i}\theta_k}$, one has[36]

$$(f,f) = \frac{1}{\pi^n} \int |f(z)|^2 \mathrm{e}^{-|z|^2} \mathrm{d}^n z = \sum_{m,m'} \bar{\alpha}_{[m]} \alpha_{[m']} \Theta_{mm'} \tag{6.8.8}$$

where

$$\Theta_{mm'} = \prod_k \left[\frac{1}{\pi} \int_0^{2\pi} \mathrm{e}^{\mathrm{i}\theta(m'_k - m_k)} \mathrm{d}\theta \int_0^\infty \mathrm{e}^{-r^2} r^{m_k + m'_k + 1} \mathrm{d}r \right]$$

$$= \prod_k \left[2\delta_{m_k m'_k} \int_0^\infty r^{2m_k+1} \mathrm{e}^{-r^2} \mathrm{d}r \right] = \prod_k \left(\delta_{m_k m'_k} m_k! \right). \tag{6.8.9}$$

So

$$(f,f) = \sum_{m_k} |\alpha_{m_1 m_2 \cdots m_n}|^2 m_1! m_2! \cdots m_n!. \tag{6.8.10}$$

Thus every set of coefficients for which the sum on the right hand side of the above quation converges defines an entire function in the Hilbert space! The simplest orthonormal set of basis vectors in the Hilbert space is given by $\{z^{[m]}/\sqrt{[m!]}\} = \left\{ \prod_k z_k^{m_k}/\sqrt{m_k!} \right\}$.

6.8.2 2-photon Quantum Rabi Model (两光子量子 Rabi 模型)

The Hamiltonian of the 2-photon Rabi model reads[21]

$$H_{2p} = \omega b^\dagger b + \Delta \sigma_z + g \sigma_x \left[(b^\dagger)^2 + b^2 \right]. \tag{6.8.11}$$

Let us make a canonical Bogoliubov transformation from $b, b^\dagger$ to the squeezed bosons $a, a^\dagger$[21],

$$b = \frac{a + \tau a^\dagger}{\sqrt{1 - \tau^2}}, \quad b^\dagger = \frac{\tau a + a^\dagger}{\sqrt{1 - \tau^2}}, \tag{6.8.12}$$

6.8 3-term Recurrence Relation and Continued Fractions

where $|\tau| < 1$ is a real parameter. In terms of the squeezed bosons, the Hamiltonian (6.8.11) takes the form

$$\tilde{H}_{2p} = \Delta \sigma_z + \frac{1}{1-\tau^2} \{[\omega\tau + g\sigma_x(1+\tau^2)][(a^\dagger)^2 + a^2]$$
$$+ [\omega(1+\tau^2) + 4g\tau\sigma_x]a^\dagger a + \omega\tau^2 + 2g\tau\sigma_x\}. \quad (6.8.13)$$

Introduce the operators $K_\pm, K_0$

$$K_+ = \frac{1}{2}(a^\dagger)^2, \quad K_- = \frac{1}{2}a^2, \quad K_0 = \frac{1}{2}\left(a^\dagger a + \frac{1}{2}\right). \quad (6.8.14)$$

Then (6.8.13) becomes

$$\tilde{H}_{2p} = \Delta \sigma_z + \frac{1}{1-\tau^2} \{2[\omega\tau + g\sigma_x(1+\tau^2)](K_+ + K_-)$$
$$+ 2[\omega(1+\tau^2) + 4g\tau\sigma_x]K_0\} - \frac{1}{2}\omega. \quad (6.8.15)$$

The operators $K_\pm, K_0$ form the $su(1,1)$ Lie algebra. Its quadratic Casimir, $C = K_+K_- - K_0(K_0-1)$, takes the particular values $C = \frac{3}{16}$ in the representation (6.8.14). This is the well-known infinite-dimensional unitary irreducible representation $\mathcal{D}^+(q)$ of $su(1,1)$ with $q = \frac{1}{4}, \frac{3}{4}$. Thus the Fock-Hilbert space decomposes into the direct sum of two subspaces $\mathcal{H}^q$ labeled by $q = 1/4, 3/4$.

Using the representation $a^\dagger = w$ and $a = \frac{d}{dw}$, it can be shown [22] that in the Bargmann space with basis vectors given by the monomials in $z = w^2$, $\{z^n/\sqrt{[2(n+q-1/k^2)]!}\}$, the operators $K_\pm, K_0$ (6.8.14) have the single-variable 2nd order differential realization

$$K_0 = z\frac{d}{dz} + q, \quad K_+ = \frac{z}{2}, \quad K_- = 2z\frac{d^2}{dz^2} + 4q\frac{d}{dz}. \quad (6.8.16)$$

The Bargmann space is the Hilbert space of entire functions on the complex plane if the inner product

$$(f, g) = \int \overline{f(w)}\, g(w)\, d\mu(w) \quad (6.8.17)$$

is finite for an appropriate measure $d\mu(w)$. Take the measure $d\mu(w) = \frac{1}{\pi}|w|^{4(q-1/4)} e^{-|w|^2}\, dx\, dy$[23]. Then,

$$(z^m, z^n) = (w^{2m}, w^{2n}) = [2(n+q-1/4)]!\,\delta_{mn}. \quad (6.8.18)$$

Thus the monomials $\{z^n/\sqrt{[2(n+q-1/k^2)]!}\}$ form an orthonormal basis of the BH space. Note in passing that the standard measure $\dfrac{1}{\pi}e^{-|w|^2}\,dx\,dy$ is no longer appropriate here. It is now not difficult to see that if $f(z) = \sum_{n=0}^{\infty} c_n z^n$ then

$$\|f\|^2 = \sum_{n=0}^{\infty} |c_n|^2\, [2(n+q-1/4)]! \tag{6.8.19}$$

and $f(z)$ is entire if the sum on the right hand side converges.

Using this differential realization, working in a representation defined by σ_x diagonal and choosing τ such that $\omega\tau + g(1+\tau^2) = 0$, i.e.①

$$\tau = -\frac{\omega}{2g}(1-\Omega), \quad \Omega = \sqrt{1 - \frac{4g^2}{\omega^2}}, \tag{6.8.20}$$

where $\left|\dfrac{2g}{\omega}\right| < 1$, then the transformed 2-photon Rabi Hamiltonian becomes the matrix differential operator

$$\tilde{H}_{2p} = \begin{pmatrix} 2\omega\Omega\left(z\dfrac{d}{dz}+q\right) - \dfrac{1}{2}\omega & \Delta \\ \Delta & -\dfrac{8g}{\Omega}z\dfrac{d^2}{dz^2} + \dfrac{2}{\Omega}[\omega(2-\Omega^2)z - 8gq]\dfrac{d}{dz} \\ & -\dfrac{2g}{\Omega}z + \dfrac{2\omega(2-\Omega^2)q}{\Omega} - \dfrac{1}{2}\omega \end{pmatrix}. \tag{6.8.21}$$

In terms of two-component wavefuntion $\psi(z) = (\psi_+(z), \psi_-(z))^{\mathrm{T}}$, the time-independent Schrödinger equation, $\tilde{H}_{2p}\psi(z) = E\psi(z)$, yields a system of coupled differential equations,

---

① The quadratic equation has two roots $\tau = -\dfrac{\omega}{2g}(1\mp\Omega)$. The root given in (6.8.20) is real and obeys $|\tau| < 1$, the requirement of τ from the Bogoliubov transformation, provided that $\left|\dfrac{2g}{\omega}\right| < 1$. This is seen as follows. Assume $|\tau| = \left|\dfrac{\omega}{2g}\right|\left[1 - \sqrt{1-\left(\dfrac{2g}{\omega}\right)^2}\right] \geqslant 1$, i.e. $1 - \left|\dfrac{2g}{\omega}\right| \geqslant \sqrt{1-\left(\dfrac{2g}{\omega}\right)^2}$. Then we would have $\sqrt{1-\left|\dfrac{2g}{\omega}\right|} \geqslant \sqrt{1+\left|\dfrac{2g}{\omega}\right|}$, which is impossible for the non-trivial case $g \neq 0$.

6.8 3-term Recurrence Relation and Continued Fractions

$$\left[2\omega\Omega\left(z\frac{d}{dz}+q\right)-\frac{1}{2}\omega-E\right]\psi_+ + \Delta\psi_- = 0, \qquad (6.8.22)$$

$$\left\{8gz\frac{d^2}{dz^2}+\left[-2\omega(2-\Omega^2)z+16gq\right]\frac{d}{dz}\right.$$
$$\left.+2gz-2\omega(2-\Omega^2)q+\left(\frac{1}{2}\omega+E\right)\Omega\right\}\psi_- - \Omega\Delta\psi_+ = 0. \qquad (6.8.23)$$

This is a system of differential equations of Fuchsian type. Solutions to these equations must be analytic in the whole complex plane if E belongs to the spectrum of $\tilde{H}_{2p}$. So we are seeking solutions of the form

$$\psi_+(z) = \sum_{n=0}^{\infty} \mathcal{K}_n^+(E)\, z^n, \quad \psi_-(z) = \sum_{n=0}^{\infty} \mathcal{K}_n^-(E)\, z^n, \qquad (6.8.24)$$

which converge in the entire complex plane, i.e. solutions which are entire.

Substituting (6.8.24) into (6.8.22), we obtain[23]

$$\mathcal{K}_n^+ = \frac{\Delta}{E+\frac{1}{2}\omega-(2n+2q)\omega\Omega}\mathcal{K}_n^-. \qquad (6.8.25)$$

So $\mathcal{K}_n^+$ is not analytic in E but has simple poles at

$$E = -\frac{1}{2}\omega + (2n+2q)\omega\Omega, \quad n = 0, 1, \cdots. \qquad (6.8.26)$$

The energies (6.8.26) appear for special values of model parameters[18, 21] and correspond to the exceptional solutions of the 2-photon Rabi model. If (6.8.26) is satisfied, the infinite series expansions (6.8.24) truncate and reduce to polynomials in z but only if the system parameters satisfy certain constraints. Majority part of the spectrum of the 2-photon Rabi model is regular for which (6.8.26) is not satisfied. The regular spectrum of the model is given by the zeros of the transcendental function $F(E)$ obtained below. Thus similar to the Rabi case, the spectrum of the 2-photon Rabi model consists of two parts, the regular and the exceptional spectrum.

From (6.8.23), we obtain the 3-step recurrence relation for $\mathcal{K}_n^-$ [23]:

$$\mathcal{K}_1^- + A_0 \mathcal{K}_0^- = 0,$$
$$\mathcal{K}_{n+1}^- + A_n \mathcal{K}_n^- + B_n \mathcal{K}_{n-1}^- = 0, \quad n \geqslant 1, \qquad (6.8.27)$$

where

$$\begin{cases} A_n = \dfrac{1}{8g(n+1)(n+2q)}\left[-(2n+2q)\omega(2-\Omega^2)\right.\\ \left.\qquad + \left(E+\dfrac{1}{2}\omega - \dfrac{\Delta^2}{E+\dfrac{1}{2}\omega-(2n+2q)\omega\Omega}\right)\Omega\right], \\ B_n = \dfrac{1}{4(n+1)(n+2q)}. \end{cases} \qquad (6.8.28)$$

The coefficients A_n, B_n have the behavior as $n \to \infty$,

$$A_n \sim a n^\alpha, \quad B_n \sim b n^\beta \qquad (6.8.29)$$

with

$$a = -\frac{\omega}{4g}(2-\Omega^2), \quad \alpha = -1, \quad b = \frac{1}{4}, \quad \beta = -2. \qquad (6.8.30)$$

Thus the asymptotic structure of solutions to the 2nd equation of (6.8.27) depends on the Newton-Puiseux diagram formed with the points $P_0(0,0), P_1(1,-1)$, $P_2(2,-2)$ [35]. Let γ be the slope of $\overline{P_0 P_1}$ and δ the slope of $\overline{P_1 P_2}$ so that $\gamma = \alpha$ and $\delta = \beta - \alpha$. Then we have $\gamma = \delta = \alpha$. The characteristic equation of the $n \geqslant 1$ part of (6.8.27) reads $t^2 + at + b = 0$ with a, b given in (6.8.30). It has two roots $t_1 = \dfrac{\omega}{4g}, t_2 = \dfrac{g}{\omega}$. Remembering the condition $\left|\dfrac{2g}{\omega}\right| < 1$, we have $|t_2| < |t_1|$. Applying the Perron-Kreuser theorem, we conclude that the two linearly independent solutions $\mathcal{K}^-_{n,1}$ and $\mathcal{K}^-_{n,2}$ of the $n \geqslant 1$ part (i.e. the truly 3-term part) of (6.8.27) satisfy

$$\lim_{n\to\infty} \frac{\mathcal{K}^-_{n+1,r}}{\mathcal{K}^-_{n,r}} \sim t_r\, n^{-1}, \quad r = 1, 2. \qquad (6.8.31)$$

So $\mathcal{K}^-_{n,2}$ is a minimal solution and $\mathcal{K}^-_{n,1}$ is a dominant one. By (6.8.19), we can see that the infinite power series in (6.8.24) with expansion coefficients $\mathcal{K}^-_{n,r}$ is entire if the sum

$$\sum_{n=0}^{\infty} |\mathcal{K}^-_{n,r}|^2 [2(n+q-1/4)]! \qquad (6.8.32)$$

converges. Using the asymptotic form (6.8.31) we get

$$\lim_{n\to\infty} \frac{|\mathcal{K}^-_{n+1,r}|^2 [2(n+1+q-1/4)]!}{|\mathcal{K}^-_{n,r}|^2 [2(n+q-1/4)]!} = 4|t_r|^2 \qquad (6.8.33)$$

6.8 3-term Recurrence Relation and Continued Fractions

which is less than 1 for $r = 2$ and greater than 1 for $r = 1$. Thus by the ratio test, the sum (6.8.32) converges for the minimal solution $\mathcal{K}_n^{\min} \equiv \mathcal{K}_{n,2}^-$ and diverges for the dominant solution $\mathcal{K}_{n,1}^-$. It follows that the infinite power series expansions $\psi_\pm^{\min}(z)$, obtained by substituting $\mathcal{K}_n^{\min}$ for the $\mathcal{K}_n^-$'s in (6.8.25) and (6.8.24), converge in the whole complex plane, i.e. they are entire.

We now present the results obtained in [23] for the energy eigenvalues E corresponding to the minimal solution $\mathcal{K}_n^{\min}$ (and thus to the entire wavefunctions $\psi_\pm^{\min}(z)$). The procedure used in that work is similar to that in [15], [37], [38] which uses the relationship between minimal solutions and infinite continued fractions[35].

By the Pincherle theorem, the ratio of successive elements of the minimal solution sequence $\mathcal{K}_n^{\min}$ is expressible as continued fractions. Proceeding in the direction of increasing n, we have

$$R_n = \frac{\mathcal{K}_{n+1}^{\min}}{\mathcal{K}_n^{\min}} = -\frac{B_{n+1}}{A_{n+1}-}\frac{B_{n+2}}{A_{n+2}-}\frac{B_{n+3}}{A_{n+3}-}\cdots, \qquad (6.8.34)$$

which for $n = 0$ gives

$$R_0 = \frac{\mathcal{K}_1^{\min}}{\mathcal{K}_0^{\min}} = -\frac{B_1}{A_1-}\frac{B_2}{A_2-}\frac{B_3}{A_3-}\cdots. \qquad (6.8.35)$$

Note that the ratio $R_0 = \frac{\mathcal{K}_1^{\min}}{\mathcal{K}_0^{\min}}$ involves $\mathcal{K}_0^{\min}$, although the above continued fraction expression is obtained from the 2nd equation of (6.8.27), i.e the recurrence (6.8.27) for $n \geq 1$. However, for single-ended sequences such as those appearing in the infinite series expansions (6.8.24), the ratio $R_0 = \frac{\mathcal{K}_1^{\min}}{\mathcal{K}_0^{\min}}$ of the first two terms of a minimal solution is unambiguously fixed by the first equation of the recurrence (6.8.27), namely,

$$R_0 = -A_0 = \frac{1}{16gq}\left[2q\omega(2-\Omega^2) - \left(E + \frac{1}{2}\omega - \frac{\Delta^2}{E + \frac{1}{2}\omega - 2q\omega\Omega}\right)\Omega\right]. \qquad (6.8.36)$$

In general, the R_0 computed from the continued fraction (6.8.35) can not be the same as that from (6.8.36) for arbitrary values of recurrence coefficients A_n and B_n. As a result, general solutions to the recurrence (6.8.27) are dominant and are usually generated by simple forward recursion from a given value of $\mathcal{K}_0^-$. Physical meaningful solutions are those that are entire in the BH space [39]. They can be obtained if

E can be adjusted so that equations (6.8.35) and (6.8.36) are both satisfied. Then the resulting solution sequence $\mathcal{K}_n^-(E)$ will be purely minimal and the power series expansion (6.8.24) will converge in the whole complex plane.

Therefore, if we define the transcendental function $F(E) = R_0 + A_0$ with R_0 given by the continued fraction in (6.8.35), then the zeros of $F(E)$ correspond to the points in the parameter space where the condition (6.8.36) is satisfied. In other words, $F(E) = 0$ is the eigenvalue equation of the 2-photon Rabi model, which may be solved for E by standard nonlinear root-search techniques. Only for the denumerable infinite values of E which are the roots of $F(E) = 0$, do we get entire solutions of the differential equations (6.8.22) and (6.8.23).

6.8.3 Two-Mode Quantum Rabi Model (双模量子 Rabi 模型)

We consider the Hamiltonian of the two-mode Rabi model introduced in [18]:

$$H_{2m} = \omega(b_1^\dagger b_1 + b_2^\dagger b_2) + \Delta \sigma_z + g\, \sigma_x (b_1^\dagger b_2^\dagger + b_1 b_2), \tag{6.8.37}$$

where we assume that the boson modes are degenerate with the same frequency ω. Introduce the two-mode Bogoliubov transformation[23]

$$b_1 = \frac{a_1 + \sigma a_2^\dagger}{\sqrt{1-\sigma^2}}, \quad b_1^\dagger = \frac{\sigma a_2 + a_1^\dagger}{\sqrt{1-\sigma^2}}, \quad b_2 = \frac{a_2 + \sigma a_1^\dagger}{\sqrt{1-\sigma^2}}, \quad b_2^\dagger = \frac{\sigma a_1 + a_2^\dagger}{\sqrt{1-\sigma^2}}. \tag{6.8.38}$$

Here $|\sigma| < 1$ is a real parameter and $a_1, a_2, a_1^\dagger, a_2^\dagger$ are squeezed bosons satisfying the canonical commutation relations $[a_i, a_i^\dagger] = 1$, $[a_i, a_j] = [a_i, a_j^\dagger] = [a_i^\dagger, a_j^\dagger] = 0$, $i,j = 1, 2$. As far as I know, the above two-mode Bogoliubov transformation is a new result. In terms of the 2-mode squeezed bosons, the Hamiltonian (6.8.37) has the form

$$\begin{aligned}\tilde{H}_{2m} = \frac{1}{1-\sigma^2}\Big\{ & \left[2\omega\sigma + g\sigma_x(1+\sigma^2)\right](a_1^\dagger a_2^\dagger + a_1 a_2) \\ & + \left[\omega(1+\sigma^2) + 2g\sigma\,\sigma_x\right](a_1^\dagger a_1 + a_2^\dagger a_2) \\ & + 2\omega\sigma^2 + 2g\sigma\,\sigma_x \Big\} + \Delta\sigma_z. \end{aligned} \tag{6.8.39}$$

Introduce the operators $K_\pm, K_0$

$$K_+ = a_1^\dagger a_2^\dagger, \quad K_- = a_1 a_2, \quad K_0 = \frac{1}{2}(a_1^\dagger a_1 + a_2^\dagger a_2 + 1). \tag{6.8.40}$$

Then (6.8.39) becomes

$$\begin{aligned}\tilde{H}_{2m} = \Delta\sigma_z + \frac{1}{1-\sigma^2}\Big\{ & \left[2\omega\sigma + g\sigma_x(1+\sigma^2)\right](K_+ + K_-) \\ & + 2\left[\omega(1+\sigma^2) + 2g\sigma\sigma_x\right]K_0 \Big\} - \omega. \end{aligned} \tag{6.8.41}$$

6.8 3-term Recurrence Relation and Continued Fractions

The operators $K_\pm, K_0$ form the $su(1,1)$ Lie algebra. Its quadratic Casimir, $C = K_+K_- - K_0(K_0-1)$, takes the particular values $C = \kappa(1-\kappa)$ in the representation (6.8.40), where $\kappa = 1/2, 1, 3/2, \cdots$. This is the well-known infinite-dimensional unitary irreducible representation of $su(1,1)$ known as the positive discrete series $\mathcal{D}^+(\kappa)$. Thus the Fock-Hilbert space decomposes into the direct sum of infinite subspaces $\mathcal{H}^\kappa$ labeled by $\kappa = 1/2, 1, 3/2, \cdots$.

Using the representation $a_i^\dagger = w_i$ and $a_i = \dfrac{d}{dw_i}$, $i = 1, 2$, it has been shown in [22] that in the Bargmann space with basis vectors $\left\{z^n/\sqrt{n!(n+2\kappa-1)!}\right\}$ given by the monomials in $z = w_1 w_2$, the operators $K_\pm, K_0$ (6.8.40) have the single-variable differential realization

$$K_0 = z\frac{d}{dz} + \kappa, \quad K_+ = z, \quad K_- = z\frac{d^2}{dz^2} + 2\kappa\frac{d}{dz}, \tag{6.8.42}$$

where $\kappa = 1/2, 1, 3/2, \cdots$. The Bargmann space is the Hilbert space of entire functions on $\mathbf{C}^2$ if the inner product

$$(f, g) = \iint \overline{f(w_1, w_2)}\, g(w_1, w_2)\, d\mu(w_1, w_2) \tag{6.8.43}$$

is finite for an appropriate measure $d\mu(w_1, w_2)$. It can be easily seen that if we choose $d\mu(w_1, w_2) = \dfrac{1}{\pi^2}|w_1|^{2(2\kappa-1)}\, e^{-|w_1|^2 - |w_2|^2}\, d^2w_1\, d^2w_2$, then

$$(z^m, z^n) = ((w_1w_2)^{2m}, (w_1w_2)^{2n}) = n!\,(n+2\kappa-1)!\,\delta_{mn}. \tag{6.8.44}$$

Thus the monomials $\left\{z^n/\sqrt{n!(n+2\kappa-1)!}\right\}$ form an orthonormal basis of the Bargamann-Hilbert space. It is now not difficult to see that if $f(z) = \sum_{n=0}^{\infty} c_n z^n$ then

$$\|f\|^2 = \sum_{n=0}^{\infty} |c_n|^2\, n!\,(n+2\kappa-1)! \tag{6.8.45}$$

and $f(z)$ is entire if the sum on the right hand side converges.

Using this differential realization, working in a representation defined by σ_x diagonal and choosing σ such that $2\omega\sigma + g(1+\sigma^2) = 0$, i.e.[1]

$$\sigma = -\frac{\omega}{g}(1 - \Lambda), \quad \Lambda = \sqrt{1 - \frac{g^2}{\omega^2}}, \tag{6.8.46}$$

---

[1] By arguments similar to the 2-photon case we can show that σ given by (6.8.46) obeys the requirements that it is real and $|\sigma| < 1$, provided that $\left|\dfrac{g}{\omega}\right| < 1$.

where $\left|\frac{g}{\omega}\right| < 1$, then the transformed Hamiltonian (6.8.41) becomes a matrix differential operator

$$\tilde{H}_{2m} = \begin{pmatrix} 2\omega\Lambda\left(z\dfrac{\mathrm{d}}{\mathrm{d}z} + \kappa\right) - \omega & \Delta \\ \Delta & -\dfrac{2g}{\Lambda}z\dfrac{\mathrm{d}^2}{\mathrm{d}z^2} + \dfrac{2}{\Lambda}\left[\omega(2-\Lambda^2)z - 2g\kappa\right]\dfrac{\mathrm{d}}{\mathrm{d}z} \\ & -\dfrac{2g}{\Lambda}z + \dfrac{2\omega(2-\Lambda^2)\kappa}{\Lambda} - \omega \end{pmatrix}. \quad (6.8.47)$$

In terms of two-component wavefuntion $\phi(z) = (\phi_+(z), \phi_-(z))^{\mathrm{T}}$, the time-independent Schrödinger equation, $\tilde{H}_{2m}\phi(z) = E\phi(z)$, yields a system of coupled differential equations,

$$\left[2\omega\Lambda\left(z\frac{\mathrm{d}}{\mathrm{d}z} + \kappa\right) - \omega - E\right]\phi_+ + \Delta\phi_- = 0, \quad (6.8.48)$$

$$\left\{2gz\frac{\mathrm{d}^2}{\mathrm{d}z^2} + \left[-2\omega(2-\Lambda^2)z + 4g\kappa\right]\frac{\mathrm{d}}{\mathrm{d}z}\right.$$
$$\left. + 2gz - 2\omega(2-\Lambda^2)\kappa + (E+\omega)\Lambda\right\}\phi_- - \Lambda\Delta\,\phi_+ = 0. \quad (6.8.49)$$

This is a system of differential equations of Fuchsian type. Solutions to these equations must be analytic in the whole complex plane if E belongs to the spectrum of $\tilde{H}_{2m}$. Similar to the 2-photon Rabi case, we seek solutions of the form

$$\phi_+(z) = \sum_{n=0}^{\infty} \mathcal{Q}_n^+(E)\,z^n, \quad \phi_-(z) = \sum_{n=0}^{\infty} \mathcal{Q}_n^-(E)\,z^n, \quad (6.8.50)$$

which converge in the entire complex plane.

Substituting (6.8.50) into (6.8.48), we obtain[23]

$$\mathcal{Q}_n^+ = \frac{\Delta}{E + \omega - (2n+2\kappa)\omega\Lambda}\mathcal{Q}_n^-. \quad (6.8.51)$$

So $\mathcal{Q}_n^+$ is not analytic in E but has simple poles at

$$E = -\omega + (2n + 2\kappa)\omega\Lambda, \quad n = 0, 1, \cdots. \quad (6.8.52)$$

The energies (6.8.52) appear for special values of model parameters[18] and correspond to the exceptional solutions of the two-mode Rabi model. If (6.8.52) is satisfied, the infinite series expansions (6.8.50) truncate and reduce to polynomials in z but only if the model parameters obey certain constraints. Majority part of the spectrum of the

6.8 3-term Recurrence Relation and Continued Fractions

two-mode Rabi model is regular spectrum which does not have the form (6.8.52). The regular spectrum of the model is given by the zeros of the transcendental function $G(E)$ obtained below. Thus again, the spectrum of the two-mode Rabi model consists of two parts, the regular and the exceptional spectrum.

From (6.8.49), we obtain the 3-step recurrence relation for $\mathcal{Q}_n^{-}$ [23]:

$$\begin{cases} \mathcal{Q}_1^{-} + C_0 \mathcal{Q}_0^{-} = 0, \\ \mathcal{Q}_{n+1}^{-} + C_n \mathcal{Q}_n^{-} + D_n \mathcal{Q}_{n-1}^{-} = 0, \quad n \geqslant 1, \end{cases} \quad (6.8.53)$$

where

$$\begin{cases} C_n = \dfrac{1}{2g(n+1)(n+2\kappa)} \left\{ -(2n+2\kappa)\omega(2-\Lambda^2) \right. \\ \qquad\qquad \left. + \left[E + \omega - \dfrac{\Lambda^2}{E+\omega-(2n+2\kappa)\omega\Lambda} \right] \Lambda \right\}, \\ D_n = \dfrac{1}{(n+1)(n+2\kappa)}. \end{cases} \quad (6.8.54)$$

The coefficients C_n, D_n have the behavior as $n \to \infty$,

$$C_n \sim c n^{\mu}, \quad D_n \sim d n^{\rho} \quad (6.8.55)$$

with

$$c = -\frac{\omega}{g}(2-\Lambda^2), \quad \mu = -1, \quad d = 1, \quad \rho = -2. \quad (6.8.56)$$

By analysis similar to the 2-photon case, we see that the two linearly independent solutions $\mathcal{Q}_{n,1}^{-}$ and $\mathcal{Q}_{n,2}^{-}$ of the $n \geqslant 1$ part of the recurrence (6.8.27) obey

$$\lim_{n\to\infty} \frac{\mathcal{Q}_{n+1,r}^{-}}{\mathcal{Q}_{n,r}^{-}} \sim t_r n^{-1}, \quad r = 1, 2, \quad (6.8.57)$$

where $t_1 = \dfrac{\omega}{g}$, $t_2 = \dfrac{g}{\omega}$ and $|t_2| < |t_1|$ (from the condition $\left|\dfrac{g}{\omega}\right| < 1$). Thus $\mathcal{Q}_{n,2}^{-}$ is a minimal solution and $\mathcal{Q}_{n,1}^{-}$ is a dominant one. Using (6.8.45) and by similar analysis to the 2-photon case, we can conclude that the infinite power series expansions $\phi_{\pm}^{\min}(z)$ generated by substituting the minimal solution $\mathcal{Q}_n^{\min} \equiv \mathcal{Q}_{n,2}^{-}$ for the $\mathcal{Q}_n^{-}$'s in (6.8.51) and (6.8.50), converge in the whole complex plane.

Similar to the 2-photon case, let

$$S_0 = \frac{\mathcal{Q}_1^{\min}}{\mathcal{Q}_0^{\min}} = -\frac{D_1}{C_1-}\frac{D_2}{C_2-}\frac{D_3}{C_3-}\cdots \quad (6.8.58)$$

be the continued fraction computed from the coefficients of the 2nd equation of the three-term recurrence relation (6.8.53). Define the function

$$G(E) = S_0 + C_0,$$

where

$$C_0 = -\frac{1}{4g\kappa}\left[2\kappa\omega(2-\Lambda^2) - \left(E + \omega - \frac{\Delta^2}{E + \omega - 2\kappa\omega\Lambda}\right)\Lambda\right] \quad (6.8.59)$$

is the ratio of the first two terms of the minimal solution unambiguously fixed by the $n = 0$ part of the recurrence (6.8.53). Then the zeros of transcendental function $G(E)$ correspond to the points in the parameter space where the the power series solutions will converge in the whole complex plane. In other words, $G(E) = 0$ yield the eigenvalue equations which give the regular energies for the two-mode Rabi model.

6.9 Solutions to QES Differential Equations: Heine-Stieltjes Polynomials (准精确可解差分方程的解: Heine-Stieltjes 多项式)

Extended Heine-Stieltjes (HS) polynomials have been applied to solve a set of algebraic equations of interesting physical systems (e.g. [40, 41]). HS polynomials are polynomial solutions of the 2nd-order Fuchsian equation of the form,

$$\left[A(z)\frac{\mathrm{d}^2}{\mathrm{d}z^2} + B(z)\frac{\mathrm{d}}{\mathrm{d}z} + V(z)\right]y(z) = 0, \quad (6.9.1)$$

where $A(z)$ is a polynomial of degree k which can be written as $A(z) = \prod_{i=1}^{k}(z - a_i)$, $B(z)$ is a polynomial of degree $k-1$ such that for a set of real positive parameters γ_i,

$$\frac{B(z)}{A(z)} = \sum_{i=1}^{k}\frac{\gamma_i}{z - a_i} \quad (6.9.2)$$

and $V(z)$ is the so-called Van Vleck polynomial of degree at most $k - 2$ such that the above differential equation has polynomial solutions. The Van Vleck polynomial $V(z)$ is allowed to depend on the solutions $y(z)$.

6.9 Solutions to QES Differential Equations: Heine-Stieltjes Polynomials

The Fuchian differential equation can be recast into the more familiar form

$$\frac{d^2y}{dz^2} + \left(\sum_{i=1}^{k} \frac{\gamma_i}{z-a_i}\right)\frac{dy}{dz} + \frac{V(z)}{\prod_{i=1}^{k}(z-a_i)} y = 0. \tag{6.9.3}$$

The $k=2$ and $k=3$ cases correspond to the hypergeometric and Heun equations, respectively. If $A(z)$ and $B(z)$ are algebraically independent, i.e. they do not satisfy any algebraic equations with integer coefficients, then for every integer $\mathcal{N}$ there exist at most $d(k,\mathcal{N}) = (\mathcal{N}+k-2)!/(k-2)!\mathcal{N}!$ different Van Vleck polynomials $V(z)$ such that $y(z)$ is a polynomial solution of degree $\mathcal{N}$.

Let

$$y(z) = \prod_{i=1}^{\mathcal{N}}(z-z_i) \tag{6.9.4}$$

be a HS polynomial solution and $\{z_i,\ i=1,2,\cdots,\mathcal{N}\}$ be its zeros, called Stieltjes zeros. We can easily check that

$$\begin{cases} y'(z) = \sum_{i=1}^{\mathcal{N}} \prod_{j\neq i}(z-z_j), \\ y''(z) = \sum_{i=1}^{\mathcal{N}} \sum_{l=1,l\neq i}^{\mathcal{N}} \prod_{j\neq i,l}(z-z_j) = 2\sum_{i<l}\prod_{j\neq i,l}(z-z_j). \end{cases} \tag{6.9.5}$$

Therefore at a zero z_r of $y(z)$, we have

$$\begin{cases} y'(z_r) = \prod_{j\neq i}(z_r-z_j), \\ y''(z_r) = s\sum_{i=1}^{\mathcal{N}}\prod_{j\neq i,r}(z_r-z_j) = y'(z_r)\sum_{i\neq r}\frac{2}{z_r-z_i}, \end{cases} \tag{6.9.6}$$

so that

$$\frac{y''(z_r)}{y'(z_r)} = \sum_{i\neq r}^{\mathcal{N}} \frac{2}{z_r-z_i}. \tag{6.9.7}$$

In general, we have

$$\begin{cases} y'(z) = y(z)\sum_{i=1}^{\mathcal{N}}\frac{1}{z-z_i}, \\ y''(z) = y'(z)\sum_{i=1}^{\mathcal{N}}\frac{1}{z-z_i} + y(z)\sum_{i=1}^{\mathcal{N}}\frac{-1}{(z-z_i)^2} \\ \qquad = y(z)\sum_{i=1}^{\mathcal{N}}\frac{1}{z-z_i}\sum_{j\neq i}\frac{2}{z_i-z_j}. \end{cases} \tag{6.9.8}$$

Substituting into the ODE, we get

$$-V(z) = A(z) \sum_{i=1}^{N} \frac{1}{z-z_i} \left(\sum_{j \neq i}^{N} \frac{2}{z_i - z_j} + \sum_{\mu=1}^{k} \frac{\gamma_\mu}{z - a_\mu} \right). \tag{6.9.9}$$

Note that at any zeros z_r of $y(z)$, the ODE becomes $y''(z_r)/y'(z_r) + B(z_r)/A(z_r) = 0$ which gives rise to

$$\sum_{j \neq i}^{N} \frac{2}{z_r - z_j} + \sum_{\mu=1}^{k} \frac{\gamma_\mu}{z_r - a_\mu} = 0, \quad r = 1, 2, \cdots, \mathcal{N}. \tag{6.9.10}$$

These (non-linear) algebraic equations are called the Bethe Ansatz equations (BAEs).

Using the BAEs, we get

$$V(z) = -A(z) \sum_{i=1}^{N} \frac{1}{z - z_i} \left(-\sum_{\mu=1}^{k} \frac{\gamma_\mu}{z_i - a_\mu} + \sum_{\mu=1}^{k} \frac{\gamma_\mu}{z - a_\mu} \right)$$

$$= A(z) \sum_{\mu=1}^{k} \frac{1}{z - a_\mu} \sum_{i=1}^{N} \frac{\gamma_\mu}{z_i - a_\mu}. \tag{6.9.11}$$

This shows that the zeros $\{\tilde{z}_l^{(\zeta)}, \, l = 1, 2, \cdots, k-2\}$ of $V^{(\zeta)}(z)$ related to the ζ-th HS polynomial $y^{(\zeta)}(z) = \prod_{i=1}^{N}(z - z_i^{(\zeta)})$ are determined by $(k-2)$ algebraic equations

$$\sum_{\mu=1}^{k} \frac{1}{\tilde{z}_l^{(\zeta)} - a_\mu} \sum_{i=1}^{N} \frac{\gamma_\mu}{z_i^{(\zeta)} - a_\mu}, \quad l = 1, 2, \cdots, k-2. \tag{6.9.12}$$

The roots $\{\tilde{z}_l^{(\zeta)}, \, l = 1, 2, \cdots, k-2\}$ are called Van Vleck zeros related to the ζ-th HS polynomial $y^{(\zeta)}(z)$.

Once the Van Vleck zeros are determined from the above $(k-2)$ algebraic equations, $V^{(\zeta)}(z)$ can be expressed explicitly as

$$V^{(\zeta)}(z) \sim \prod_{l=1}^{k-2}(z - \tilde{z}_l^{(\zeta)}). \tag{6.9.13}$$

If a_μ ($\mu = 1, 2, \cdots, k$) are real, then according to the Stieltjes results, one may give an electrostatic interpretation of the location of zeros of the extended HS polynomials $y(z)$ as follows. Put k negative fixed charges $-\gamma_\mu/2$, $\mu = 1, 2, \cdots, k$, along a real line, respectively, and allow $\mathcal{N}$ positive unit charges to move freely on the two dimensional complex plane. Then, up to a constant, the total energy of the system may be written as

$$U(z_1, z_2, \cdots, z_N) = \frac{1}{2} \sum_{i=1}^{N} \sum_{\mu=1}^{k} \gamma_\mu \ln|z_i - a_\mu| - \sum_{1 \leq i \neq j \leq N} \ln|z_i - z_j|. \tag{6.9.14}$$

The BAEs given in (6.9.10) imply that there are $d(k, \mathcal{N})$ different configurations for the position of the $\mathcal{N}$ positive charges $\{z_1^{(\zeta)}, z_2^{(\zeta)}, \cdots, z_l^{(\zeta)}\}$ with $\zeta = 1, 2, \cdots, d(k, \mathcal{N})$, corresponding to global minimums of the total energy $U(z_1, z_2, \cdots, z_N)$.

Note: By applying the spirit of the above electrostatic intepretation, the author in [42] gave a nice connection of the roots of the Jacobi polynomials to the solutions of the special cases of the scattering equations[43]

$$\sum_{j \neq i}^{n} \frac{k_i \cdot k_i}{\sigma_i - \sigma_j} = 0, \quad i = 1, 2, \cdots, n. \tag{6.9.15}$$

These equations connect the space of kinematic invariants of n massless particles with momentum k_i in arbitrary dimensions with the positions σ_i of n points on a Riemann sphere.

6.10 Solutions to QES Systems: Functional Bethe Ansatz Method (准精确可解系统的解: Bethe Ansatz 方法)

Consider the generic 2nd-order ordinary differential equation (ODE) involving the differential operator (6.3.5),

$$\mathcal{H}S(z) = 0, \tag{6.10.1}$$

We are interested in the following two problems:

Problem 1. Given a pair of polynomials $X(z)$, $Y(z)$ and a positive integer n,

(a) Find all polynomials $Z(z)$ such that the ODE (6.10.1) has a polynomial solution $S(z)$ of degree n.

(b) Find the polynomial solutions $S(z)$.

Problem 2. If the coefficients of $X(z)$ and $Y(z)$ do satisfy some algebraic relations with integer coefficients, i.e. are algebraically dependent, then how many polynomials $Z(z)$ are there which lead to degree n polynomial solutions of the ODE (6.10.1)?

Problem 1 is answered by the theorem[44].

Theorem 5. Given a pair of polynomials $X(z)$ and $Y(z)$, then the values of the coefficients c_2, c_1, c_0 of polynomial $Z(z)$ such that the differential equation (6.10.1) has degree n polynomial solution

$$S(z) = \prod_{i=1}^{n}(z - z_i) \qquad (6.10.2)$$

with distinct roots $z_1, z_2, \cdots, z_n$ are given by

$$c_2 = -n(n-1)a_4 - nb_3, \qquad (6.10.3)$$

$$c_1 = -[2(n-1)a_4 + b_3]\sum_{i=1}^{n} z_i - n(n-1)a_3 - nb_2, \qquad (6.10.4)$$

$$c_0 = -[2(n-1)a_4 + b_3]\sum_{i=1}^{n} z_i^2 - 2a_4\sum_{i<j}^{n} z_i z_j$$

$$- [2(n-1)a_3 + b_2]\sum_{i=1}^{n} z_i - n(n-1)a_2 - nb_1, \qquad (6.10.5)$$

where the roots $z_1, z_2, \cdots, z_n$ satisfy the Bethe Ansatz equations,

$$\sum_{j \neq i}^{n} \frac{2}{z_i - z_j} + \frac{Y(z_i)}{X(z_i)} = 0, \quad i = 1, 2, \cdots, n. \qquad (6.10.6)$$

The above equations (6.10.3)~(6.10.6) give all polynomials $Z(z)$ such that the ODE (6.10.1) has polynomial solution (6.10.2).

Remark 3. Theorem 5 gives all polynomials $Z(z)$ such that (6.10.1) is QES as well as the corresponding polynomial solutions of (6.10.1). These include the case in which $\mathcal{H}$ ha no $sl(2)$ algebraization.

Remark 4. In contrast to the results of Heine-Stieltjes in the last section, this theorem determines the explicit expressions of all Van Vleck polynomials $Z(z)$ without the need of solving the algebraic equations (6.9.12) in the last section to find the zeros of $Z(z)$.

We here provide the proof of Theorem 5 given in [44] by using the functional Bethe Ansatz (BA) method[45, 46]. We first prove some identities.

Lemma 1. Let $\{z_i, i = 1, 2, \cdots, n\}$ be any set of n real or complex numbers. Then we have

6.10 Solutions to QES Systems: Functional Bethe Ansatz Method

I1. $\displaystyle\sum_{i=1}^{n}\sum_{j\neq i}^{n}\frac{1}{z_i-z_j}=0,$

I2. $\displaystyle\sum_{i=1}^{n}\sum_{j\neq i}^{n}\frac{z_i}{z_i-z_j}=\frac{1}{2}n(n-1),$

I3. $\displaystyle\sum_{i=1}^{n}\sum_{j\neq i}^{n}\frac{z_i^2}{z_i-z_j}=(n-1)\sum_{i=1}^{n}z_i,$

I4. $\displaystyle\sum_{i=1}^{n}\sum_{j\neq i}^{n}\frac{z_i^2 z_j}{z_i-z_j}=\sum_{i<j}^{n}z_i z_j,$

I5. $\displaystyle\sum_{i=1}^{n}\sum_{j\neq i}^{n}\frac{z_i^3}{z_i-z_j}=(n-1)\sum_{i=1}^{n}z_i^2+\sum_{i<j}^{n}z_i z_j.$

Proof. (I1) is obvious from the anti-symmetry of the denominator z_i-z_j.
I2. Writing the numerator z_i as $z_i-z_j+z_j$, we obtain

$$\text{l.h.s.}=n(n-1)-\sum_{j=1}^{n}\sum_{i\neq j}^{n}\frac{z_j}{z_j-z_i}, \tag{6.10.7}$$

which gives the result, as required.
I3. Writing the numerator z_i^2 as $z_i^2=z_i(z_i-z_j)+z_i z_j$, we have

$$\text{l.h.s.}=\sum_{i=1}^{n}\sum_{j\neq i}^{n}z_i+\sum_{i=1}^{n}\sum_{j\neq i}^{n}\frac{z_i z_j}{z_i-z_j}, \tag{6.10.8}$$

The 2nd term above is vanishing due to anti-symmetry of $z_i z_j/(z_i-z_j)$.
I4. Noting $z_i^2 z_j=z_i(z_i-z_j)z_j+z_i z_j^2$, we get

$$\text{l.h.s.}=2\sum_{i<j}^{n}z_i z_j-\sum_{j=1}^{n}\sum_{i\neq j}^{n}\frac{z_j^2 z_i}{z_j-z_i}, \tag{6.10.9}$$

which leads to the result.
I5. Writing the numerator z_i^3 as $z_i^3=z_i^2(z_i-z_j)+z_i^2 z_j$ and using the result in (I4), we can easily check the identity (I5). □

Now let

$$S(z)=\prod_{i=1}^{n}(z-z_i) \tag{6.10.10}$$

be a degree n polynomial with undetermined, distinct roots $z_1,z_2,\cdots,z_n$. We will be looking for the values of the coefficients c_2,c_1,c_0 of $Z(z)$ and the roots z_i, $1\leqslant i\leqslant n$, such that (6.10.10) is a solution of (6.10.1) with fixed coefficients of $X(z)$ and $Y(z)$. Substituting $Z(z)$ into the ODE and dividing on both sides by $Z(z)$ gives rise to[44]

$$-c_0 = X(z_i) \sum_{i=1}^{n} \frac{1}{z - z_i} \sum_{j \neq i}^{n} \frac{2}{z_i - z_j} + Y(z_i) \sum_{i=1}^{n} \frac{1}{z - z_i} + c_2 z^2 + c_1 z. \qquad (6.10.11)$$

The left hand side is a constant and the right hand side is meromorphic function with simple poles at $z = z_i$ and singularity at $z = \infty$. The residues of $-c_0$ at the simple poles $z = z_i$ are

$$\text{Res}(-c_0)_{z=z_i} = X(z_i) \sum_{j \neq i}^{n} \frac{2}{z_i - z_j} + Y(z_i). \qquad (6.10.12)$$

It can then be shown[44] that

$$-c_0 - \sum_{i=1}^{n} \frac{\text{Res}(-c_0)_{z=z_i}}{z - z_i} = \sum_{i=1}^{n} \frac{X(z) - X(z_i)}{z - z_i} \sum_{j \neq i}^{n} \frac{2}{z_i - z_j}$$

$$+ \sum_{i=1}^{n} \frac{Y(z) - Y(z_i)}{z - z_i} + c_2 z^2 + c_1 z$$

$$= [n(n-1)a_4 + nb_3 + c_2] z^2$$

$$+ \left[(2(n-1)a_4 + b_3) \sum_{i=1}^{n} z_i + n(n-1)a_3 + nb_2 + c_1 \right] z$$

$$+ [2(n-1)a_4 + b_3] \sum_{i=1}^{n} z_i^2 + 2a_4 \sum_{i<j}^{n} z_i z_j$$

$$+ [2(n-1)a_3 + b_2] \sum_{i=1}^{n} z_i + n(n-1)a_2 + nb_1,$$

where we have used identities (I1)$\sim$(I5). That is

$$-c_0 = [n(n-1)a_4 + nb_3 + c_2] z^2$$

$$+ \left[(2(n-1)a_4 + b_3) \sum_{i=1}^{n} z_i + n(n-1)a_3 + nb_2 + c_1 \right] z$$

$$+ \sum_{i=1}^{n} \frac{\text{Res}(-c_0)_{z=z_i}}{z - z_i}$$

$$+ [2(n-1)a_4 + b_3] \sum_{i=1}^{n} z_i^2 + 2a_4 \sum_{i<j}^{n} z_i z_j$$

$$+ [2(n-1)a_3 + b_2] \sum_{i=1}^{n} z_i + n(n-1)a_2 + nb_1. \qquad (6.10.13)$$

6.10 Solutions to QES Systems: Functional Bethe Ansatz Method

The right hand side of (6.10.13) is a constant if and only if the coefficients of z^2 and z as well as all the residues at the simple poles are equal to zero, respectively. This gives the results in the theorem and thus completes the proof. □

From (6.10.3)~(6.10.6), we have, as special cases of the theorem 5, the following corollary which provides the Answer to Problem 2[44]:

Corollary 1. (a) If for an arbitrary integer n the coefficients a_4 and b_3 in the ODE (6.10.1) are algebraically dependent,

$$2(n-1)a_4 + b_3 = 0, \qquad (6.10.14)$$

then there are $n+1$ polynomials $Z(z)$ with coefficients $c_2 = n(n-1)a_4$, $c_1 = -n[(n-1)a_3 + b_2]$ and

$$c_0 = -n\left[(n-1)a_2 + 2b_1\right] - [2(n-1)a_3 + b_2]\sum_{i=1}^{n} z_i - 2a_4 \sum_{i<j}^{n} z_i z_j,$$

such that (6.10.1) has degree n polynomial solution $S(z)$ (6.10.2), where the roots z_i satisfy (6.10.6) with $b_3 = -2(n-1)a_4$.

(b) If $a_4 = b_3 = c_2 = 0$ and if for an arbitrary integer n the coefficients a_3 and b_2 are algebraically dependent,

$$2(n-1)a_3 + b_2 = 0, \qquad (6.10.15)$$

then there is 1 polynomial $Z(z) = -n[(n-1)a_3 + b_2]z - n(n-1)a_2 - nb_1$ such that the corresponding ODE has degree n polynomial solution $S(z)$ (6.10.2) with the roots z_i determined by (6.10.6) with $a_4 = b_3 = 0$, $b_2 = -2(n-1)a_3$.

Proof [44]. Part (b) is obvious. We now prove part (a). By Proposition 2, If $b_3 = -2(n-1)a_4$, then the ODE (6.10.1) has a hidden $sl(2)$ algebra structure provided that $c_2 = n(n-1)a_4$, $c_1 = -n[(n-1)a_3 + b_2]$. Namely if we rewrite the ODE (6.10.1) with such coefficients c_2, c_1 as the form

$$\mathcal{H} S(z) \equiv (H + c_0) S(z) = 0, \qquad (6.10.16)$$

then by Proposition 2, H is an element of the enveloping algebra of Lie-algebra $sl(2)$

$$\begin{aligned} H = &\, a_4 J^+ J^+ - a_3 J^+ J^0 + a_2 J^0 J^0 + a_1 J^0 J^- + a_0 J^- J^- \\ &- \left(\frac{3n-2}{2} a_3 + b_2\right) J^+ + [(n-1)a_2 + b_1] J^0 \\ &+ \left(\frac{n}{2} a_1 + b_0\right) J^- + \frac{n}{2}\left[\left(\frac{n}{2} - 1\right) a_2 + b_1\right]. \end{aligned} \qquad (6.10.17)$$

$-c_0$ is the eigenvalue of H with polynomial eigenfunction $S(z) = \prod_{i=1}^{n}(z - z_i)$, given by (from theorem 5)

$$-c_0 = n\left[(n-1)a_2 + 2b_1\right] + \left[2(n-1)a_3 + b_2\right]\sum_{i=1}^{n} z_i + 2a_4 \sum_{i<j}^{n} z_i z_j, \qquad (6.10.18)$$

and the roots z_i of the eigenfunction $S(z)$ are determined by the BAEs (6.10.6) with $b_3 = -2(n-1)a_4$. It is well known that the solution space of differential operators with a $sl(2)$ algebraization is $n+1$ dimensional[3, 4]. Applying this to the H above, we conclude that the above Bethe ansatz equations have $n+1$ sets of solutions and thus there $n+1$ eigenvalues $-c_0$, i.e. $n+1$ polynomials $Z(z)$. This completes the proof. □

6.11 Examples of Solutions of QES Models (准精确可解模型的解的一些例子)

6.11.1 Bose-Hubbard Dimer with Local M-body Interaction (Bose-Hubbard 二聚体中的局域多体相互作用)

The Bose-Hubbard (BH) model describing cold atomic gases in optical lattices is defined by the hamiltonian

$$\mathcal{H} = -\sum_{i=1}^{L}(a_i^\dagger a_{i+1} + \text{h.c.}) + \frac{U}{2}\sum_{i=1}^{L} n_i(n_i - 1) + \frac{V}{M!}\sum_{i=1}^{L}\prod_{k=0}^{M-1}(n_i - k), \qquad (6.11.1)$$

where $n_i = a_i^\dagger a_i$ are the number operators. The last term in the hamiltonian represents the local N-body interaction. In terms of $gl(L)$ generators $E_{ij} = a_i^\dagger a_j$, $i,j = 1, 2, \cdots, L$, the hamiltonian can be written as

$$\mathcal{H} = -\sum_{i=1}^{L-1}(E_{i,i+1} + E_{i+1,i}) + \frac{U}{2}\sum_{i=1}^{L} E_{ii}(E_{ii} - 1) + \frac{V}{M!}\sum_{i=1}^{L}\prod_{k=0}^{M-1}(E_{ii} - k). \qquad (6.11.2)$$

We will focus on the $L = 2$ case which corresponds to the BH dimer. The hamiltonian in this case can be expressed in terms of the generators of $sl(2)$ algebra,

$$J^+ \equiv E_{12} = a_1^\dagger a_2, \quad J^- \equiv E_{21} = a_2^\dagger a_1, \quad J^0 \equiv \frac{1}{2}(E_{22} - E_{11}) = \frac{1}{2}(n_2 - n_1). \qquad (6.11.3)$$

6.11 Examples of Solutions of QES Models

It is easily seen that $N = n_1 + n_2$ is a central element of $sl(2)$, i.e. $[N, J^{\pm,0}] = 0$, and

$$n_1 = \frac{1}{2}(N - 2J^0), \quad n_2 = \frac{1}{2}(N + 2J^0). \tag{6.11.4}$$

It follows that the BH dimer hamiltonian can be recast into the form

$$\mathcal{H} = J^+ - J^- + U\left[J^0 J^0 + \frac{N}{2}\left(\frac{N}{2} - 1\right)\right]$$

$$+ \frac{V}{M!}\left[\prod_{k=0}^{M-1}\left(\frac{N}{2} - 2J^0 - k\right) + \prod_{k=0}^{M-1}\left(\frac{N}{2} + 2J^0 - k\right)\right]. \tag{6.11.5}$$

Using the differential realization of $sl(2)$, we obtain

$$\mathcal{H} = Uz^2\frac{d^2}{dz^2} + [z^2 - U(n-1)z - 1]\frac{d}{dz} - nz + \frac{U}{2}n(n-1)$$

$$+ \frac{V}{M!}\left[\prod_{k=0}^{M-1}\left(\frac{3n}{2} - 2z\frac{d}{dz} - k\right) + \prod_{k=0}^{M-1}\left(2z\frac{d}{dz} - \frac{n}{2} - k\right)\right]. \tag{6.11.6}$$

It is easily seen that $\mathcal{H}$ preserves a $(n+1)$-dimensional invariant subspace $\mathcal{V}_{n+1} = \text{span}\{1, z, z^2, \cdots, z^n\}$. Thus the Bose-Hubbard dimer with local M-body interaction is QES and the eigenfunctions ψ of the eigenvalue equation $\mathcal{H}\psi = E\psi$ are given by polynomials in z of degree n

$$\psi(z) = \prod_{i=1}^{n}(z - z_i), \quad n = 0, 1, 2, \cdots. \tag{6.11.7}$$

The roots z_i of the polynomial ψ and the eigenvalue E can be obtained by applying the functional BA method.

As an example, let us consider the $V = 0$ case, the Bose-Hubbard dimer without local M-body interaction. For this case, we derive, by theorem 5 in the last section, the energy eigenvalues E

$$E = \frac{1}{2}n(n-1)U + \sum_{i=1}^{n} z_i \tag{6.11.8}$$

and the algebraic equations satisfied by the roots z_i,

$$\sum_{j \neq i}^{n} \frac{2}{z_j - z_i} + \frac{z_i^2 - (n-1)Uz_i - 1}{Uz_i^2} = 0, \quad i = 1, 2, \cdots, n. \tag{6.11.9}$$

We can convert the dimer hamiltonian into the Schrödiner form. That is we seek a change of variable $z = z(x)$ and a gauge factor $W(x)$ such that $e^{W(x)}\mathcal{H}e^{-W(x)} =$

$-\dfrac{\mathrm{d}^2}{\mathrm{d}x^2} + V(x)$. Comparing the hamiltonian with that in section 6.4, we have

$$P(z) = -Uz^2, \quad Q(z) = -z^2 + nUz + 1, \quad R(z) = nz - \dfrac{U}{2}n(n-1). \quad (6.11.10)$$

Thus

$$x = x(z) = \pm \int^z \dfrac{1}{\sqrt{-U}y}\mathrm{d}y = \pm \dfrac{1}{\sqrt{-U}}\ln z \longrightarrow z = z(x) = \mathrm{e}^{\pm\sqrt{-U}x}, \quad (6.11.11)$$

$$\begin{aligned}W(x) &= \int^z \dfrac{-y^2 + nUy + 1}{-2Uy^2}\mathrm{d}y = \dfrac{1}{2U}(z + z^{-1}) - \dfrac{n}{2}\ln z \\ &= \mp\dfrac{n}{2}\sqrt{-U}x + \dfrac{1}{U}\cosh(\pm\sqrt{-U}x).\end{aligned} \quad (6.11.12)$$

It follows that the potential $V(x)$ is given by

$$V(x) = -\dfrac{1}{U}\cosh^2(\sqrt{-U}x) - (n+1)\cosh(\sqrt{-U}x) + \dfrac{1}{U} + \dfrac{n}{2}\left(\dfrac{n}{2} - 1\right)U. \quad (6.11.13)$$

6.11.2 BA Solutions of the Driven Rabi Model (驱动 Rabi 模型的 Bethe Ansatz 解)

In section 6.5.5 we have seen that $\mathcal{H}$ and $\tilde{\mathcal{H}}$ have hidden algebraic symmetry and preserve the $(n+1)$-dimensional representation space $\mathcal{V}_{n+1}$ of $sl(2)$ when the energies are given by (6.5.41) and (6.5.47), respectively. Here $\mathcal{V}_{n+1} = \mathrm{span}\{1, z, z^2, \cdots, z^n\}$. In other words, the driven Rabi model is QES and can be exactly diagonalized in $\mathcal{V}_{n+1}$ for certain models parameters. In this section we solve the driven Rabi model by means of the functional BA method. In addition to the exact spectrum, we are also able to obtain the closed form expressions for the allowed model parameters and the polynomial wavefunctions in terms of the roots of a set of algebraic equations.

We first derive exact solutions to the differential equation (6.5.37). Obviously they are polynomials in z of degree n, which can be written as

$$\phi_+(z) = \prod_{i=1}^n (z - \alpha_i), \quad n = 1, 2, \cdots, \quad (6.11.14)$$

where α_i are the roots of the polynomial $\phi_+(z)$ to be determined. Applying the results in theorem 5, we obtain the constraints on the model parameters

$$\Delta^2 + 2ng^2 + 2\omega g\sum_{i=1}^n \alpha_i = 0, \quad (6.11.15)$$

6.11 Examples of Solutions of QES Models

where $\{\alpha_i\}$ are determined by the BAEs

$$\sum_{j\neq i}^{n}\frac{2}{\alpha_i-\alpha_j}=\frac{2\omega g\alpha_i^2+[(2n-1)\omega+2\delta]\omega\alpha_i+g[\omega+2(\delta-g^2/\omega)]}{(\omega\alpha_i-g)(\omega\alpha_i+g)},$$

$$i=1,2,\cdots,n. \qquad (6.11.16)$$

Here use has been made of the E values given in (6.5.41). The corresponding wavefunction component $\psi_+(z)$ of the model is then given by

$$\psi_+(z)=e^{-\frac{g}{\omega}z}\prod_{i=1}^{N}(z-\alpha_i), \qquad (6.11.17)$$

and the component $\psi_-(z)=e^{-gz/\omega}\phi_-(z)$ with $\phi_-(z)$ determined by the first equation of (6.5.37) for $\Delta\neq 0$. Because the differential operator in the l.h.s. of (6.5.37) preserves $\mathcal{V}_{n+1}$ for any system parameters, $\phi_-(z)$ automatically belongs to the same invariant subspace as $\phi_+(z)$.

Similarly set

$$\varphi_-(z)=\prod_{i=1}^{n}(z-\beta_i),\quad n=1,2,\cdots, \qquad (6.11.18)$$

with β being the roots of $\varphi_-(z)$. Then we obtain

$$\Delta^2+2ng^2-2\omega g\sum_{i=1}^{n}\beta_i=0, \qquad (6.11.19)$$

where $\{\beta_i\}$ are determined by the BAEs

$$\sum_{j\neq i}^{n}\frac{2}{\beta_i-\beta_j}=\frac{-2\omega g\beta_i^2+[(2n-1)\omega-2\delta]\omega\beta_i+g[-\omega+2(\delta+g^2/\omega)]}{(\omega\beta_i-g)(\omega\beta_i+g)},$$

$$i=1,2,\cdots,n. \qquad (6.11.20)$$

Here we have used the formula (6.5.47) for E. The corresponding wavefunction component $\psi_-(z)$ of the model is then given by

$$\psi_-(z)=e^{\frac{g}{\omega}z}\prod_{i=1}^{N}(z-\beta_i), \qquad (6.11.21)$$

and the component $\psi_+(z)=e^{gz/\omega}\varphi_+(z)$ with $\varphi_+(z)$ determined by the second equation of (6.5.43) for $\Delta\neq 0$. Because the differential operator in the l.h.s. of (6.5.43) preserves $\mathcal{V}_{n+1}$ for any system parameters, $\varphi_+(z)$ automatically belongs to the same invariant subspace as $\varphi_-(z)$.

Remark: See [18] for BA solutions of the 2-photon and two-mode Rabi models (see also [47]).

6.11.3 BA Solutions of Two Electrons on a Sphere (球面上两电子的 Bethe Ansatz 解)

Consider the Hamiltonian in section 6.6.4 describing two electrons, interacting via a Coulomb potential, but constrained to remain on the surface of a D-dimensional sphere of radius R[28]. The Hamiltonian differential equation can be written as

$$\left\{\frac{d^2}{dz^2} + \left[\frac{1/\gamma}{z} + \frac{\frac{1}{2}(\delta - 1/\gamma)}{z+1} + \frac{\frac{1}{2}(\delta - 1/\gamma)}{z-1}\right]\frac{d}{dz} + \frac{-4R^2Ez + 2R}{z(z+1)(z-1)}\right\}\Psi = 0. \quad (6.11.22)$$

This ODE has the form of the Heun equation. It follows from our general results that this equation has polynomial solutions of degree $n = 1, 2, \cdots$,

$$\Psi(z) = \prod_{i=1}^{n}(z - z_i), \quad (6.11.23)$$

where z_i are the roots of the above polynomial to be determined, provided that E and R take the values given by[44]

$$E = \frac{n}{4R^2}(n + \delta - 1), \quad (6.11.24)$$

$$R = -\frac{1}{2}[2(n-1) + \delta]\sum_{i=1}^{n} z_i, \quad (6.11.25)$$

where $z_1, z_2, \cdots, z_n$ obey the Bethe Ansatz equations

$$\sum_{j \neq i}^{n} \frac{2}{z_i - z_j} + \frac{1/\gamma}{z_i} + \frac{\frac{1}{2}(\delta - 1/\gamma)}{z_i + 1} + \frac{\frac{1}{2}(\delta - 1/\gamma)}{z_i - 1} = 0, \quad i = 1, 2, \cdots, n. \quad (6.11.26)$$

For $n = 1$, we have, from the Bethe Ansatz equations, $z_1 = \pm\frac{1}{\sqrt{\delta\gamma}}$. Then $2R = -\delta z_1 = \sqrt{\frac{\delta}{\gamma}}$ (by choosing the negative root $z_1 = -\frac{1}{\sqrt{\delta\gamma}}$ so that the radius R is non-negative), $E = \gamma$ and the wave function is $\Psi = z + \frac{1}{\sqrt{\delta\gamma}} = \frac{1}{\sqrt{\delta\gamma}}(1 + \gamma u)$.

For $n = 2$, we find

$$\begin{cases} z_1 = \frac{1}{2(\delta + 2)}\left[-\sqrt{2(\delta+2) + \frac{4\delta + 6}{\gamma}} \pm \sqrt{2(\delta+2) - \frac{2}{\gamma}}\right], \\ z_2 = \frac{1}{2(\delta + 2)}\left[-\sqrt{2(\delta+2) + \frac{4\delta + 6}{\gamma}} \mp \sqrt{2(\delta+2) - \frac{2}{\gamma}}\right]. \end{cases} \quad (6.11.27)$$

The other root satisfies $z_1 = -z_2$ which leads to $R = 0$ and therefore is discarded. The radius R and the energy E are

$$\begin{cases} R = -\dfrac{1}{2}(\delta+2)(z_1+z_2) = \dfrac{1}{2}\sqrt{2(\delta+2)+\dfrac{4\delta+6}{\gamma}}, \\ E = \dfrac{\gamma(\delta+1)}{\gamma(\delta+2)+2\delta+3}, \end{cases} \quad (6.11.28)$$

and the wave function is

$$\Psi = (z-z_1)(z-z_2) = \frac{1}{\gamma(\delta+2)}\left[1+\gamma u + \frac{\gamma^2(\delta+2)}{2\gamma(\delta+2)+4\delta+6}u^2\right]. \quad (6.11.29)$$

6.11.4 BA Solutions of the Inverse Sextic Power Potential (逆六次势的 Bethe Ansatz 解)

The exact ground state solution of the Schrödinger equation with inverse sextic power potential (6.6.51) was previously obtained in [48]. Here we present the general exact solutions of the system.

It is shown in section 6.6.5 that the differential equation for the transformed wavefunction $v(z)$ can be written as

$$z^2 v'' + \left[-\omega z^2 + \left(2+\frac{e}{\sqrt{2d}}\right)z+\sqrt{2d}\right]v' + \frac{z}{2}\left[E-\omega\left(2+\frac{e}{\sqrt{2d}}\right)\right]v$$
$$= \frac{1}{4}\left[2\omega\sqrt{2d}+\left(\ell+\frac{1}{2}\right)^2-\left(\frac{e}{\sqrt{2d}}+1\right)^2\right]v. \quad (6.11.30)$$

It follows that from our general results that this equation has the degree n polynomial solutions

$$v = \prod_{i=1}^{n}(z-z_i), \quad v \equiv 1 \text{ for } n=0 \quad (6.11.31)$$

with distinct roots z_i provided that the potential parameters satisfy certain constraints. The closed-form expressions for the energies and the wave functions are[30]

$$E_n = \omega\left(2n+2+\frac{e}{\sqrt{2d}}\right),$$

$$\Psi_n(r) = r^{3/2+e/\sqrt{2d}}\left[\prod_{i=1}^{n}(r^2-z_i)\right]\exp\left(-\frac{\omega}{2}r^2-\frac{\sqrt{2d}}{2}\frac{1}{r^2}\right) \quad (6.11.32)$$

and the constraint for the potential parameters reads

$$2\omega\left(\sqrt{2d}+2\sum_{i=1}^{n}z_i\right)+\left(\ell+\frac{1}{2}\right)^2=4n\left(n+1+\frac{e}{\sqrt{2d}}\right)+\left(\frac{e}{\sqrt{2d}}+1\right)^2. \quad (6.11.33)$$

Here the roots $\{z_i\}$ are determined by the BAEs,

$$\sum_{j\neq i}^{n}\frac{2}{z_i-z_j}=\frac{\omega z_i^2-\left(2+\dfrac{e}{\sqrt{2d}}\right)z_i-\sqrt{2d}}{z_i^2}, \quad i=1,2,\cdots,n. \quad (6.11.34)$$

As examples of the above general expressions for the exact solutions, we study the ground and first excited states of the system. The $n=0$ case gives the ground state energy and wave function

$$E_0=\omega\left(2+\frac{e}{\sqrt{2d}}\right),$$

$$\Psi_0(r)=r^{3/2+e/\sqrt{2d}}\exp\left[-\frac{\omega}{2}r^2-\frac{\sqrt{2d}}{2}\frac{1}{r^2}\right] \quad (6.11.35)$$

with the potential parameters constrained by

$$2\omega\sqrt{2d}+\left(\ell+\frac{1}{2}\right)^2=\left(\frac{e}{\sqrt{2d}}+1\right)^2. \quad (6.11.36)$$

The first excited state solution corresponds to the $n=1$ case of the general expressions Eqs. (6.6.53) and (6.11.31)~(6.11.34). The energy and wave function are given respectively by

$$E_1=\omega\left(4+\frac{e}{\sqrt{2d}}\right),$$

$$\Psi_1(r)=r^{3/2+e/\sqrt{2d}}\left(r^2-z_1\right)\exp\left[-\frac{\omega}{2}r^2-\frac{\sqrt{2d}}{2}\frac{1}{r^2}\right], \quad (6.11.37)$$

where the root z_1 is determined by the Bethe Ansatz equation,

$$\omega z_1^2-\left(2+\frac{e}{\sqrt{2d}}\right)z_1-\sqrt{2d}=0$$

$$\Rightarrow z_1=\frac{1}{2\omega}\left[2+\frac{e}{\sqrt{2d}}\pm\sqrt{\left(2+\frac{e}{\sqrt{2d}}\right)^2+4\omega\sqrt{2d}}\right] \quad (6.11.38)$$

6.11 Examples of Solutions of QES Models

and the potential parameters obey the constraint,

$$\frac{1}{4}\left[\frac{e^2}{2d} - 2\omega\sqrt{2d} + 5 + \frac{4e}{\sqrt{2d}} - \left(\ell + \frac{1}{2}\right)^2\right]^2 = \left(2 + \frac{e}{\sqrt{2d}}\right)^2 + 4\omega\sqrt{2d}. \qquad (6.11.39)$$

Here (6.11.38) has been used in deriving this equation from (6.11.33).

6.11.5 Kink Stability Analysis of the ϕ^6-type Field Theory (ϕ^6 型场论的扭结稳定性)

In this section, we provide a highly non-trivial example of QES systems which has no algebraization and therefore is non Lie-algebraic!

Consider the ϕ^6-type field theory in 1+1 dimensions characterized by the Lagrangian

$$\mathcal{L} = \frac{1}{2}\partial_\lambda\phi\partial^\lambda\phi - \frac{\mu^2}{8g^2(1+\epsilon^2)}\left(g^2\phi^2 + \epsilon^2\right)\left(1 - g^2\phi^2\right)^2, \qquad (6.11.40)$$

where ϵ is real dimensionless constant and μ has the dimension of mass. As is well-known, the field equation of this theory has a kink solution. When one performs stability analysis around the kink solution one arrives at the Schrödinger equation[49, 50],

$$\left[-\frac{d^2}{dx^2} + V(x)\right]\psi(x) = E\psi(x), \qquad (6.11.41)$$

where $E \geqslant 0$ and the potential $V(x)$ is given by

$$V(x) = \mu^2 \frac{8\sinh^4\frac{\mu x}{2} - \left(\frac{20}{\epsilon^2} - 4\right)\sinh^2\frac{\mu x}{2} + 2\left(\frac{1}{\epsilon^2} + 1\right)\left(\frac{1}{\epsilon^2} - 2\right)}{8\left(1 + \frac{1}{\epsilon^2} + \sinh^2\frac{\mu x}{2}\right)^2}. \qquad (6.11.42)$$

The Schrödinger equation can be transformed into the form of general ODE[44]. Let

$$\psi = \left(1 + \frac{1}{\epsilon^2} + \sinh^2\frac{\mu x}{2}\right)^{-\frac{3}{2}} y. \qquad (6.11.43)$$

Then it can be shown that y satisfies the ODE,

$$-y'' + 3\mu\frac{\sinh\frac{\mu x}{2}\cosh\frac{\mu x}{2}}{1 + \frac{1}{\epsilon^2} + \sinh^2\frac{\mu x}{2}} y' + \frac{\frac{3}{2}\mu^2\left(1 + \frac{1}{\epsilon^2}\right)}{1 + \frac{1}{\epsilon^2} + \sinh^2\frac{\mu x}{2}} y = \left(E + \frac{5}{4}\mu^2\right) y. \qquad (6.11.44)$$

Make a change of variable,
$$z = \cosh\frac{\mu x}{2}. \tag{6.11.45}$$

Then the above equation becomes
$$\left[z^4 + \left(\frac{1}{\epsilon^2} - 1\right)z^2 - \frac{1}{\epsilon^2}\right]y'' - \left[5z^3 - \left(\frac{1}{\epsilon^2} + 6\right)z\right]y'$$
$$+ \left[\left(\frac{4E}{\mu^2} + 5\right)z^2 + \frac{4E}{\epsilon^2\mu^2} - \frac{1}{\epsilon^2} - 6\right]y = 0. \tag{6.11.46}$$

From our general results, the ODE (6.11.46) has polynomial solutions of degree $n = 1, 2 \cdots$,
$$y(z) = \prod_{i=1}^{n}(z - z_i), \tag{6.11.47}$$

where z_i are the roots of the above polynomial to be determined, provided that E, ϵ satisfy the relations[44]
$$E = \frac{\mu^2}{4}(n-1)(5-n), \tag{6.11.48}$$

$$\sum_{i=1}^{n} z_i = 0, \tag{6.11.49}$$

$$\frac{6(n-1)}{\epsilon^2} = (n-1)(n-6) + [5 - 2(n-1)]\sum_{i=1}^{n} z_i^2 - 2\sum_{i<j}^{n} z_i z_j, \tag{6.11.50}$$

and the Bethe Ansatz equations,
$$\left[z_i^4 + \left(\frac{1}{\epsilon^2} - 1\right)z_i^2 - \frac{1}{\epsilon^2}\right]\sum_{j\neq i}^{n}\frac{2}{z_i - z_j} = 5z_i^3 - \left(\frac{1}{\epsilon^2} + 6\right)z_i, \quad i = 1, 2, \cdots, n.$$
$$\tag{6.11.51}$$

For $n = 1$, we have $E = 0$ and $z_1 = 0$ so that $y(z) = z$. There is no constraint on ϵ, and
$$\psi(x) = \frac{\cosh\frac{\mu x}{2}}{\left(1 + \frac{1}{\epsilon^2} + \sinh^2\frac{\mu x}{2}\right)^{\frac{3}{2}}} \tag{6.11.52}$$

which is the ground state of the system with energy eigenvalue $E = 0$.

For $n = 2$, we have $E = \frac{3}{4}\mu^2$ and equations,

6.11 Examples of Solutions of QES Models

$$\frac{6}{\epsilon^2} = -4 + 3(z_1^2 + z_2^2) - 2z_1z_2, \quad z_1 = -z_2, \tag{6.11.53}$$

$$\left[z_1^4 + \left(\frac{1}{\epsilon^2} - 1\right)z_1^2 - \frac{1}{\epsilon^2}\right]\frac{2}{z_1 - z_2} = 5z_1^3 - \left(\frac{1}{\epsilon^2} + 6\right)z_1, \tag{6.11.54}$$

$$\left[z_2^4 + \left(\frac{1}{\epsilon^2} - 1\right)z_2^2 - \frac{1}{\epsilon^2}\right]\frac{2}{z_2 - z_1} = 5z_2^3 - \left(\frac{1}{\epsilon^2} + 6\right)z_2, \tag{6.11.55}$$

which give $z_1 = -z_2 = \sqrt{2}$ and $\frac{1}{\epsilon^2} = 2$. Thus $y(z) = (z - \sqrt{2})(z + \sqrt{2}) = z^2 - 2$ and

$$\psi(x) = \frac{\cosh^2\frac{\mu x}{2} - 2}{\left(1 + \frac{1}{\epsilon^2} + \sinh^2\frac{\mu x}{2}\right)^{\frac{3}{2}}}. \tag{6.11.56}$$

This gives the first excited state eigenfunction. The two energy eigenvalues and eigenfunctions above for $n = 1, 2$ reproduce those obtained in [49] (from a completely different analysis). Here, we have obtained the closed form expressions for all eigenvalues and eigenfunctions of the system. From (6.11.48), we see that there are only five non-negative energy solutions to the Schrödinger equation with energy eigenvalues $E \leqslant \mu^2$. All other analytic solutions (corresponding to $n \geqslant 6$) have negative energy eigenvalues. These negative energy solutions correspond to unstable modes.

Remark: If we rewrite the ODE (6.11.46) as the form $\mathcal{H}y = \mathcal{E}y$ with $\mathcal{E} = -4E/\epsilon^2\mu^2$ and

$$\mathcal{H} = \left[z^4 + \left(\frac{1}{\epsilon^2} - 1\right)z^2 - \frac{1}{\epsilon^2}\right]\frac{d^2}{dz^2}$$

$$- \left[5z^3 - \left(\frac{1}{\epsilon^2} + 6\right)z\right]\frac{d}{dz} + \left[\left(\frac{4E}{\mu^2} + 5\right)z^2 - \frac{1}{\epsilon^2} - 6\right], \tag{6.11.57}$$

then it can be easily seen that the criteria in Proposition 2 is not satisfied. This means that the operator $\mathcal{H}$ has no hidden $sl(2)$ algebraic structure (i.e. not $sl(2)$ algebraic). So this model provides a non-trivial 1D example which is QES but has no $sl(2)$ algebraization.

It should be pointed out that there are, in fact, many known QES systems which have no Lie algebraization (see e.g. [44] and references therein for other examples). Thus the previous claim in the literature by some researchers that every QES system is Lie algebraic does not seem justified-the above system gives a non-trivial counterexample to such claim.

6.12 Realization of $gl(M)$ in Fock Spaces and Bargmann-Hilbert Spaces ($gl(M)$ 在 Fock 空间和 Bargmann-Hilbert 空间中的表示)

In our lectures we mainly concentrated on QES linear operators in single variable or QES systems in one dimension. The generalization of the Lie algebraic approach to higher dimensions are possible by means of the differential realizations of $gl(M)$ algebra in Bargamann-Hilber spaces. We did not cover this topic in our lectures in the Summer School. Nevertheless let us briefly discuss the realization which students might find useful for further study.

Proposition 5. Let $a_i^\dagger$ and a_i, $i = 1, 2, \cdots, M$, be a set of boson creation and annihilation operators which satisfy the commutation relations

$$[a_i, a_j^\dagger] = \delta_{ij}, \quad [a_i, a_j] = 0 = [a_i^\dagger, a_j^\dagger].$$

Then the operators

$$E_{ij} = a_i^\dagger a_j, \quad i, j = 1, 2, \cdots, M$$

form the $gl(M)$ algebra and give an M-mode boson realization of $gl(M)$.

Proof. By means of the commutation relations above, we have

$$[E_{ij}, E_{kl}] = [a_i^\dagger a_j, a_k^\dagger a_l] = a_i^\dagger [a_j, a_k^\dagger a_l] + [a_i^\dagger, a_k^\dagger a_l] a_j$$
$$= a_i^\dagger [a_j, a_k^\dagger] a_l + a_k^\dagger [a_i^\dagger, a_l] a_j = \delta_{jk} a_i^\dagger a_l - \delta_{il} a_k^\dagger a_j$$
$$= \delta_{jk} E_{il} - \delta_{il} E_{kj}.$$

This completes the proof. □

The Fock space associated with the boson operators has vacuum $|0\rangle$ defined by $a_i|0\rangle = 0$, $i = 1, 2, \cdots, M$ and is spanned by states

$$|n_1, n_2, \cdots, n_M\rangle = \frac{a_1^{\dagger n_1} a_2^{\dagger n_2} \cdots a_M^{\dagger n_M}}{\sqrt{n_1! n_2! \cdots n_M!}} |0\rangle, \quad n_i = 0, 1, 2, \cdots. \tag{6.12.1}$$

Note that $E_{11} + \sum_{k=2}^{M} E_{ll}$ is a central element of $gl(M)$ and it thus takes constant value, $E_{11} + \sum_{k=2}^{M} E_{ll} = n$, where n is a complex parameter. Thus there is a different

6.12 Realization of $gl(M)$ in Fock Spaces and Bargmann-Hilbert Spaces

realization of $gl(M)$ in terms of $(M-1)$ pairs of boson operators $\{b_i^\dagger, b_i | i = 2, 3, \cdots, M\}$[51]:

Proposition 6. The explicit realization of $gl(M)$ generators in terms of the $(M-1)$ pairs of boson operators are given by

$$E_{i1} = b_i, \quad E_{1i} = b_i^\dagger \left(n - \sum_{l=2}^{M} b_l^\dagger b_l \right), \quad i = 2, 3, \cdots, M,$$

$$E_{11} = n - \sum_{l=2}^{M} b_l^\dagger b_l, \quad E_{ij} = b_i^\dagger b_j, \quad i, j = 2, 3, \cdots, M.$$

The generators E_{ij}, $i, j = 2, 3, \cdots, M$, span the $gl(M-1)$ algebra.

Proof. It is obvious that $[E_{ij}, E_{kl}] = \delta_{jk} E_{il} - \delta_{il} E_{kj}$ for $i, j, k, l = 2, 3, \cdots, M$. Now

$$[E_{1i}, E_{i1}] = \left[b_i^\dagger \left(n - \sum_{l=2}^{M} b_l^\dagger b_l \right), b_i \right] = [b_i^\dagger, b_i] \left(n - \sum_{l=2}^{M} b_l^\dagger b_l \right) - b_i^\dagger \left(\sum_{l=2}^{M} b_l^\dagger b_l, b_i \right)$$

$$= -\left(n - \sum_{l=2}^{M} b_l^\dagger b_l \right) + b_i^\dagger b_i = E_{ii} - E_{11}.$$

Other relations can be similarly checked. □

By the definition of vacuum $|0\rangle$: $b_i |0\rangle = 0$, $i = 2, 3, \cdots, M$. Then, $E_{11} |0\rangle = n |0\rangle$, $E_{ii} |0\rangle = 0$, $i = 2, 3, \cdots, M$. So the above boson realization provides a representation of $gl(M)$ with highest weight $(n, 0, 0, \cdots, 0)$ and highest weight state $|n, 0, 0, \cdots, 0\rangle$. If n is a non-negative integer, then the boson realization becomes a finite dimensional representation of $gl(M)$ acting on the space of polynomials

$$\mathcal{V} = \mathrm{span}\left\{ b_2^{\dagger n_2} b_3^{\dagger n_3} \cdots b_M^{\dagger n_M} \mid 0 \leqslant \sum_{l=2}^{M} n_l \leqslant n \right\}.$$

This is the symmetric representation of $gl(M)$ represented by the Young tableau with n horizontal boxes. The dimension of this representation is $(M+n-1)!/(M-1)!n!$.

By means of the Bargmann correspondence $b_i^\dagger = z_i$, $b_i = \dfrac{\partial}{\partial z_i}$, $i = 2, 3, \cdots, M$, one obtains the differential operator realization of $gl(M)$ in the Bargmann-Hilbert space.

Proposition 7. When n is non-negative integer, the generators of $gl(M)$ have the differential realization in $\mathcal{V} = \text{span}\left\{z_1^{n_1}, z_2^{n_2}, \cdots, z_M^{n_M} \mid 0 \leqslant \sum_{l=2}^{M} n_l \leqslant n\right\}$ as

$$E_{1i} = z_i\left(n - \sum_{l=2}^{M} z_l \frac{\partial}{\partial z_l}\right), \quad E_{i1} = \frac{\partial}{\partial z_i},$$

$$E_{11} = n - \sum_{l=2}^{M} z_l \frac{\partial}{\partial z_l}, \quad E_{ij} = z_i \frac{\partial}{\partial z_j}, \quad i, j = 2, 3, \cdots, M.$$

This differential realization has been applied to QES systems in higher dimensions or with many-body interactions (see. e.g. [52, 53] and references therein), which were not covered in our lectures.

I would like to thank the organizers of the Summer School for invitation to present these lectures, and students in the class for their seemly-keen interest and many impressive questions. I also thank the Institute of Theoretical Physics, Chinese Academy of Sciences, where these notes were written up and typed, for kind hospitality and support. My research has been partly supported by the Australian Research Council through Discovery Projects grant DP140101492.

参考文献

[1] Turbiner A. Comm. Math. Phys., 1988, 118: 467.
[2] Ushveridze A G. Quasi-exactly Solvable Models in Quantum Mechanics. Bristol: Institute of Physics Publishing, 1994.
[3] Gonzárez-López A, Kamran N, Olver P. Comm. Math. Phys., 1993, 153: 117.
[4] Turbiner A. In CRC Handbook of Lie Group Analysis of Differential Equations, Vol. 3: New Trends in Theoretical Developments and Computational Methods, Chapter 12, CRC Press, N. Ibragimov (ed.), 1995: 331-366.
[5] Sasaki R, Takasaki K. J. Phys., A: Math. Gen., 2001, 34: 9533.
[6] Ho C L. Ann. Phys., 2008, 323: 2241; ibid 2009, 324: 1095.
[7] Kalnins E G, Miller Jr W, Pogosyan G S. J. Math. Phys., 2006, 47: 033502.
[8] Yu Morozov A, Perelomov A M, Rosly A A, et al. Int. J. Mod. Phys. A, 1990, 5: 803.
[9] Gorsky A. JETP Lett., 1991, 54: 289.
[10] Shifman M A. QES spectral problems and CFTs, preprint. Theoretical Physics Inst. Univ of Minnesota, 1992.
[11] Zhang Y Z. Ann. Phys., 2016, 375: 460.
[12] Bender C M, Boettcher. J. Phys. A: Math. Gen., 1998, 31: L273.

[13] Braak D. Phys. Rev. Lett., 2011, 107: 100401.
[14] Chen Q H, Wang C, He S, et al. Phys. Rev. A, 2012, 86: 023822.
[15] Moroz A. Europhys. Lett., 2012, 100: 60010.
[16] Zhong H, Xie Q, Batchelor M T, et al. J. Phys. A: Math. Theor., 2013, 46: 415302.
[17] Moroz A. Ann. Phys., 2013, 338: 319; ibid, 2014, 340: 252.
[18] Zhang Y Z, Math J. Phys., 2013, 54: 102104.
[19] Reik H G, Nusser H, Ribeiro L A. J. Phys. A: Math. Gen., 1982, 15: 3431.
[20] Kus M. J. Math. Phys., 1985, 26: 2792.
[21] Emary C, Bishop R F. J. Math. Phys., 2002, 43: 3916; J. Phys. A: Math. Gen., 2002, 35: 8231.
[22] Zhang Y Z. J. Phys. A: Math. Theor., 2013, 46: 455302.
[23] Zhang Y Z. arVix:1304.7827v2 [quant-ph]; Rev. Math. Phys., 2017, 29: 1750013.
[24] Li Z M, Batchelor M T. J. Phys. A: Math. Theor., 2015, 48: 454005.
[25] Taut M. Phys. Rev. A, 1993, 48: 3561; J. Phys. A: Math. Gen., 1995, 28: 2081.
[26] Turbiner A V. Phys. Rev. A, 1994, 50: 5335.
[27] Chiang C M, Ho C L. Phys. Rev. A, 2001, 63: 062105.
[28] Loos P F, Gill P M W. Phys. Rev. Lett., 2009, 103: 123008.
[29] Pais A, Wu T T. J. Math. Phys., 1964, 5: 799.
[30] Agboola D, Zhang Y Z. J. Math. Phys., 2012, 53: 042101; Ann. Phys., 2013, 330: 246.
[31] Lanczos C, Res J. NBS, 1950, 45: 255.
[32] Szego G. Orthogonal polynomials. American Mathematical Society, 1993.
[33] Bender C M, Dunne G V. J. Math. Phys., 1996, 37: 6.
[34] Krajewska A, Ushveridze A, Walczak Z. arXiv:hep-th/9601088v1.
[35] Gautschi W. SIAM Rev., 1967, 9: 24.
[36] Bargmann V. Comm. Pure Appl. Math., 1961, 14: 187.
[37] Leaver E W. J. Math. Phys., 1986, 27: 1238.
[38] Zhang Y Z. Ann. Phys., 2014, 347: 122.
[39] Schweber S. Ann. Phys., 1967, 41: 205.
[40] Pan F, Bao L, Zhai L, et al. J. Phys. A: Math. Theor., 2011, 44: 395305.
[41] Pan F, Li B, Zhang Y Z, et al. Phys. Rev. C, 2013, 88: 034305.
[42] Kalousios C. J. Phys. A: Math. Theor., 2014, 47: 215402.
[43] Cachazo F, He S, Yuan E Y. Phys. Rev. D, 2014, 90: 065001.
[44] Zhang Y Z. J. Phys. A: Math. Theor., 2012, 45: 065206.
[45] Wiegmann P B, Zabrodin A V. Phys. Rev. Lett., 1994, 72: 1890; Nucl. Phys. B, 1995, 451: 699.
[46] Sasaki R, Yang W L, Zhang Y Z. SIGMA, 2009, 5: 104.
[47] Tomka M, El Araby O, Pletyukhov M, et al. Phys. Rev. A, 2014, 90: 063839.
[48] Kaushal R S. Ann. Phys. (NY), 1991, 206: 90.

[49] Christ N H, Lee T D. Phys. Rev. D, 1975, 12: 1606.
[50] Jatkar D P, Kumar C N, Khare A. Phys. Lett. A, 1989, 142: 200.
[51] Turbiner A. Lie algebras in Fock space. arXiv: q-alg/9710012v1.
[52] Hou X, Shifman M. Int. J. Mod. Phys. A, 1999, 14: 2993.
[53] Tanaka T. Phys. Lett. B, 2003, 567: 100; Ann. Phys., 2004, 309: 239.

《21世纪理论物理及其交叉学科前沿丛书》
已出版书目

(按出版时间排序)

1. 真空结构、引力起源与暗能量问题　　王顺金　　　　　　2016 年 4 月
2. 宇宙学基本原理（第二版）　　　　　龚云贵　　　　　　2016 年 8 月
3. 相对论与引力理论导论　　　　　　　赵　柳　　　　　　2016 年 12 月
4. 纳米材料热传导　　　　　　　　　　段文晖，张　刚　　2017 年 1 月
5. 有机固体物理（第二版）　　　　　　解士杰　　　　　　2017 年 6 月
6. 黑洞系统的吸积与喷流　　　　　　　汪定雄　　　　　　2018 年 1 月
7. 固体等离子体理论及应用　　　　　　夏建白　　　　　　2018 年 6 月
8. 量子色动力学专题　　　　　　　　　黄　涛，王　伟　等　2018 年 6 月
9. 可积模型方法及其应用　　　　　　　杨文力，杨战营　等　2019 年 4 月